Range Management

RANGE MANAGEMENT
Principles and Practices

Jerry L. Holechek

Professor, Range Science

Rex D. Pieper

Professor, Range Science

Carlton H. Herbel

Range Scientist (Retired)
Agricultural Research Service,
U.S. Department of Agriculture and
Adjunct Professor, Range Science

Department of Animal and Range Sciences
New Mexico State University
Las Cruces, NM 88003

Prentice Hall
Upper Saddle River, New Jersey 07458

Library of Congress Cataloging-in-Publication Data

Holechek, Jerry.

    Range management: principles and practices/Jerry L. Holechek,
Rex D. Pieper, Carlton H. Herbel. — 3rd ed.

      p.   cm.

    Includes bibliographical references and index.

    ISBN 0-13-626988-5

    1. Range management.   2. Range management—United States.
I. Pieper, Rex D.   II. Herbel, Carlton H.,   III. Title.

SF85.H56   1998

636.08'45—dc21   98-21991

                                          CIP

Acquisitions Editor: *Charles Stewart*
Production Editor: *Dawn M. Sitzmann*
Production Liaison: *Eileen M. O'Sullivan*
Managing Editor: *Mary Carnis*
Director of Manufacturing and Production: *Bruce Johnson*
Manufacturing Manager: *Ed O'Dougherty*
Production Manager: *Marc Bove*
Marketing Manager: *Melissa Bruner*
Editorial Assistant: *Kimberly Yehle*
Cover Artist: *John N. Smith*
Formatting/page make-up: *Carlisle Publishers Services*
Printer/Binder: *RR Donnelley & Sons Company*

© 1998 by Prentice-Hall, Inc.
Simon & Schuster / A Viacom Company
Upper Saddle River, New Jersey 07458

Printed in the United States of America

10 9 8 7 6 5 4 3

ISBN: 0-13-626988-5

Prentice-Hall International (UK) Limited, *London*
Prentice-Hall of Australia Pty. Limited, *Sydney*
Prentice-Hall Canada Inc., *Toronto*
Prentice-Hall Hispanoamericana, S.A., *Mexico*
Prentice-Hall of India Private Limited, *New Delhi*
Prentice-Hall of Japan, Inc., *Tokyo*
Simon & Schuster Asia Pte. Ltd., *Singapore*
Editora Prentice-Hall do Brasil, Ltda., *Rio de Janeiro*

We dedicate this book to authors of previous range management textbooks that have contributed tremendously to our profession and this textbook.

This book is also dedicated to the many managers, researchers, and graduate students who have contributed so much to the development of range management.

Finally, we dedicate this book to our families and colleagues who have supported and encouraged our various endeavors in range management.

Contents

Chapter 3
RANGE MANAGEMENT HISTORY   47

Chapter 4
DESCRIPTION OF RANGELAND TYPES   66

Chapter 5
RANGE PLANT PHYSIOLOGY   110

Chapter 6
RANGE ECOLOGY   130

Chapter 7
RANGE INVENTORY AND MONITORING   163

Chapter 8
CONSIDERATIONS CONCERNING STOCKING RATE   190

Chapter 9
SELECTION OF GRAZING METHODS   228

Chapter 10
METHODS OF IMPROVING LIVESTOCK DISTRIBUTION   269

Chapter 11
RANGE ANIMAL NUTRITION   286

Chapter 12
RANGE LIVESTOCK PRODUCTION   331

FOREWORD

This is a good book and it is sorely needed. Its chapters are arranged in a traditional and useful way, each capturing the fundamentals and perspectives of the key subjects in the field of range management. It has always seemed rather sad for us in range management to feel so strongly about the importance of the field, yet to have so few textbooks that bring the concepts of rangeland management into clear focus and establish important fundamentals for the profession. This text does that quite well, and makes an important addition to the few books available in range management.

The central theme of this book is managerial. It is a textbook intended primarily for midlevel university students studying to become managers of rangelands or students who wish to broaden their overall education in natural resource use and management. A textbook cannot satisfy all needs, but this one is ideal for students and readers seeking a sound overview of range management principles and practices. The book covers more topics than do the texts presently available. The range science literature has been properly applied to support the management concepts and principles presented. Overall, it is well written, clear, and distinct, and the level of difficulty will challenge most readers.

Although rangelands of the world have been used by the livestock of pastoralists for thousands of years, information about the land's response to grazing is relatively recent. In fact, research of any magnitude into the use and management of rangelands did not commence until after 1900. Adequate scientific knowledge to undergird principles and theories sufficiently was not accumulated until after 1950. Trial and error was the only method available to the grazier to improve management prior to about 1930.

Until very recently, range researchers and teachers have been too busy researching and teaching to write books that range management professionals have needed. Drs. Holechek, Pieper, and Herbel, in addition to the demands of their full-time research and teaching positions, have written an excellent book that integrates insights gained through their experiences with the information provided in the scientific literature.

I am very pleased with this text. Its appearance is timely and its contribution, especially to teachers, will be considerable.

DON D. DWYER
PROFESSOR EMERITUS
UTAH STATE UNIVERSITY
LOGAN, UTAH

PREFACE

The purpose of the third edition of this book is to introduce students to the science of range management coupling the latest concepts and technology with proven traditional approaches. We hope that our audience continues to include employed range managers on public and private lands, ranchers, wildlife biologists, soil scientists, and the growing segment of the public interested in natural resource management. We have tried to improve the text for those concerned with range management, not only in the United States but in other parts of the world as well.

Our approach has involved tempering fundamental topics, such as range plant physiology, range plant ecology, stocking-rate considerations, and grazing system selection, with the most recent research. Some traditional range management concepts have been altered since the second edition as the result of new findings. This is particularly true in the subject areas of grazing management, range nutrition, and range plant ecology. We have placed greater emphasis on economics and multiple use in this edition along with improving the quality of Tables and Figures.

Although approaches to range management change, the basic objectives of range management remain essentially the same as in the past. These are to provide society with meat, water, wildlife, and recreational opportunities on a sustained basis from lands unsuited for permanent cultivation. In recent years, the relative importance to society of these products has shifted on many rangelands in the United States. However, with modern range management practices, most rangelands can be made to yield near their potential of each product simultaneously. Although multiple-use has been practiced for over 20 years on federal rangelands in the United States, it is now being more widely practiced on private rangelands as ranchers find the sale of recreational opportunities on their land to be increasingly profitable. We have tried to emphasize range management practices oriented toward multiple-use wherever possible.

Range management is distinguished from other land management disciplines in that it involves manipulation of grazing by large domestic or wild animals. Since control of grazing is the foundation of any range management program, this still receives primary emphasis in this third edition of our text. We have restricted our coverage of manipulation of vegetation by practices other than grazing to fundamental concepts since several other good texts are available that deal exclusively with this subject.

We received both encouragement and helpful criticism from many of our colleagues. Those who received and provided valuable suggestions on our manuscript include Dr. Billie E. Dahl, Dr. Richard M. Hansen, Dr. Kris Havstad, Dr. Don D. Dwyer, Dr. David L. Scarnecchia, Dr. Sam L. Beasom, Dr. Jack L. Butler, and Dr. Randy Rosiere.

JERRY L. HOLECHEK
REX D. PIEPER
CARLTON H. HERBEL

Conversion from a Metric Unit to the English Equivalent[a]

METRIC UNIT	ENGLISH UNIT EQUIVALENT
LENGTH	
kilometer (km)	0.621 mile (mi)
meter (m)	1.094 yards (yd)
meter (m)	3.281 (ft)
centimeter (cm)	0.394 inch (in.)
millimeter (mm)	0.039 inch (in.)
AREA	
kilometer^2 (km^2)	0.386 mile^2 (mi^2)
kilometer^2 (km^2)	247.1 acres
hectare (ha)	2.471 acres
VOLUME	
meter^3 (m^3)	1.308 cubic yards (yd^3)
meter^3 (m^3)	35.316 cubic feet (ft^3)
hectoliter (hl)	3.532 cubic feet (ft^3)
hectoliter (hl)	2.838 bushels (U.S.) (bu)
liter (l)	1.057 quarts (U.S. liq.) (qt)
MASS	
ton (t)	1.102 tons
quintal (q)	220.5 pounds (lb)
kilogram (kg)	2.205 pounds (lb)
gram (g)	0.00221 pound (lb)
YIELD OR RATE	
ton/hectare (t/ha)	0.446 ton/acre
hectoliter/hectare (hl/ha)	7.013 bushels/acre
kilogram/hectare (kg/ha)	0.892 pound/acre
quintal/hectare (q/ha)	89.24 pounds/acre
quintal/hectare (q/ha)	0.892 hundredweight/acre
PRESSURE	
bar (10^6 dynes/cm^2)	14.5 pounds/inch^2 (psi)
bar (bar)	0.9869 atmosphere (atm)
atmosphere (atm or atmos)	14.7 pounds/inch^2 (psi)
(an atmosphere may be specified in metric or English units)	
TEMPERATURE	
Celsius (°C)	$1.80°C + 32 = $ Fahrenheit (F)
kelvin (K) $= °C + 2.73.15$	

[a]Additional information may be found in the 1979 Standard for Metric Practice. *Annual Book of ASTM Standards,* p. 41. Designation 380-79, pp. 504–545. Philadelphia: American Society for Testing Materials (ASTM).

CHAPTER 1

RANGELAND AND MAN

RANGELAND DEFINED

All areas of the world that are not barren deserts, farmed, or covered by bare soil, rock, ice, or concrete can be classified as rangelands. Therefore, rangelands include deserts and forests and all natural grasslands. Dyksterhuis (1955), Hedrick (1966), Society for Range Management (1989), and Stoddart et al. (1975) provide slightly different definitions of rangeland. However, all these sources are consistent in considering rangelands to be uncultivated and capable of providing habitat for domestic and wild animals. Therefore, in this book *rangeland* is defined as uncultivated land that will provide the necessities of life for grazing and browsing animals (Figure 1.1). *Grazing* refers to the consumption of standing forage (edible grasses and forbs) by livestock or wildlife, while *browsing* is the consumption of edible leaves and twigs from woody plants (trees and shrubs) by the animal.

Herbivory is the essential process that characterizes rangelands. Range managers may be interested in a whole host of activities and products, but the main focus is on grazing and browsing animals.

Pasturelands are distinguished from rangelands by the fact that periodic cultivation is used to maintain introduced (nonnative) forage species, and agronomic inputs such as irrigation and fertilization are applied annually. At one time, the intensity of fencing was useful to distinguish rangeland and pasturelands. However, during the past two decades,

(a)

(b)

range management has become much more intensive and many rangelands are now heavily fenced. At the turn of the century, rangeland could be defined as supporting native vegetation. However, many introduced (exotic) plants such as crested wheatgrass (*Agropyron cristatum*), buffelgrass (*Cenchrus ciliaris*), and Lehman lovegrass (*Eragrostis lehmanniana*) have remained established without periodic cultivation in many parts of the western United States. Presently, stands of introduced forages that are maintained without annual cultivation and irrigation and are harvested by grazing animals are considered as rangeland.

(c)

(d)

***Figure* 1.1** Native perennial grass-
land (a), desert shrubland (b), coniferous forest (c), and annual grassland (d) are all considered rangeland.

Because economic and social values change constantly, the amount of rangeland in the
world varies from year to year. For example, large areas of land in the central Great Plains region of the United States have been shifted between rangeland and cropland several times during the past 100 years, depending on economic conditions (the price of beef versus the price of
grain). Therefore, it is erroneous to define all rangelands as being unsuitable for cultivation.
However, most rangelands in the world will not sustain cultivation, because of low precipitation, thin soils, rugged topography, and/or cold temperatures.

Forage for livestock is the primary contribution to humans of rangelands in developing countries. This is true in most parts of Africa and South America. In affluent societies such as the United States, where food surpluses have been a problem in recent years, if the current trends continue, water, wildlife, and recreation may have greater monetary value on many rangelands than forage for livestock.

Ecosystem is a commonly used word when rangelands are discussed. A rangeland ecosystem is an area with similar ecological characteristics on which humans have placed boundaries for management purposes. An ecosystem includes both the biotic (plants and animals) and abiotic (soils, topography, and climate) components of the defined area. A particular plant and animal community is associated with each ecosystem. The components of rangeland ecosystems and the usable products they produce for humans are shown in Figure 1.2.

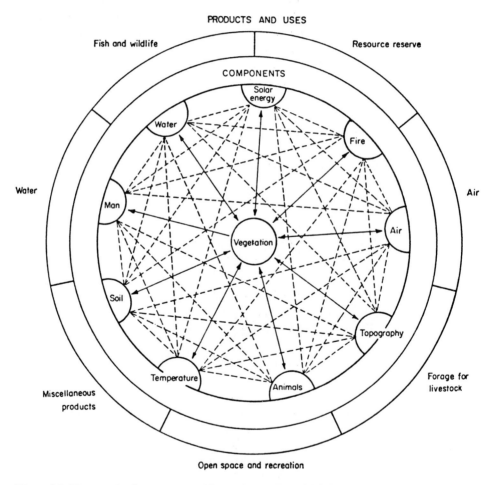

Figure **1.2** The rangeland ecosystem and its products. (From Blaisdell et al. 1970.)

RANGE MANAGEMENT DEFINED

Range management is the manipulation of rangeland components to obtain the optimum combination of goods and services for society on a sustained basis. These goods and services are shown in Figure 1.2. Range management has two basic components: (1) protection and enhancement of the soil and vegetation complex, and (2) maintaining or improving the output of consumable range products, such as red meat, fiber, wood, water, and wildlife.

The range management profession is unique among agricultural vocations in that it deals with the plant and animal interface rather than dealing with plants or animals in isolation. The distinguishing feature of range management is that it deals with manipulating the grazing activities by large herbivores so that both plant and animal production will be maintained or improved. The science and art of range management was born in the western United States in the early twentieth century. In this period, increased flooding, declining forage production, and heavy livestock deaths due to poisonous plants were recognized as being the result of soil and vegetation degradation from uncontrolled livestock grazing (Smith 1899). Early rangeland management practices concerned manipulation of livestock grazing intensity, timing, and frequency to ameliorate adverse grazing impacts on the soil and vegetation. More recently, range management has become broadened to include manipulation of many components of rangeland ecosystems other than livestock, such as fire, wildlife, and human activities. Still the focal point of range management is the control of livestock grazing.

Range management is based on five basic concepts:

1. Rangeland is a renewable resource.
2. Energy from the sun can be captured by green plants that can only be harvested by the grazing animal.
3. Rangelands supply humans with food and fiber at very low energy costs compared to those associated with cultivated lands. Ruminant animals are best adapted to use range plants. Unlike human beings, ruminants have microbes in their digestive systems that efficiently break down fiber, which is quite high in most range plants.
4. Rangeland productivity is determined by soil, topographic, and climatic characteristics.
5. A variety of "products" including food, fiber, water, recreation, wildlife, minerals, and timber are harvested from rangelands.

Range science is the organized body of knowledge upon which range management is based. In the past 20 years, a tremendous amount of scientific information was accumulated that can be applied in the management of rangelands. However, range management remains as much an art as a science because every piece of rangeland has a different set of physical and biological characteristics. This requires the manager to synthesize information from various other rangelands and apply it to the present situation. Experience coupled with scientific information is much more valuable than either scientific information or experience in isolation. The statement by Stoddart et al. (1975) that "the feel for resource is the hallmark of the rangeman" remains applicable today and will probably be applicable 100 years from now.

Range Management Information

The Society for Range Management, which originated in 1948, is the primary professional organization that deals with rangelands and range management on a global basis. It is head-quartered at 1839 York Street, Denver, Colorado 80206. It publishes two journals, (*Journal of Range Management* and *Rangelands*) and has published several books that are primary sources of information for range managers and scientists. These journals and books serve as primary sources of information for this textbook.

Over the past 10 years the information available on range management in the United States and other countries has increased tremendously. While we will provide an overview of past and current information; we recognize that range management is a large subject area. Throughout this book we will provide the reader with references that offer more detailed coverage of specific subject areas. At the same time, we refer readers who wish to broaden their coverage of range management to basic textbooks by Stoddart et al. (1975), Vallentine (1989, 1990), Heitschmidt and Stuth (1991), and Heady and Child (1994).

RELATIONSHIP OF RANGE MANAGEMENT TO OTHER DISCIPLINES

Range management is distinct from other disciplines in that it integrates into a unified system knowledge from several disciplines (Figure 1.3). The basic components of range management can be categorized into biological, physical, and anthropological factors. Initially, range managers were concerned primarily with the biological component, particularly plants. Responses of plant communities to grazing were studied because plant communities are the primary producers of food for grazing animals. However, it was soon recognized that a plant community's responses to grazing could not be understood without knowledge of individual plant physiological processes. Because plant productivity depends on the interaction between climate and soils as well as grazing influences, the additional need to understand the physical environment became apparent.

Different animal species affect rangeland ecosystems in different ways, and animal productivity is not consistent among rangeland ecosystems. Therefore, the range manager must have a thorough understanding of the behavior and nutritional requirements of domestic and wild animals if their productivity is to be maximized. During the past 20 years, range science has focused increasingly on the range animal. At the same time, range management has changed from a protection-oriented vocation to one of production orientation. This has resulted from a rapidly expanding human population with ever-increasing demands on a finite resource.

Since range management is geared toward producing products usable to man, social, economic, cultural, and technological considerations are a critical part of the range management decision-making process. As an example, range management on some private lands in the United States is becoming increasingly oriented to the production of harvestable wildlife. In the United States food costs are low, much of the human population has con-

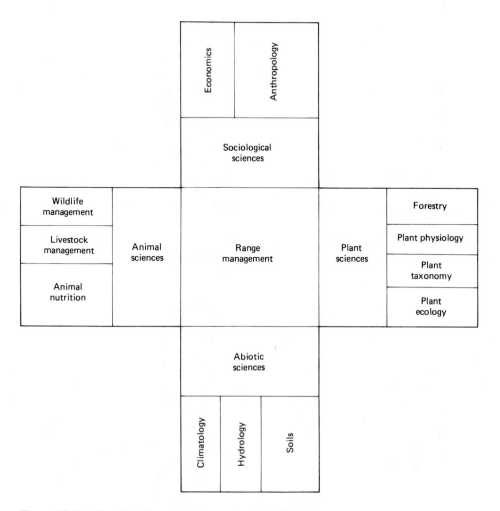

***Figure* 1.3** Relationship of range management to other disciplines.

siderable time for recreational pursuits, growth in the human population is relatively low, and the country is characterized by a high level of affluence. In contrast, most countries of Africa have rapidly expanding human populations and low levels of affluence, and food costs account for a major portion of the per capita income. In such countries, maximization of livestock production is the primary orientation of range management. However, because most of the livestock in Africa is owned by nomadic herders, range management practices useful in the United States are often difficult to apply in Africa and other developing areas.

Ecology, defined as the study of the relationship between an organism or group of organisms and their environment, has been and will continue to be the foundation for range management. Early rangemen were trained primarily in plant ecology. However, during the past 20 years, animal ecology has become an important part of the range curriculum.

Because the range manager often is dealing with several resources in an area at one time, he or she must have basic knowledge of watershed, forest, wildlife, and recreation management; and will often depend on professionals in these disciplines for advice. The harmonious use of native rangeland for more than one purpose, such as livestock, recreation, wildlife, and water production, is defined as *multiple use* (Society for Range Management 1989).

THE IMPORTANCE OF RANGELANDS TO HUMANS

The Human Population Explosion

Many scientists consider the most pressing problem confronting humankind to be the tremendous increase in the human population expected in the next 50 years. According to anthropologists, about 8000 B.C. the human population was around 1 million (Chrispeels and Sadava 1977). By the time of Christ it had reached 250 million, and around 1850 there were 1 billion people in the world. The next doubling time took only 80 years. Even more impressive is the fact that between 1930 and 1976 the world's population doubled, reaching 4 billion. The world human population in 1993 was 5.6 billion (USDC 1996). The annual growth rate in 1996 was 1.4%, which is a drop in human population growth rate that started in the 1980s. However, the world population could reach 11 billion before stabilizing.

The falloff in the population growth rate has come almost exclusively from developed countries. Birthrates remain high in the developing South American, Central American, and African countries (Table 1.1). Religious beliefs, family labor needs, and lack of education on birth control largely explain the high birthrates in developing countries as compared to those of developed countries.

In developed countries such as the United States, food production has kept far ahead of the population growth rate. Since the early 1980s food surpluses have been one of the major problems confronting agriculture in the United States (Schiller 1991). However, food shortages remain an important problem in many developing countries, and this is expected to continue for many future decades.

Trends in human population growth and economic development will have considerable influence on how rangelands in various countries will be used in coming years. Although the emphasis may shift among rangeland products, the rapid expansion in the human population will undoubtedly make rangelands more important to humankind then ever before.

Land Area and Forage Production

Regardless of whose figures are used, rangelands are the primary land type in the world. Based on Food and Agriculture Organization (FAO) (1995) data, 11% of the land in the world is farmed; 24% is in permanent pasture; 31% is forest or woodland; and deserts, glaciated areas, high mountain peaks, and urbanized or industrialized land comprise the other 34% (Table 1.2). Although exact estimates vary, total land covered by concrete (houses, cities, and highways, etc.) is between 3% and 4%.

TABLE 1.1   Growth and Density of the Human Population in Various Parts of the World in 1996

	PERCENT NET GROWTH PER YEAR	DOUBLING TIME (YEARS)	POPULATION PER SQUARE MILE	POPULATION PER SQUARE KILOMETER
World	1.4	51	114	45
Africa	2.7	27	66	26
Kenya	2.4	30	128	50
Sudan	2.9	25	34	13
China	1.0	72	336	131
India	1.7	42	829	325
Europe	0.3	240	270	106
North America	1.2	60	40	16
Canada	1.2	60	8	3
Mexico	1.9	38	129	51
U.S.A.	1.0	72	75	29
Oceania	1.5	48	9	4
Australia	1.1	65	6	2
New Zealand	1.1	65	34	13
South America	1.7	42	48	19
Argentina	1.1	65	33	13
Brazil	1.2	60	50	20
Former USSR	0.3	—	22	9

Sources: USDC 1996 and FAO 1995.

The FAO (1995) defines *permanent pasture* as land used for 5 years or more for herbaceous forage crops, either cultivated or growing wild. However, these data do not include large tracts of land classed as forest or woodland used for grazing, or the deserts and tundra that fall into the *other land* category that are in many cases grazed by nomadic herders. When all the land resources presently receiving grazing by domestic animals are taken into account, they allow for about 50% of the world's land area. If all the uncultivated land with the potential to support grazing by domestic animals is taken into account, rangelands comprise about 70% of the world's land area. Rangelands are the major land type found on all continents (see Table 1.2).

In most developing African and South American countries, rangelands provide over 85% of the total feed needs of domestic ruminants (cattle, sheep, and goats) (ARPAC 1975). Rangelands in the United States provide domestic ruminants with between 50% and 65% of their total feed needs, depending on how rangeland is defined. On a worldwide basis,

TABLE 1.2 Percentages of Farmland, Grassland, Forest Land, and Rangeland for Selected Countries and Regions in 1994

	FARMLAND	PERMANENT PASTURELAND	FOREST AND WOODLAND	RANGELAND (TOTAL ESTIMATED POTENTIALLY GRAZABLE LAND)
World	11	25	29	70
Africa	6	30	23	69
Kenya	4	7	4	91
Sudan	5	24	20	87
China	10	42	13	77
India	52	4	20	33
Europe	28	17	32	53
North America	13	17	32	66
Canada	5	3	35	64
Mexico	13	38	21	70
U.S.A.	20	26	31	61
Oceania	5	56	18	66
Australia	6	60	14	69
New Zealand	2	53	26	90
South America	6	28	46	74
Argentina	10	51	21	75
Brazil	7	22	58	77
Former USSR	10	17	41	62

Source: Based on FAO (1995) data.

rangelands contribute about 70% of the feed needs of domestic ruminants. Rangelands provide wild ruminants with over 95% of their feed needs in the United States and worldwide. Throughout the world, rangelands are the major source of feed for both domestic and wild ruminant animals.

Production of Animal Products

Rangelands play a major role in supplying human populations with animal products in all the land regions of the world not covered by ice. India has the largest percentages of the world's cattle and goat populations, while Australia has the highest percentage of sheep

TABLE 1.3   Percentages of the World's Cattle, Sheep, and Goat
Populations for Selected Countries and Regions in 1994

	CATTLE	SHEEP	GOATS
Africa	15	19	29
Kenya	1	4	1
Sudan	2	2	3
China	6	10	17
India	15	4	19
Europe	9	12	2
North America	13	2	2
Canada	1	<1	<1
Mexico	2	1	2
U.S.A.	9	1	<1
Oceania	3	17	2
Australia	2	12	<1
New Zealand	1	5	<1
South America	22	9	4
Argentina	4	2	1
Brazil	12	2	2

Source: Based on FAO (1995) data.

(Table 1.3). The United States is the leading producer of beef, while China leads in mutton production worldwide (Table 1.4). Australia is the leading producer of wool.

Although India has the largest cattle production in the world, it has the lowest output of meat per animal. Religious custom in India opposes the slaughter of cattle for meat. However, cattle in India are used as draft animals and they provide some milk. Although Africa has nearly five times as many cattle as the United States, three times more beef is marketed in the United States than in Africa. In many parts of Africa, nomadic herders consider cattle a source of wealth; many animals in the herd contribute nothing to meat production and are generally not used for draft purposes.

However, it is important to recognize that 80% to 90% of the food energy consumed by nomadic African herders comes from meat, milk, and blood supplied by their livestock. These animals do serve as a cash crop that can be used to buy other food. A large herd also helps to ensure that some animals will be left to restock ranges after cessation of drought. The American Indians, prior to settlement of the United States, viewed horses as a source of wealth. Maintaining large numbers of nonproducing animals by nomadic cultures has led

TABLE 1.4   Percentages of the World's Beef, Mutton, and Wool
Production from Selected Countries and Regions in 1994

	BEEF	MUTTON	WOOL
Africa	6	13	9
Kenya	4	<1	<1
Sudan	<1	1	1
China	4	12	10
India	3	2	1
Europe	19	18	9
North America	28	3	2
Canada	2	<1	<1
Mexico	3	<1	1
U.S.A.	22	2	1
Oceania	5	16	38
Australia	4	9	28
New Zealand	1	7	10
South America	16	4	9
Argentina	5	1	3
Brazil	6	1	1

Source: Based on FAO (1995) data.

to rangeland deterioration. Large increases in both nomad and livestock populations, as a result of interventions from Western culture in Africa, necessitate a change in the practice of keeping nonproducing livestock as a form of wealth.

Between 1980 and 1994, cattle populations in the world increased about 6% while sheep populations showed no change (Table 1.5). Developing countries with rapidly increasing human populations, such as Sudan and Brazil, have experienced large-scale increases in cattle and sheep populations, whereas developed countries with low human population growth rates, such as the United States and many European nations, have had declines or have reached stability in cattle and sheep populations.

In developing countries, range livestock numbers in the next 25 years are expected to increase further as more and more herders share a declining land base due to conversion of rangeland to cropland. This will place tremendous pressure on rangelands in these countries and necessitate major changes in grazing practices to prevent widespread rangeland degradation.

The drop in cattle and sheep numbers in the developed countries has been attributed to a poor market due to meat supply exceeding demand. This situation is expected to continue

TABLE 1.5   Percentage Change in Cattle and Sheep Populations
Between 1980 and 1994 for Selected Countries and Regions

	CATTLE	SHEEP
World	+ 6	0
Africa	+ 12	+ 16
Kenya	+ 10	+ 8
Sudan	+ 18	+ 30
China	+ 73	+ 10
India	+ 3	0
Europe	− 20	+ 6
North America	− 5	− 17
Canada	+ 2	+ 44
Mexico	+ 11	− 9
U.S.A.	− 9	− 24
Oceania	− 2	− 10
Australia	− 5	− 2
New Zealand	+ 6	− 26
South America	+ 17	− 9
Argentina	− 10	− 41
Brazil	+ 30	+ 11

Source: FAO 1995.

at least into the late 1990s. In the United States, recreation, water production, and wildlife are expected to receive increasing emphasis as rangeland products. However, meat will probably still be the primary product produced by rangelands for human use.

Wildlife

Rangelands are the primary habitat for nearly all the land-dwelling wild animals highly valued for meat, hunting, and aesthetic viewing. The economic value of wildlife on rangelands is becoming increasingly recognized in developed and developing countries. In parts of the western United States, Texas in particular, income generated from selling hunting privileges can exceed that from livestock on some ranches (Ramsey 1965; Henson et al. 1977; Payne 1988). In certain African countries, such as Kenya, income from tourists viewing wildlife is of critical importance to the national economy. Rangeland wildlife has potential as a source of meat for human consumption in many African countries (Small 1988).

If agricultural technology keeps ahead of population growth in the United States, wildlife will probably become greater in economic importance than livestock on many rangelands. However, if agricultural technology lags, wildlife populations may be adversely affected by more intensive use of rangelands for meat production.

Water

In some parts of the western United States, such as Arizona, New Mexico, California, and Texas, where human populations are rapidly growing but arid to semiarid conditions prevail, water is becoming of greater importance than forage as a rangeland product. In the western states, forested and alpine rangelands are the primary source of water for agricultural, industrial, and domestic use. The condition of the soil and vegetation complex on which precipitation falls has a major influence on the quality and quantity of water available for human use. Range management practices can affect flooding of streams and rivers; silting rates of reservoirs; coliform bacteria and sediment counts in reservoirs, streams, and rivers; flow rates of springs, seeps, streams, and rivers; and the quality of water from overland flow that can be trapped in reservoirs (Branson et al. 1981).

By the year 2000, water will be the major problem confronting California, Arizona, New Mexico, and Nevada (USDA 1989). This is due not only to rapid growth in the human population, but also to the depletion of groundwater reserves at an ever-increasing rate (Figure 1.4). In the Southwest, this has led in recent years to an increase in rangeland at the expense of farmland (Cox et al. 1983). On most arid to semiarid rangelands, range man-

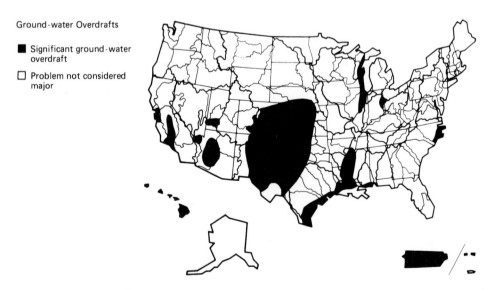

Ground-water Overdrafts

■ Significant ground-water overdraft

□ Problem not considered major

Figure **1.4** Regions in the United States where groundwater depletion is an important problem. (From USDA 1989.)

agement practices have been and will continue to be geared toward improving water retention on the site, thus reducing sedimentation and improving recharge of the water table. On forested and alpine rangelands, considerable potential exists for increasing yields from overland flow by vegetation modification through logging and grazing practices. However, little potential generally exists to increase water yields from overland flow on desert and grassland ranges.

Recreational Products

The large human population increases in the United States since the 1940s have made rangelands increasingly important as places for people to engage in outdoor recreational pursuits. Hiking, camping, trail biking, picnicking, hunting, fishing, and rock hounding are some of the important recreational uses of rangelands. The importance of open space, scenery, and aesthetic values from rangelands in the United States is difficult to quantify. However, the value of this outdoor product was demonstrated by considerable public pressure in the late 1960s and 1970s for legislation that now requires federal land management agencies to draft statements and consider public input on any major action affecting lands under their control. During the 1970s and 1980s, large areas in the United States were set aside as wilderness areas, where human activities other than sightseeing are tightly restricted.

Ranchers are finding more and more opportunity to market recreational values from rangelands. Marketable recreational products include hunting, fishing, and camping privileges; horseback riding; and homesites. There are many "dude ranches" in the United States that provide a vacationing visitor with the opportunity to experience the life and culture of the American cowboy.

Plant Products

Rangelands produce a wide variety of plants that could be very important in meeting our future needs. Salt-tolerant shrubs such as fourwing saltbush (*Atriplex canescens*) have considerable potential to be productive forage species on lands with prolonged drought and excessive salinity. Many rangeland shrubs are being developed and used for landscaping purposes. Some ranchers in the southwestern United States have obtained substantial income from the sale of shrubs, particularly cactus (*Opuntia* sp.), agave (*Agave* sp.), and ocotillo (*Fouqueria* sp.), for use as ornamentals in landscaping (Steger and Beck 1973). The desert shrub jojoba (*Simmondsia chinensis*), endemic in the Sonoran desert in the southwestern United States and northern Mexico, may produce a suitable replacement for sperm whale oil as an industrial lubricant. Guayule (*Parthenium argentatum*), another desert shrub, contains rubber that can be used for tires, medical supplies, and other items. Some of the desert forbs and shrubs contain substances that may inhibit cancer growth and have other medicinal properties. Many rangeland plants have the potential to be developed into valuable domestic food and forage species using new genetic engineering techniques.

Wood and Minerals

Wood and minerals are important products from some rangelands. Decisions regarding the use of minerals are usually made by geologists, and grazing management has little effect on this rangeland resource. Timber, livestock, and wildlife production occur concurrently on many lands in the western and southeastern United States. Grazing by livestock can influence timber production. It is therefore essential that both the forester and range manager understand interrelationships between grazing and timber management.

Open Space

Loss of open space is emerging as one of the biggest environmental problems that will occur in the twenty-first century in developed countries. During the 1990s the western United States experienced a massive building boom. Large numbers of people left the crowded cities along the Atlantic and Pacific coastal areas for the open spaces of Utah, Nevada, New Mexico, Arizona, Idaho, and Colorado. The population of these states has increased by more than 25% over the last 10 years. Thousands of acres of rangeland are rapidly being converted into ranchettes, retirement communities, industrial parks, and housing projects. As much as 1.5 million acres of rangeland per year may be lost to urbanization in the United States. There is growing concern that many parts of the West will resemble a ghetto with extended tracts of former rangeland being occupied by cheap, scattered housing in a haphazard arrangement. The problem of loss of open space is being increasingly recognized by ranchers, environmental groups, and politicians. More and more it is being recognized that rangelands have a esthetic value beyond the physical products they provide. Not only are large expanses of open terrain in a natural condition pleasing to the human spirit, but they play an important role in air and water purification and serve as a source of habitat for many rare and unique plant and animal species. If present trends continue the biggest challenge confronting tomorrow's range managers in the United States could be to find economic ways to maintain large areas of rangeland in a fairly natural condition free from the clutter of urban sprawl (Huntsinger and Hopkinson 1996).

DESERTIFICATION

Desertification is the formation of desertlike conditions, largely through human actions, in areas that do not have desert climates. Biological productivity declines while the prevailing climatic conditions are thought to remain constant. Human activities implicated as causes of desertification include uncontrolled livestock grazing, burning, woodcutting, temporary cultivation, and abandonment of semiarid to arid lands. Africa has been the focal point of concern over desertification during the past 20 years because of continued droughts in the Sahel region. These droughts have caused tremendous losses of livestock and human hardship. A world conference was held on desertification in Nairobi, Kenya, in 1977. Comprehensive reviews on the subject of desertification are provided by United Nations (1977), Glantz (1977), Postel (1989), and in Mouat and Hutchinson (1995).

Past climatological data show that drought has been a recurring phenomenon in the Sahel region (Wallen and Gwynne 1978; Winstanley 1983). However, the effects of drought on the vegetation have been magnified in recent years because of rapidly increasing human and livestock populations. There is also some evidence that recent droughts have been more severe than those in the past (Winstanley 1983).

As human populations increase in semiarid to arid areas, desertification may become an important problem in other parts of the world as well as Africa. Application of range management practices has considerable potential to reduce or reverse the desertification problem in many areas.

CHANGES IN THE AMOUNT OF RANGELAND

The amount of rangeland in the world is expected to decline substantially in the next 30 years. Large amounts of rangeland in Africa and South America are presently being converted to farmland, and this trend is expected to continue until most of the potentially farmable land is put under cultivation. Rapid increases in the human population will necessitate the farming of all available lands on these continents. The expected rangeland-to-farmland conversion could decrease the amount of rangeland by 20% to 30% in Africa and as much as 40% in South America in the next 50 years. In some instances this conversion will be temporary and will cause degradation of the land resource.

In the United States, large amounts of rangeland in the central Great Plains, particularly in Colorado, South Dakota, Nebraska, and Kansas, were converted to farmland during the 1970s (USDA 1989). Low prices for beef relative to those for wheat caused this trend (Huszar and Young 1984). This trend was reversed by the 1985 Farm Bill and low grain prices in the middle 1980s through the early 1990s. In the Southwest, particularly Arizona and New Mexico, large areas of farmland reverted back to rangeland in the 1980s because of increased irrigation costs, loss of water to urbanization, and lowering of the water table (Cox et al. 1983). Now much of this same land is being converted into low-density housing.

In the early 1980s, there was considerable concern over the conversion of Great Plains rangeland to farmland (Huszar and Young 1984). Attempts were made to farm this area in the late 1800s through the 1920s. The great drought and Dust Bowl of the 1930s resulted in a switch of much farmland back to rangeland. The concern that the Dust Bowl era could be repeated was one reason for passage of the 1985 Farm Bill. In parts of Colorado and Nebraska there were reports during the early 1980s of severe wind and water erosion degrading both range and farmland, and driving inhabitants from their homes (Huszar and Young 1984). The 1985 Farm Bill has been successful in providing economic incentives for landowners to convert back to grasslands highly erodible lands in the Great Plains.

Center-pivot irrigation systems were responsible for conversion of rangeland to cropland in the Great Plains, particularly in Colorado, Nebraska, Kansas, and Texas in the 1960s and 1970s (Supalla et al. 1982). Most of the land in these states is irrigated from the huge Ogalalla aquifer, which extends from the Texas panhandle to northern Nebraska. This aquifer is rapidly being depleted, and already, much irrigated land has been converted back to rangeland due to a falling water table. Depletion of this aquifer has resulted in considerable farmland conversion to rangeland in the 1980s and 1990s. California, Arizona, Nevada, Utah, and New Mexico are other states where, due to falling water tables, large areas of farmland have been converted back to rangeland.

In general, it appears that the amount of rangeland in the western United States will probably decrease in the next 20 to 30 years. In the Intermountain region large scale human immigration is causing rapid loss of rangeland to urbanization around major cities such as Albuquerque, Phoenix, Salt Lake City, Denver, Las Vegas, and El Paso. Unless there are major changes in land use planning, it appears that the vast open spaces that have long characterized many western states could easily disappear over the next 20–40 years (Huntsinger and Hopkinson 1996). In most developing countries of Africa and South America, considerable reduction will probably occur in the amount of rangeland. Regardless of these changes, rangeland will continue to be the major type of land in the world.

Range Management Challenges in the Twenty-First Century

We believe the late 1990s are a period of transition for the range management profession. It has been characterized by Kennedy et al. (1995) as a period in which the commodity production-goal-oriented approaches of the past are giving way to new approaches that center around sustainability, diversity, holism, and integrated natural resource planning.

We consider the primary challenges to range managers in the twenty-first century to be as follows:

1. Sustaining ranching as an occupation and way of life.
2. Preservation of open space.
3. Prevention and resolution of social conflicts over usage and management of natural resources.
4. Maintaining and improving the health of rangeland ecosystems.
5. Preservation of threatened and endangered species.
6. Expansion of supply of rangeland products as follows:
   - rangeland products
   - recreation
   - wildlife
   - water
   - esthetics
   - other

These will be discussed in detail throughout the book.

RANGE MANAGEMENT PRINCIPLES

- Rangeland is a renewable resource.
- Rangelands supply humans with food and fiber at very low energy costs compared to cultivated lands.

■ Rangeland productivity is determined by soil, topographic, and climatic characteristics.

■ Rangelands provide society with a variety of products that include food, fiber, water, wildlife, recreation, minerals, timber, and open space.

■ Social, economic, cultural, and technological considerations are all a part of the range management decision-making process.

■ Many rangelands in the United States are still characterized by vast expanses of open space dominated by natural vegetation. In the future, the value of these areas for esthetics and preservation of biological diversity may be far greater than the value of the commodities they can produce.

Literature Cited

Agricultural Research Policy Advisory Committee (ARPAC). 1975. *Research to meet U.S. and world food needs.* Report of Working Conference, Kansas City, MO, p. 208.

Blaisdell, J. P., V. L. Duvall, R. W. Harris, R. D. Lloyd, and E. H. Reid. 1970. Range research to meet new challenges and goals. *J. Range Manage.* 22:227–234.

Branson, F. A., G. F. Gifford, K. G. Renard, and R. F. Hadley. 1981. *Rangeland hydrology.* Kendall/Hunt Publishing Company, Dubuque, IA.

Chrispeels, M. J., and D. Sadava. 1977. *Plants, food, and people.* W. H. Freeman and Company, Publishers, San Francisco, CA.

Cox, J. R., H. L. Morton, J. T. Labaume, and K. G. Renard. 1983. Reviving Arizona's rangelands. *J. Soil Water Conserv.* 38:342–346.

Dyksterhuis, E. J. 1955. What is range management? *J. Range Manage.* 8:193–196.

Food and Agriculture Organization (FAO). 1995. *Production yearbook. United Nations FAO Statistics Series No. 48.* Rome, Italy.

Glantz, M. H. (Ed.). 1977. *Desertification: Environmental degradation in and around arid lands.* Westview Press, Inc., Boulder, CO.

Heady, H. F., and D. Child. 1994. *Rangeland ecology and management.* Westview Press, San Francisco, CA.

Hedrick, D. W. 1966. What is range management? *J. Range Manage.* 19:111.

Heitschmidt, R. K., and J. W. Stuth (Eds.). 1991. *Grazing management: An ecological perspective.* Timber Press, Portland, OR.

Henson, J., F. Sprague, and G. Valentine. 1977. Soil Conservation Service assistance in managing private lands in Texas. *Trans N. Am. Wildl. Nat. Resour. Conf.* 42:264–270.

Huntsinger, L., and P. Hopkinson. 1976. Viewpoint: Sustaining rangeland landscapes: a social and ecological process. *J. Range Manage.* 49:167–174.

Huszar, P. C., and J. E. Young. 1984. Why the great Colorado plowout? *J. Soil Water Conserv.* 39:232–235.

Kennedy, J. J., B. L. Fox, and T. D. Osen. 1995. Changing social values and images of public rangeland management. *Rangelands* 17:127–132.

Mouat, D. A., and C. F. Hutchinson (Eds.). 1995. *Desertification in developed countries.* Kluwer Academics Publishers, London.

Payne, J. 1988. Wildlife enterprise opportunities on a limited land base. *Proc. Intnl. Wildl. Ranching Symp.* 1:82–94. Las Cruces, NM.

Postel, S. 1989. Halting land degradation. In *State of the world 1989. World Watch Institute Report.* W. W. Norton & Co., NY.

Ramsey, C. W. 1965. Potential economic returns from deer as compared with livestock in the Edwards Plateau region of Texas. *J. Range Manage.* 18:247–250.

Schiller, B. R. 1991. *The economy today.* 5th ed. McGraw-Hill, Inc., NY.

Small, C. P. 1988. Overview of wildlife ranching in Africa. *Proc. Intl. Wildl. Ranching Symp.* 1:8–20, Las Cruces, NM.

Smith, J. G. 1899. Grazing problems in the southwest and how to meet them. *U.S. Dep. Agric. Div. Agrost. Bull.* 16:1–7.

Society for Range Management. 1989. *A glossary of terms used in range management.* 3d ed. Society of Range Management, Denver, CO.

Steger, R. E, and R. F. Beck. 1973. Range plants as ornamentals. *J. Range Manage.* 26:72–74.

Stoddart, L. A., A. D. Smith, and T. W. Box. 1975. *Range management.* 3d ed. McGraw-Hill, Inc., NY.

Supalla, R. J., N. R. Lansford, and N. R. Gallehon. 1982. Is the Ogalalla going dry? *J. Soil Water Conserv.* 37:310–315.

United Nations (UN). 1977. *Desertification: Its causes and consequences.* Pergamon Press Ltd., Oxford.

United States Department of Agriculture (USDA). 1989. *The Second RCA Appraisal: Soil, water, and related resources on nonfederal land in the United States.* U.S. Department of Agriculture, Soil Conservation Service.

United States Department of Commerce (USDC). 1996. *Statistical Abstract of the United States, 116th Edition.* U.S. Department of Commerce, Bureau of Census.

Vallentine, J. F. 1989. *Range development and improvement.* 3d ed. Brigham Young University Press, Provo, UT.

Vallentine, J. F. 1990. *Grazing management.* Academic Press, Inc., NY.

Wallen, C. L., and M. D. Gwynne. 1978. Drought—a challenge to rangeland management. *Proc. Int. Rangel. Congr.* 1:21–32.

Winstanley, D. 1983. Desertification: A climatological perspective. In S. G. Wells and D. R. Naragan (Eds.). *Origin and evolution of deserts,* 185–213. University of New Mexico Press, Albuquerque, NM.

CHAPTER 2

RANGELAND PHYSICAL CHARACTERISTICS

A thorough knowledge of rangeland physical characteristics is essential for an understanding of range management problems. These physical characteristics include climate, soil, and topography. Physical characteristics determine the type of vegetation and its productivity in any area. The type of vegetation and the terrain on a piece of rangeland govern the types of livestock and wildlife it will support.

PRECIPITATION

Precipitation is the most important single factor determining the type and productivity of vegetation in an area. Table 2.1 shows average annual precipitation and forage production for selected ranges in the United States. Forage production increases rapidly as precipitation increases up to about 500 mm per year. Above 500 mm of precipitation per year, soil characteristics can assume much greater importance than precipitation in determining forage production. Critical characteristics of precipitation that affect vegetation are the total amount, the distribution, the relative humidity, the form, and the annual variability.

Most rangelands are characterized by low total precipitation (less than 500 mm per year). In the 11 conterminous western states of the United States, 80% of the area receives less than 500 mm of annual average precipitation (Figure 2.1). In Africa about 50% of the area receives less than 500 mm of annual average precipitation and about 20% receives less

TABLE 2.1 Average Annual Precipitation, Forage Production, and Stocking Rate for Various U.S. Ranges in Good Condition

LOCATION	AVERAGE ANNUAL PRECIPITATION IN	MM	PATTERN OF PRECIPITATION	TYPE OF RANGE	FORAGE PRODUCTION (DRY MATTER) LB/ACRE	KG/HA	NUMBER OF HECTARES (ACRES) REQUIRED TO SUPPORT A 450-KG (1000-LB) COW FOR A YEAR ACRES	HA	REFERENCE
California	60	1520	Pacific	Annual grassland	2759	3100	3–5	2	Bartolome et al. (1980)
	28	710	Pacific	Annual grassland	1958	2200	7	3	Bartolome et al. (1980)
	20	510	Pacific	Annual grassland	1157	1300	11	5	Bartolome et al. (1980)
	8	200	Pacific	Annual grassland	400	450	30	14	Bartolome et al. (1980)
Northeastern Oregon	20	510	Great Basin	Coniferous forest	250	280	82	37	Skovlin et al. (1976)
	20	510	Great Basin	Palouse bunchgrass	330	370	62	25	Skovlin et al. (1976)
Southwestern Utah	7	180	Great Basin	Salt desert (winterfat)	213	239	100	45	Hutchings and Stewart (1953)
Southwestern Oregon	11	290	Great Basin	Sagebrush grassland	322	360	64	26	Hyder (1953)
Eastern Oregon	16	400	Great Basin	Sagebrush grassland	270	302	77	31	Miller et al. (1980)
Central Utah	13	330	Great Basin	Crested wheatgrass	442	495	25	11	Frischknecht and Harris (1968)
Northern Utah	24	610	Great Basin	Mountain shrubland	870	974	24	10	Laycock and Conrad (1981)
Southcentral New Mexico	8	219	Southwestern	Desert grassland	276	309	74	30	Herbel and Gibbens (1996)
Southeastern Arizona	14	340	Southwestern	Desert grassland	343	384	60	24	Martin and Ward (1976)
Northeastern Colorado	12	310	Plains	Shortgrass prairie	1260	1411	14	6	Uresk et al. (1975)
Western Nebraska	13	330	Plains	Shortgrass prairie	1294	1449	13	6	Burzlaff and Harris (1969)
Southeastern Alberta	13	330	Plains	Northern mixed prairie	345	388	52	21	Smoliak (1986)

Location			Region		Vegetation				Reference
Western North Dakota	16	410	Plains	700	Northern mixed prairie	784	26	10	Rogler and Lorenz (1957)
Western South Dakota	14	390	Plains	2100	Northern mixed prairie	2352	7	3	Hanson et al. (1970)
Northcentral Texas	9	710	Plains	1424	Southern mixed prairie	1595	13	5	Kothmann et al. (1978)
Northwest Texas	20	410	Plains	1800	Southern mixed prairie	2016	10	4	Brown and Schuster (1969)
Edwards Plateau Texas	24	610	Plains	1331	Oak savannah	1490	14	5	Reardon and Merrill (1976)
Eastern Kansas	33	840	Plains	4100	Tallgrass prairie	4592	3–4	1–2	Owensby et al. (1973)
Northeastern Oklahoma	32	810	Plains	3100	Tallgrass prairie	3441	4–5	2	Baker and Powell (1978)
Southeast Texas	34	860	Plains	2772	Coastal prairie	3115	5–6	2–3	Scifres et al. (1982)
Central Louisiana	58	1470	Eastern	2000	Southern pine forest	2240	7	3	Pearson and Whitaker (1974)
Southern Illinois	50	200	Eastern	1750	Bluegrass bluestem	1960	8	3–4	Voigt (1959)
Southcentral Florida	60	1520	Florida		Southern pine forest				Kalmbacher et al. (1984)
				988	Pine-palmetto	1110	14	6	
				3296	Fresh marsh	3703	4–5	2	

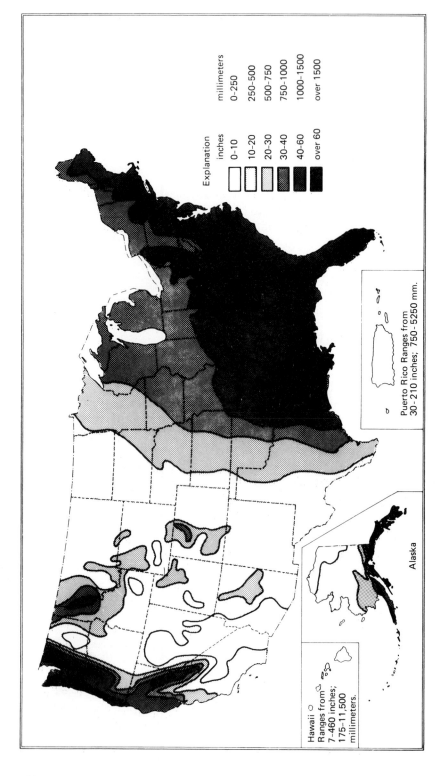

Figure **2.1** Mean annual precipitation in the United States. (From USWRC 1978.)

Explanation

inches		millimeters
0–10		0–250
10–20		250–500
20–30		500–750
30–40		750–1000
40–60		1000–1500
over 60		over 1500

Puerto Rico Ranges from
30–210 inches; 750–5250 mm.

Hawaii
Ranges from
7–460 inches;
175–11,500
millimeters.

Alaska

than 250 cm. About 70% of the area in Australia receives less than 500 mm of annual average precipitation and about 33% receives less than 250 mm (Rumney 1968). In Mexico 75% of the area receives less than 500 mm of annual average precipitation and 40% receives less than 250 mm. South America differs in that between 65% and 75% of the area receives over 500 mm of annual average precipitation.

Orographic Influences

Tremendous differences in the geographic distribution of precipitation occur over short distances in the 11 conterminous western U.S. states. However, in the eastern and central states, precipitation decreases gradually with movement inland from the Atlantic Ocean. This is explained by the relatively flat terrain in the eastern compared to the western United States.

Precipitation distribution is influenced by topography and distance from oceans. Inland areas have lower precipitation than coastal areas because air masses gradually lose water evaporated from the oceans as they move inland. Topography influences precipitation because air masses cool as they move upward over mountains. This is called the *orographic effect.* Condensation and precipitation occur because cool air can hold less moisture than can warm air. Areas on the leeward side of mountains are usually quite dry. This is because air masses descending mountain slopes lose much of their moisture during ascent, and they can hold more moisture as they warm with decreased elevation. The Great Basin area of the western United States is quite arid because it lies between the Rocky Mountains to the east and the Sierra-Cascade Mountains to the west. San Francisco, on the Pacific coast, receives 500 mm of rainfall compared to 180 mm for Reno, Nevada, which is 142 km inland. The much lower precipitation at Reno results from the fact that air masses lose most of their moisture while crossing the Coast and the Sierra Mountain ranges (Figure 2.2). Reno is at the base of the Sierra Mountains in the zone of lowest precipitation, which is called the "rain shadow." The higher-elevation areas on the western slopes of the Sierras, 25 km west of Reno, receive over 750 mm of annual precipitation. Throughout the western United States, large differences in precipitation occur over short distances due to the sharp changes in topography. In the state of Washington there is 10 to 14 times more precipitation in the Cascade Range than in the Columbia Basin, which is 62 km to the east. Because of the orographic effect, precipitation generally increases with elevation. Therefore, low-elevation ranges typically produce much less vegetation than do those at high elevations. The annual average precipitation for New Mexico's weather stations below 1220 m altitude is 250 mm, those between 1220 and 1820 m receive 360 mm, and those above 1820 m receive over 410 mm.

The lower amount of precipitation that characterizes most rangelands is made more severe by the great year-to-year variability (Figure 2.3). The variability in precipitation increases rapidly as the annual total drops below 450 mm per year (Conrad and Pollak 1950). Even slight reductions from normal precipitation can cause severe reductions in plant yield in areas below 300 mm of precipitation, while much greater reductions in precipitation may have no influence on plant yield in areas with over 800 mm of precipitation (Klages 1942).

Annual variability in timing of precipitation can be more important than variability in the total amount that occurs. In 1980, total annual precipitation on desert grassland ranges around

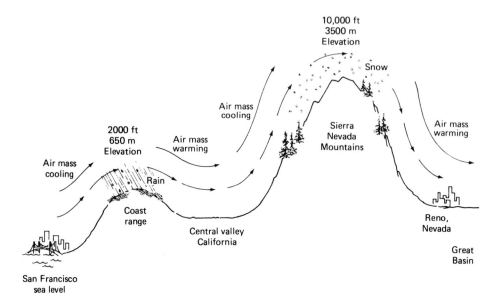

Figure 2.2 As air masses move inland from the Pacific Ocean, they rapidly lose their moisture due to cooling as a result of elevation increases from the Coast Ranges and Sierra Nevada Mountains. This is called *orographic effect*. (Drawing by John N. Smith)

Las Cruces, New Mexico, was near average. However, this precipitation occurred primarily in the fall and winter when temperatures were too low for good growth of the native perennial grasses. Range forage production was severely depressed in 1980 compared to 1979, when total precipitation was well below average but arrival occurred primarily in the summer.

Drought is defined as prolonged dry weather, generally when precipitation is less than 75% of average annual amount (Society for Range Management 1989). Using this criterion, over the past 40 years (1944–1984) drought in the United States has occurred in 43% of the years in the Southwest, 13% of the years in the Northwest, 21% of the years in the northern Great Plains, and 27% of the years in the southern Great Plains. The most severe and widespread drought in the Great Plains region of the United States in recorded history occurred between 1933 and 1935. In the southwestern United States, the most severe drought occurred between 1951 and 1956.

These droughts all had a tremendous influence on rangeland vegetation. Vegetation recovery in the Great Plains from the 1930s drought involved 5 to 15 years even under conditions of light to no grazing, depending on how severely drought affected particular areas. On the Jornada Range in central New Mexico, annual precipitation between 1951 and 1956 averaged 55% of the predrought average (Herbel et al. 1972). Cover of black grama (*Bouteloua eriopoda*), the dominant grass on upland sites, was reduced over 60%, and honey mesquite (*Prosopis glandulosa*) heavily invaded many areas supporting black grama. After 40 years black grama stands on deep sands on the Jornada have not recovered even though grazing has been light (Herbel and Gibbens 1996).

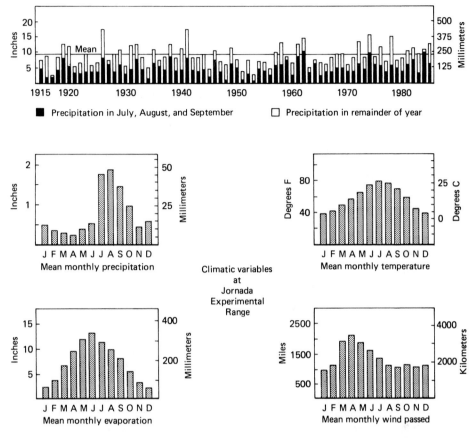

***Figure* 2.3** Annual variation in precipiation at the Jornada Experimental Range in south-central New Mexico. (From USDA 1987.)

Two or more consecutive years of drought have far more impact on vegetation than 1 year of drought followed by normal or above-normal precipitation. One of the most severe droughts on record occurred in the northwestern United States in 1977. Precipitation in central and eastern Oregon was only 49% of the annual average. This area had above-average precipitation in 1978. Ganskopp and Bedell (1981) reported little mortality of important forage species resulting from this drought. They found that grasses, which were lightly to moderately grazed during the drought, produced as much forage as did ungrazed plants. Weaver and Albertson (1936) observed that moderately grazed plants withstood drought better than did ungrazed plants in the Great Plains, but heavy grazing reduced plant cover and productivity. Although light to moderate grazing may improve a plant's ability to withstand drought, heavy grazing during and after drought years reduces plant cover and productivity (Young 1956; Pieper and Donart 1975).

WIND

Wind is caused by uneven heating of land and water areas as the earth turns on its axis. Wind can substantially reduce precipitation effectiveness by increasing soil evaporation losses and increasing plant transpiration. Wind has little influence on soil moisture below 20 cm to 30 cm (Veihmeyer 1938). The highest wind velocities occur in flat terrain with few trees. The Great Plains in the central United States are characterized by high wind velocities, particularly in the Texas Panhandle, southeastern Wyoming, and North Dakota. Low wind velocities occur in most of Washington, Oregon, and California, where topographic barriers are at a maximum. Hot summer winds in the Great Plains during dry years greatly magnify the effect of drought by increasing both soil moisture losses and plant transpiration. Cold, windy springs in the Great Basin region of eastern Oregon, northern Nevada, and southern Idaho generally result in poor forage production even though winter and spring precipitation may be above average. Much of the precipitation may be lost from the soil surface before temperatures are warm enough for significant plant growth. Spring winds in southern New Mexico greatly reduce the effectiveness of winter precipitation.

TEMPERATURE

Temperatures for different seasons and years vary substantially on rangelands in temperate zones. In tropical areas, monthly and yearly temperatures show little variation. In the mountainous areas of the northwestern United States, temperature can have as much influence on annual variation in forage production as does precipitation. Nearly all forage species are cool-season plants in this region. They make limited growth during intermittent warm periods in the winter, but most of their growth occurs in the spring. Years with above-average spring temperatures are usually characterized by above-average forage production even though total precipitation may be below average. However, years when below-freezing temperatures occur late into the spring often have low forage production regardless of precipitation. This is due to the fact that much of the moisture in the soil may be evaporated by wind before temperatures are warm enough for high rates of growth. When high wind velocities occur, considerable snow may be evaporated directly from the snowpack on high-elevation areas without ever reaching the soil. In most years summers are too dry for plant growth in this region. In the northern portions of the Great Plains, particularly Montana and North Dakota, spring temperatures vary considerably from year to year and can have a substantial influence on forage production of cool-season species.

Temperatures above average usually occur concurrently with drought in the Great Plains. This increases evaporation of the limited soil moisture and greatly magnifies the effect of the drought.

FROSTFREE PERIOD

On high mountainous areas (over 2,000 m elevation) in the western United States, the frost-free period at many locations is less than 100 days. This gives plants only a very short period to complete their growth cycle. The low temperatures on these ranges are a much bigger constraint on forage production than is precipitation.

HUMIDITY

Humidity refers to the amount of moisture in the air. It is usually expressed as *relative humidity*, which is the percentage of the maximum quantity of moisture that the air can hold at the prevailing temperature. Cold air can hold less moisture than can warm air. Water evaporation from the soil and plant transpiration losses increase as the relative humidity decreases. Therefore, areas with high humidity give greater plant growth per unit of precipitation than do areas where humidity is low. The lowest relative humidity averages in the United States during the summer occur at low elevations in Nevada, Arizona, New Mexico, and California. Relative humidities increase with movement eastward across the Great Plains toward the Atlantic Ocean. The highest humidities occur in the southeastern states, particularly Florida.

CLIMATE TYPES

Timing of precipitation, total annual precipitation, and the annual average temperature regime are the primary factors determining the climatic type of an area. We will discuss the climatic types of the United States following the classification and descriptions of Humphrey (1962) (Figure 2.4).

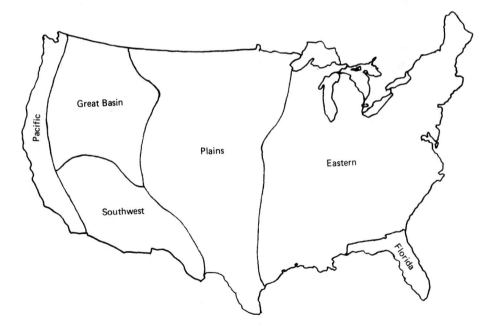

Figure **2.4** Climatic types in the United States. (Based on the classification of Humphrey 1962.)

The Pacific Climate

The Pacific climate characterizes the Pacific coastal areas of California, Oregon, Washington, and British Columbia. It extends inland to the Cascade and Sierra Nevada Mountains. Summers are quite dry and winters are wet (Figure 2.5). The percentage of precipitation occurring during the winter increases from north to south, although total precipitation decreases. Precipitation also decreases rapidly with movement inland. The percentage of the total precipitation that falls during summer ranges from about 10% in northern Washington to 1% in most of California. The temperatures associated with this pattern are mild near the coast but become progressively colder with movement inland. The vegetation of the southern half, in California and southwestern Oregon, is primarily chaparral and grassland. In Washington, British Columbia, and northwestern Oregon, dense coniferous forest occurs as a result of the high total rainfall (900 mm to 2500 mm). Because this entire area is characterized by little summer rainfall, the herbaceous species are almost entirely cool-season. Grasslands occur in low-elevation areas with 250 mm to 460 mm of annual precipitation. The warmer temperatures associated with these areas permit growth of the cool-season grasses in the winter and spring when moisture is available. Other parts of the world with this pattern of climate and similar vegetation include southern Europe around the Mediterranean Sea and the coastal areas of South Africa, Australia, and southwestern South America.

The Great Basin Climate

The Great Basin climate occurs largely in the states of Washington, Oregon, Idaho, Nevada, Utah, and portions of eastern California. In California, Oregon, and Washington, this pattern occurs east of the Sierra-Cascade Mountains. The Great Basin region has much less precipitation than the Pacific region, but the pattern of precipitation is similar. Like the Pacific climate, summers are the driest period. However, more precipitation falls in the spring, and summers are less droughty than in the Pacific region. Total precipitation ranges from 500 mm in northeastern Oregon and eastern Washington to 100 mm in Nevada. Low precipitation is largely in reflection of the rain-shadow effect of the Sierra-Cascade and Rocky Mountains. Winter temperatures over much of this region are cold, with the exception of palouse prairie in Oregon and Washington. Shrubs of the genera *Artemisia, Atriplex,* and *Juniperus* dominate most of this region because they can utilize moisture stored deep in the soil profile during the summer dry period. Grasses in the region are generally all of cool-season type because of the lack of summer moisture. Portions of the Middle East, the former Soviet Union, South Africa, Australia, and western South America have this pattern of climate and similar associated vegetation.

The Southwestern Climate

The southwestern climate occurs in New Mexico, Arizona, southern Utah and Nevada, and southwestern Texas. This region has basically a biomodal precipitation pattern with winter precipitation, spring drought, summer precipitation, and fall drought. Pacific air moving in

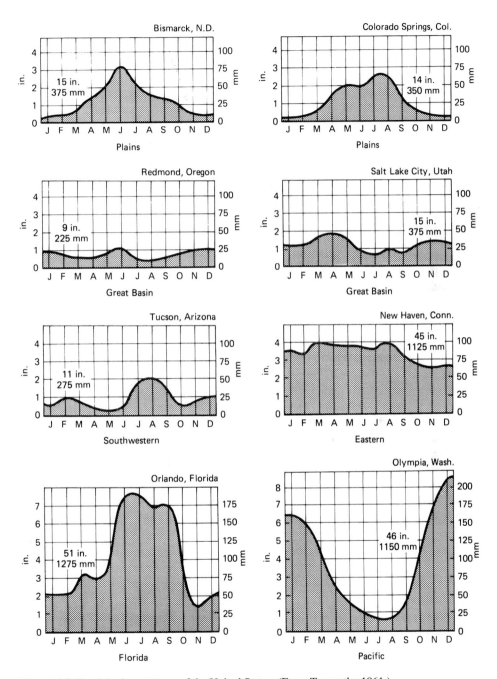

***Figure* 2.5** Precipitation patterns of the United States. (From Trewartha 1961.)

from the west causes the winter rains, which are primarily frontal storms. Summer rains result from air moving in from the Gulf of Mexico, and they are primarily convectional storms. These storms are local in nature and usually of short duration and very high intensity. Their effectiveness for forage growth is much lower than that of the more frequent, light-intensity storms in the central Great Plains. Shrubs in the genera *Prosopis, Larrea, Opuntia*, and *Acacia* dominate many areas because they are well adapted to warm temperatures and long dry periods. Nearly all of the perennial grasses in this region are warm-season plants and grow as the result of summer rainfall. Winter precipitation favors cool-season annual forbs. This biomodal pattern has maximum development in Arizona. Parts of Africa and South America also have this type of precipitation pattern. Temperatures are warm throughout this region during the entire year. Total precipitation ranges from 150 mm to 460 mm.

The Plains Climate

The plains precipitation pattern characterizes the central United States. Maximum precipitation delivery occurs in the late spring and summer, with a moderate to light amount in the fall. The least amount of precipitation occurs during the winter. It is of considerable importance that most of the precipitation in the central plains area comes as frequent, light-intensity rains during the spring-summer growing season, which maximizes its effectiveness. This in large part explains why the central United States is a grassland. Because precipitation occurs during both the spring and summer growing season, this region is dominated by a mixture of warm- and cool-season grasses. Total precipitation ranges from 250 mm to 900 mm. Other parts of the world with this type of climate and similar vegetation include the lee side of the Andes in South America and continental, midlatitude Europe and Asia.

The Eastern Climate

The eastern climate covers all of the eastern United States with the exception of the Florida peninsula. It is distinguished by uniformity in amount of precipitation during each month. Over most of the east the precipitation ranges from 900 mm to 1400 mm, with greater amounts in the southern portion and the mountains. With movement northward, an increasing percentage of the precipitation comes as snow. A large amount of precipitation with even distribution in a temperate climate provides favorable conditions for deciduous trees. This explains why much of the natural vegetation is deciduous forest. Northern Europe, including the British Isles, has this pattern of climate and similar vegetation.

The Florida Climate

The Florida climate is restricted to the Florida peninsula. This area receives very heavy rainfall (250 mm or more per month) during the summer period from June to October, but the period November to May is quite dry. Annual total precipitation for this area ranges from 1270 mm to 1500 mm. Because of the heavy rainfall and warm climate, this type is the most productive of all regions for forage. However, the rainfall also leaches the soils, and several minerals are deficient in the forage.

The Tropical Wet and Dry Climate

In addition to the climates described previously, there is the tropical wet and dry climate, which characterizes rangelands in southeastern Asia, central Africa, India, northern Australia, and parts of South America. This pattern is described by Wallen and Gwynne (1978) and Heady and Heady (1982). It is characterized by high temperatures throughout the year, with sharply defined wet and dry periods. Considerable variability exists in precipitation between years in most areas receiving this pattern. Temperatures below freezing occur only on high mountain areas. Tropical wet and dry climates with 130 mm to 250 mm of precipitation typically support desert shrubs, grasses, and cactus. Where annual rainfall is between 250 mm and 500 mm, an open savanna of grasses and trees occurs. A dry forest with large trees and scattered shrubs occurs in the 500- to 1140-mm precipitation zone. Tropical forests dominate most landscapes that have over 1140 mm of precipitation.

TOPOGRAPHY

The wide difference in topography between the eastern and western United States largely explains the difference in climate and vegetation. Hopkins (1938) related altitude and latitude and longitude to vegetation in a law known as *Hopkin's bioclimatic law.* The interpretation of Hopkin's law is that a 305-m increase in altitude will result in essentially the same phenological (relationship between climate and biological phenomena) changes that would be encountered traveling 107 km north with no increase in elevation. This law applies to many parts of the western United States, where extremes in forage development during the spring may occur within a few kilometers due to elevation. In April new growth may be available for grazing at low-elevation ranges in the Great Basin region of the United States, but the surrounding mountains, only a few kilometers away, will still be covered with snow. During the summer, with seasonal advance, livestock must be removed from high-elevation ranges in the fall because of the onset of severe cold and snow.

Aspect

Aspect refers to the directional orientation of slopes. Temperature on slopes in North America increases as aspect changes from north to east to west to south. The aspect of the slopes has considerable influence on vegetation they support and their use by grazing animals. In the spring the warmer south- and west-facing slopes support forage species that are more advanced in growth than those on the cooler north- and east-facing slopes (Figure 2.6). During the winter, grazing animals usually prefer south- and west-facing slopes because of the warmer temperatures. In the summer north- and east-facing slopes are preferred by grazing animals because they are cooler, forage is less advanced in growth, and shade is more available.

***Figure* 2.6** South-facing slopes in northeastern Oregon support grassland and are best used in the spring. North-facing slopes are best used in summer because there is more shade and forage species are less advanced in growth.

Degree of Slope

Degree of slope is of considerable importance in range management because it affects both vegetation productivity and use by range animals. In range surveys, slope is commonly expressed in percent. As slope increases, vegetation productivity declines per unit precipitation because less water enters the soil and more runs off as overland flow. Livestock use, particularly that of cattle, decreases with increasing slope because of the difficulty the animals have in climbing. The influence of slope on range utilization by livestock is discussed in more detail in Chapters 9 and 10.

SOILS

Soil is defined as the dynamic, natural body of the surface of the earth in which plants grow (Brady and Weil 1996). The most severe consequence of rangeland mismanagement or overgrazing is loss of the soil profile. This is because soil is the primary factor determining the potential for forage production of an area within a particular climate. Soil formation is a very slow process; a thousand years or more are required to form an inch of soil. However under poorly controlled grazing this same inch of soil can be lost to erosion within a few years (Figure 2.7). A critical part of any range management plan is to maintain enough vegetation cover to protect the soil profile from erosion.

Knowledge of soil characteristics and classification is essential for the rangeland manager. Soil is comprised of minerals, organic materials, and living forms. Soil is distin-

***Figure* 2.7** This heavily stocked range in the Nebraska sandhills shows severe erosion from depletion of vegetation cover.

guished from regolith (unconsolidated rock and materials) by a higher organic matter content, more intense weathering, the presence of horizontal layers, and the presence of living organisms (Figure 2.8).

Soil characteristics of importance to the range manager include texture, structure, depth, pH, organic matter content, and mineral status (fertility). The interaction between these six factors with climate and topography determines the type and quantity of vegetation that an area is capable of supporting.

Texture

Soil *texture* refers to the size of the mineral particles comprising the soil. Soil textural sizes from the smallest to largest are clay (less than 0.002 mm), silt (0.002 mm to 0.02 mm), fine sand (0.02 mm to 0.2 mm), coarse sand (0.2 mm to 2.0 mm), fine gravel (2 mm to 5 mm), and coarse gravel (more than 5 mm). Based on the proportions of these particle sizes, a soil can be classed as sand, loamy sand, sandy loam, silt loam, clay loam, or clay (Figure 2.9).

To a considerable degree, soil texture determines the fertility of the soil. Soils with a high clay content retain nutrients such as nitrogen, phosphorus, and potassium compounds much better than do sands. This is because the small particles have a much greater total surface area for attracting and binding nutrients per unit volume than do large particles.

Soil texture also plays an important role in soil moisture status. Water enters coarse, sandy soils much more rapidly than fine clay because there is more space between particles

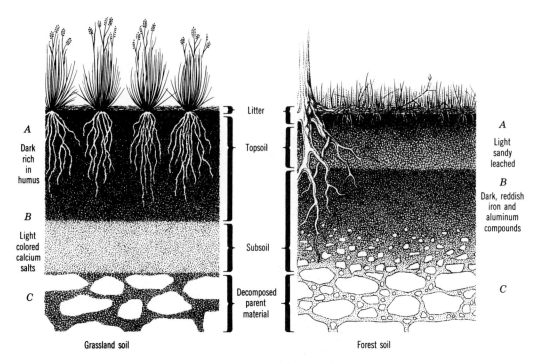

Figure 2.8 Profiles of grassland and forest soils. (From R. F. Dasmann, 1976, *Environmental Conservation*, 4th ed. Copyright © 1976, John Wiley & Sons, Inc., New York. Reprinted by permission of John Wiley & Sons, Inc.)

(Figure 2.10). On the other hand, clay soils retain water much better than do sandy soils. The best balance between moisture infiltration and retention is obtained with loamy soils, which have a mixture of sand, silt, and clay.

In arid regions such as the western United States, sandy soils provide a more favorable habitat for plants, particularly grasses, than do clay soils. This is because the limited precipitation that occurs rapidly infiltrates the soil rather than moving overland. In the higher-rainfall areas (over 900 mm of annual average precipitation) east of the Mississippi River, moisture and nutrient retention by soils has more influence on vegetation productivity than rate of water infiltration. In these areas soils high in clay are more productive than those high in sand.

Structure

Soil *structure* refers to how soil particles are arranged. Soil structure is quite important since it determines the rate at which water can enter the soil. There are six basic types of soil structure: platelike, prismlike, blocklike, spheroidal, single-grained, and massive

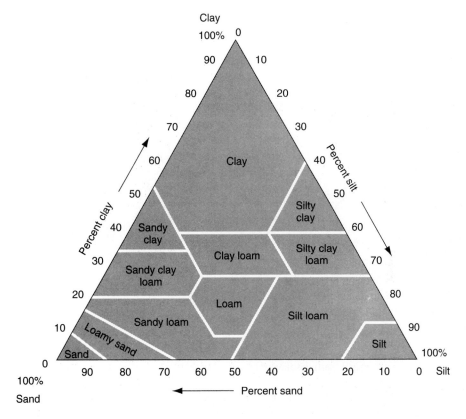

***Figure* 2.9** Soil texture designation based on percentages of sand, silt, and clay. (From Owen and Chiras 1995.)

(Figure 2.11). A massive soil structure is tightly compacted with little or no space between soil particles for infiltration. This structure type often occurs on overgrazed areas. Single-grained structure refers to a loose arrangement of individual soil particles that do not stick together, such as sands. Platelike structure refers to soil particles aggregated together into leafy plates. This type of structure can result naturally from parent material or soil-forming forces. Infiltration is minimized with this type of structure. Prismlike structure refers to soil particles aggregated in vertical columns. This type of structure is most common in arid areas. When the tops are round, this type of structure is referred to as *columnar*. Columnar structured soils in the west are often associated with a high sodium content and unfavorable soil-water relations. Soils with blocklike structure have a cubical aggregation. The blocks have sharp edges and are found in the subsoil. The spheroidal structured soils have the best characteristics for water infiltration and storage. The aggregates are round and easily shaken apart. Soils high in organic matter usually have a spheroidal structure. This type of structure is associated with grassland areas.

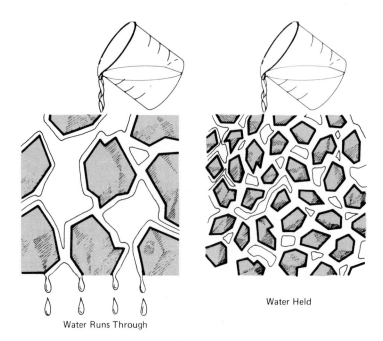

Water Held

Water Runs Through

Figure **2.10** Soil water-holding capacity. The water-holding capacity of soil increases as particle size decreases because water clings to surfaces and small particles have relatively the most surface. (From Nebel 1981.)

Depth

Soil depth has considerable influence on range productivity since it determines how much moisture the soil can hold. A deep soil in an area with moderate precipitation will often produce more forage than a thin soil receiving heavy precipitation. Conversely, in some arid areas, grasses on thin, sandy soils underlain with a hard impermeable layer (caliche) are more productive than grasses on deeper sandy soils (Herbel et al. 1972). The impermeable layer restricts moisture to the portion of the soil profile near the surface where it can readily be used by the fibrous roots of the grasses. Depth usually means distance from the soil surface to bedrock or the unconsolidated material. Soil depth, in conjunction with texture and structure, largely determines the potential of a site for cultivation.

pH

Soil pH indicates the status of the soil in regard to exchangeable mineral ions. The pH value expresses whether a soil is basic or acidic. The proportional weight of the free hydrogen ions in the soil solution is given by the actual numerical value. The pH value is a logarithmic expression and is demonstrated as follows: pH 1 = one hydrogen ion per 10,000 ions, pH

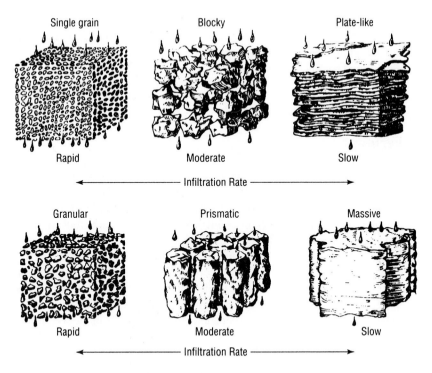

Figure 2.11 Note the variety of shapes occurring in soil aggregates and the relative rates at which water can move through them. (From Owen and Chiras 1995.)

2 = one hydrogen ion per 100,000 ions, pH 3 = one hydrogen ion per 1 million ions, and so on. A high-acid soil would have a pH of 4, while a highly basic soil would have a pH of 10 or more. The scale ranges from 1 to 14. Minerals are most available to plants in soils with pH values between 6 and 8. A pH value of 7 is considered ideal for growth of most plants. However, many species of plants grow best where soils are either highly acidic or basic.

Most soils in the western United States are slightly basic because they are derived from calcareous parent material and receive low precipitation. In contrast, most eastern soils are slightly acidic. The low rainfall in the west results in little leaching of soil minerals and the associated base losses. *Leaching* refers to the downwashing of soil minerals from precipitation. Soils of the southeast are the most heavily leached in the United States. South America has large areas of soils called Oxisols that are the most heavily leached soils in the world.

Organic Matter

Soil organic matter represents an accumulation of partially decayed and partially synthesized plant and animal residues. It represents a relatively small part of the soil (<1% to 6%) and is found primarily in the upper 30 cm of soil. Partially decomposed organic matter that has been incorporated into the soil is called *humus*. Humus provides a constant, although small supply of nutrients to plants. More important, it serves as a binding site for cations and prevents them from leaching out of the soil profile. Three other important functions of humus include binding of soil particles together, increasing soil moisture-holding capacity, and providing food for microorganisms. For these reasons, it is important to maintain as much humus in the soil as possible. Grassland soils are generally much higher in organic matter than are forest soils.

Fertility

Next to water, fertility is the second most limiting factor to forage production in arid rangelands with less than 500 mm precipitation. On high-rainfall rangelands (over 1,000 mm of annual precipitation) soil fertility is usually the most limiting factor to forage production. Nitrogen is the most deficient element in rangeland soils from the standpoint of plant growth. Most rangeland soils are also deficient in phosphorus and some in potassium and sulfur. Rangeland soils in high-rainfall areas are heavily leached and have low levels of elements required by animals, such as copper, cobalt, magnesium, sodium, and zinc. Forages growing on these soils often have inadequate levels of these elements to meet livestock requirements. Selenium and molybdenum are two elements that cause both toxicity and deficiency problems for livestock depending on their levels in the soil. Low soil fertility is a major problem limiting forage and livestock production in tropical rangelands in many parts of Africa, Asia, Central America, and South America. Forest soils are generally much more leached and less fertile than grassland soils.

Soil Classification

The soil classification system presently in use was developed by the Soil Survey Staff to the U.S. Department of Agriculture. It is called the "Comprehensive Soil Survey System." The major feature of this system is that it is based on the characteristics of soils as they are found in the field. It classifies soils rather than soil-forming processes. Soil names give the major physical characteristics. The levels of classification in this system are as follows: (a) Order, (b) Suborder, (c) Great Group, (d) Sub Group, (e) Family, and (f) Series. This system uses diagnostic surface and subsurface horizons as the basis of soil classification.

The ten orders of the present soil classification system are discussed below. Different types of vegetation are associated with different soil orders (Figure 2.12).

Entisols –   Soils found in the Rocky Mountains area of the United States. They are the youngest of soil orders. They are lacking in horizon development.

Inceptisols – These soils are slightly more advanced in profile development than Entisols. They are formed from volcanic ash and found primarily in the Pacific Northwest.

Aridisols –   Desert soils found in the Southwest with little profile development. They are dry more than 6 months of the year.

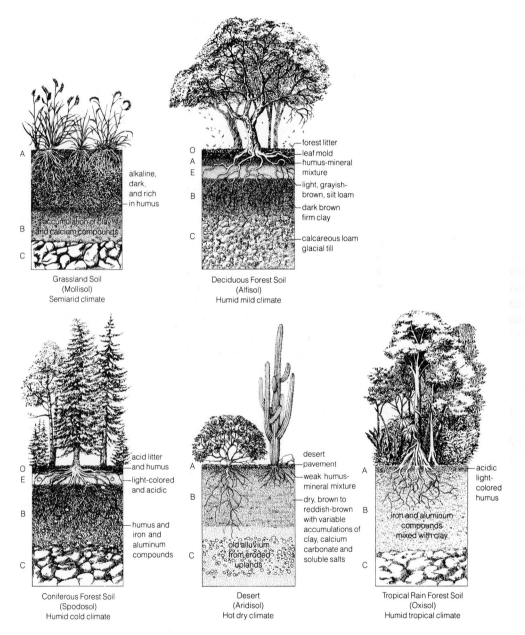

Figure 2.12 Soil profiles of the major soil orders typically found in five different biomes. (From Miller 1990.)

Vertisols – These soils have a high content of swelling clay that shrinks and cracks when wet. They are found mostly in southcentral United States.

Mollisols – These are the natural grassland or prairie soils. They typically are deep, have a high organic matter content, a high base supply, and moderate profile development.

Alfisols – Alfisols are similar to mollisols but have a higher organic matter content. These soils are associated with the eastern deciduous forest.

Spodosols – These are leached mineral soils with distinct organic horizon and a developed B horizon but with less accumulation of clays than Ultisols. These soils are found primarily in the northeastern and northcentral United States, and generally support coniferous forest.

Ultisols – Ultisols are the most highly leached soils in the United States. They are found primarily in the Southwest. These soils are moist much of the time and have a B horizon of clay accumulation, a leached A horizon, and a low base supply. They usually support forest or savannah woodland.

Oxisols – These are the most developed of all soils. They occur in tropical areas and are almost absent from the contiguous United States. Oxisols have the lowest fertility of all orders. They are the least suited for farming.

Histosols – These soils have a very high organic matter content and are often referred to as peat. They are found primarily in Minnesota and Louisiana, in lowland areas with swamps and bogs.

INFLUENCE OF RANGELAND PHYSICAL CHARACTERISTICS UPON RANGE ANIMALS

The interaction between climate, soil, and topography determines the potential of a particular range to support livestock and wildlife. Warm wet rangelands produce high quantities of forage for livestock. Parasites and diseases are important problems on these rangelands but are usually of minor importance on temperate arid rangelands where forage production is much lower. Parasites and diseases have been major obstacles to development of a range livestock industry on many of the more productive rangelands in Africa (Petrides and Swank 1958; Pratt and Gwynne 1977). Native animals that evolved with these parasites and diseases are much less affected than are introduced domestic animals. Range forages in high-rainfall areas are often deficient in many minerals required by range animals, such as phosphorus, potassium, copper, cobalt, magnesium, sodium, and zinc, because of severe soil leaching. In contrast, these minerals are usually nearly adequate to adequate in forage from arid areas.

In temperate areas, cold temperatures and heavy snow can have important adverse impact on both range livestock and wildlife. Large losses of livestock and game animals occurred in the western United States during the winter of 1948–1949 and 1983–1984 due to prolonged periods of heavy snow and cold temperature. Supplemental feeding of livestock

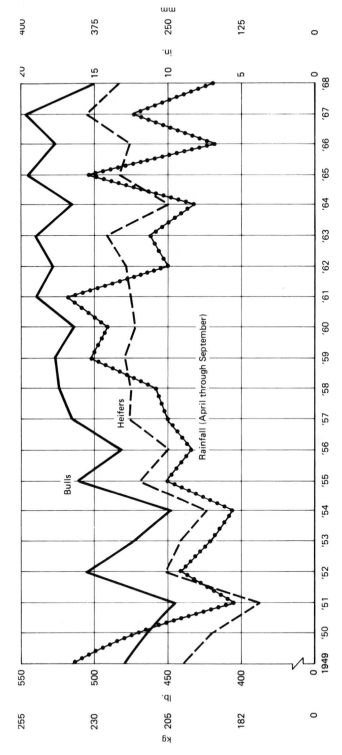

Figure 2.13 Weights of cattle and growing season rainfall on desert grassland range in southcentral New Mexico. (From Knox and Ellis 1969.)

43

during the winter is necessary on many temperate rangelands, which greatly increases production costs.

Annual weather variation has considerable influence on range animal productivity. Figure 2.13 shows the relationship between yearly total precipitation and cattle weights on a southcentral New Mexico range. Large decreases in weights were associated with the dry years. In Africa the drought in the Sahel region between 1969 and 1974 was responsible for loss of over 65% of the cattle production (King 1983).

Both climate and topography are important considerations in the selection of type or types of livestock for a particular range. European cattle and sheep are well suited to the cold, windy conditions associated with temperate ranges, while zebu cattle and goats can best handle the parasites, disease, and heat problems associated with tropical ranges. Yaks, musk oxen, llamas, and alpacas can withstand the severe cold of alpine-tundra ranges due to their heavy coat of hair. Camels efficiently use hot desert ranges because of their low water requirements and ability to browse shrubby vegetation. Sheep and goats make better use of steep, rocky ranges than do cattle, bison, or water buffalo. Cattle and bison are best adapted to flat, open ranges.

It is important to recognize that variation in rangeland physical characteristics necessitates that management practices vary widely from area to area. Climatic conditions, particularly precipitation, vary considerably both within and between years on most rangelands. This variability has a major influence on both plant and animal productivity. Successful range management depends on a good knowledge of the interactions among physical, plant, and animal factors.

RANGE MANAGEMENT PRINCIPLES

- Precipitation is the most important single factor determining the type and productivity of vegetation in an area. Differences in amount, timing, and frequency of precipitation among rangeland types necessitate different management strategies.

- Soil is the primary factor determining forage production within a particular climate. A thousand years or more are required to form an inch of soil. A critical part of range management is to maintain enough vegetation cover to protect the soil profile.

Literature Cited

Baker, R. L., and J. Powell. 1978. Oklahoma tallgrass prairie responses to atrazine and 2,4-D and fertilizer. *Proc. Int. Rangel. Congr.* 1:681–684.

Bartolome, J. W., M. C. Stroud, and H. F. Heady. 1980. Influence of natural mulch on forage production on differing California annual rangelands. *J. Range Manage.* 33:4–8.

Brady, N. C., and R. R. Weil. 1996. *The nature and properties of soils.* 11th ed. Prentice-Hall, Inc., Upper Saddle River, NJ.

Brown, J. W. and J. L. Schuster. 1969. Effects of grazing on hardland site in the southern high plains. *J. Range Manage.* 22:418–423.

Burzlaff, D. F., and L. Harris. 1969. Yearling steer gains and vegetation changes of western Nebraska rangeland under three rates of stocking. *Univ. Nebr. Agric. Exp. Stn. Bul.* 50.

Conrad, V., and L. W. Pollak. 1950. *Methods in climatology.* Harvard University Press, Cambridge, MA.

Dasmann, R. F. 1976. *Environmental conservation.* John Wiley & Sons, Inc., NY.

Frischknecht, N. C, and L. E. Harris. 1968. Grazing intensities and systems on crested wheatgrass in central Utah: Response of vegetation and cattle. *U.S. Dep. Agric. Tech. Bull.* 1388.

Ganskopp, D. C., and T. E. Bedell. 1981. An assessment of vigor and production of range grasses following drought. *J. Range Manage.* 34:137–141.

Hanson, C. L., A. R. Kuhlman, C. J. Erickson, and J. K. Lewis. 1970. Grazing effects on runoff and vegetation on western South Dakota rangeland. *J. Range Manage.* 23:418–420.

Heady, H. F., and E. B. Heady. 1982. *Range and wildlife management in the Tropics.* Longman, Inc., NY.

Herbel, C. H., F. N. Ares, and R. A. Wright. 1972. Drought effects on a semidesert grassland range. *Ecology* 53:1084–1093.

Herbel, C. H., and R. P. Gibbens. 1996. Post-drought vegetation dynamics on arid rangelands in southern New Mexico. *N. Mex. Agric. Exp. Sta. Bull.* 776.

Hopkins, A. D. 1938. Bioclimatics: A science of life and climatic relations. *U.S. Dep. Agric. Misc. Publ.* 280.

Humphrey, R. R. 1962. *Range ecology,* p. 234. The Ronald Press Company, NY.

Hutchings, S. S., and G. Stewart. 1953. Increasing forage yields and sheep production on intermountain winter ranges. *U.S. Dep. Agric. Circ.* 925.

Hyder, D. N. 1953. Grazing capacity as related to range condition. *J. For.* 51:206.

Kalmbacher, R. S., K. R. Long, M. K. Johnson, and F. G. Martin. 1984. Botanical composition of diets of cattle grazing south Florida rangeland. *J. Range Manage.* 37:334–340.

King, J. M. 1983. Livestock needs in pastoral Africa in relation to climate and forage. *International Livestock Center, Africa (ILCA) Res. Rep.* 7.

Klages, K. H. W. 1942. *Ecological crop geography.* The Macmillan Company, NY.

Knox, J. H., and G. F. Ellis. 1969. Cattle improvement on the Bell Ranch. *N. Mex. Agric. Exp. Stn. Mem. Ser.* 3.

Kothmann, M. M., W. J. Waldrip, and G. W. Mathis. 1978. Rangeland vegetation of the Texas rolling plains; response to grazing management and weather. *Proc. Int. Rangel. Congr.* 1:606–609.

Laycock, W. A., and P. W. Conrad. 1981. Responses of vegetation and cattle to various systems of grazing on seeded and native mountain rangelands in eastern Utah. *J. Range Manage.* 35:52–59.

Martin, S. C., and D. E. Ward. 1976. Perennial grasses respond inconsistently to alternate year seasonal rest. *J. Range Manage.* 29:346.

Miller, G. T. 1990. *Resource conservation and management.* Wadsworth Publishing Co., Belmont, CA.

Miller, R. F., R. R. Findley, and J. Alderfer-Findley. 1980. Changes in mountain big sagebrush habitat types following spray release. *J. Range Manage.* 33:278–282.

Nebel, B. J. 1981. *Environmental science.* Prentice-Hall, Inc., Englewood Cliffs, NJ.

Owen, O. S., and D. Chiras. 1995. *Natural resource conservation.* 6th ed. Prentice-Hall, Inc., Englewood Cliffs, NJ.

Owensby, C. E., E. F. Smith, and K. L. Anderson. 1973. Deferred-rotation grazing with steers in the Kansas Flint Hills. *J. Range Manage.* 26:393–395.

Pearson, H. A., and L. B. Whitaker 1974. Forage and cattle responses to different grazing intensities on the southern pine ridges. *J. Range Manage.* 27:444–446.

Petrides, G. A., and W. G. Swank. 1958. Management of the big game resource in Uganda, East Africa. *Trans. N. Am. Wildl. Nat. Resour. Conf.* 23:461–477.

Pieper, R. D., and G. B. Donart. 1975. Drought on the range: Drought and southwestern range vegetation. *Rangeman's J.* 2:176–178.

Pratt, D. J., and M. D. Gwynne. 1977. *Rangeland management and ecology in East Africa.* R. E. Krieger Publishing Co., Inc., Huntington, NY.

Reardon, P. O., and L. B. Merrill. 1976. Vegetative response under various grazing management systems in the Edwards Plateau of Texas. *J. Range Manage.* 29:195–198.

Rogler, G. A., and R. J. Lorenz. 1957. Nitrogen fertilization of northern Great Plains rangelands. *J. Range Manage.* 10:156–159.

Rumney, G. R. 1968. *Climatology and the world's climates.* The Macmillan Company, NY.

Scifres, C. J., J. L. Mutz, R. E. Whitson, and D. L. Drawe. 1982. Interrelationships of huisache canopy cover with range forage on coastal prairie. *J. Range Manage.* 35:558–563.

Skovlin, J. M., R. W. Harris, G. A. Strickler, and G. A. Garrison. 1976. Effects of cattle grazing methods on ponderosa pine-bunchgrass range in the Pacific northwest. *U.S. Dept. Agric. Tech. Bull.* 1531.

Smoliak, S. 1986. Influence of climatic conditions on production of *Stipa-Bouteloua* prairie over a 50-year period. *J. Range Manage.* 39:100–103.

Society for Range Management. 1989. *A glossary of terms used in range management,* 3d ed. Society for Range Management, Denver, CO.

Trewartha, G. 1961. *The Earth's problem climates.* University of Wisconsin Press, Madison, WI.

United States Department of Agriculture (USDA). 1987. The Jornada Experimental Range, Las Cruces, New Mexico. *U.S. Dept. Agric. Misc. Publ.* 689–479.

United States Water Resources Council (USWRC). 1978. *The nation's water resources, 1975–2000. Second national water assessment. Vol. I. Summary.* Supt. Doc. 052-045-00051-7. U.S. Government Printing Office, Washington, D.C.

Uresk, D. W., P. L. Sims, and D. A. Jameson. 1975. Dynamics of blue grama within a short grass ecosystem. *J. Range Manage.* 28:205–208.

Veihmeyer, F. J. 1938. Evaporation from soils and transpiration. *Am. Geophys. Union Trans.* 1938:612–619.

Voigt, J. W. 1959. Ecology of a southern Illinois bluegrass-broomsedge pasture. *J. Range Manage.* 12:175–179.

Wallen, C. C., and M. D. Gwynne. 1978. Drought—a challenge to rangeland management. *Proc. Int. Rangel. Congr.* 1:21–32.

Weaver, J. E., and F. W. Albertson. 1936. Effects of the great drought on the prairies of Iowa, Nebraska, and Kansas. *Ecology.* 17:567–639.

Young, V. A. 1956. The effect of the 1949–1954 drought on the ranges of Texas. *J. Range Manage.* 9:139–142.

CHAPTER 3

RANGE MANAGEMENT HISTORY

ORIGINATION OF RANGE SCIENCE

The exact time when the profession of range management began is unknown. However, concern over the influences of livestock grazing practices on rangeland health and productivity date back to the 1890s in the United States. People such as Smith (1895) in west Texas, Colville (1898) in Oregon, Nelson (1898) in Wyoming, and Kennedy and Doten (1901) were among the first to define the problems of uncontrolled livestock grazing on western rangelands. Smith (1899), in west Texas, provided a description of range destruction by uncontrolled livestock that can be summarized as follows:

1. Reduction in grazing capacity
2. Replacement of desirable forages with unpalatable plants
3. Compaction of soil by livestock
4. Decreased soil fertility due to loss of plant cover
5. Decreased absorption of rainfall by soil
6. High loss of soil during periods of torrential rain
7. Rapid increase in prairie dogs and jackrabbits

Smith (1899) was among the first to recommend control of livestock numbers, range rest periods, water development, brush control, and seeding as potential means of range improvement. These proposals form the backbone of present-day range management. The first range research studies were initiated in the 1890s near Abilene and Chandler, Texas. Results of these studies were reported by Bentley (1902). Between 1910 and 1915, Arthur Sampson conducted the first American grazing system experiments in northeastern Oregon. He reasoned that deferment of grazing until seed maturity would allow for seedling establishment and replenishment of carbohydrate reserves. Because of these and other range studies, Arthur Sampson is considered the father of range management.

Published accounts of livestock grazing problems and scientific studies on rangelands from other parts of the world before 1900 are generally lacking. Therefore, it appears that the science of range management was established in the United States. However, it is important to recognize that pastoral tribes in Asia and Africa have grazed their livestock on rangelands for thousands of years. Until interference by Europeans in the mid-nineteenth century, a system was maintained in which animals and the forage resource were in balance. The pastoral herders in some areas used grazing rotations similar to some of the more sophisticated systems being applied in the United States today. Although Americans have developed the science of range management, many modern range management practices in the United States have been applied for centuries in other parts of the world. Primary sources for our discussion of range management history are Sampson (1952), Lewis (1969), Stoddart et al. (1975), Schickedanz (1980), and Holechek (1981).

CHRONOLOGICAL HISTORY OF LIVESTOCK GRAZING

Grazing by Native Animals

Countless numbers of grazing animals were present on the North American continent prior to the arrival of European man. Stoddart et al. (1975), using Seton (1927) estimates, calculated that over 67 million animal units were present, which nearly equals the livestock grazing pressure on the area today.

Accounts by early explorers indicate that bison (Figure 3.1) grazing on the Great Plains may have been abusive, at least in some areas. England and De Vos (1969) reviewed considerable information from journals kept by fur trappers that allude to high bison populations and mention poor range conditions. Records of the Hudson Bay Fur Company described overgrazing by bison in western Canada to the point where no forage existed for their horses. Meriwether Lewis, in his 1814 account of the Lewis and Clark exposition, described cactus (*Opuntia* sp.) patches so thick that travel was impeded in many parts of the Montana prairie country. He also mentioned traveling through large areas dominated by thick stands of big sagebrush (*Artemisia tridentata*). Both species are associated with severely overgrazed range in the area discussed. Kirsch and Kruse (1972) reviewed substantial information indicating that bison may have heavily overgrazed some parts of the prairie country of Kansas. Their review suggests that prairie grouse and waterfowl populations were kept at low levels because of lack

***Figure* 3.1** Bison were the primary native grazing animal of the central Great Plains. Accounts by early explorers indicate bison grazing was damaging to the range in some areas.

of cover due to heavy bison grazing. It does appear that once the bison left an area, they did not return for several years, which gave the range time to recover (England and De Vos 1969).

There is little information available on the impact of native large herbivore on the area west of the Rocky Mountains. However, reviews of accounts by early settlers indicate that large herbivores had little influence on the vegetation in most locations (Hull and Hull 1974; Vale 1975). It is important to recognize that bison made very little use of this area, probably due to poor water distribution, dry summers, and rugged terrain. Elk probably caused range degradation at some locations, but accounts of elk grazing impact prior to the arrival of the white man are generally unavailable.

Early Livestock Grazing

Hernando Cortez, conqueror of the Aztec Indians in Mexico, brought the first livestock into Mexico in 1515. Between 1515 and 1530, he imported many cattle and sheep into Mexico and was instrumental in developing the livestock industry. The first cattle, sheep, and horses were brought into the United States from Mexico by Coronado in 1540. Several animals escaped from this expedition and began stocking ranges in New Mexico, Arizona, Texas, and Colorado. Numerous Spanish settlements were established in the Southwest in the seventeenth century. Cattle and horses escaping from these settlers multiplied rapidly under the mild climatic and abundant forage conditions in the Southwest. Wild horses were common in the Great Plains and Southwest by the mid-seventeenth century. Horses were brought into eastern Oregon and Idaho by Indians in the early eighteenth century, but cattle and sheep did not come into the Northwest until the late eighteenth century.

By the 1590s, sheep were well established in northeastern Arizona and central New Mexico (Schickedanz 1980). Don Juan de Onate, who moved 4,000 sheep, 1,000 goats, and 1,000 cattle from southern Texas to northern New Mexico in the 1590s, was a key figure in establishing the livestock industry in the southwestern United States.

The missions were also important in the development of the livestock industry in the Southwest. During the seventeenth century the Spanish concentrated on Christianizing Indians near their missions. In addition, they attempted to teach the Indians about farming and animal husbandry. These missions were established in chains along the major rivers in New Mexico, Arizona, and Texas (Schickedanz 1980). Each mission had its own flock of sheep and herd of cattle. There was considerable conflict between the Indians and Spaniards in the late seventeenth century. Indians attacking the missions would often drive stolen livestock back to their pueblos. By the eighteenth century nearly all the Indian pueblos in the Southwest had large flocks of sheep, some numbering 30,000 or more (Carlson 1969).

During the early to mid-nineteenth century, New Mexico was the primary supplier of sheep to other parts of the Southwest. A system had developed in which *patrones,* receiving large Spanish and later Mexican (after Mexico's independence in 1820) land grants, held huge herds of cattle and flocks of sheep. Possession of a quarter- to half-million sheep was not uncommon for these landowners (Schickedanz 1980). Shepherds or peons were in charge of individual flocks of sheep. Part of the increase in the flock of sheep was used to reward the peons for their efforts and expenses. However, the peon was also responsible for losses. The losses due to adverse weather, predators, and Indians usually kept the peons in constant debt to the *patrones.*

The discovery of gold in California in 1849 triggered a huge influx of people. Local sheep flocks in California could not begin to meet the demand for meat. Large sheep flocks numbering 5,000 or more were driven from New Mexico to California. Over one-half million sheep were trailed into California from New Mexico from 1849 to 1860 (Carlson 1969). Sheep flocks from New Mexico were also trailed to the Mormon farms in Utah, silver mining settlements in Nevada, and the gold mining settlements in Colorado in this period.

The livestock industry in the northwestern United States developed during the 1830s through 1850s. Small herds of cattle were present on many farms in the Willamette Valley of western Oregon by the early 1830s. In 1836, Marcus Whitman brought several heads of cattle to the Walla Walla Valley in southeastern Washington. These cattle served as the nucleus for herds that developed in the area in the 1840s. Large numbers of people poured into the Oregon-Washington country in the 1840s because of reports of the mild climate coupled with abundant grass, water, and timber. By the 1850s large herds of cattle were present in the palouse country of eastern Washington and Oregon (Galbraith and Anderson 1971). The discovery of gold in the Kamloops region of British Columbia created a lucrative market for herds of palouse cattle in the 1850s. The development of the cattle industry in the Northwest in the 1850s marked the beginning of the era of the cattle baron and range exploitation.

The 1865–1900 Period

The end of the Civil War in 1865 triggered the real beginning of the livestock industry in the West. The completion of the railroad into Kansas in 1866 provided a market for longhorn cattle that were present in vast numbers on Texas ranges. The early drives north were accomplished with great difficulty because the cattle were generally wild, with little previous herding experience. There were numerous rivers to cross, which resulted in considerable drowning of cattle; there were large numbers of hostile Indians on the plains; and white settlers on the plains generally resented the Texas livestock crossing through their lands. Cattle drives to Kansas from Texas peaked around 1872 and had nearly ended by 1880 because of completion of the railroad into Texas. Many of the drives from Texas to Kansas lasted 2 to 3 years, with the cattle being grown and fattened along the way. By 1885, when the Texas drives had ended, more than 5 million cattle had been trailed north from Texas.

The growth of the cattle industry on the northern and central Great Plains began in the early 1870s. Most of the cattle used for base herds on the Plains area came from Texas and the palouse country of Oregon and Washington. The United States was in an inflationary period during the 1870s. Because cattle were selling at high prices in eastern markets and pamphlets were in wide circulation advertising the enormous profits, millions of dollars from the eastern United States and Europe were invested in western cattle operations. This boom caused huge increases in western cattle numbers during the 1870s. In 1870 there were an estimated 4.6 million cattle in the 17 western states, compared to 35 to 40 million when cattle reached peak numbers in 1884 (USDC 1943). During the 1880s, range degradation in the western United States was at a maximum. Cattle numbers were greatly reduced on the northern Great Plains and palouse prairie by the severe winter of 1885–1886. Severe drought during the summers of 1891 and 1892 resulted in heavy cattle losses from starvation in the Southwest. In 1890, cattle numbers in the 17 western states were estimated at 27 million (USDC 1943).

The sheep industry also expanded rapidly in the West during the 1880s and 1890s, peaking about 1910. Trail herding of sheep was nearly as spectacular as that of cattle. Large numbers were trailed from New Mexico, California, and Oregon to railways for marketing and for stocking other ranges. Considerable conflict occurred between cattlemen and sheepmen in the 1890s and early 1900s. This was because sheepmen were quite mobile and grazed through ranges that cattlemen claimed because of past use. Lack of mobility on the part of the cattlemen made overstocking much more critical to their operation than to the sheepmen, whose operation was based on movement from one range to another. Government intervention beginning in 1905 gradually ended the conflict.

Between 1870 and 1900 there was a great reduction in the rangeland resource. This reduction was caused by overgrazing and conversion of rangeland to farmland. During this period, the railroads brought into the West large numbers of people who were seeking land provided by the Homestead Act of 1862 (Table 3.1). Much of the tallgrass, palouse, and California prairies were put under cultivation as a result of this act.

The deterioration of western ranges in the late nineteenth century did not go unnoticed. Range abuse on the palouse prairie was reported in a Walla Walla, Washington newspaper as early as 1862. Numerous reports of overgrazing were made by livestock associations and reporters in practically all parts of the West in the 1880s and 1890s.

The 1900–1930 Period

Government intervention into grazing problems on the western range began in 1898 when the Department of Interior granted grazing permits to limit the number of livestock on federal lands.

TABLE 3.1   Summary of Important Congressional Acts Influencing Rangeland in the United States

---

Homestead Act (1862): Granted 160 acres after 5 years' residence; encouraged large numbers of people to move west for farming purposes.

Morrill Act (1862): Set up the land-grant colleges (provided land for these institutions).

Transcontinental Railroad Act (1862): Large acreages of land were granted to railroads to provide incentives for railroad construction in the West; this resulted in rapid transportation for large numbers of people to come west.

Forest Reserves Act (1891): Set aside forested areas for timber; grazing privileges were allocated to ranchers on these areas.

Enlarged Homestead Act (1909): Granted 320 acres if one-fourth was cultivated; designed to promote farming on remaining federal land; caused rangeland destruction by cultivation of land not suited for farming.

Stockraising Homestead Act (1916): Granted 640 acres raising 50 cows; caused rangeland destruction because 640 acres would not support 50 cows in most areas.

Taylor Grazing Act (1934): Allocated grazing privileges on unsold government lands in the West on the basis of the ranchers' ability to provide water (Southwest) or hay (Northwest); this act was passed as the result of actions by ranchers concerned about range deterioration.

Soil Erosion Act (1935): Set up Soil Conservation Service to deal with soil erosion problems on private lands.

Multiple Use Act (1960): Mandated that Forest Service lands be managed for several purposes rather than single use.

National Environmental Policy Act (1969): Required government and private agencies to draft Environmental Impact Statements on proposed actions that would affect federal lands.

Endangered Species Act (1973): Required federal agencies to protect listed wildlife species.

Federal Lands Policy and Management Act (1976): Established rationale for keeping lands covered by Taylor Grazing Act in public ownership (Bureau of Land Management lands) and set up multiple use guidelines for these lands.

Rangeland Improvement Act (1978): Set aside 50% of grazing fee receipts from federal lands for range improvement on these lands.

1985 Farm Bill: Government payments were provided to private landowners who planted their erodible lands to perennial grasses and retired them for 10 years with no grazing or haying. This program converted 35 million acres of cropland back to grassland.

In 1905 the Forest Service was set up in the Department of Agriculture and the process of forage allotment was initiated. Between 1910 and 1920 grazing laws were put into effect on national forest lands. Because the price of cattle was high during World War I, a period of very severe overgrazing took place between 1915 and 1920. National forest ranges began to improve after World War I because controlled grazing practices were once again implemented. However, private and other government-owned rangelands in the West continued to deteriorate.

In the 1920s, the discipline of range management flowered and developed. By 1925 approximately 15 colleges were offering courses in range management. During this period many basic ecological concepts were developed by Frederick Clements and John Weaver, both professors at the University of Nebraska. Clements developed theories on plant succession; at the same time, Weaver initiated some of the classic ecological studies involving the tallgrass prairie. In 1915, Clements addressed the problems of range abuse. He recommended that rest, controlled season of use, reduced livestock numbers, noxious plant control, and seeding be used to stop range deterioration and initiate range improvement. The first textbook on range management was written in 1923 by Arthur Sampson. Much of the research of the 1920s involved the effects of different stocking intensities on forage and livestock production. Some of the first research examining plant composition changes under different grazing intensities was published during this period.

The 1930–1960 Period

During the early 1930s, the United States suffered the consequences of the past 60 years of range exploitation. Between 1931 and 1936 severe drought occurred throughout the western United States. This drought was particularly severe in the Great Plains region. Huge dust storms occurred over vast acreages of cultivated and overgrazed lands in the Great Plains. Effects from these dust storms were felt in eastern cities such as New York and Washington. During the 1910s and 1920s vast areas of the Great Plains were plowed. High grain prices and favorable climatic conditions had encouraged farming where it could not be sustained. Although the Dust Bowl brought tremendous suffering to inhabitants of the Great Plains, it forced Americans to confront the consequences of their past 50 years of land abuse. During the mid-1930s the nation developed a conservation ethic. Millions of hectares of cropland in the Great Plains were returned to rangeland. In 1933 the Soil Erosion Service was established in the Department of Agriculture and renamed the Soil Conservation Service in 1935.

The biggest event in the 1930s in range management was the passage of the Taylor Grazing Act of 1934, which placed administration of remaining public lands under the Grazing Service, which later became the Bureau of Land Management. The Taylor Grazing Act was passed largely because of the actions of concerned ranchers, particularly those in New Mexico. This act resulted in the allocation of grazing privileges on unsold federal land in the West. In northern parts of the West, where severe winters require the feeding of hay to livestock, forage allocations were made on the basis of ownership of private land capable of providing hay for livestock in the winter. These are called *land-based allotments*. In the Southwest, where ranges can be grazed year-round, grazing privileges were allocated to those who owned watering points. These are called *water-based allotments*. Generally, the cattlemen benefited from the Taylor Grazing Act, but many roving sheep operators were forced out of business because they had no base property. Forage allocations were generally made on the basis of ocular estimates, which were often inaccurate and resulted in further degradation. It was soon recognized that better methods were needed for evaluating range vegetation. Because of this need, some very good research was reported concerning methods for quantifying vegetation productivity, utilization, and composition in the 1930s and 1940s.

The most important event during the 1940s was the formation of the Society for Range Management in 1948. This society grouped people working in rangeland research and management into one discipline and provided a mechanism for publishing scientific findings concerning rangelands (the *Journal of Range Management*). In 1949, E. J. Dyksterhuis published a classic paper in the *Journal of Range Management* called "Condition and Management of Rangeland Based on Quantitative Ecology." This paper (Dyksterhuis 1949) proposed that range condition be evaluated on the basis of the amount of climax vegetation remaining on a site. It introduced the terms *decreasers, increasers,* and *invaders,* which have become key words in describing plant responses to grazing. In the early 1940s auxins that regulate plant growth were discovered. This led to the development of 2,4-D, which made large-scale control of noxious plants practical and provided a tool for improving brush-infested rangelands. The first studies examining the nutrition of range animals were conducted in Utah in the late 1940s by C. Wayne Cook and Lorin Harris. The first edition of *Range Management* was published in 1943 by L. A. Stoddart and A. D. Smith. This book was revised in 1955 and again in 1975. This book organized the knowledge of range science into the framework that forms the basis for the present-day range profession.

The 1950s were years of terrific improvement in the range resource on public lands. These improvements resulted from water development, brush control, seeding, stocking rate adjustments, and grazing period adjustments. Considerable range research was directed toward range improvement by the use of herbicides. Watershed and range nutrition problems also received much attention. More range research was conducted in the 1950s than in all years prior to 1950 put together.

The 1960–1997 Period

The 1960s were characterized by considerable change in the philosophy of range management. The Multiple Use Act of 1960 mandated that Forest Service lands be managed for several uses, such as grazing, wildlife, timber, and recreation, rather than for a single use such as grazing or timber. Previously, range research and management had been geared toward producing forage for livestock. This change in philosophy had considerable influence on how range management practices such as brush control were applied. The Federal Lands Policy and Management Act of 1976 provided a similar multiple use mandate for Bureau of Land Management lands. Methodologies for range nutrition and ecology studies were greatly improved. George Van Dyne made valuable contributions in range nutrition and vegetation sampling techniques. He developed computer modeling as a tool for simulating and managing range ecosystems. During the 1960s public concern over the environment and natural resources accelerated. There was growing opposition to livestock grazing on federal lands. The National Environmental Policy Act of 1969 was passed by Congress in response to pressure from environmentally concerned groups. This led to the requirement for environmental impact statements by federal and private agencies on actions that would affect federal lands. During the 1960s several other countries developed range research and management programs based on methodology and concepts derived in the United States.

The 1970s were characterized by considerable change in regard to the emphasis of both range management and research. There was a considerable shift to nonconsumptive use of public rangelands. Livestock grazing on federal ranges was reduced 25% between 1960 and 1992 (Table 3.2), and this trend is expected to continue through the 1990s. Range improvement on private lands accelerated during the 1980s because of improved information and education programs by state and federal agencies. Another factor of considerable importance was that the 1980s and 1990s rancher was much better educated than were those of previous periods. Watershed and wildlife aspects of range management attracted more public attention and research dollars than in the 1960s. It appears that range watershed management will receive more and more emphasis in the future because of restricted water supplies. Fire has received considerable attention as a management tool in the 1990s and will probably receive greater use in the future because some segments of the public find herbicides objectionable.

Considerable progress has been made in the past 50 years. This progress is reflected by the major improvements in federal (Tables 3.3 and 3.4) and private ranges (Figure 3.2). Despite a general improvement in range conditions on both private and government-owned

TABLE 3.2   Trend (1960–1992) in Actual Animal Unit Months (thousands) of Livestock Grazing on Bureau of Land Management and Forest Service Lands

TYPE OF LIVESTOCK	BUREAU OF LAND MANAGEMENT				
	1960	1970	1980	1990	1993
Cattle	10,277	13,368	8,984	9,327	8,486
Horses & burros	198	400	72	72	62
Sheep & goats	4,790	2,741	1,351	1,445	1,211
Total AUMs	15,265	16,209	10,407	10,844	9,759
	FOREST SERVICE				
Cattle	8,084	—	7,594	—	6,784
Sheep	1,083	—	893	—	805
Horses & burros	210	—	176	—	130
Total AUMs	9,377	—	8,663	—	7,719

Sources: U.S. Department of Interior annual reports and U.S. Forest Service annual reports.

TABLE 3.3   Comparative Percentages of Bureau of Land Management
Rangelands in Excellent, Good, Fair, and Poor Condition between 1936 and 1993

YEAR	EXCELLENT (CLIMAX)	GOOD (LATE SERAL)	FAIR (MID SERAL)	POOR (EARLY SERAL)
1936	1.5	14.3	47.9	36.3
1966	2.2	16.7	51.6	29.5
1975	2.0	15.0	50.0	33.0
1984[a]	5.0	31.0	42.0	18.0
1993[a]	4.0	33.0	38.0	14.0

Source: USDI 1984, 1994.

[a]Less than 100% totals because some lands have not been rated as to range condition.

TABLE 3.4   Comparative Percentages of Bureau of Land Management Rangelands in
Different Ecological Condition Categories in 1993 for 10 Western States[a]

STATE	EXCELLENT (CLIMAX)	GOOD (LATE SERAL)	FAIR (MID SERAL)	POOR (EARLY SERAL)
Arizona	6	38	39	9
California	3	19	42	30
Colorado	3	18	36	22
Idaho	2	22	34	39
Montana	6	62	23	1
Nevada	2	29	34	9
New Mexico	1	35	46	14
Oregon	1	25	58	16
Utah	11	29	42	13
Wyoming	5	44	34	6
Bureauwide	4	33	38	14

Source: USDI 1994.

[a]Less than 100% in some states where some lands have not been rated as to range condition.

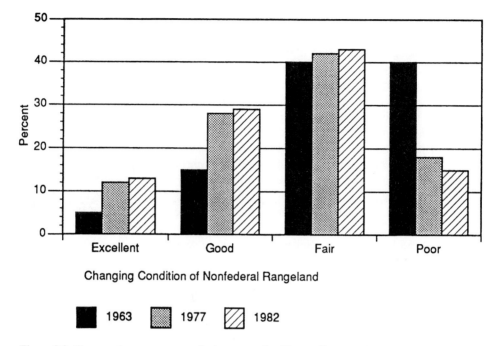

Changing Condition of Nonfederal Rangeland

■ 1963   ▨ 1977   ▨ 1982

***Figure* 3.2** Comparative percentages of private rangeland in excellent, good, fair, and poor condition between 1963 and 1982 for 17 western states. (From USDA 1980, 1989.)

rangelands, range improvement has been a slow process. There are five reasons for this as follows:

1. Because of the low precipitation (less than 300 mm) in much of the western United States, range recovery from grazing abuse is very slow. It is important to recognize that precipitation is the driving force in natural recovery from grazing.

2. Range ecosystems are highly complex and have considerable variation. Long time periods are generally required before conclusions can be drawn from these studies.

3. Government agencies such as the Bureau of Land Management have been forced to allocate much of their monetary and personnel resources toward developing environmental impact statements since the 1970s. This has slowed the implementation of range management practices on public lands.

4. Until the last 20 years, range management principles have not been widely applied on private lands.

5. Government policies, programs, and subsidies have often created disincentives rather than incentives for sustainable grazing practices on both privately and publicly owned rangelands (see discussions by Anderson and Leal 1991; Holechek 1993; Holechek and Hess 1994; Hess and Holechek 1995).

The Future

Looking into the future it appears that rangelands in the United States will become increasingly important in providing red meat, water, wildlife, energy, and recreation needed by the American public. If present trends continue, livestock production will receive greater emphasis on Great Plains and eastern rangelands. However, livestock production will be deemphaiszed in the Intermountain area and the Southwest. During the past 15 years many ranches in the West have subdivided into ranchettes of 20 ha or less, and livestock are no longer produced. It is expected that human populations will continue to expand in the West and more ranches will subdivide. Public pressure has increasingly forced government agencies to emphasize recreation and wildlife on federal lands. As urbanization of the western states continues, this pressure will probably increase rather than decrease. The potential for increased livestock production in the West compared to the Great Plains and the East is quite low (Holechek and Hawkes 1993). Therefore, it is expected that these areas will be emphasized as future sources of red meat. In some parts of the West, game ranching will become more and more common. Water will undoubtedly become the most important resource derived from rangeland in the Southwest.

How rangelands are used in this country in the future will depend on what happens in other countries as well as in the United States. As long as the world population grows at a rapid rate, there will be greater pressure to increase livestock production on rangelands in developing countries. Hopefully, in the future the public will become better informed on range management principles and more supportive of range management programs.

Many range scientists are beginning to question the sustainability of traditional range livestock production systems in the present era of low livestock prices and increasing environmental sensitivity (Holechek et al. 1994; Box 1995; Heitschmidt et al. 1996; Hess and Holechek 1995; Holechek 1996a; Vavra 1996). Sustaining western ranching as a viable economic enterprise, developing more sustainable range livestock production systems, maintaining open space, and better educating the public on range management issues will be serious challenges for range managers of the twenty-first century in the United States (Brunson and Steel 1996; Huntsinger and Hopkinson 1996; Vavra 1996).

GOVERNMENT LAND POLICIES

To the early settlers in the United States, land resources seemed unlimited. Until the 1930s, government policies were concerned with disposing of the vast land resource. These policies were oriented toward converting undisposed land into farmland. It was generally unrecognized that large areas of the West were unsuited for farming and suited only for grazing. The vastness of the land resource coupled with the failure of the government to recognize that most of the West could not sustain farming resulted in tremendous land damage and economic loss.

Prior to the Forest Reserves Act and the Taylor Grazing Act, there was no control over how federal lands were used. If one person did not exploit the forage in an area, his or her neighbor probably would. This is commonly referred to as the "tragedy of the commons" (Figure 3.3; see also Chapter 16). This resulted in a constant struggle in the Intermountain

Figure **3.3** "Tragedy of the commons." (Redrawn by John N. Smith, based on Bernard J. Nebel, *Environmental Science,* 3d ed., Copyright © 1987, p. 485. Reprinted by permission of Prentice-Hall, Inc., Englewood Cliffs, NJ.)

area to get on spring ranges as soon as snowmelt and to stay on them as long as possible. This competition between stockmen to gain and maintain use of the best forage resulted in total disregard of prudent grazing practices.

The homestead era occurred between 1862 and 1934. The primary goal during this era was settlement of the West. The Homestead Act of 1862 allowed the settler to obtain 160-acre tracts after living on the land for 5 years and making specified improvements. The act also contained a clause that allowed the settler to purchase the land for $1.25 per acre after 6 months of residence. This act led to rapid settlement of the midwestern states (Illinois, Iowa, Minnesota, and Missouri) in the 1860s and 1870s.

By 1900 only semiarid lands of low productivity remained unsettled. The Enlarged Homestead Act of 1909 increased the homestead size to 320 acres and provided that one-fourth must be cultivated. This act was passed to encourage settlement of lands in the nine westernmost states and the western portions of the Great Plains. The major problem with this act was that 320 acres often was not enough to sustain a family (Hibbard 1924). This act resulted in the plowing of vast areas of rangeland that would not sustain cultivation, particularly in the western portions of the Great Plains. This act played a key role in causing the Dust Bowl of the 1930s.

The Stockraising Homestead Act of 1916 resulted in severe overgrazing in the arid portions of the nine westernmost states. This act granted stockmen 640 acres which was thought to be enough to support 50 head of cattle. In reality very little of the remaining rangeland in the West could support 50 head of cattle on 640 acres. This act resulted in severe overgrazing and often abandonment before the homestead could be established.

The Transcontinental Railroad Act in 1862 had a tremendous influence on the land resource in the western United States. This act was designed to encourage the spread of the railroad across the United States, and hence to provide a mechanism for large numbers of people to come west. It also permitted western crops and livestock to be shipped east. It granted alternate sections of land for several miles on each side of the railroad tracks to the railroad company. This economic incentive resulted in completion of a railroad across the United States in 1868. By the 1890s nearly all parts of the West were accessible by rail.

Much of the land in state ownership in the western states is the result of school grants (Hibbard 1924). After 1848, two sections in each township were given to the states for school establishment. Utah and New Mexico were granted four sections. The Morrill Act of 1862 granted two townships (40,000 acres) to each state for the establishment of what are presently known as the "land-grant colleges." The Morrill Act resulted in a highly efficient system of agricultural education and research.

The Taylor Grazing Act triggered the era of land retention by the federal government. It placed under management the vast, arid areas of the West that were suffering from severe overgrazing. With the passage of this act, the government formally recognized that unsold arid western lands were unsuited for cultivation and that grazing on these lands must be controlled. Prior to the Taylor Grazing Act, little attention was given to the custodianship of unsold western lands.

The present status of public land is shown in Tables 3.5 and 3.6. Presently, about 37% of land in the United States (Alaska included) is in public ownership. Federally owned lands account for 29% of the total land area in the United States (Alaska included). Lands in state, county, and city ownership account for the other 8% of public lands.

Much of the rangeland in the western states is in federal ownership (Table 3.5). Federally owned rangelands provide about 10% of the feed requirements of livestock in the United States (CAST 1986). Since the Taylor Grazing Act, there have been several movements by private interest groups to transfer remaining federal lands into private ownership. The most recent attempt, known as the "sagebrush rebellion," occurred in the late 1970s and early 1980s.

TABLE 3.5   Custodianship of Federal Lands in the United States in 1992

	ACRES (MILLIONS)	PERCENT OF TOTAL
Total land in the United States	2,271	100
Not owned by federal government	1,621	71
Owned by federal government	650	29
Bureau of Land Management	268	12
Forest Service	191	6
Fish and Wildlife Service	86	4
National Park Service	73	3
Department of the Army	12	1
Bureau of Reclamation	4	<1
Department of the Air Force	<1	<1
Corps of Engineers	9	<1
Bureau of Indian Affairs	3	<1
Department of the Navy	3	<1
Other Agencies	4	<1

Sources: USDI 1983, 1993; USDA 1992.

TABLE 3.6   Percent of Land in the 18 Western States in Federal Ownership in 1990[a]

STATE	PERCENTAGE	STATE	PERCENTAGE
Alaska	68	New Mexico	31
Arizona	47	North Dakota	4
California	44	Oklahoma	2
Colorado	36	Oregon	52
Idaho	62	South Dakota	6
Kansas	<1	Texas	1
Montana	28	Utah	64
Nebraska	1	Washington	24
Nevada	83	Wyoming	48

Source: USDI 1993.

[a]28% of the United States is in federal ownership.

These movements have been unsuccessful due to heavy public opposition. The majority of citizens in the United States evidently consider the large tracts of public lands in the West to be an important part of their natural heritage and favor retaining these lands in federal ownership. Most ranchers grazing livestock on federal lands would probably not be able to compete with large corporations if these lands were auctioned off. Provided that grazing privileges are maintained and grazing fees are kept reasonable, it is advantageous to most ranchers for these lands to remain in federal ownership. Taxes, costs of maintaining physical structures (fences, corrals, water developments, and roads), and interest on land purchase money would make grazing uneconomical for most ranchers if they were forced to buy federal grazing lands.

Historically ranchers on federal rangelands could only derive income from these lands through livestock production. Permitting ranchers to derive income from converting livestock grazing privileges to recreational or wildlife use has been advocated by some reformists (Anderson and Leal 1991; Holechek and Hess 1994, 1996; Hess and Holechek 1995).

Importance of Federal Lands

The issue of livestock grazing on federal lands has been quite controversial during the 1990s (Wuerthner 1990). Therefore we have included a brief discussion of its importance (Holechek and Hess 1994). At present 30,600 permittees graze cattle on federal lands. This is about 2% of the nation's ranchers or approximately 7% of the ranchers in the 11 western states. Bureau of Land Management and Forest Service rangelands provide forage for about 10.8 and 7.7 million animal unit months (AUMs), respectively, for a total of 18.5 million AUMs. This represents 1.54 million animal units (AUs) or 3.76% of the nation's beef cattle herd (41 million AUs). At an average fair market value of $80 per AUM, the total value of federal land grazing permits is roughly $1.48 billion dollars.

Although federal rangelands provide only a small part of total livestock forage requirements, they are seasonally important in the production process. Around 22% of the yearling cattle in the United States spend a portion of their lives on federal rangelands. Federal rangelands play an even bigger role in sheep production. They support about 20% of the nation's stock sheep from which about 21% of the nation's wool is shorn.

It is doubtful that discontinuation of federal land grazing would have much impact on the price of meat to the consumer. Increases on beef production on private lands in the Great Plains and Southeast would likely compensate for any reduction on federal lands in the West.

Discontinuation of federal land grazing would severely harm some local economies. Negative impacts on wildlife populations would be likely if private land holdings associated with federal land grazing are subsequently subdivided into ranchettes. Further, many water points on federal lands would no longer be maintained. Those watering points play a crucial role in supporting many wildlife populations. In some areas wildfire would become a serious problem due to a build-up of vegetation residues. We consider livestock grazing a legitimate and important use on federal lands. However, we acknowledge that situations do exist where livestock grazing on public lands should be discontinued. The economic effectiveness of livestock grazing becomes doubtful when forage production drops below 100 lbs per acre and/or slopes over 60% dominate the area (Stoddart and Smith 1943; Holechek

1996b). Approaches that deal with equity problems when livestock grazing conflicts with other federal rangeland uses are discussed in detail by Anderson and Leal (1991), Holechek (1993), and Holechek and Hess (1994). In our opinion public land ranchers should receive fair market monetary compensation when their grazing privileges are compromised so the land can be more fully used for wildlife, recreation, or some other purpose. An analysis by Holechek and Hess (1996) indicated that a sum of $400 million dollars would solve most of the major conflicts between livestock and recreation if this approach were used. Many ranchers on these lands would welcome the opportunity of selling their grazing privileges in these situations. This type of program may seem costly, but it must be considered in the context of established public policy on equity rights and the future costs of administration, monitoring, and litigation as conflict increases. Further, these costs are relatively small in the context of the total budget for the Bureau of Land Management ($1.1 billion dollars in 1996) and Forest Service ($5.2 billion dollars in 1996).

DEVELOPMENT OF RANGE MANAGEMENT IN OTHER COUNTRIES

In the 1960s, range principles and practices developed in the United States were spread to other countries, particularly those in Africa. Harold Heady, an American range scientist and educator, played a key role in initiating range research and developing range management programs in Africa in the 1950s. He has written two books on African range management (Heady 1960; Heady and Heady 1982). A book on African range management is also available by Pratt and Gwynne (1977), and one on Pakistan range management from Quraishi et al. (1993).

During the late 1960s, the government of Australia established a range research program (Perry 1967). In the late 1970s, the Australian Society for Range Management was formed. Mexico developed a range research program during the 1970s and the first International Rangeland Congress was held in 1978.

Although the science of range management has developed in other countries, rangeland degradation continues to be a problem, particularly in African countries, where nomadic systems exist. This is discussed in Chapter 16.

RANGE MANAGEMENT PRINCIPLES

- Public policies directed toward rangelands in the United States have been characterized by both successes and failures. Knowledge of the outcome of these policies will permit managers and policymakers to avoid repeating past mistakes and to build on past successes.

- Government policies that provide individual public rangeland users with security of their grazing privileges and opportunities to benefit from sound management practices are essential in preventing rangeland degradation based on history in the United States and other parts of the world.

Literature Cited

Anderson, J. L., and D. L. Leal. 1991. *Free market environmentalism.* Westview Press, Boulder, CO.

Bentley, H. L. 1902. Experiments in range improvements in central Texas. *U.S. Bur. Plant Ind. Bull.* 13.

Box, T. W. 1995. A viewpoint: Range management and the tragedy of the commons. *Rangelands* 17:83–85.

Brunson, M. W., and B. S. Steel. 1996. Sources of variation in attitudes and beliefs about federal rangeland management. *J. Range Manage.* 49:69–75.

Carlson, A. W. 1969. New Mexico sheep industry, 1850–1900: Its role in the history of the territory. *N. Mex. Hist. Rev.* XLIV:1.

Colville, F. J. 1898. Forest growth and sheep grazing in the Cascade Mountains of Oregon. *U.S. Dep. Agric. For. Div. Bull.* 15.

Council for Agricultural Science and Technology (CAST). 1986. *Forages: Resources of the future.* Rep. 108, Ames, IA.

Dyksterhuis, E. J. 1949. Condition and management of rangeland based on quantitative ecology. *J. Range Manage.* 2:104–115.

England, R. E., and A. De Vos. 1969. Influence of animals on pristine conditions on the Canadian grasslands. *J. Range Manage.* 22:87–94.

Galbraith, W. A., and E. W. Anderson. 1971. Grazing history of the northwest. *J. Range Manage.* 24:6–13.

Heady, H. F. 1960. *Range management in East Africa.* Government Printer, Nairobi, Kenya.

Heady, H. F., and E. B. Heady. 1982. *Range and wildlife management in the Tropics.* Longman, Inc., NY.

Heitschmidt, R. K., R. E. Short, and E. E. Grings. 1996. Ecosystems, sustainability and animal agriculture. *J. Anim. Sci.* 74:1395–1405.

Hess, K., Jr., and J. L. Holechek 1995. Beyond the grazing fee: An agenda for rangeland reform. *Cato Inst. Paper No. 234,* Washington, DC.

Hibbard, B. H. 1924. *A history of the public land policies.* The Macmillan Company, NY.

Holechek, J. L. 1981. A brief history of range management in the United States. *Rangelands* 3:16–18.

Holechek, J. L. 1993. Policy changes on federal rangelands: A perspective. *J. Soil & Water Cons.* 48:166–174.

Holechek, J. L. 1996a. Drought and low cattle prices: Hardship for New Mexico ranchers. *Rangelands* 18:11–14.

Holechek, J. L. 1996b. Financial returns and range condition on southern New Mexico ranches. *Rangelands* 18:52–56.

Holechek, J. L., and J. Hawkes. 1993. Desert and prairie ranching profitability. *Rangelands* 15:104–109.

Holechek, J. L., J. Hawkes, and T. Darden. 1994. Macro-economics and cattle ranching. *Rangelands* 16:118–123.

Holechek, J. L., and K. Hess. 1994. Free market policy for public land grazing. *Rangelands* 16:63–67.

Holechek, J. L., and K. Hess, Jr. 1996. Market forces would benefit rangelands. *Forum for Appl. Res. and Public Policy* 2:5–15.

Hull, A. C., Jr., and M. K. Hull. 1974. Presettlement vegetation of Cache Valley, Utah, and Idaho. *J. Range Manage.* 27:27–30.

Huntsinger, L., and P. Hopkinson. 1996. Viewpoint: Sustaining rangeland landscapes: Social and ecological process. *J. Range Manage.* 49:167–173.

Kennedy, P. B., and S. B. Doten. 1901. A preliminary report on the summer ranges of western Nevada sheep. *Nev. State Univ. Agric. Exp. Stn. Bull.* 51.

Kirsch, L. M., and A. D. Kruse. 1972. Prairie fires and wildlife. *Proc. Tall Timbers Fire Ecol. Conf.* 12:289–305.

Lewis, J. K. 1969. Range management viewed in the ecosystem framework. In G. M. Van Dyne (Ed.). *The ecosystem concept in natural resource management.* Academic Press, Inc., NY.

Nebel, B. J. 1987. *Environmental science: The way the world works.* 3rd ed. Prentice-Hall, Inc., Englewood Cliffs, NJ.

Nelson, A. 1898. The red desert of Wyoming and its forage resources. *U.S. Dep. Agric. Div. Agrost. Bull.* 13.

Perry, R. A. 1967. The need for rangelands research in Australia. *Proc. Ecol. Soc. Aust.* 2:1–4.

Pratt, D. J., and M. D. Gwynne. 1977. *Rangeland ecology and management in East Africa.* R. E. Krieger Publishing Co., Inc., Huntington, NY.

Quraishi, M. A., G. S. Khan, and M. S. Yaqoob. 1993. *Range management in Pakistan.* Kazi Publications. Univ. Agr., Faisalabad, Pakistan.

Sampson, A. W. 1952. *Range management principles and practices.* John Wiley & Sons, Inc., NY.

Schickedanz, J. G. 1980. History of grazing in the southwest. In *Proc. grazing management systems for Southwest rangelands symposium.* New Mexico State University, Las Cruces, NM.

Seton, E. T. 1927. *Lives of game animals, Vol. III. Hoofed animals.* Doubleday & Company, Inc., Garden City, NY.

Smith, J. G. 1895. Forage conditions of the prairie region. In *U.S. Department of Agriculture Yearbook.* U.S. Government Printing Office, Washington, DC.

Smith, J. G. 1899. Grazing problems in the southwest and how to meet them. *U.S. Dep. Agric. Div. Agrost. Bull.* 16:1–47.

Stoddart, L. A., and A. D. Smith. 1943. *Range management.* McGraw-Hill Book Company, NY.

Stoddart, L. A., A. D. Smith, and T. W. Box. 1975. *Range management.* 3d ed. McGraw-Hill Book Company, NY.

United States Department of Agriculture (USDA). 1980. *America's soil and water: Condition and trends.* U.S. Department of Agriculture. Soil Conservation Service. Washington, DC.

United States Department of Agriculture (USDA). 1989. *The second RCA appraisal: Soil, water, and related resources on nonfederal land in the United States.* U.S. Department of Agriculture. Soil Conservation Service. Washington, DC.

United States Department of Agriculture (USDA). 1992. *Report of the Forest Service: Fiscal year 1992.* U.S. Department of Agriculture. Forest Service. Washington, DC.

United States Department of Commerce (USDC). 1943. *Sixteenth Census of the United States, 1940.* Agric. (Gen. Rep.) 3:9–1092.

United States Department of Interior (USDI). 1983. *Public land statistics.* U.S. Department of Interior. Bureau of Land Management.

United States Department of Interior (USDI). 1984. *Fifty years of public land management.* U.S. Department of Interior. Bureau of Land Management.

United States Department of Interior (USDI). 1993. *Public land statistics.* U.S. Department of Interior. Bureau of Land Management.

United States Department of Interior (USDI). 1994. *Public land statistics.* U.S. Department of Interior. Bureau of Land Management.

Vale, T. R. 1975. Presettlement vegetation in the sagebrush-grass area of the intermountain west. *J. Range Manage.* 28:32–37.

Vavra, M. 1996. Sustainability of animal production systems: An ecological perspective. *J. Anim. Sci.* 74:1418–1423.

Wagner, F. H. 1978. Livestock grazing and the livestock industry. In *Wildlife and America symposium.* Council for Environmental Quality, Washington, DC.

Wuerthner, G. 1990. The price is wrong. *Sierra* 25:38–48.

CHAPTER 4

DESCRIPTION OF RANGELAND TYPES

TYPES OF RANGELAND

Grasslands, desert shrublands, savanna woodlands, forests, and tundra are the basic rangeland types of the world. Each of these types is comprised of several plant associations that support a slightly different biota due to variations in climate, soils, and human influences. Activities such as logging, cultivation, grazing, and industrialization have substantially altered the natural biota in all the rangeland types. The major vegetation types of the world are shown in Figures 4.1 through 4.3.

Grasslands

Grasslands are the most productive rangelands in the world when forage production for wild and domestic animals is the major consideration. Grasslands are typically free of woody plants (shrubs and trees) and are dominated by plants in the family Gramineae (grasses). Grasslands occur from sea level to 5,000 m but are most common on relatively flat, inland areas at elevations from 1,000 m to 2,000 m.

Grasses, Forbs, and Shrubs Defined. *Grasses* are distinguished by having hollow, jointed stems; fine, narrow leaves with large parallel veins; and fibrous root systems (Figure 4.4). Many grasslands have a high component of sedges (*Carex* sp.) and rushes (*Juncus* sp.).

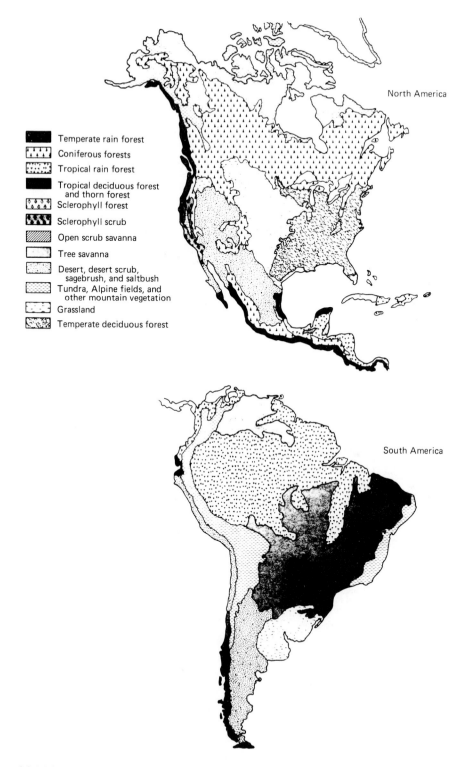

Legend:
- Temperate rain forest
- Coniferous forests
- Tropical rain forest
- Tropical deciduous forest and thorn forest
- Sclerophyll forest
- Sclerophyll scrub
- Open scrub savanna
- Tree savanna
- Desert, desert scrub, sagebrush, and saltbush
- Tundra, Alpine fields, and other mountain vegetation
- Grassland
- Temperate deciduous forest

North America

South America

***Figure* 4.1** Major vegetation types of North and South America. "Schlerophyll" refers to wood plants with broad, hard leaves. (From R. F. Dasmann, 1976, *Environmental Conservation,* 4th ed. Copyright© 1976, John Wiley & Sons, Inc., New York. Reprinted by permission of John Wiley & Sons, Inc.)

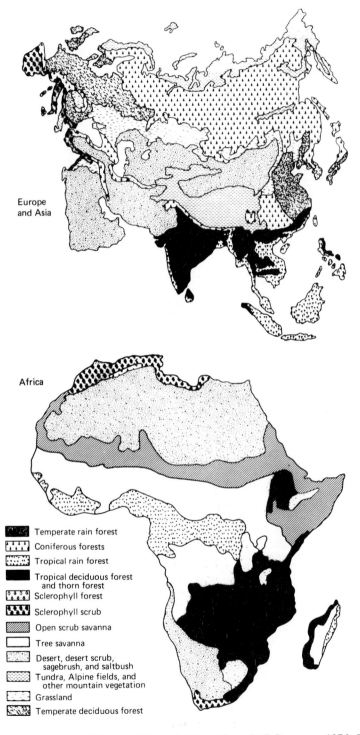

Europe
and Asia

Africa

Temperate rain forest
Coniferous forests
Tropical rain forest
Tropical deciduous forest
 and thorn forest
Sclerophyll forest
Sclerophyll scrub
Open scrub savanna
Tree savanna
Desert, desert scrub,
 sagebrush, and saltbush
Tundra, Alpine fields, and
 other mountain vegetation
Grassland
Temperate deciduous forest

***Figure* 4.2** Major vegetation types of Europe, Asia, and Africa. (From R. F. Dasmann, 1976, *Environmental Conservation,* 4th ed. Copyright © 1976, John Wiley & Sons, Inc., New York. Reprinted by permission of John Wiley & Sons, Inc.)

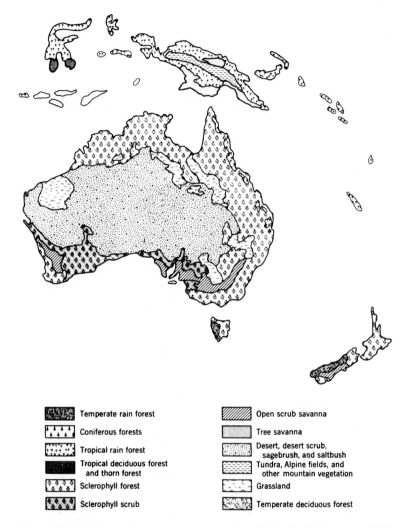

Temperate rain forest		Open scrub savanna
Coniferous forests		Tree savanna
Tropical rain forest		Desert, desert scrub, sagebrush, and saltbush
Tropical deciduous forest and thorn forest		Tundra, Alpine fields, and other mountain vegetation
Sclerophyll forest		Grassland
Sclerophyll scrub		Temperate deciduous forest

Figure **4.3** Major vegetation types of Australia. (From R. F. Dasmann, 1976, *Environmental Conservation,* 4th ed. Copyright © 1976, John Wiley & Sons, Inc., New York. Reprinted by permission of John Wiley & Sons, Inc.)

These grasslike plants have leaves and fibrous roots like true grasses but differ in having nonjointed, solid stems. Forbs are also an important component on many grasslands. Forbs are nongrasslike plants with tap roots, generally broad leaves with netlike veins, and solid nonjointed stems. Shrubs, which are a minor component of most grasslands, have woody stems that branch near the base, and long, coarse roots. Trees differ from shrubs in that they have a definite trunk that branches well above ground. Shrubs and trees have stems that remain alive during dormant periods, but the aboveground parts of most grasses and forbs die back to the crown during winter in temperate areas or during the dry season in the tropics.

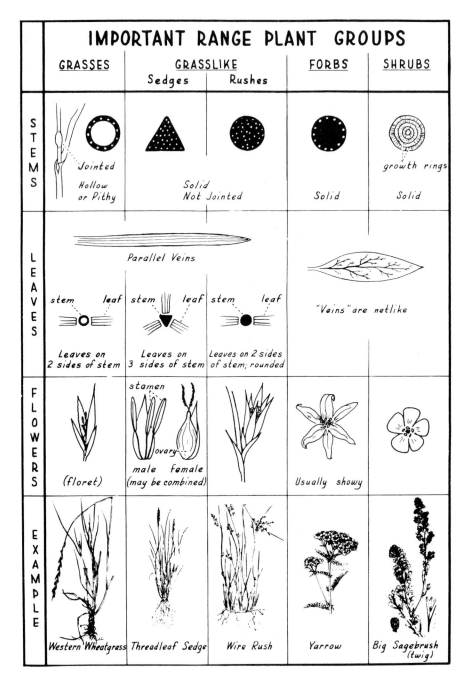

Figure 4.4 Characteristics of the major range plant groups. (From Gay 1965.)

Grassland Climate and Soils. Grasslands generally occur in areas receiving between 250 mm and 900 mm annual precipitation (Figure 4.5). This precipitation generally occurs as frequent light rains over an extended period (90 days or more). In temperate areas, extended light rains during the summer favor grassland over shrubland because the shallow, fibrous roots of grasses use moisture near the soil surface more efficiently than do the long, coarse roots of trees and shrubs (Figure 4.6). Winter rainfall with dry summers generally favors shrubland or woodland over grassland in temperate areas because moisture levels near the soil surface are low during the summer period when temperatures are high enough (at least 10°C) for plant growth. During the summer dry period, the long, coarse roots of shrubs and trees can use moisture stored deep (over 1 m) in the soil profile more efficiently than can the shorter, fibrous grass roots. In the central valley of California and in the Mediterranean region of Europe and northern Africa, grasslands occur in a wet winter, dry summer type of climate. However, winter and spring day temperatures in these areas are above freezing during most days and are warm enough to permit growth by cool season grasses (grasses that grow between 30°C and 35°C). Warm season grasses dominate tropical areas and temperate areas with dry winters and wet summers.

Soils associated with grasslands are usually deep (over 2 m), loamy textured, high in organic matter, and very fertile (Mollisols). These characteristics make them highly suitable for cultivation. In sandy arid areas (less than 300 mm of annual precipitation), soils (less than 600 mm deep) often support grassland, while deeper soils on the surrounding area support shrubland. The thin soils retain the limited moisture near the soil surface that can readily be used by the fibrous grass roots. In contrast, moisture quickly percolates through the deeper sandy soils and is more readily available to the longer tap roots of the shrubs.

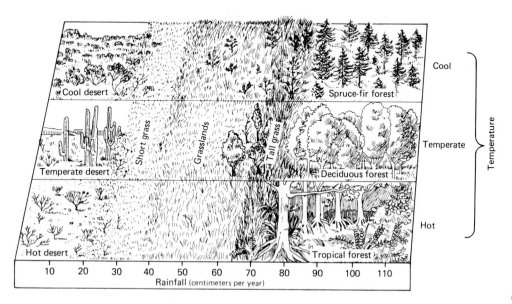

Figure **4.5** Abiotic factors determine the type of ecosystem that can develop in an area. (From Nebel 1981.)

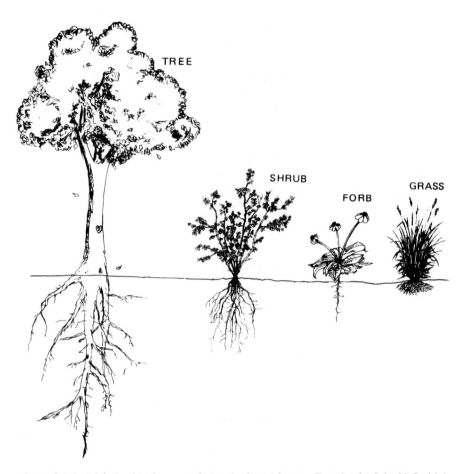

***Figure* 4.6** Rooting depths of grasses, forbs, shrubs, and trees. (Drawing by John N. Smith.)

In areas with over 250 mm of annual precipitation, heavy clay soils often support grassland while surrounding loamy or sandy soils support forest. The fine clay soils permit less moisture infiltration and retain more moisture near the soil surface than do loams or sands. This results in more favorable conditions for fibrous-rooted than for tap-rooted species.

Desert Shrublands

Desert shrublands are the driest of the world's rangelands and cover the largest area. Woody plants less than 3 m in height with a sparse herbaceous understory characterize vegetation of this type. Desert shrublands have received the greatest degradation by heavy grazing of the rangeland biomes and, as a result, show the slowest recovery. In some cases, desert shrublands have been created through degradation of arid grasslands by heavy livestock grazing.

Desert shrublands generally receive less than 250 mm of annual precipitation. The amount of precipitation varies much more from year to year than in the other biomes. In hot desert shrubland areas, precipitation occurs as infrequent, high-intensity rains during a short period (less than 90 days) of the year. This results in long periods where the water content of the soil surface is below the permanent wilting point. This provides highly unfavorable conditions for short, fibrous-rooted plants (grasses). Coarse-rooted plants (shrubs) can collect moisture from a much greater portion of the soil profile than can those with short, fibrous roots near the soil surface. Desert shrub roots extend considerable distances laterally as well as downward. The sparse spacing of desert shrubs permits individual plants to collect moisture over a large area. This explains why they can survive long, dry periods much better than grasses. In temperate areas with high winter snowfall and dry summers, considerable moisture is available deep in the soil profile during the summer growing season. Shrubs can use this moisture much better than grasses because of their longer roots.

Sandy- to loamy-textured soils of variable depth are typical of desert shrublands. Coarse-textured soils permit deep water infiltration and retain little moisture near the soil surface unless there is a restrictive layer. Heavy clay soils and sandy soils, with a shallow restrictive layer in desert areas, typically show a much higher grass component than do surrounding areas. Desert shrubland soils are mainly in the order Aridisol, although some are Entisols.

Savanna Woodlands

Savanna woodlands are dominated by scattered, low-growing trees (less than 12 m tall). They have a productive herbaceous understory if not excessively grazed. Heavy grazing usually results in loss of the understory grasses and an increase in the density of the trees and shrubs. Typically, savanna woodlands occur as a transition zone between grassland and forest. Shifts toward grassland or forest take place continually in this biome, depending on grazing intensity, fire control, logging, and drought. Shrub and tree densities on many savanna woodlands have increased substantially due to fire suppression and heavy grazing of the understory.

Considerable potential exists for conversion of savanna woodlands to grasslands when they occur on flat, nonrocky soils over 1 m in depth. Rocky, thin soils favor woodlands in grassland climatic zones because the long, coarse roots of woody plants can grow down into cracks in the rocky layer where moisture is collected. Further, many woody species have long lateral roots that can absorb moisture over a large area of very thin, rocky soil. Without periodic fire most of the wetter portion of the tallgrass type with loamy to sandy soils is quickly invaded by trees and shrubs, because considerable moisture reaches that portion of the soil profile below 2 m.

Forests

Forests are distinguished from savanna woodlands by having trees over 12 m in height that are closely spaced (less than 10 m apart). In many areas forests are managed primarily for timber production and are too dense to have any grazing value. However, they can produce considerable forage for both livestock and wildlife when thinned by logging or fire, or when in open stands.

Forests generally occur in high-rainfall areas (over 500 mm). Under high rainfall that portion of the soil profile below 3 m has a high water content during most of the year. Much larger quantities of moisture are needed to support the higher biomass of trees compared to grasses and shrubs. Forests occur under as little as 450 mm of annual precipitation in temperate areas of the western United States that have high snowfall in the winter and dry summers. In temperate areas with a fairly even distribution of precipitation during the year, such as the eastern United States, at least 800 mm of annual precipitation is required for forest vegetation. Forests can occur in tropical areas with as little as 500 mm of annual precipitation when soils are coarse textured and the wet season is short (less than 150 days). Coarse-textured and/or thin, rocky soils often favor forest over grassland because they retain low amounts of moisture near the soil surface but store considerable moisture deep in the soil profile and/or rocky crevices. Forest soils are low in fertility, due to greater leaching as compared to grassland soils. Temperate, deciduous forests in most parts of the world are characterized by soils in the order Alfisol. Coniferous forests usually occur on either Spodosols or Ultisols. Oxisols, the most highly leached soils, support tropical forests.

Tundra

Tundra refers to a level, treeless plain in arctic or high-elevation (cold) regions. It is estimated that arctic tundra covers about 5% of the world's surface. Large areas of arctic tundra occur in North America, Greenland, northern Europe (the Soviet Union), and northern Asia. Low-growing, tufted perennial plants and lichens dominate this type. Shrubs in the genus *Salix* are the main type of woody plants. The tundra is frozen for over 7 months of the year and this permafrost restricts the growth of trees. Precipitation is low (250 mm to 500 mm) over much of the area. Strong winds further increase the severity of the area for plant life.

Alpine tundra occurs at high elevations (over 2800 m) above timberline in mountainous portions of the western United States and Canada, Europe, South America, and New Zealand. Because of its limited area, rough terrain, high aesthetic value, and short grazing season (less than 90 days), alpine tundra is of little importance as livestock grazing type. An exception is Peru, where alpacas and llamas make heavy use of this type.

RANGELANDS OF THE UNITED STATES

There are 15 basic rangeland types in the United States that are economically important when forage production and/or total area are considered (Table 4.1). Important historical, climatological, vegetational, and managerial aspects of each type will be discussed. Primary references used for our discussion include USDA (1936), Stoddart and Smith (1943), USDA (1972), Stoddart et al. (1975), USDA (1977), Branson (1985), and Shiflet (1994).

TABLE 4.1   Present Extent and Productivity of the 15 Rangeland Types in the United States

ITEM	FORAGE PRODUCTION (KG/HA) DRY MATTER	AREA (MILLION HA)	PERCENT IN FEDERAL OWNERSHIP
Grasslands			
Tallgrass prairie	1,500–3,500	15[a]	1
Southern mixed prairie	1,000–2,500	20	5
Northern mixed prairie	1,000–2,000	30	25
Shortgrass prairie	600–1,000	20	5
California annual grassland	400–3,500	3	6
Palouse prairie	300–1,000	3	15
Desert shrublands			
Hot desert	150–500	26	55
Cold desert	100–500	73	77
Woodlands			
Piñon-juniper woodland	100–600	17	65
Mountain shrubland	800–2,500	13	70
Western coniferous forest	300–1,000	59	62
Southern pine forest	2,000–4,000	81	6
Eastern deciduous forest	1,200–3,500	100	7
Oak woodland	700–2,500	16	4
Tundra			
Alpine	500–700	4[b]	99

Source: USDA 1972, 1977.

[a]Includes sandhills and coastal prairie.

[b]Does not include Alaskan tundra.

Tallgrass Prairie

The tallgrass prairie is located in the central United States east of the mixed and shortgrass prairies and west of the deciduous forest (Figure 4.7). The major remaining range areas of tallgrass prairie are the Flint Hills of eastern Kansas and the Osage Hills of Oklahoma. These two areas are contiguous and remain because of thin, rocky soils. The Nebraska sandhills and the Texas coastal prairie are considered to be subunits of the tallgrass prairie,

TALLGRASS PRAIRIE

SHORTGRASS OR MIXED PRAIRIE

Figure **4.7** Distribution of shortgrass, mixed, and tallgrass prairies. This and following maps of vegetation represent the approximate original boundaries of major vegetation types. (Adapted from USDA 1936 and USDA-SCS 1954 by Branson 1985.)

although they differ in certain vegetational respects. In the northern United States, very little tallgrass prairie remains. The principal areas remaining include the Waubin prairie in
Minnesota and the Fort Pierre National Grasslands in central South Dakota. Recently, there
has been considerable interest by conservation groups in buying up relict areas of tallgrass
prairie, particularly in the more eastern states of Iowa, Indiana, Minnesota, Wisconsin, and
Illinois.

The climate of the tallgrass prairie is subhumid, mesic, and temperate. Along the southeastern boundary, the mean annual precipitation is almost 1,000 mm. Precipitation for the
northeastern boundary averages 600 mm, with 500 mm occurring on the northwestern boundary. On the southwest, the average is 760 mm. More precipitation is required for tallgrass
prairie in the south than in the north because evaporation is higher. Throughout the tallgrass
prairie most of the precipitation (75%) comes during the summer growing season. This is one

reason why a grassland is favored. Another reason is that summer drought is periodic, and causes considerable mortality of young tree seedlings. It appears that under pristine conditions most of the tallgrass prairie burned every 3 to 4 years, which further favored the grasses.

The soils of the tallgrass prairie are primarily Mollisols. These soils are deep, very fertile, and largely free of rocks. They support cultivated grasses such as corn (*Zea mays*) and wheat (*Triticum aestivum*) very well. Generally, the soil profile is not leached enough to impair fertility. The topography of the tallgrass prairie in the north tends to be monotonous rolling hills, while it is very flat in the south.

The tallgrass prairie evolved under grazing by wild ungulates (primarily, the American bison). It is one of the most grazing-resistant range types, due to the high amount and favorable timing of precipitation (Table 4.2).

Four species characterize the tallgrass prairie. These include little bluestem (*Schizachyrium scoparium*), which dominates the uplands, and big bluestem (*Andropogon gerardii*), which dominates the lowlands (see Figure 1.1a, also Weaver 1954). These two species comprise about 80% by weight of tallgrass prairie climax composition. Yellow indiangrass (*Sorghastrum nutans*) and switchgrass (*Panicum virgatum*) are the other two major grasses. Saline areas are dominated by saltgrass (*Distichlis spicata*). On heavy soils on lowland areas, western wheatgrass (*Agropyron smithii*) is dominant while porcupine grass (*Stipa spartea*) is found on eroded areas. With overgrazing, the tallgrass prairie moves from tall to mid to short grasses. Severely overgrazed areas are dominated by the annual sunflower (*Helianthus annuus*), Kentucky bluegrass (*Poa pratensis*), and blue grama (*Bouteloua gracilis*). Ecotones (areas where two communities come together) are gradual in the true prairie, although communities can be separated by dominants. Ecotones will shift in wet and dry years. In dry years the upland species will move down the hill, while the reverse is true in years of above-average precipitation (Weaver 1954). Several

TABLE 4.2   Grazing Resistance of the Rangeland Types[a]

HIGH RESISTANCE	MODERATE RESISTANCE	LOW RESISTANCE
Southern pine forest	Coniferous forest	Piñon-juniper
Tallgrass prairie	Mountain shrub	Cold desert
Shortgrass prairie	Oak woodland	Hot desert
Eastern deciduous forest	Bunchgrass	Alpine tundra
Northern mixed prairie	California annual grassland	
Southern mixed prairie		

[a]High resistance: Will recover from overgrazing within 3 to 10 years.

Moderate resistance: Will recover from overgrazing in 10 to 30 years.

Low resistance: Over 30 years before any substantial recovery from overgrazing; some locations show no recovery after 50 years. These areas require brush control, and often seeding, for recovery to occur.

genera of forbs are found in the tallgrass prairie. Leadplant (*Amorpha canescens*) and scurfpea (*Psoralea* sp.) are important legumes with nitrogen-fixing abilities. The *Psoraleas* are relatively unpalatable for sheep and cattle but receive some use by wildlife. Leadplant is a highly palatable legume half-shrub that fixes considerable nitrogen in the true prairie ecosystem. Buckbrush (*Symphoricarpos orbiculatus*) is the primary shrub on the tallgrass prairie. It is not particularly palatable to livestock, but it provides valuable food and cover for wildlife. On the eastern edge and along the waterways, several other woody plants are important. These include oaks (*Quercus*), cottonwoods (*Populus*), elms (*Ulmus*), and roses (*Rosa*).

The tallgrass prairie is some of the finest rangeland in the world for summer and fall use. However, the tall grasses are coarse, unpalatable, and low in nutritive content in winter. Rotation grazing systems have given better results than has season-long grazing (Herbel and Anderson 1959; Owensby et al. 1973). Spring burning is necessary every 2 to 3 years to prevent excessive mulch accumulations (Anderson et al. 1970). When yearling cattle are grazed, a modification of season-long grazing called intensive-early stocking has given best results (Smith and Owensby 1978; McCollum et al. 1990) (also see Chapter 9).

Southern Mixed Prairie

The southern mixed prairie is the most important of the western range types for livestock production. It extends from eastern New Mexico to eastern Texas and from southern Oklahoma to northern Mexico. The precipitation varies from 300 mm in eastern New Mexico to 700 mm in central Texas. Over most of the area the frost-free period is at least 180 days. Soils are primarily Mollisols, Entisols, and Aridisols. Because of the wide range of soil and climatic conditions, both productivity and vegetation are variable. We consider four basic subtypes to occur in the southern mixed prairie. These include true mixed prairie, desert prairie, high plains bluestem, and oak savannah. Much of the true mixed prairie and high plains bluestem communities are now under cultivation, although sizable portions still remain as native range in eastern New Mexico and the Texas Panhandle. The desert prairie and oak savannah subtypes exist primarily as native range.

The southern mixed prairie has a long history of grazing, dating back to the early seventeenth century when the Spaniards started bringing livestock into the area. This land type also received considerable use by the buffalo. Most of the grasses associated with this type evolved with grazing and are relatively grazing resistant. Important grasses occurring over the entire type include blue grama, buffalograss (*Buchloe dactyloides*), little bluestem, various threeawn species (*Aristida* sp.), silver bluestem (*Bothriochloa saccharoides*), vine mesquite (*Panicum obtusum*), and sideoats grama (*Bouteloua curtipendula*). In the desert prairie, scotone tobosa (*Hilaria mutica*) and blue grama are the primary grass species. However, as one moves into Texas and climatic conditions become more mesic, silver bluestem and Texas wintergrass (*Stipa leucotricha*) become common.

Areas with sandy, deep soils occur throughout the southern mixed prairie. These areas are characterized by tall grasses, primarily big bluestem, silver bluestem, and little bluestem. In southeastern New Mexico and adjacent Texas these plants grow in association with sand sagebrush (*Artemisia filifolia*) and shinnery oak (*Quercus harvardii*). Collectively, the area supporting these tall grasses is referred to as the high plains bluestem subtype.

In central Texas the oak savannah occurs on what is known as the Edwards Plateau. Originally, this land type supported mostly midgrasses with little bluestem, Texas wintergrass, vine mesquite, silver bluestem, and sideoats grama dominating the composition. However, this type has been heavily overgrazed, with threeawn and curly mesquite (*Hilaria belangeri*) replacing the midgrasses. Texas wintergrass is of major importance since it grows during the winter and provides high-quality feed for livestock when other grasses are dormant.

The southern true mixed prairie occurs primarily in far eastern New Mexico, western Oklahoma, and northwestern Texas. Under climax conditions, this subtype is dominated by little bluestem. Texas wintergrass is a very important species in this type since it is the only cool-season grass of significance to grow in the southern mixed prairie. It provides valuable winter and spring forage when the warm season grasses are dormant.

Mesquite (*Prosopis* sp.) and several other noxious species create serious problems to livestock producers over most of the southern mixed prairie, particularly when overgrazing occurs (Figure 4.8). Ranchers have found that they can increase animal production and improve their ranges with "common use" grazing (Taylor 1985). This involves grazing two or more types of animals on the same range, such as cattle, sheep, goats, and deer. Game animals have been given much attention in this type, particularly in Texas, because many ranchers have found that they can generate more income from fee hunting than from livestock (White 1987). Under controlled grazing coupled with brush control, maximum returns are achieved by a mixture of domestic and wild animals (Taylor 1985; White 1987; Taylor and Kothmann 1993). The Merrill three-herd and four-pasture grazing system has proven superior to continuous grazing when soils, vegetation, livestock, and wildlife are considered (Merrill and Taylor 1975; Whitson et al. 1982; Wood and Blackburn 1984; Heitschmidt et al. 1990).

***Figure* 4.8** Mesquite invasion is heavy in the southern mixed prairie of Texas.

Northern Mixed Prairie

The northern mixed prairie is that portion of the Great Plains extending northward from the Nebraska-South Dakota state line. This type includes the western half of North and South Dakota, the eastern two-thirds of Montana, the northeastern one-fourth of Wyoming, and the southeastern part of Alberta and southern Saskatchewan in Canada. This is the second most important western range type from the standpoint of livestock production.

The climate of the northern mixed prairie is characterized by long, severe winters with warm summers. Precipitation ranges from 300 mm to 650 mm, with two-thirds of it coming during the summer and most of the other third in the spring. Over most of this land type, the peak period of precipitation is June. The average frost-free period ranges from about 140 days in the south to less than 100 days in Canada. The first killing frost usually occurs between September 1 and 10 in the fall and the last freeze generally comes from May 10 to June 10 in the spring.

The major soils associated with the type are in the order Mollisol. Large areas of Entisols also occur throughout the region. Although much of the soil is suitable for cultivation, severe winters, a short growing season, periodic drought, and low precipitation produce conditions that are not favorable for crop production. Therefore, most of the northern mixed prairie is still rangeland, although there are some large areas of productive wheatland in central Montana. Much of the northern mixed prairie was farmed in the period between 1900 and 1933, but drought resulted in this area being returned to range. In western South Dakota heavy clay soils in the order Vertisol severely limit farming. These soils produce almost pure stands of western wheatgrass when in good condition.

The northern mixed prairie supports the highest diversity of grasses of all the western range types. It has short, mid, and tall grasses as well as both cool- and warm-season grasses. Under climax conditions the cool-season mid grasses dominate the composition. This type also supports a very diverse composition of shrubs and forbs. Because of the great diversity in vegetation, the northern mixed prairie is one of the best of all types for both wildlife and livestock from a nutritional standpoint. The shrubs and blue grama provide excellent winter feed. However, snow periodically covers the grass in winter, necessitating winter feeding of livestock. Cool-season grasses such as bluebunch wheatgrass (*Agropyron spicatum*) and various bluegrasses (*Poa* spp.) provide good early spring feed. Green needlegrass (*Stipa viridula*), needleandthread (*Stipa comata*), western wheatgrass, and various forbs provide excellent late spring feed. Little bluestem, blue grama, and sideoats grama provide high-quality summer and fall forage.

Several different vegetation subtypes occur within the northern mixed prairie. The southern part is dominated by the species of *Stipa, Agropyron*, and *Buchloe;* the central and northern parts support the genera of *Bouteloua* and *Stipa;* and scattered through the central and northern parts are areas dominated by *Festuca.* Because of temperature, buffalograss does not extend past southern Montana. Western wheatgrass dominates the western part of South Dakota and southwestern North Dakota. Rough fescue (*Festuca scabrella*) occurs through parts of central Montana, eastern Alberta, and Saskatchewan. The fescue grasslands are some of the best in the world for livestock, due to high plant diversity and the fact that rough fescue is one of the most palatable and nutritious of all range grasses. If one species was used to characterize the northern mixed prairie, it would probably be green needlegrass. This species is a decreaser occurring throughout the type but is relatively uncommon in other range types. Needleandthread, blue grama, and western wheatgrass are important species over the entire type.

The northern mixed prairie has perhaps the widest variety of shrubs of all the western range types. This is a primary reason why it supports an abundant and diverse wildlife population. Important shrubs include silver sagebrush (*Artemisia cana*), found on heavy soils in the lowlands, big sagebrush (*Artemisia tridentata*), found on well-drained soils in the more xeric portions, skunkbrush sumac (*Rhus trilobata*) on the hillsides, western snowberry (*Symphoricarpos occidentalis*) and various species of rose (*Rosa* spp.) in the creek bottoms, and fourwing saltbush (*Atriplex canescens*) and black greasewood (*Sarcobattus vermiculatus*) in the more saline areas. Scattered areas of ponderosa pine (*Pinus ponderosa*) occur in the central portions of the northern mixed prairie. Areas with ponderosa pine usually have sandy soils and support species such as prairie sandreed (*Calamovilfa longifolia*), big bluestem, and soapweed yucca (*Yucca glauca*). In years of severe drought and with overgrazing, the mid grasses are replaced with blue grama. Invasion by annual bromes (*Bromus tectorum* and *japonicus*), Russian thistle (*Salsola iberica*), and prickly pear cactus (*Opuntia polyacantha*) also occurs. In the xeric portions, big sagebrush may increase in density and canopy cover.

Crested wheatgrass (*Agropyron cristatum*) and Russian wildrye (*Elymus junceus*) are introduced grasses that provide excellent early spring forage for cattle and allow deferred and lighter use of native range (Smoliak and Slen 1974). Large acreages of these species are not recommended because they are low in nutritive value during the summer and winter. In addition, seedings over 160 ha are detrimental to wildlife. Nitrogen fertilization can be used to improve plant palatability and range productivity, but heavy rates over extended periods are not recommended because cool-season bunchgrasses and warm-season grasses tend to be replaced by western wheatgrass and annual bromes (Lodge 1959; Lorenz and Rogler 1972). The biggest benefit from fertilization may be improved livestock distribution. Pitting and furrowing have been very effective in many areas in improving forage productivity and reducing clubmoss (*Selaginella densa*) (Ryerson et al. 1970). Although the northern mixed prairie has traditionally been cattle country, common use of cattle and sheep would probably be more efficient than either class of animal alone because of the diversity in forage. However, this could adversely affect many important wildlife species, such as pronghorn, mule deer, and sage grouse, since they depend heavily on forbs and shrubs that receive light use by cattle. The most limiting factors to both livestock and wildlife in the northern mixed prairie are drought and severe winters. These two problems can be minimized by preventing excessive range use and maintaining vegetation diversity. Because they are available above snow during the winter and are green during drought, shrubs provide insurance against each of these two climatic problems.

Shortgrass Prairie

The shortgrass prairie extends from northern New Mexico into northern Wyoming. Pieces of this type are scattered through central Wyoming, western South Dakota, and southern Montana. Because of low precipitation, much of this type remains as rangeland. It is the third most important western range type from the standpoint of livestock production.

The climate of the area is characterized by cool winters and warm summers. Annual precipitation ranges between 300 mm and 500 mm, with 60% to 75% of the precipitation coming from slight rains fairly evenly distributed over the summer. This type of climate is very favorable to warm-season grasses such as blue grama, which have shallow but extensive root systems.

Soils of this type are largely Mollisols. However, sandy soils in the order Entisol and clay soils in the order Vertisol are scattered through the area. Mid grasses such as little bluestem occupy the sandy soils, while the heavy clay soils are dominated by western wheatgrass. Medium-textured soils support primarily blue grama and buffalograss.

The vegetation of the shortgrass prairie is relatively simple since blue grama and buffalograss comprise 70% to 90% of the composition by weight. The third most important grass species associated with this type is western wheatgrass. Because blue grama and buffalograss evolved with grazing pressure by the American bison, they have morphological and physiological characteristics which make them quite resistant to grazing. An important shrub associated with this type is winterfat (*Ceratoides lanata*). This shrub is palatable to both domestic and wild ungulates. It is an important food of the pronghorn antelope, which reaches its highest numbers in this type (Figure 4.9). Scarlet globemallow (*Sphaeralcea*

***Figure* 4.9** Pronghorn antelope reach their highest numbers in the shortgrass prairie.

coccinea) is an abundant forb heavily used by cattle, sheep, and pronghorn. Heavy grazing causes the grasses to be replaced by cactus (*Opuntia* sp.), snakeweed (*Gutierrezia* sp.), Russian thistle, and fringed sagewort (*Artemisia frigida*). During periods of drought, buffalograss tends to replace blue grama.

Specialized rotation grazing systems have not proven superior to continuous grazing (see Chapter 9). Nitrogen fertilizer has proven to be an effective means of increasing forage production and livestock performance (Hyder et al. 1975). Over most of the shortgrass prairie, livestock are grazed throughout the year without provision of hay. During the winter months, a protein supplement can substantially reduce livestock weight losses (Bellido et al. 1981).

Unfortunately, repeated attempts have been made to cultivate large areas of the shortgrass prairie. In recent years, this has been a particular problem in eastern Colorado. Past experience has shown that the shortgrass prairie will not sustain cultivation without irrigation. During the drought of the 1930s, severe wind erosion on cultivated land created conditions known as the dust bowl and vast tracts of cultivated land were abandoned. These tracts were slowly revegetated back to rangeland during the 1940s and 1950s. However, these tracts were again cultivated in the 1970s. In the early 1980s, there was much concern that the dust bowl might be repeated. The 1985 Farm Bill has encouraged the return of these highly erodible lands back to grassland.

Because of flat terrain, good distribution of water, long growing season, and the high nutritional quality of the short grasses, this range type is well suited for grazing by both cattle and sheep.

California Annual Grassland

The California annual grassland is found primarily west of the Sierra Nevada Mountains (Figure 4.10). A subtype occurs in Oregon west of the Cascade Mountains as a savannah with an overstory of Oregon white oak (*Quercus garryana*) but with understory species the same as the California annual type.

The climate of the area is Mediterranean, characterized by mild, wet winters and long, hot, dry summers. Rainfall varies from about 200 mm in the southern foothills to almost 1,000 mm in some areas near the coast. Most of the precipitation comes between October and May, with a peak occurring in January. There is almost no precipitation during the summer and the high evaporation rate quickly vaporizes that which does fall. Summer days are sunny and clear with maximum temperatures frequently above 40°C. The frost-free period ranges from 200 to 260 days.

Soils of the California annual type are quite variable. Many have prairielike profiles (Mollisols), while others display desert characteristics (Aridisols). In the western Oregon subtype they are almost entirely from volcanic ash (Enceptisols). The more prairielike soils occur near the coast, while the drier types with low organic matter are found in the foothills and uncultivated portions of the interior valley. Some of the grassland soils are deep, medium textured, high in base saturation, and have a porous structure. These make excellent cropland and

CALIFORNIA GRASSLAND

Figure **4.10** Approximate original distribution of California annual grassland. (Adapted from USDA 1936 by Branson 1985.)

are in most cases farmed. Those that are shallow, extremely heavy or light, and low in organic matter remain as rangeland.

The California annual grassland has one of the longest livestock grazing histories of the western range types. Spanish settlements were made in California during the seventeenth century and by the eighteenth century, much of the valley and coastal areas were being grazed. The original vegetation was comprised mostly of cool-season bunchgrasses in the genus *Stipa*, which did not evolve with grazing by large herbivores. Under pristine conditions, the California grasslands were quite beautiful, but they have suffered severe degradation since the arrival of the white man. Practically all the native perennial grasses are gone. They have been replaced by introduced, cool-season, annual bromes and forbs. These cool-season annuals have nearly ideal environmental conditions in California. This is because they have adequate moisture and temperature for growth and reproduction in winter while their seeds remain dormant during the dry summer period. Today almost 400 species of introduced plants from many lands are found in California. Presently, less than 5% of the herbaceous cover is comprised of perennials (Sampson et al. 1951).

The original vegetation of the California annual grasslands was comprised of cool-season bunchgrasses dominated by mid grasses such as purple needlegrass (*Stipa pulchra*), nodding needlegrass (*Stipa cernua*), prairie junegrass (*Koeleria pyramidata*), pine blue-grass (*Pos scabrella*), California melicgrass (*Melica imperfecta*), creeping wildrye (*Elymus triticoides*), and California oatgrass (*Danthonia californica*). The most important of these species under pristine conditions were purple and nodding needlegrasses. It appears that introduced annuals may have dominated the area for over 100 years. Numerous fires and overgrazing at an early date probably account for the change from perennial to annual grasses. Today the vegetation is dominated by slender oat (*Avena barbata*), wild oat (*Avena fatua*), soft brome (*Bromus mollis*), ripgut brome (*Bromus rigidus*), foxtail brome (*Bromus rubens*), and little barley (*Hordeum pusillum*) (see Figure 1.1d).

As one moves up from the valley floor into the Coast Range on the west or Sierra Nevada foothills on the east, conditions become more and more favorable for trees and shrubs. Because of climatic conditions several evergreen shrubs with small thick leaves known as chaparral occupy the foothill ranges. Important species include pointleaf manzanita (*Arctostaphylos pungens*), wedgeleaf Ceanothus (*Ceanothus cuneatus*), hollyleaf buckthorn (*Rhamnus crocea*), poison oak (*Rhus toxicodendron*), chamise (*Adenostoma fasciculatum*), California scrub oak (*Quercus dumosa*), blue oak (*Quercus douglasii*), and interior live oak (*Quercus wislizenii*). Chamise and scrub oak are the dominant chaparral species in California. Other important genera are *Prunus* and *Holodiscus*.

Because the grasslands of California are dominated by annuals they have some unique problems. Generally, they are not very responsive to grazing intensity, although heavy grazing does cause erosion problems, some change in species composition, and reduced forage production (Bartolome et al. 1980; Rosiere 1987). Season-long grazing has been superior to specialized systems (see Chapter 9). These ranges should be grazed lightly during plant germination and establishment in the fall, but grazing intensity can be increased substantially in the early spring. Primary problems in this type include inadequate forage quantity in the fall and winter and inadequate forage quality in the summer (Sampson et al. 1951; Rosiere and Vaughn 1986). The second problem can be solved by supplementation with protein and phosphorus. The first problem requires access to perennial pasture or drylot feeding of harvested forages. A major disadvantage of annual range is that forage quantity fluctuates drastically from year to year (Swanson and Sellars 1978; Rosiere 1987). This means that operators must be able to adjust their stocking rates rapidly or have a good reserve of harvested forage. A large operator can do this much more efficiently than can a small one. Nitrogen and sulfur fertilizer can greatly increase forage production in this type in average and wet years but response is low in dry years (Jones 1967). Fertilization is most economical for operators who can easily change their stocking rates. Fertilizer application also allows ranchers to extend the grazing season.

Brush invasion is a real problem in many areas. Fire and common use grazing can both be utilized to control brush. Generally, a single fire favors brush because it increases the germination of many brush species and prepares a seedbed (Bentley 1967). However, two fires in close succession retard brush invasion because the second fire destroys the seedlings. Browsing by goats, sheep, and deer also retards brush establishment. Small burns of 15 ha and 35 ha in the chaparral are effective for enhancement of black-tailed deer and other wildlife species, since this increases both food and edge effect (Biswell et al. 1952). Guzzlers have been useful in increasing quail populations in this type.

Palouse Prairie

The palouse prairie, also referred to as the northwest bunchgrass prairie, has had the highest percentage conversion into farmland of all the western range types. Today it is used primarily for wheat production. The palouse prairie occurs primarily in eastern Washington, north-central and northeastern Oregon, and western Idaho (Figure 4.11). Certain areas in northern Utah and western Montana are very similar to the palouse prairie. Although most of the contiguous palouse prairie in Oregon, Washington, and Idaho has been plowed, large tracts of native grassland remain in these states. The Blue Mountain region of northeastern Oregon and southeastern Washington is an elevated plateau containing considerable area of open palouse grassland that is unfarmed because of thin soils and a short growing season. Northcentral Oregon has the largest area of remaining true palouse prairie.

The palouse prairie grasses evolved with little grazing pressure from large herbivores. The American bison did not use this type because barriers such as the Rocky Mountains and Snake River Canyon restricted access. Both bluebunch wheatgrass and Idaho fescue

Figure **4.11** Approximate original distribution of palouse prairie. (Adapted from Ross and Hunter 1976 and USDA 1936 by Branson 1985.)

(*Festuca idahoensis*) have very low grazing resistance. For this reason the palouse prairie responds rapidly to overuse. Cattle grazing was initiated on this type in the late 1840s, and by 1863 overgrazing was reported by a Walla Walla, Washington, newspaper. The palouse prairie was the first of the northern ranges to support a livestock industry (Galbraith and Anderson 1971). It is notable that cattle from the palouse were used to stock northern mixed prairie ranges in the 1870s.

Rainfall in the area ranges from 30 cm to 64 cm annually, with approximately 65% to 70% falling during the winter months. Peak precipitation occurs in December or January. The months of July and August have the lowest precipitation, although these months are wetter than in the California annual type. The growing season lasts from 140 to 160 days, extending from May 10 to October 10. Winters in the palouse country are relatively mild, and summer temperatures are seldom over 35°C.

The soils of the palouse country are primarily loessial dunes in the order Mollisol. They have excellent textural, structural, and chemical properties for agriculture. In many areas the hills or dunes are 100 m deep. Soils on the western edge are heavily mixed with volcanic ash and fall into the order of Inceptisol. The topography of the type is rolling with biscuits and swales. Deep canyons bisect the prairie but are generally unnoticeable as one looks across it.

The palouse prairie is one of the most productive and most beautiful grasslands in the world. The major characteristics distinguishing the palouse prairie from other North American grasslands is that the climax vegetation is dominated by bluebunch wheatgrass, or bluebunch wheatgrass is a codominant with either Idaho fescue or Sandberg bluegrass (*Pos sandbergii*) (Figure 4.12). Unlike the central prairie of North America, where the grasses form a sod, the palouse prairie grasses grow in bunches with open interspaces between plants (Figure 4.13). The palouse prairie is like the California grasslands in that the

Figure **4.12** The palouse bunchgrass prairie dominated by bluebunch wheatgrass and Idaho fescue in northeastern Oregon. Most of the palouse is now farmed due to excellent soils and climate for wheat production.

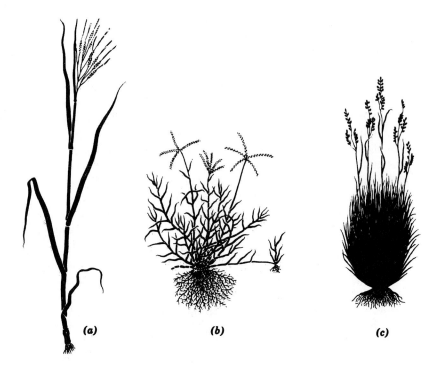

Figure 4.13 Growth forms of grasses: (a) annual, (b) sod farming, (c) bunch. (From R. F. Dasmann, 1976, *Environmental Conservation,* 4th ed. Copyright © 1976, John Wiley & Sons, Inc., New York. Reprinted by permission of John Wiley & Sons, Inc.

grasses are almost entirely cool-season bunchgrasses because of the dry summers. With over-grazing, the invader downy brome (cheatgrass) (*Bromus tectorum*) replaces the perennial grasses and Sandberg bluegrass increases (Daubenmire 1940; Tisdale 1961). In some areas the unpalatable medusahead rye (*Taeniatherum asperum*) has replaced downy brome after re-peated long-term heavy grazing. The palouse prairie supports a wide variety of forbs. Most of these forbs are composites. Under climax conditions arrowleaf balsamroot (*Balsamorhiza sagittata*) is dominant. Western yarrow (*Achillea millefolium*) becomes more dominant when range condition declines. Several shrubs occur in the type. Big sagebrush occupies the more xeric sites and increases with heavy grazing. Rose species (*Rosa* spp.) are found on the more mesic sites. Crested wheatgrass and several other wheatgrasses can be seeded on the more productive sites to provide early spring grazing and allow deferment of native range.

Hot Desert

The hot desert is one of the largest western range types. However, because of low precipita-tion and high temperatures, it rates relatively low from the standpoint of livestock produc-tion compared to other types. This type is found in southern California, southern Nevada,

Arizona, New Mexico, southwestern Texas, and northern Mexico. It contacts piñon-juniper range to the north, chaparral on the west, southern mixed prairie on the east, and the southern Mexico mountains to the south. Elevations for the type range from 925 m to 1,400 m.

A desert climate characterizes the type. Precipitation varies from 130 mm to 500 mm and increases with elevation above sea level. Precipitation occurs primarily during winter and summer with the wettest months in July, August, and September. A much smaller precipitation peak occurs in January. May and June are extremely dry. Summer rains occur as convectional storms caused by solar heating. The winter rains are a result of general frontal movement. Because the storms during the summer are convectional with short duration and high intensity, they are often not very effective. Temperatures are quite high, with long periods of greater than 38°C expected during the summer months. Bright sunshine is characteristic of the semiarid regions, with some stations in Arizona reporting 90% of the possible days to be cloudless. Relative humidity is generally low. The frost-free period is over 200 days during the year. Many areas may go 2 to 3 years without a killing frost. On the average, the last killing frost in the spring occurs between February 28 and March 30, with the first frost occurring between October 30 and November 30.

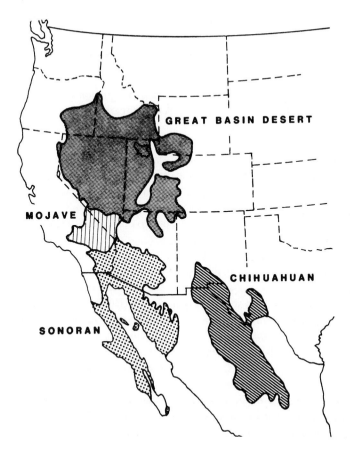

***Figure* 4.14** Deserts of North America. The Great Basin desert is the cold desert. (After Shreve 1942 by Blaisdell and Holmgren 1984.)

TABLE 4.3 Some Climatic Features of the Four Desert Regions of North America

	COLD DESERT (GREAT BASIN)	MOJAVE	SONORAN	CHIHUAHUAN
Area (km^2)	409,000	140,000	275,000	453,000
Annual precipitation (mm)	100–300	100–200	50–300	150–300
Precipitation falling in summer (percent of total)	30	35	45	65
Snowfall (cm; 10 cm snow = 1 cm rain)	150–300	25–75	Trace	Trace
Winter mean max/min temperatures (°C)	+8/−8	+15/0	+18/+4	+16/0
Hours of frost (percent of total hours per year)	5–20	2–5	0–1	2–5
Summer mean max/min temperatures (°C)	34/10	39/20	40/26	34/19
Elevation (m)	>1,000	Variable	<600	600–1,400

Source: Barbour et al. 1987.

Soils of the area are mixed and difficult to categorize. The major soil order is Aridisol. Some areas that are well drained have a calcium carbonate (caliche) layer at the surface or under a shallow profile.

Three general desert areas occur within this type: the Mojave, Sonoran, and Chihuahuan deserts (Figure 4.14). The Mojave desert, found in southeastern California, southern Nevada, and northwestern Arizona, is the driest of these three types (Table 4.3). It has the least diversity of vegetation and is primarily a shrubland. The Chihuahuan desert, which is found in southwestern Texas and southcentral New Mexico, has less species diversity than the Sonoran desert in southern Arizona due to colder weather in the winter and less winter precipitation. The Sonoran desert is quite rich in species because it is virtually frost free and most areas have nearly 300 mm of precipitation. Warmer winter temperatures explain the wider diversity of cactus found in Arizona compared to New Mexico.

There is considerable evidence that originally the Sonoran and Chihuahuan deserts were an open grassland or grassland scattered with shrubs (Buffington and Herbel 1965). Presently, much of this area supports a mixture of shrubs and grasses, with the shrubs more or less dominant. Mesquite (*Prosopis* sp.), catclaw (*Acacia* sp.), and creosotebush (*Larrea tridentata*) are common throughout the area. Several theories have been advanced for the increase in shrubs during the past 100 years. Overgrazing, cessation of fire, climatic change, and seed dissemination by domestic animals have all been suggested as possible causes. Probably a combination of these factors explains the increase. Regardless of the cause, brush covers a large area that was originally grassland.

The vegetation of the hot desert type did not evolve with grazing by large herbivores. However, this area was one of the first types in the United States to receive grazing by do-

Figure 4.15 Chihuahuan desert grassland in excellent condition dominated by black grama. This New Mexico range provides high-quality forage for cattle throughout the year due to high nutritive value of forbs and shrubs when grasses are dormant in winter.

mestic livestock. The Spaniards brought cattle and sheep into the area early in the seventeenth century (Schickedanz 1980). Black grama (*Bouteloua eriopoda*), which apparently dominated much of the type, particularly southcentral New Mexico, under pristine conditions is not highly resistant to grazing (Figure 4.15). Both severe overuse and lack of adaptation to grazing probably explain the large-scale decline of black grama and other climax grasses in the past 100 years.

Almost all the plants found in the hot desert type are warm-season. The true climax dominants are almost completely of the genera *Bouteloua, Aristida*, and *Hilaria*. They begin growth early in the spring and grow rapidly any time moisture is available, until cessation by frost. On areas where soil moisture relationships are favorable, mid grasses such as a sideoats grama, silver bluestem, green sprangletop (*Leptochloa dubia*), tanglehead (*Heteropogon contortus*), and Arizona cottontop (*Digitaria californica*) are dominant. These areas are referred to as either a post-disclimax or an edaphic climax, depending on the ecologist. Intensive grazing in the desert plains results in a disclimax of low-producing perennials such as burrograss (*Scleropogon brevifolius*) mixed with annuals in the genera of *Bouteloua* and *Aristida*. In many areas the grasslands are reduced to almost pure stands of annuals. Broom snakeweed (*Gutierrezia sarothrae*) is an unpalatable half shrub that has severely infested many heavily grazed desert grassland ranges.

Because of low forage production and grazing resistance, livestock grazing is not practical in much of the Mojave and Sonoran deserts. Because of the dry, sunny climate, large numbers of people have moved into these deserts during the past 20 years. Tourism, wildlife, water, and recreation have become far more important products than forage for livestock. In the Chihuahuan desert, livestock grazing is and will probably continue to be an important rangeland use. Because of the low potential, attempts to control brush species such as mesquite and creosotebush have generally not been economically effective (Holechek 1992, Warren et al. 1996). Management of livestock focuses on water development to improve livestock distribution and carefully adjusting stocking rates to forage availability (Holechek 1991, 1992). Because of the low precipitation, recovery from heavy grazing is often very slow to nonexistent (little recovery after 30 years) (Smith and Schmutz 1975; Beck and Tober 1985). Therefore, the consequences of abusive grazing are much more severe than in most of the other types.

There is growing concern about large tracts of the Sonoran desert that are being converted into housing and other urban uses in Arizona. If this continues within 30 to 50 years less than 50% of this type may remain as rangeland. Rangeland loss to urbanization is less severe in the Mojave and Chihuahuan deserts, but still a problem. Maintaining open space and natural plant communities could be one of the biggest challenges confronting rangeland managers in the southwestern United States in the twenty-first century.

Cold Desert

The cold desert, sometimes called the Great Basin, is comprised of the sagebrush grassland and the salt desert. These two rangeland types intermingle with each other over vast portions of the intermountain United States. The sagebrush grassland generally occurs at higher elevations than does the salt desert. It is characterized by higher precipitation and less saline soils. Although the frost-free period for both types is typically less than 180 days, the salt desert has warmer temperatures, due to its occurrence at low elevations. Differences in vegetation and managerial components between the two cold deserts warrant a separate discussion of each.

Sagebrush Grassland. The sagebrush grassland, or shrub steppe as it is sometimes called, is one of the most extensive of the western range types (Figure 4.16). There are approximately 39 million hectares of this range type, of which about 65% is in federal control and 35% is in private ownership. Most of the federal land is controlled by the Bureau of Land Management. This range type occurs primarily in Oregon, Idaho, Nevada, Utah, Montana, and Wyoming.

The climate of the sagebrush grassland is semiarid. Precipitation ranges between 200 mm and 500 mm, with an average of 250 mm. In the northern half about 50% to 60% of the precipitation comes in the late fall, winter, and early spring as snow. Summers are quite dry. In the southern part most of the precipitation also comes in the winter, but slightly more precipitation comes in the summer. The sagebrush grassland occurs between the Cascade-Sierra Nevada Mountains and the Rocky Mountains. Many smaller mountain ranges occur through the area. These mountains result in considerable rain-shadow effect and account for the dryness of this type. Temperatures in the area are extreme, dropping as low as −37°C in winter and climbing up to 38°C in the summer. Most of the area is characterized by about a 100-day growing season. This type occurs at elevations over 1,235 m.

Soils of the area are primarily volcanic materials with much sand and little clay. They are mainly Aridisols. Depth varies from very shallow (less than 1 m) to quite deep and topography is highly variable. There are many level plains dominated by sagebrush extending to rough foothills. Central Oregon, Idaho, and Nevada have considerable lava.

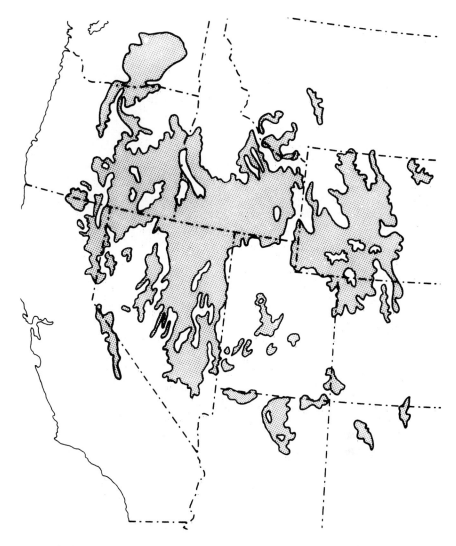

***Figure* 4.16** Distribution of the big sagebrush type. (Modified from Branson et al. 1967, by Branson 1985.)

The sagebrush grassland is characterized by big sagebrush (Figure 4.17). Important grasses of the sagebrush grassland are bluebunch wheatgrass, bottlebrush squirreltail (*Sitanion hystrix*), Idaho fescue, western wheatgrass, Indian ricegrass (*Oryzopsis hymenoides*), needleandthread, and great basin wildrye (*Elymus cinereus*). Bottlebrush squirreltail is an important grass over the entire area. Bluebunch wheatgrass dominates the understory in the northern half, with western wheatgrass dominating the understory in the southern half if grazing has not been abusive. Originally, the sagebrush grassland had an open stand of big sagebrush, with native wheatgrasses comprising most of the total aboveground

Figure **4.17** The sagebrush grassland is one of the largest types in the western United States. The high levels of volatile oils associated with the various sagebrushes make them unpalatable to livestock. Sagebrush does provide native big-game animals such as muledeer and pronghorn with important winter feed.

vegetation. However, with overgrazing, the big sagebrush becomes more prominent and the invader, downy brome, replaces the wheatgrass in the understory.

There are several other shrubs associated with this type that are also important. Black sagebrush (*Artemisia nova*) occurs on rocky and shallow soils in the northern and central areas; low sagebrush (*Artemisia arbuscula*) is found on soils with a poor water supply at higher elevations in the northern half; rabbitbrush (*Chrysothamnus* sp.) occupies sandy soils with a low salt content in the central and northern part; antelope bitterbrush (*Purshia tridentata*) occurs on sandy, rocky soils at the higher elevations; and Mormon tea (*Ephedra* sp.) occurs on coarse soils in the central and southern areas.

Creased wheatgrass has been successfully seeded on large areas of sagebrush grassland after control of big sagebrush (Holechek 1981). This species provides high-quality spring forage. Crested wheatgrass seeding allows delay of grazing on native rangeland until late spring or early summer. By this time the native cool-season grasses have completed most of their growth and can withstand grazing much better than during the spring. Sagebrush grassland ranges generally recover very slowly from overgrazing (Holechek and Stephenson 1983). Control of big sagebrush can greatly speed rate of recovery. Both burning and herbicidal control can be economically effective (Heady and Bartolome 1977; Nielsen 1977; Holechek and Hess 1994).

Salt Desert Shrubland. The salt desert occurs primarily in the states of Utah and Nevada, although smaller pieces of this type occur in Wyoming, Montana, Idaho, Oregon, Colorado, and New Mexico (Figure 4.18). It is one of the least productive of the range types in the United States because of climate and soils. There are approximately 35 million hectares of the salt desert shrub, of which about 85% is in public ownership and 15% is privately owned. Most of the salt-desert shrub type is controlled by the Bureau of Land Management.

The salt desert occurs between the Cascade-Sierra Nevada Mountains and the Rocky Mountains. It occurs in many areas as a mosaic with the sagebrush grassland. The salt desert

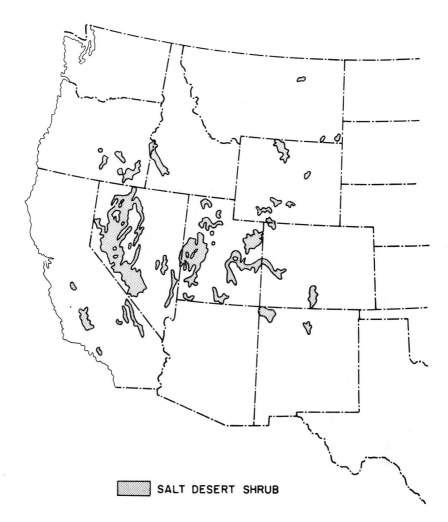

Figure **4.18** Distribution of the salt-desert shrub type. (Revised from Kuchler 1964 by Branson 1985.)

occurs on the lower areas of the Great Basin, where drainage is often restricted and the water table is high. Evaporation causes salts to accumulate at the soil surface. Sagebrush grassland occupies the upper areas free of salts and receiving more precipitation.

The climate of the salt desert is very xeric (it has the lowest precipitation of all types, with the exception of the Mojave desert) and this condition is made more severe by the high salt content of the soil. Precipitation for this type ranges from 80 mm to 250 mm, with an average of 120 mm. About one-half the precipitation is snow and one-half is spring or fall rain. Summers are quite dry in much of this type. Precipitation is highly variable from year to year, and there is very little vegetation growth in the dry years. As with the sagebrush grassland, the low rainfall is caused by the rain-shadow effect of the mountains. The growing season in this type is approximately 200 days. Temperatures can be quite cold in the winter and quite hot in the summer.

Soils of the area are primarily Aridisols. They have varying degrees of alkalinity and salinity. Soils where water collects have the highest salt concentrations. Although both clays and sand occur, sands are more common.

The vegetation of the salt desert is typically characterized by a few species of low, spiny, grayish, and widely spaced microphyllous (small-leaved) shrubs in the Chenopodiaceae and Asteraceae families. The vegetation canopy cover is usually less than 10%. Under climax conditions the percent composition by weight is about 75% browse, 24% grasses, and 1% forbs (Hutchings and Stewart 1953). With deterioration, the area is dominated almost entirely by shrubs or low palatability. Shadscale saltbush (*Artiplex confertifolia*) is the shrub species dominating the largest area, with winterfat being the second most common (see Figure 1.1b). Shadscale is relatively unpalatable to game and livestock, but winterfat is very palatable. Important understory species include inland saltgrass, Indian ricegrass, bottle-brush squirreltail, and galleta grass (*Hilaria jamesii*). With deterioration the understory is occupied by Russian thistle, halogeton (*Halogeton glomeratus*), and downy brome. The overstory species associated with deterioration include spiny hopsage (*Grayia spinosa*), shadscale saltbush, gray horsebrush (*Tetradymia canescens*), and rubber rabbitbrush (*Chrysothamnus nauseosus*). Much of the salt desert has been severely overgrazed. Vegetation recovery is slow (Turner 1971; Rice and Westoby 1978) but Yorks et al., 1992, reported meaningful improvement over a 56-year period under moderate livestock grazing pressure.

The salt desert has traditionally been used as winter range for sheep. During the summer sheep are trailed to the surrounding mountains. Many areas of the salt desert have been invaded by halogeton, an accidentally introduced annual from Asia (Branson 1985). Halogeton is quite toxic to sheep, particularly in the fall, and control of this plant has proven difficult. The best insurance against halogeton is a dense stand of perennial plants (Frischknecht 1967). Generally, control of undesirable shrubs in the salt desert has not been economically practical. However, on some of the wetter sites with a low salt content, tall wheatgrass (*Agropyron elongatum*) seedings after brush control have been successful. Tall wheatgrass is an introduced, salt-tolerant perennial from Asia that provides good spring forage.

Piñon-Juniper Woodland

The piñon-juniper type is one of the most widely distributed range types in the western United States (Figure 4.19). This type occurs from the state of Washington to 220 km north of Mexico City. Although it occurs on the eastern side of the Cascades and Sierras, most of it is found in the states of Utah, Colorado, New Mexico, and Arizona. North of Utah this type consists entirely of juniper (*Juniperus* sp.) because juniper is more cold resistant and can tolerate lower precipitation than can piñon pine. This type is found at elevations from 100 m to 2300 m. However, in New Mexico it is found as high as 2400 m. Its upper limits contact Gambel oak (*Quercus gambelii*) and/or ponderosa pine, while the lower limits may contact

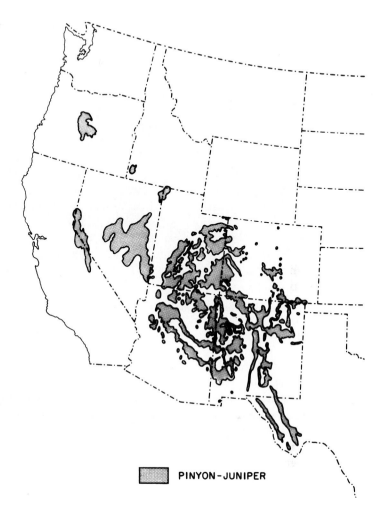

Figure **4.19** Distribution of the piñon-juniper type. The type in Oregon consists mainly of a single tree species, western juniper (*Juniperus occidentalis*). (From Franklin and Dyrness 1969; modified from Clary et al. 1974 and USDA 1936 by Branson 1985.)

☐ **PINYON-JUNIPER**

hot desert, shortgrass prairie, chaparral, or sagebrush grassland. The amount of this type has increased over the past 100 years because of juniper invasion into surrounding grassland.

Piñon-juniper range is commonly referred to as woodland rather than forest since the trees are small and below sawtimber size. The climate of this type is relatively harsh for tree growth. It is characterized by low precipitation, hot summer, high wind, low relative humidity, high evaporation rates, and much clear weather and intense sunlight. The annual precipitation varies from 300 mm to 450 mm, with local areas receiving up to 500 mm. For most of the type, precipitation averages 380 mm to 420 mm. The frost-free period is variable and ranges from 91 to 205 days.

Soils of this type are poorly developed and are primarily of the orders Entisol and Aridisol. Prior to widespread settlement of the West, piñon-juniper stands were confined

Figure **4.20** The piñon-juniper type has been severely overgrazed in most locations where it occurs. Natural recovery is very low in this type due to low precipitation and thin soils.

largely to the rocky ridges or more level sites with shallow soils. As in many other parts of the West, heavy grazing in the late nineteenth century depleted much of this type. Due to a combination of overgrazing, absence of fires, dissemination of seeds by mammals and birds, and possibly climatic change, trees have encroached on the grasslands and original stands have become more dense (Johnsen 1962). Overgrazing of understory species has reduced protective soil cover and resulted in severe soil erosion in much of this type (Figure 4.20).

The understory herbaceous vegetation varies with grazing history, physical features of the site, and density of the trees. Juniper ranges north of Utah have entirely cool-season grasses because of lack of summer rainfall. In Utah, Arizona, Colorado, and New Mexico this type supports a mixture of warm- and cool-season grasses. Spring grazing has been detrimental to many of the cool-season grasses. As the tree overstory increases, perennial grasses and forbs decrease because of shading and increased competition from the trees. Germination of some grasses may be decreased by foliage extract of the juniper.

This is one of the most depleted range types occurring in the United States, especially in northern areas. Very few areas presently support a good grass understory. Recovery from overgrazing is considered to be slow to nonexistent in most areas without control of the trees. However in western Utah Yorks et al. (1994) found meaningful increases in perennial grass cover occurred over a 56-year period in response to lower grazing pressure. Presently wood from this type often has more economic value than the forage it provides for domestic animals. Partial removal of trees can greatly increase forage for both livestock and wildlife.

Mountain Browse

The mountain browse range type occurs primarily in the Rocky Mountains and Sierra-Cascade Mountains of the western United States. It is a narrow intermittent strip between the uppermost reaches of the grasslands (northern mixed prairie, palouse prairie, sagebrush grassland, and California annual grassland) and the coniferous forest types. It occurs above the piñon-juniper type in the intermountain area. This type is most prevalent in Colorado, Utah, Oregon, and Idaho.

Climate is intermediate between that favoring grassland and forest. There is not quite enough precipitation for forest. Annual precipitation averages around 460 cm to 500 cm. Temperatures range from 35°C in the summer to −34°C in the winter. The growing season lasts 100 to 120 days.

Soils are mostly Entisols and Inceptisols. Topography is variable and includes mostly ridges and dry, rocky slopes. Elevations range from 1200 m to 2800 m.

The vegetation is dominated by shrubs 1 m to 10 m tall (most 2 m to 4 m tall). Important species occurring throughout the type are chokecherry (*Prunus virginiana*) and several species of buckbrush (*Ceanothus*). However, about any western shrub requiring mesic conditions may occur in this type, depending on location. Gambel oak and true mountain mahogany (*Cercocarpus montanus*) are two of the most important shrubs associated with this type in the southwest. In northern areas antelope bitterbrush is an important forage species.

This type is a very important big-game winter range. When in public ownership, management should probably be directed toward big-game animals. Shrub plants require moderate browsing to maintain productivity. They tend to stagnate and grow out of reach of browsing animals when protected (Tueller and Tower 1979). Late spring and early summer grazing of cattle and sheep has been beneficial to mule deer because they graze primarily the understory species (Smith and Doell 1968; Jensen et al. 1972). This reduces competition to shrub species and results in more browse being available to wintering mule deer. Gambel oak is very important forage for mule deer on southwestern mountain browse ranges. However, stands often grow out of reach of game animals and retard movement. Herbicide application has been effective in killing top growth and stimulating lateral growth (Kufeld 1977). In the northwestern half of the western United States bitterbrush is the primary mountain browse shrub, particularly in Oregon, Idaho, and Utah. This shrub is extremely palatable to both big game and livestock. However, it tends toward apical dominance and stagnation when only lightly or not grazed. Antelope bitterbrush stands have been effectively rejuvenated by cutting off the top part of the plant with a chain saw (Ferguson 1972). This results in the plant growing laterally rather than apically.

Conversion of the mountain browse type into summer homes has become a serious concern to wildlife managers in the 1990s. This has resulted in critical loss of big-game wintering range throughout the West. Another problem caused by the increase in housing is fire control. Much of the mountain browse type has historically burned periodically. These fires play an important role in keeping the shrub growth vigorous and within reach of livestock and game animals. The increase in housing in the type has led to strong public pressure for fire suppression. This causes excessive fuel loads to accumulate and results in wildfires that are much more damaging than those that occurred under natural conditions.

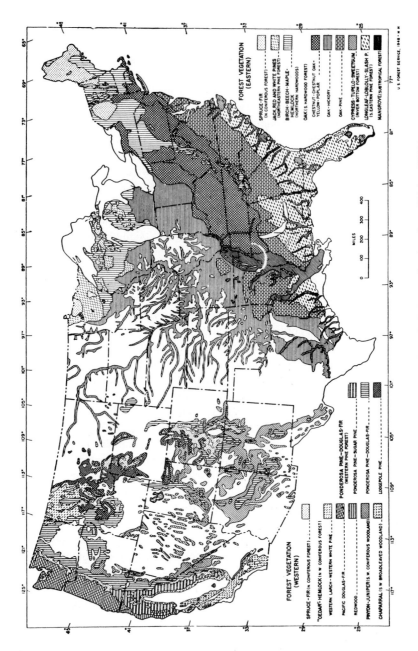

Figure 4.21 Natural forest regions of the United States. (From U.S. Forest Service, from Shirley 1973.)

Western Coniferous Forest

The western coniferous forest range type is comprised primarily of areas dominated by ponderosa pine (16 million hectares) and Douglas fir (*Pseudotsuga menziesii*) (17 million hectares). This type is found in all of the interior western United States (Figure 4.21). Much of this type exists in seral stages.

Ponderosa pine ranges are the largest and most xeric of the true forest types in the western United States (see Figure 1.1c). Ponderosa pine is found between the piñon-juniper of various brush ranges and the Douglas fir zone. These ranges occur from 242 km to 322 km inland from the Pacific Ocean to Nebraska and the Dakotas and from southern Alberta and British Columbia south into northern Mexico. Approximately 55% of the ponderosa pine subtype is under the jurisdiction of the Forest Service. Precipitation ranges from 450 mm to 650 mm, with most as snow in the north and as rain in the south. The growing season ranges from 105 to 140 days, and frost can occur in any month. Soils are primarily Entisols, with Enceptisols on the benchlands and ridges and Mollisols dominating areas with gentle topography. In the southwest ponderosa pine is found at from 2,000 m to 2,500 m, while it may occur as low as 1100 m in northern areas.

The Douglas fir-aspen zone occurs directly above the ponderosa pine zone. Douglas fir-aspen ranges occur mostly in Colorado, Idaho, Wyoming, Montana, Oregon, and Washington. They occur in isolated areas in Arizona and New Mexico. The bulk of this type is in seral stages, with about 8 million hectares in lodgepole pine (*Pinus contorta*), 5 million hectares in quaking aspen (*Populus tremuloides*), 2 million hectares in western larch (*Larix occidentalis*), and 1.2 million hectares in brush. Approximately 62% of this type is owned by the Forest Service. This zone receives 640 mm to 900 mm of precipitation per year, mostly as snow. The growing season lasts from 100 to 125 days. The soils are primarily Alfisols, Entisols, and Inceptisols. The topography is rarely level. In the southwest it is found between 2,500 m and 3,100 m in elevation. However, it occurs much lower in the north. Quaking aspen is adapted to fire but intolerant of shade. Douglas fir, in contrast, is shade tolerant but cannot survive fire.

The coniferous forest is in relatively good condition compared to other western range types because of high precipitation and a long history of controlled grazing. The main cause of grazing capacity loss is from regrowth of timber after fires and logging, which shade out the understory species (Skovlin et al. 1976). Multiple use is very important in this type. Recreational, watershed, wildlife, and scenic values are more important than livestock production in many areas. However, when properly controlled, all these uses are usually compatible.

Southern Pine Forest

The southern pine type is of the largest and most important range types in the United States (Figure 4.22). It is the most important of all range types for livestock production. Precipitation for this type averages 1,250 mm or more per year. The fall period is dry, while precipitation distribution is fairly even during the rest of the year. The frost-free period ranges from 200 to 365 days. Temperatures are high, with a yearly average of 21°C. Soils

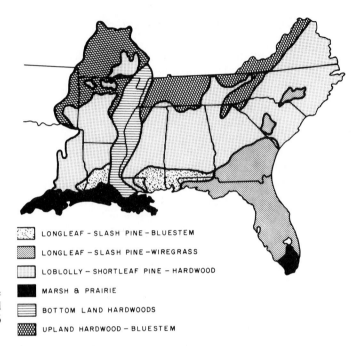

***Figure* 4.22** Southern pine forest rangeland. (Adapted from Byrd and Lewis 1976 by Grelen 1978.)

LONGLEAF – SLASH PINE – BLUESTEM

LONGLEAF – SLASH PINE – WIREGRASS

LOBLOLLY – SHORTLEAF PINE – HARDWOOD

MARSH & PRAIRIE

BOTTOM LAND HARDWOODS

UPLAND HARDWOOD – BLUESTEM

of this area are highly leached and primarily in the order of Ultisol. Acid soils are a primary limitation to vegetation production. Vegetation productivity is quite high due to warm temperatures and the large amount of precipitation.

The climax vegetation for most of this area is oak-hickory (*Carya* sp.) hardwoods. Presently most of it is in a pine [longleaf pine (*Pinus palustris*), shortleaf pine (*Pinus echinata*), loblolly pine (*Pinus taeda*)] seral stage (Figure 4.23). The pines are grown for lumber and other uses. Open and cutover forests provide abundant forage, particularly in the coastal plain areas (Grelen 1978). However, grazing has only a limited place in upland hardwoods and shortleaf-loblolly pine types, where low grazing value does not justify the risks of damage to tree reproduction. Important forage species of the type include grasses in the genera *Andropogon, Panicum, Aristida, Paspalum, Sporobolus,* and *Cynodon.*

Periodically prescribed burning in the winter is practiced to eliminate mulch accumulations and old growth. Native grasses provide nutritious forage for only about 3 months in the spring and a few weeks in the autumn. Heavy use of improved pasture and supplemental feed is necessary because protein and phosphorus are especially deficient (Kalmbacher 1983; Long et al. 1986). Cattle are the most important range animal throughout this area. Hogs are used as a range animal in some areas, but feral hogs are pests at other locations. Sheep are seldom used. Grazing is second to timber as a land use in the Southeast. Demand for fee hunting is rapidly increasing in this type, and consequently, wildlife are receiving more consideration in management programs. Cattle grazing and wildlife (white-tailed deer, wild turkey, bobwhite quail) are highly compat-

***Figure* 4.23** Southern pine ranges are highly productive for wood, wildlife, and livestock. Prescribed burning is used to eliminate excessive accumulation of vegetation and control hardwoods.

ible in this type. Without cattle grazing many important wildlife forbs are shaded out by heavy vegetation accumulation of the tall rank grasses (Reid 1954; Moore and Terry 1981).

Eastern Deciduous Forest

The eastern deciduous forest is becoming increasingly important for livestock production. Large areas of this type occur in Missouri, Indiana, Ohio, Kentucky, Virginia, and Wisconsin. Annual precipitation ranges between 800 mm and 2,000 mm, which is uniformly distributed throughout the year. Snow and frost are common in the winter. The growing season lasts 120 to 240 days. Soils are primarily in the order Alfisol. Most of the eastern deciduous forest has been heavily modified by farming, logging, and industrialization.

Important genera of the eastern deciduous forest trees include the maples (*Acer* sp.), birches (*Betula* sp.), oak (*Quercus* sp.), hickories (*Carya* sp.), beeches (*Fagus* sp.), and basswood (*Tilia* sp.). Forage production is low in climax eastern deciduous forest because of heavy shading by the overhead forest canopy. However, this type is very productive when the trees have been partially or completely removed. Grasses in the genera *Andropogon* (bluestem), *Festuca* (fescue), *Phleum* (timothy), *Poa* (bluegrass), *Bromus* (brome), *Lolium* (ryegrass), and *Dactylis* (orchardgrass) are important as forages.

Rotational grazing schemes have generally been superior to continuous grazing in maximizing livestock production per unit area (Matches and Burns 1985). Nitrogen and phosphorus fertilization are economically effective in improving forage quantity and quality since soil fertility is much more limiting than precipitation.

Because of the high precipitation, seedings of introduced forage grasses (fescues, bromes, ryegrasses, timothy) are widely used. Legumes, particularly clovers (*Trifolium* sp.), are often grown with the pasture grasses (Carlson et al. 1985). Because they have nitrogen-fixing bacteria, clovers (*Trifolium* sp.) greatly improve soil fertility and productivity of the grasses.

Oak Woodland

This type comprises the rangelands dominated by several different species of oak (*Quercus* sp.). There are three basic subdivisions: the shinnery oak type found in southeastern New Mexico, adjoining Texas, and Chihuahua, Mexico; Gambel oak in the central and southern Rockies; and open savannah dominated by tree oaks in California, Oregon, southern Arizona, and central Texas.

The climate of the oak brush ranges is variable depending on location. The post oak savannah of central Texas has from 660 mm to 800 mm of precipitation per year, which comes mostly in the late spring (May) and early fall (September). The chaparral ranges of southern Arizona and New Mexico average about 360 mm per year, with two peaks which occur in January and late July. Gambel oak ranges which occupy the central and southern Rocky Mountains average from 380 mm to 500 mm of precipitation per year. Precipitation distribution is fairly even during the year in the central Rocky Mountain zone, but the southern Rocky Mountains have the biomodal pattern of summer and winter precipitation. The California chaparral ranges receive from 650 mm to 1,000 mm per year, with most of it arriving during the winter. Oaks are sensitive to winter cold and do not occur much farther north than northcentral Oregon. One species, Oregon white oak (*Quercus garryana*) does occur as far north as southwestern British Columbia. The maximum temperature of the oak brush ranges is about 36°C and the minimum is about −34°C, although oaks may endure even greater extremes during short periods. Soils of the oak brush type are varied but usually well drained. Topography is characteristically rolling uplands and foothills, although the shinnery oak type in west Texas and southeastern New Mexico is quite flat.

As one moves from east to west in the southern Great Plains, the oak species become smaller and smaller as the result of reduced precipitation. In eastern Texas the rather tall-growing post oak (*Quercus stellata*) is dominant; in central Texas and Oklahoma the small tree forms of common live oak (*Quercus virginiana*) are most common; and in western Oklahoma, Texas, and southeastern New Mexico the shinnery oak is the primary species. Oak brush range in Utah and Colorado is comprised of Gambel oak. A variety of oaks occur in Arizona and New Mexico.

Oak woodland is important to wildlife. The acorns of oaks provide valuable food for many game species, such as the black-tailed deer, white-tailed deer, mule deer, Rocky Mountain elk, collared peccary, wild turkey, band-tailed pigeon, lesser prairie chicken, and bobwhite quail.

Most of the oaks have limited forage value for livestock. The new growth of deciduous oak species such as shinnery and Gambel oak can be poisonous, especially to cattle, in the spring. Goats readily browse most oak species without experiencing poisoning problems. In Texas, goats have been used effectively to control common live oak and associated species (Merrill and Taylor 1976).

Figure **4.24** Foothill ranges of the Rocky Mountains are dominated by dense stands of Gambel oak. This plant is toxic to livestock in the spring, but provides important food and cover for deer.

Gambel oak provides valuable browse for deer and elk in Colorado, Utah, Arizona, and New Mexico, although mature stands have little value because the forage is mostly out of reach and animal movement is restricted (Figure 4.24). Herbicidal spraying and burning have been used effectively to kill top growth and stimulate lateral growth (Kufeld 1977). Deer and elk have responded positively to such treatments. Practically all oak species are rapid sprouters and very difficult to kill with either herbicides or fire. Herbicidal treatment has potential for reducing dense shinnery oak stands in the Southwest, and consequently, improving lesser prairie chicken habitat and forage production for livestock (Pettit 1979; Taylor and Guthery 1980). Complete control of oak over large areas in the western United States is not recommended because of wildlife habitat loss and the fact that oak increases soil stability on steep and sandy areas. In addition, oaks have considerable esthetic value and provide forage variety and shade for livestock.

Alpine Tundra

The alpine tundra is the highest range type in altitude. It occurs above the spruce-fir type. Alpine ecosystems occupy those mountain areas above timberline that are characterized by short, cool growing seasons and long, cold winters. The vegetation is characteristically dominated by low-growing (20 cm or less in height), perennial, herbaceous, shrubby vascular plants, extensive mats of crytograms [mosses (*Selaginella* sp.), lichens (*Cladonia* sp.), etc.] and the complete absence of trees due to permafrost. Alpine ecosystems are found primarily in Alaska, Colorado, Washington, Montana, and California, but Oregon, Idaho, Utah, New Mexico, Arizona, and Wyoming have small amounts of this type. Alaska and Colorado are the states with the largest amounts.

The alpine tundra receives 1,000 mm to 1,500 mm of precipitation, most of which occurs as snow. The alpine tundra is seldom calm, with wind blowing most of the time. The overriding environmental attribute of the alpine tundra is cold temperature. The mean growing-season air temperature is often at or near the freezing point. The high winds and low temperatures result in temperature that is very stressful to plant growth. Plants must be adapted to a short growing season and high ultraviolet radiation of high altitudes. Alpine tundra soils range from shallow, rocky, and gravelly Entisols to boggy Histosols. Soils in the pockets and valleys are well developed, but soils in upper areas have poor development.

Compared to floras of other range types, the alpine tundra flora is species poor. Usually, there are no more than 200 to 300 species present, and most of these are common in all alpine areas. Members of the bluegrass (Poaceae) and sedge (Cyperaceae) families occur throughout alpine areas.

Additional families with wide alpine distribution are the saxifrage (Saxifragaceae), rose (Rosaceae), mustard (Brassicaceae), buckwheat (Polygonaceae), and pink (Caryophyllaceae). Many of the shrub species are members of the willow (Salicaceae) and heath (Ericaceae) families.

The alpine tundra is a very important source of water in the western United States (Johnston and Brown 1979). Because of esthetics and its fragility, large tracts of alpine tundra have been turned into wilderness areas. Presently, this area is grazed primarily by sheep that are herded (Thilenius 1979). This minimizes excessive use of flat, convenient areas. Compared to water production and recreational values, livestock grazing ranks low in importance.

RANGE MANAGEMENT PRINCIPLES

■ Because the rangeland types differ considerably in precipitation, soils, and terrain, management practices that work well in one type are often poorly suited for other types.

■ The interaction of climate, soils, and topography are the primary determinants of the vegetation that naturally occurs in a particular area. Attempts to permanently convert natural shrublands or forests to grasslands usually require high management inputs to prevent return of the original vegetation.

Literature Cited

Anderson, K. L., E. F. Smith, and C. Owensby. 1970. Burning bluestem range. *J. Range Manage.* 18:311–316.

Barbour, M. G., J. H. Burk, and W. D. Pitts. 1987. *Terrestrial plant ecology,* 2d ed. The Benjamin-Cummings Publishing Co., Menlo Park, CA.

Bartolome, J. W., M. C. Stroud, and H. F. Heady. 1980. Influence of natural mulch on forage production on differing California annual range sites. *J. Range Manage.* 33:4–8.

Beck, R. F., and D. A. Tober. 1985. Vegetational changes on creosotebush sites after removal of shrubs, cattle and rabbits. *N. Mex. Agric. Exp. Stn. Bull.* 417.

Bellido, M. M., J. D. Wallace, E. E. Parker, and M. D. Finkner. 1981. Influence of breed, calving season, supplementation and year on productivity of range cows. *J. Anim. Sci.* 52:455–462.

Bentley, J. R. 1967. Conversion of chaparral areas to grassland: Techniques used in California. *U.S. Dep. Agric. Handb.* 328.

Biswell, H. H., R. D. Taber, D. W. Hedrick, and A. M. Schultz. 1952. Management of chamise brushlands for game in the north coast region of California. *Calif. Fish Game* 38:453–484.

Blaisdell, J. P., and R. C. Holmgren. 1984. Managing intermountain rangelands-salt desert shrub ranges. *U.S. Dep. Agric. For. Serv. Gen. Tech. Rep.* INT–163.

Branson, F. A. 1985. *Vegetation changes in western ranges. Range Monograph 2.* Society for Range Management, Denver, CO.

Branson, F. A., R. F. Miller, and R. S. McQueen. 1967. Geographic distribution and factors affecting the distribution of salt desert shrubs in the United States. *J. Range Manage.* 20:287–296.

Buffington, L. C., and C. H. Herbel. 1965. Vegetational changes on semidesert grassland range from 1858 to 1963. *Ecol. Monogr.* 35:139–164.

Byrd, N. A., and C. E. Lewis. 1976. Managing southern pine forests to produce forage for beef cattle. *U.S. Dep. Agric. For. Serv. Priv. For. Manage. Bul.,* Atlanta, GA.

Carlson, G. E., P. B. Gibson, and D. D. Baltensperger. 1985. White clover and other perennial clovers. In *Forages: The science of grassland agriculture.* 4th ed. Iowa State University Press, Ames, IA.

Clary, W. P., M. B. Baker, Jr., P. F. O. Connell, T. N. Johnsen, Jr., and R. F. Campbell. 1974. Effects of piñon juniper removal on natural resource products and uses in Arizona. *U.S. Dep. Agric. For. Serv. Res. Pap.* RM–128.

Dasmann, R. F. 1976. *Environmental conservation.* 4th ed. John Wiley & Sons, Inc., New York.

Daubenmire, R. F. 1940. Plant succession due to overgrazing in the *Agropyron* bunchgrass prairie of southeastern Washington. *Ecology* 21:55–65.

Ferguson, R. B. 1972. Bitterbrush topping: Shrub response and cost factors. *U.S. Dep. Agric. For. Serv. Res. Pap.* INT–125.

Franklin, J. F., and C. T. Dyrness. 1969. Vegetation of Oregon and Washington. *U.S. Dep. Agric. For. Serv. Res. Pap.* PNW–80.

Frischknecht, N. C. 1967. How far will halogeton spread? *J. Soil Water Conserv.* 22:135–139.

Galbraith, W. A., and E. W. Anderson. 1971. Grazing history of the northwest. *J. Range Manage.* 24:6–13.

Gay, C. 1965. Range management. How and why. *New Mexico State Univ. Coop. Ext. Circ.* 376, Las Cruces, NM.

Grelen, H. E. 1978. Forest grazing in the south. *J. Range Manage.* 31:244–250.

Heady, H. F., and J. Bartolome. 1977. The Vale rangeland and rehabilitation programs. The desert repaired in southeastern Oregon. *U.S. Dep. Agric. For. Serv. Res. Bull.* PNW–70.

Heitschmidt, R. K., J. R. Conner, W. E. Pinchals, J. W. Walker, and S. L. Dowhower. 1990. Cow/calf production and economic returns from yearlong continuous, deferred rotation and rotational grazing treatments. *J. Prod. Agr.* 3:92–99.

Herbel, C. H., and K. L. Anderson. 1959. Response of true prairie vegetation on major Flint Hills range sites to grazing treatment. *Ecol. Monogr.* 29:171–198.

Holechek, J. L. 1981. Crested wheatgrass. *Rangelands* 3:141–153.

Holechek, J. L. 1991. Chihuahuan desert rangeland, livestock grazing and sustainability. *Rangelands* 13:115–120.

Holechek, J. L. 1992. Financial benefits of range management practices in the Chihuahuan Desert. *Rangelands* 14:279–284.

Holechek, J. L., and K. Hess. 1994. Brush control considerations: A financial perspective. *Rangelands* 16:193–196.

Holechek, J. L., and T. Stephenson. 1983. Comparison of big sagebrush vegetation in northcentral New Mexico under moderately grazed and grazing excluded conditions. *J. Range Manage.* 36:455–457.

Hutchings, S. S., and G. Stewart. 1953. Increasing forage yields and sheep production in intermountain winter ranges. *U.S. Dep. Agric. Circ.* 925.

Hyder, D. N., R. E. Bement, E. E. Remmenga, and D. F. Hervey. 1975. Ecological responses of native plants and guidelines for management of shortgrass range. *U.S. Dep. Agric. Tech. Bull.* 1503.

Jensen, C. H., A. D. Smith, and G. V. Scotter. 1972. Guidelines for grazing sheep on rangelands used by big game in winter. *J. Range Manage.* 25:346–352.

Johnsen, T. N. 1962. One-seed juniper invasion of northern Arizona grasslands. *Ecol. Monog.* 32:187–207.

Johnston, R. S., and R. W. Brown. 1979. Hydrologic aspects related to the management of alpine areas. In *Special management needs of alpine ecosystems. Range Science Series No. 5.* Society for Range Management, Denver, CO, pp. 65–76.

Jones, M. B. 1967. Forage and protein production by subclover-grass and nitrogen fertilized California grasslands. *Calif. Agric.* 21:4–7.

Kalmbacher, R. S. 1983. Distribution of dry matter and chemical constituents in plant parts of four Florida native grasses. *J. Range Manage.* 36:298–301.

Kuchler, A. W. 1964. Potential natural vegetation of the conterminous United States. *Am. Geogr. Soc. Pub.* 36.

Kufeld, R. C. 1977. Improving Gambel oak ranges for elk and mule deer by spraying with 2,4,5-TP. *J. Range Manage.* 30:53–57.

Lodge, R. W. 1959. Fertilization of native range in the northern Great Plains. *J. Range Manage.* 12:277–279.

Long, K. R., R. S. Kalmbacher, and R. G. Martin. 1986. Diet quality of steers grazing three sites in south Florida. *J. Range Manage.* 39:389–392.

Lorenz, R. J., and G. A. Rogler. 1972. Forage production and botanical composition of mixed prairie as influenced by nitrogen and phosphorus fertilization. *Agron. J.* 64:244–249.

Matches, A. G., and J. C. Burns. 1985. Systems and grazing management. In *Forages: The science of grassland agriculture.* 4th ed. Iowa State University Press, Ames, IA.

McCollum, F. T., R. L. Gillen, D. M. Engle, and G. W. Horn. 1990. Stocker cattle performance and vegetation response to intensive early stocking of cross timbers rangeland. *J. Range Manage.* 43:99–104.

Merrill, L. B., and C. A. Taylor. 1975. Advantages and disadvantages of intensive grazing management systems near Sonora and Barnhart. *Tex. Agric. Exp. Stn.* PR–3341.

Merrill, L. B., and C. A. Taylor. 1976. Take note of the versatile goat. *Rangemans' J.* 3:74–76.

Moore, W. H., and W. S. Terry. 1981. Short-duration grazing may improve wildlife habitat in southeastern pinelands. *Proc. Annu. Conf. Southeast. Assoc. Fish Wildl. Agencies.* 33:279–287.

Nebel, B. J. 1981. *Environmental science.* Prentice-Hall, Inc., Englewood Cliffs, NJ.

Nielsen, D. B. 1977. Economics of range improvement: A rancher's handbook to economic decision-making. *Utah Agric. Exp. Stn. Bull.* 466.

Owensby, C. E., E. F. Smith, and K. L. Anderson. 1973. Deferred-rotation grazing with steers in the Kansas Flint Hills. *J. Range Manage.* 26:393–395.

Pettit, R. D. 1979. Effect of picloram and tebuthiuron pellets on sand shinnery oak communities. *J. Range Manage.* 32:196–200.

Reid, V. H. 1954. Multiple land use: Timber, cattle and bobwhite quail. *J. For.* 52:575–578.

Rice, B., and M. Westoby. 1978. Vegetative responses of some Great Basin shrub communities protected against jackrabbits or domestic stock. *J. Range Manage.* 31:28–34.

Rosiere, R. E. 1987. An evaluation of grazing intensity influences on California annual range. *J. Range Manage.* 40:160–166.

Rosiere, R. E., and C. E. Vaughn. 1986. Nutrient content of sheep diets on a serpentine barren range. *J. Range Manage.* 39:8–13.

Ross, R. L., and H. E. Hunter. 1976. *Climax vegetation of Montana based on soils and climate.* U.S. Department of Agriculture, Soil Conservation Service. Bozeman, MT.

Ryerson, D. E., J. E. Taylor, L. O. Baker, H. Houlton, and D. W. Stroud. 1970. Clubmoss on Montana rangelands: Distribution, control, range relationships. *Mont. Agric. Exp. Stn. Bull.* 645.

Sampson, A. W., A. Chase, and D. W. Hedrick. 1951. California grasslands and range forage grasses. *Calif. Agric. Exp. Stn. Bull.* 724.

Schickedanz, J. G. 1980. History of grazing in the Southwest. In *Proc. grazing management systems for Southwest Rangelands Symposium.* New Mexico State University, Las Cruces, NM.

Shiflet, T. N. (Ed.). 1994. *Rangeland cover types.* Society for Range Management, Denver, CO.

Shirley, H. L. 1973. *Forestry and its career opportunities.* 3d ed. McGraw-Hill Book Company, New York.

Shreve, F. 1942. The desert vegetation of North America. *Bot. Rev.* 8:195–246.

Skovlin, J. M., R. W. Harris, G. S. Strickler, and G. A. Garrison. 1976. Effects of cattle grazing methods on ponderosa pine bunchgrass range in the Pacific northwest. *U.S. Dep. Agric. Tech. Bull.* 1531.

Smith, A. D., and D. D. Doell. 1968. Guides to allocating forage between cattle and big game on big game winter range. *Utah State Div. Fish Game Publ.* 68–11.

Smith, E. F., and C. E. Owensby. 1978. Intensive-early stocking and season-long stocking of Kansas Flint Hills range. *J. Range Manage.* 31:14–17.

Smith, D. A., and E. M. Schmutz. 1975. Vegetation changes on protected versus grazed desert grassland ranges in Arizona. *J. Range Manage.* 28:453–458.

Smoliak, S., and S. B. Slen. 1974. Beef production on native range, crested wheatgrass and Russian wildrye pastures. *J. Range Manage.* 27:433–436.

Stoddart, L. A., and A. D. Smith. 1943. *Range management.* McGraw-Hill Book Company, New York.

Stoddart, L. A., A. D. Smith, and T. W. Box. 1975. *Range management.* 3d ed. McGraw-Hill Book Company, New York.

Swanson, J. D., and D. V. Sellars. 1978. Record drought on California's annual grass rangeland. *Proc. Int. Rangel. Congr.* 1:212–216.

Taylor, C. A., Jr. 1985. Multispecies grazing research overview. In F. H. Baker and K. R. Jones (Eds.). *Proc. Conference on Multispecies Grazing.* Winrock International Institute, Morrilton, AR.

Taylor, C. A., and M. M. Kothmann. 1993. *Managing stocking rates to achieve livestock production goals in the Edwards Plateau.* Managing Stocking Rates on Rangeland Symposia. Texas A&M Univ., College Station, TX.

Taylor, M. A., and R. S. Guthery. 1980. Status, ecology, and management of the lesser prairie chicken. *U.S. Dep. Agric. For. Serv. Tech. Rep.* RM-77.

Thilenius, D. A. 1979. Range management in the alpine zone. In *Special management needs of alpine ecosystems. Range Science Series No. 5.* Society for Range Management, Denver, CO.

Tisdale, E. W. 1961. Ecological changes in the palouse. *Northwest Sci.* 35:134–138.

Tueller, P. T., and J. D. Tower. 1979. Vegetation stagnation in three-phase big game enclosures. *J. Range Manage.* 32:258–264.

Turner, G. T. 1971. Soil and grazing influence on a salt-desert shrub range in western Colorado. *J. Range Manage.* 24:31–32.

U.S. Department of Agriculture (USDA). 1936. *The Western range.* 74th Congress, 2nd Session. Senate Document 199.

U.S. Department of Agriculture (USDA). 1972. The nation's range resources: A forest-range environmental study. *U.S. For. Serv. Resour. Rep.* 19.

U.S. Department of Agriculture (USDA). 1977. Vegetation and environmental features of forest and range ecosystems. *Forest Serv. Agric. Handb.* 475.

U.S. Department of Agriculture-Soil Conservation Service (USDA-SCS). 1954. *Map No. 4-P-7647.* Fort Worth, TX.

Warren, A., J. Holechek, and M. Cardendo, 1996. Honey mesquite influences on Chihuahuan desert vegetation. *J. Range Manage.* 49:46–52.

Weaver, J. E. 1954. *North American prairie.* Johnson Publishing Co., Chicago.

White, R. J. 1987. *Big game ranching in the United States.* Wild Sheep and Goat International Publishing Co., Mesilla, NM.

Whitson, R. E., R. H. Heitschmidt, M. M. Kothmann, and G. K. Lundgren. 1982. The impact of grazing systems on the magnitude and stability of ranch income in the Rolling Plains of Texas. *J. Range Manage.* 35:526–533.

Wood, M. K., and W. H. Blackburn. 1984. Vegetation and soil response to cattle grazing systems in the Texas Rolling Plains. *J. Range Manage.* 37:303–308.

Yorks, T. P., N. E. West, and K. M. Capels. 1992. Vegetation differences in desert shrublands of western Utah's pine valley between 1933 and 1989. *J. Range Manage.* 45:569–578.

Yorks, T. P., N. E. West, and K. M. Capels. 1994. Changes in piñon-juniper woodlands in western Utah's pine valley between 1933–1989. *J. Range Manage.* 47:359–364.

CHAPTER 5

RANGE PLANT PHYSIOLOGY

lthough range plants grow together in similar environments, different species often differ markedly in certain aspects of their physiology. Range management is often geared to community responses as a whole. However, community responses are controlled by responses of their individual grazing or browsing, water stress, and so on. Consequently, an understanding of basic physiologic processes is necessary to an understanding of ecological reactions of individual plants and groups of plants. In this chapter we provide an overview of range plant physiology. For more complete and detailed coverage of this subject we refer the reader to Bedunah and Sosebee (1995).

BASIC CONCEPTS

Range plant management is based on four fundamental concepts.

1. The plant is the only source of energy (food) for the support of grazing animals.
2. The formation of sugars, starch, proteins, and other foods is first dependent on the photosynthetic process in the leaves (seldom the stems) of plants.
3. Plants do *not* get food for their maintenance and growth from the soil. They obtain only the *raw materials* needed for photosynthesis and subsequent food elaboration. Green plants are therefore entirely dependent on green leaf tissue for their survival.
4. When leaves are removed from plants, food-producing capacity is reduced.

While these principles appear fairly simple when considered alone, they involve many complex interactions among plants, soils, herbivores, and the environment. The rest of this chapter deals with these interactions.

Photosynthesis is the process by which plants convert energy from the sun, carbon dioxide from the atmosphere, and water and minerals from the soil into food they can use for maintenance and growth (Figure 5.1). The three basic phases of photosynthesis are summarized as follows:

1. CO_2 + H_2O + energy of sunlight = simple food material

CO_2: CO_2 is supplied from the air. CO_2 is seldom limiting for photosynthesis, although only about 0.03% of the atmosphere is CO_2. Consequently, range management has little effect on the CO_2 available for photosynthesis.

H_2O: The availability of water for the photosynthetic processes is strongly affected by management. Heavy grazing can increase plant moisture stress by weakening root systems, causing excessive water runoff from soil and excessive evaporation of moisture from the soil. Three distinct photosynthetic processes have been identified in plants. Most range plants have the C_3 (cool-season species) or C_4 (warm-season species) pathway. The physiological and ecological implications of these differences are discussed in detail by Coyne et al. (1995).

2. Simple food materials produced by photosynthesis (sugars, starches, and fat) plus mineral elements from the soil are utilized in the synthesis of complex compounds such as proteins, vitamins, and other complex substances.

3. These compounds are used within the plant for the following processes:

   a. Root replacement

   b. Regeneration of leaves and stems after dormancy

   c. Respiration during dormancy

   d. Bud formation

   e. Regrowth after top removal

Many factors determine the rate of photosynthesis by the plant. These include:

   a. Area of leaf surface

   b. Intensity and quality of light

   c. Amount of carbon dioxide in the air

   d. Physiological efficiency of the plant

   e. Soil nutrients

   f. Water supply

   g. Temperature

Water availability is a primary factor determining plant growth. It is a necessary chemical constituent of photosynthesis and serves to keep the stomata open and the plant turgid.

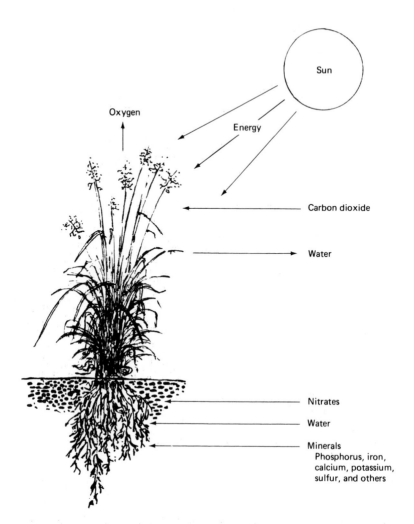

Figure **5.1** Materials used by grass plants for photosynthesis. (Drawing by John N. Smith.)

When the stomata close due to water stress, entry of carbon dioxide into the plant is halted. Water is also important in carrying minerals from the soil into the plant.

The amount of leaf area exposed to the sun is of practical significance to range managers. This is because the amount of food manufactured is proportional to the amount of foliage.

Soil nutrients can be manipulated by fertilization as discussed in Chapter 15. The potential for development of photosynthetically more efficient plants, as well as plants with more tolerance to stress through breeding and genetic engineering, is great. This is discussed in Bedunah and Sosebee (1995).

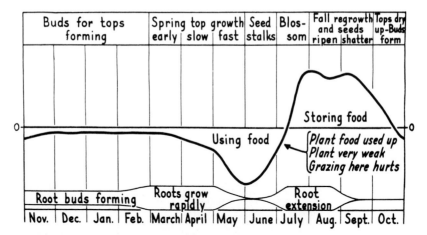

Figure 5.2 Major physiological events during the year for a typical range grass plant for an area with a cold winter and a dry summer. This diagram shows three things going on in a plant during the year: (1) top-growth (top line of diagram), (2) the rate at which the plant uses or stores food that it manufactures (curved heavy line), and (3) root growth. The rate of root growth is shown by the width of the strip just above the months of the year. Plants are most easily injured by grazing when their food storage is used up in the building of tops and roots. (From Parker 1969.)

An example of the food cycle of a grass plant on ranges having a summer dry period with a cold winter (e.g., Idaho, Oregon, Utah) is given in Figure 5.2. Although this is a typical example, it does not apply to all plants on all ranges.

Most plants can have *some* of the top material removed and still remain in productive condition (i.e., they can maintain good *vigor*). The amount that can be removed depends to a large degree on the species of the plants as well as on other environmental factors. We are *not* as much concerned, however, with the ability of individual plants to survive and produce as we are with the survival and production of the general population of desirable forage plants on a given range.

By developing grazing systems based on the food cycle, more efficient use of the range can be made. Considerations in developing grazing systems are as follows:

1. The dormant period is the least critical period for foliage removal. This is because the plant is photosynthetically inactive. However, some critical processes, such as bud formation, may occur during the dormant season.

2. The initiation of growth is intermediate relative to defoliation response. Although the plant has increasing demand for photosynthetic products, considerable opportunity exists to replace leaves removed by grazing since a long period of favorable temperature and soil moisture conditions remains. During drought years this can be the most critical period because low soil moisture can severely reduce potential photosynthetic activity in the mid and late stages of growth.

3. The most critical period for foliage removal for many plant species is from floral initiation through the seed development (post bloom). This period is critical because the plant's demand for photosynthetic products is high and opportunity for regrowth is often low, due to the approach of less favorable temperature and soil moisture conditions.

Four basic factors can be regulated by range managers in controlling grazing animal effects on plants. These include regulation of grazing intensity, timing, frequency, and selectivity (differential grazing of range plants). A wide range of studies (discussed in Chapter 8) has consistently shown that grazing intensity is of most importance (Figure 5.3). This is be-

Figure **5.3** Effects of light grazing (LG), moderate grazing (MG), and heavy grazing (HG) on top and root growth of blue grama (*Boutelou gracilis*). (From Launchbaugh 1957.)

cause intensity governs the amount of leaf area remaining for photosynthesis. Grazing can occur frequently and during critical periods if sufficient leaf area remains to sustain a high level of photosynthesis. Most plants produce more leaf area than is needed for optimum photosynthetic levels. Some of this extra leaf area may actually reduce photosynthetic efficiency by shading new leaves. Although a major objective of specialized grazing management systems has been to improve the use of forage species with low palatability (control of selectivity), increased use of secondary forages has generally resulted in destructive use of primary forages.

Many ranchers and range professionals have held the belief that residue was unimportant to key forage plants after completion of growth. However, research has shown that residue during dormancy plays a critical role in protecting plants from extreme temperatures, and destruction of the growing points in the crown by insects, rabbits, rodents, and pathogens (Sauer 1978; Sneva 1980). Further residue plays a critical role in soil protection and moisture infiltration into the soil (see Chapter 14). Heavy defoliation during dormancy reduces herbage production almost as much as during the active growth (Cook 1971; Nsinamwa 1993).

CARBOHYDRATE RESERVES

Carbohydrates in plants are often divided into structural and nonstructural components (Cook 1966). Structural carbohydrates form portions of the cell and cell wall and are complex compounds such as celluloses, hemicelluloses, and lignin. They are not reutilized by the plant in other metabolic reactions. Nonstructural carbohydrates, also called total available carbohydrates or total nonstructural carbohydrates, are translocated within the plant and used for growth, respiration, and so on. These carbohydrates include sucrose, fructosans, starch, and dextrins (Trlica 1977).

Considerable work, dating back to the 1920s, has been conducted on carbohydrate reserves (Sampson and McCarty 1930; McCarty 1935; McCarty and Price 1942). These early studies formed the basis of much of the thinking behind early grazing management. Since World War II much additional work has been conducted on carbohydrate reserves. These studies have shown that the subject is much more complicated than originally thought.

Early studies indicated that seasonal variation in carbohydrates reserves was fairly consistent for typical range plants (Figure 5.4). Starting in the spring, there is a decrease in carbohydrate content as reserves are mobilized to support new leaf growth. This depletion phase continues until sufficient leaf area is produced by photosynthesis to supply carbohydrates for growth and reserve replenishment. During the fourth or fifth leaf stage, a gradual replenishment of reserves begins and continues until a maximum is reached during formation of reproductive parts. There may be a slight drawdown of reserves as the reproductive parts of the plants are formed. Other species may show a slower rate of replenishment, with most of the storage occurring during reproductive phases (bottom diagram, Figure 5.4). Respiration probably accounts for the decrease in carbohydrates during the dormant season.

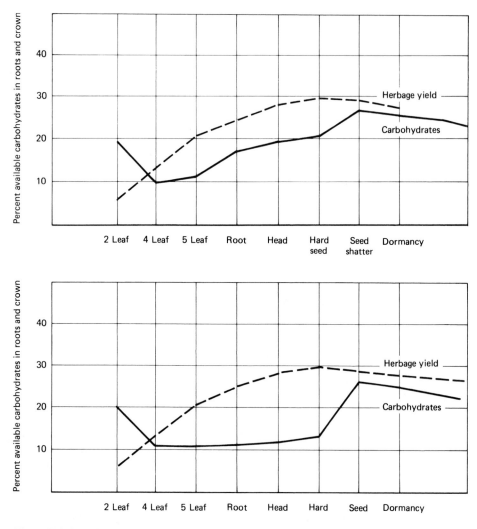

***Figure* 5.4** Carbohydrate cycles and growth curves for a typical range plant species (top) and one that replenishes its reserves late in the growing season (bottom). (From Cook 1966.)

Menke and Trlica (1981) provided additional details on the variations in carbohydrate cycles for different range species. "Typical" species exhibited a rapid decline in carbohydrate reserves in the spring as new leaves were being formed, followed by an increase during late vegetative and early reproductive phases (Figure 5.5). Fourwing saltbush (*Atriplex canescens*) and antelope bitterbrush (*Purshia tridentata*) exhibited typical V-shaped curves. Fringed sagewort (*Artemisia frigida*), scarlet globemallow (*Sphaeralcea coccinea*), and western wheatgrass (*Agropyron smithii*) had definite flattened or extended V-shaped cycles and maintained low reserves for most of the growing season (middle curve, Figure 5.5).

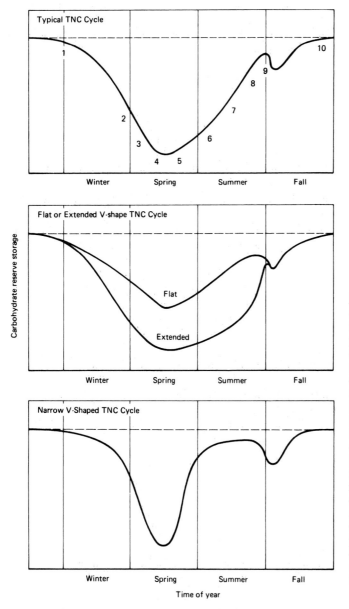

***Figure* 5.5** Generalized carbohydrate reserve cycles for three types of range plants. Numbers on top curve refer to phenological stages: (1) winter quiescence, (2) leaves regreening and apical buds swelling, (3) twigs elongating, (4) floral buds developing, (5) flowers opening (6) fruit developing, (7) seed shatter, (8) some leaves falling and most leaves brown, (9) fall regrowth, (10) fall quiescence. (From J. W. Menke and M. J. Trlica, 1981, Carbohydrate reserves, phenology, and growth cycles of nine Colorado range species. *J. Range Manage.* 34:269– 277, Fig. 3.)

Blue grama (*Bouteloua gracilis*) had a narrow V-shaped curve and replenished reserves rapidly following spring-early summer depletion (bottom curve, Figure 5.5).

Despite these studies concerning seasonal changes in reserve carbohydrate concentrations, much controversy exists concerning the interpretation of the role of carbohydrate reserves in plant growth and resistance to grazing. The nature of these contradictions will be covered in the section on grazing resistance. The reader is referred to Sosebee, (1977) Caldwell (1984), Bedunsh and Sosebee (1995) for more detailed information on carbohydrate reserves in range plants.

WATER RELATIONS

Plants require water to carry on the process of photosynthesis and for many other essential processes. However, plants usually require much more water than they actually utilize in physiological processes. Most of this water is lost through the stomata in plant leaves to the atmosphere. This transpirational movement is from the soil to the plant roots, along the vascular tissue (xylem) and out through the stomata (Brown 1977, 1995) (Figure 5.6). Factors that control evaporation also influence transpiration. The physiology of transpiration is a complex process which is only now beginning to be completely understood (Slatyer 1967; Kramer 1969; Brown 1977, 1995).

Range plants are often subjected to water stress and have evolved many modifications to reduce transpirational losses. However, there are large variations in drought tolerances among

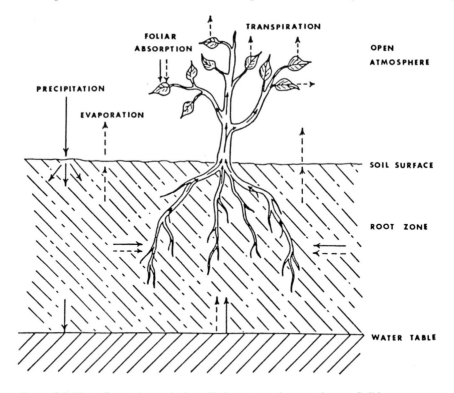

Figure **5.6** Water flux pathways in the soil-plant-atmosphere continuum. Solid arrows represent liquid water fluxes; broken arrows represent vapor fluxes. (From R. W. Brown, 1977, Water relations of range plants. In R. E. Sosebee (Ed.), *Rangeland Plant Physiology, Range Science Series, No. 4.* Society for Range Management, Denver, CO.)

range plants. Plants undergo water stress when the rate of absorption by the roots is exceeded by the rate of transpiration (Brown 1977). Ultimately, cell protoplasts collapse and the plants die. The soil water content at the point where the plants fail to respond to added water is called the permanent wilting point (Brown 1977). For many years permanent wilting was considered to be a constant for most plants when soil water was held with a pressure of -15 bar. However, more recent studies have indicated that the permanent wilting point varies considerably among different species, and more realistically is not a soil water characteristic, but a plant characteristic (Slatyer 1967). Wilting apparently occurs when leaf water potential equals osmotic potential. The ability to measure plant water potentials is a recent development and has led to a much better understanding of soil-plant-atmosphere water relations.

PLANT MORPHOLOGY

The phytomer is the basic unit of the grass plant. It consists of a leaf (sheath and blade), an internode, an axillary bud, and an internode (Hyder 1972, 1974; Dahl and Hyder 1977) (Figure 5.7). *Shoot* is a collective term applied to the stem and leaves of grasses. Shoots may be vegetative or reproductive and culmed or culmless (Dahl and Hyder 1977; Dahl 1995). Tillers are lateral vegetative shoots growing upward within the enclosed leaf sheath. Lateral shoots that grow along the soil surface are called *stolons,* and underground culms are *rhizomes.* Both rhizomes and stolons have the potential to form root and shoot systems at the node as a form of vegetative reproduction.

Plant growth results largely from cell elongation and division and is largely irreversible in nature. Development involves changes in form or structure of plant parts. Plant growth is largely a function of meristematic tissue, where cell division occurs. In grasses the primary meristematic regions are apical and intercalary. Apical meristems occur at the tips of stems and roots, while intercalary meristems lie between regions of permanent tissue, such as at the base of internodes or leaves. Buds are rudimentary shoots or portions of shoots with an active apical meristem (Dahl and Hyder 1977). Most buds are formed in leaf axils, but a few are adventitious.

Apical dominance is a physiological process enabling many range plants to respond to the removal of the apical bud by development of lateral buds into branches or tillers. Originally, the mechanism was thought to be suppression of lateral buds by hormones (mainly indoleacetic acid). Since the early studies in the 1930s, plant physiologists have discovered that the process is much more complex than simple inhibition by the action of hormones. Cline (1991) and Murphy and Briske (1992) have provided excellent reviews of the various theories advanced to explain the phenomenon. The mechanism has profound importance for grazing and browsing by range animals that remove apical growing points and influence tillering and branching patterns.

Roots of shrubby species are generally more extensive than those of herbaceous species. Some shrubs, such as mesquite (*Prosopis* spp.), have deep root systems that allow the plants to tap soil water sources at great depths. Other species, such as creosote bush (*Larrea tridentata*), have a spreading root system which enables the plants to exploit water sources for extensive areas in the soil surface around the plant.

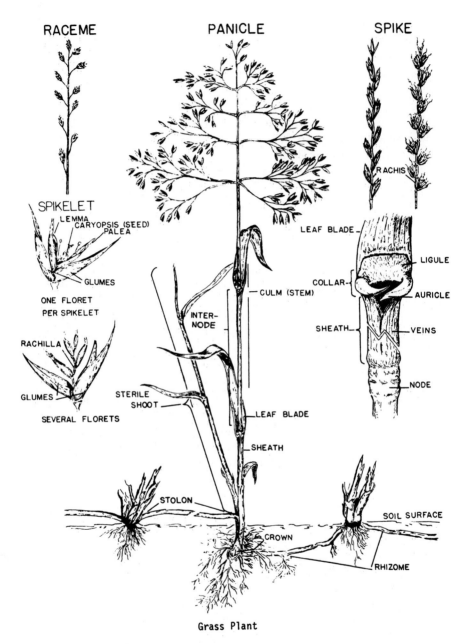

RACEME

PANICLE

SPIKE

RACHIS

SPIKELET

LEMMA
CARYOPSIS (SEED)
PALEA

GLUMES

ONE FLORET
PER SPIKELET

RACHILLA

GLUMES

SEVERAL FLORETS

LEAF BLADE

LIGULE

COLLAR

AURICLE

SHEATH

VEINS

NODE

CULM (STEM)

INTER-
NODE

STERILE
SHOOT

LEAF BLADE

SHEATH

STOLON

SOIL SURFACE

CROWN

RHIZOME

Grass Plant

Figure **5.7** The parts of a grass plant. (From Stubbendieck et al. 1986.)

REPRODUCTION

Reproduction is an important process in range plants for maintaining range plant communities in a productive and diverse status. Recruitment of plants is necessary to maintain the community in a stable condition for replacement of plants that die. Rates of replacement necessary to provide stability in terms of plant numbers varies tremendously from annuals [e.g., Russian thistle (*Salsola kali*) and downy brome (*Bromus tectorum*)] to long-lived shrubs and trees.

Sexual reproduction involves fertilization of the egg by the sperm and the development of a new individual plant. This process is one of the most complex in biology and one of the most important because the characteristics of the new plant are determined genetically at the time of fertilization. Sexual reproduction in which genetic material comes from different parents provides genetic variability, so that these populations can meet changing environmental conditions. Some grasses reproduce sexually but do not breed with other individuals since their inflorescences are enclosed in the leaf sheath. Such grasses are said to be cleistogamous. These plants are necessarily self-pollinated.

Other plants may reproduce vegetatively either by sets that root at the nodes of aboveground stolons or underground rhizomes. Black grama (*Bouteloua eriopoda*) reproduces mainly by stolons, whereas western wheatgrass reproduces mainly by rhizomes. These species produce offspring that are genetically similar to their parents. They are suited to conditions where environmental diversity is low and relatively stable. These species spread rather slowly, and once these communities are disturbed, increases in cover and production are long-term processes.

Annual plants depend on seed production every year for their survival. Plants that invest considerable energy and resources into reproduction are sometimes called "r" selected species (Colinvaux 1986). Other short-lived perennial species, such as the dropseeds (*Sporobolus* spp.) and broom snakeweed (*Gutierrezia sarothrae*), produce abundant seed to ensure their survival. Long-lived species such as black grama and shrubs and trees allocate more of their resources into vegetative growth. These are called "k" selected species.

RESISTANCE TO GRAZING

Range herbivores have been grazing and browsing on range plants for eons. Many observers believe that this coevolution of plants and animals has resulted in the development of resistance to defoliation by some species of range plants, and avoidance of defoliation in others (Table 5.1). However, defoliation, to the extent that it removes photosynthetic tissues, reduces the ability of plants to compete in natural environments (Caldwell et al. 1981).

Despite the long history of grazing or browsing to which range plants have been subjected, there are considerable differences in resistance to grazing among similar species. Early studies emphasized degree and season of defoliation. The basic idea was to determine the degree of defoliation that would minimize damage to the plants, and the time when defoliation was most damaging or rest most beneficial. In initial work, particular attention was paid to carbohydrate levels. Plants were considered most vulnerable to grazing damage when carbohydrates were at their lowest and reserves may not be sufficient to initiate re-

TABLE 5.1   Factors That Increase Grazing Resistance in Grasses, Forbs, and Shrubs

GRASSES

Higher proportion of culmless (stemless) shoots than species with low resistance.

Greater delay in elevation of the apical buds than species with low resistance.

Sprout more freely from basal buds after defoliation than species with low resistance.

Higher ratio of vegetative to reproductive stems than species with low resistance.

FORBS

Produce a large number of viable seeds.

Delayed elevation of growing points.

Poisons and chemical compounds that reduce palatability.

SHRUBS

Spines and thorns that discourage browsing.

Volatile oils and tannins that reduce palatability.

Branches make removal of inner leaves difficult.

Only current year's growth of most shrub species is palatable and nutritious.

Removal of apical meristem may stimulate axillary bud development.

growth. Since carbohydrate reserves are the lowest at about the fourth leaf stage, it was assumed that this period would be the time when plants are most susceptible to grazing (Sampson and McCarty 1930; McCarty 1935; McCarty and Price 1942). Menke and Trlica (1981, 1983), White (1973), and Buwai and Trlica (1977) found great differences in response to grazing by different species in the Great Plains. These authors related grazing tolerance to the shape of the carbohydrate reserve cycle (see Figure 5.5). Species such as blue grama, with a narrow V-shaped cycle, were more resistant to grazing than those with flat or extended cycles such as western wheatgrass, which replenished reserves slowly.

Branson (1953) first showed that plants with their growing points elevated were more susceptible to grazing than were those with basal meristems. Hyder (1972) pointed out that elevation of the growing points was partially developmental. There is also a correspondence between those plants that elevate their growing point early and those that have culmed vegetative shoots. However, environmental conditions can alter the timing of culm elongation. Species that remain largely vegetative are more resistant to grazing than are those which have a large proportion of reproductive culms (Branson 1953), partly because they are adapted for seed production rather than tolerance to grazing (Hyder 1972).

Flexibility in resource allocation can also play a role in grazing resistance. Caldwell et al. (1981) showed that crested wheatgrass (*Agropyron desertorum*) could reconstitute its canopy much faster than could bluebunch wheatgrass (*Agropyron spicatum*) (Figure 5.8).

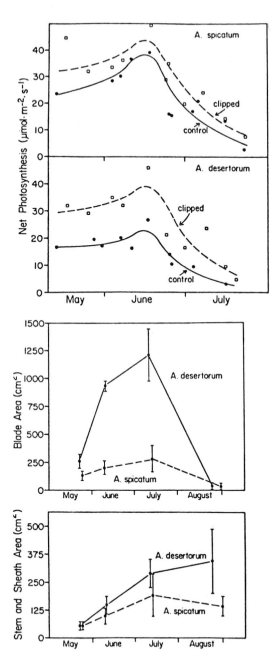

Figure **5.8** Photosynthetic capacity of clipped and control plants of bluebunch wheatgrass (*Agropyron spiecatum*) and crested wheatgrass (*Agropyron desertorum*) (top). Leaf blade and stem and sheath area for both wheatgrass species following severe defoliation (middle and bottom). (From Caldwell et al. 1981.)

Development of new tillers, leaves, and so on, was accelerated for crested wheatgrass compared to bluebunch wheatgrass, which is very sensitive to heavy defoliation. Painter and Detling (1981) found similar conditions for blue grama, which is resistant to grazing, and western wheatgrass, which is rather sensitive to defoliation. However, western wheatgrass also has early elevated growing points (Branson 1953) and a flat reserve carbohydrate curve (Menke and Trlica 1981). Several studies have shown that plants can withstand greater defoliation when competition is reduced by defoliation or removal of surrounding plant (Mueggler 1967, 1972; Archer and Detling 1984).

The various studies indicate that plant responses of defoliation are indeed complex. Plant species differ in their reactions, and multiple mechanisms are involved for each species. Field studies as well as controlled laboratory and greenhouse studies are needed to expand this important area of range plant physiology and ecology (Caldwell 1984).

Grazing resistance in plants can occur through mechanical and biochemical mechanisms as well as through physiological and morphological mechanisms (Briske 1991). Mechanical and chemical mechanisms operate through plant accessibility and palatability to specific herbivores. Tissue accessibility is primarily a function of degree of elevation of leaves and tillers above the soil surface. Species without culmed shoots such as blue grama are particularly resistant to defoliation because apical meristems are near the soil surface where accessibility is low. Spines, awns, and other epidermal characteristics (e.g., pubescence, silica, and cuticular wax) make plants unpleasant to touch and directly reduce palatability.

Biochemical compounds often referred to as secondary compounds can reduce plant palatability by interfering with herbivore metabolism (toxicity). Qualitative compounds are those that are produced at low cost to the plant and occur in low concentrations (Rhodes 1979, 1985). These compounds which include alkaloids, glucosinolates, and cyanogenic substances may increase rapidly in response to grazing. Quantitative compounds are produced in larger quantities and include tannins, lignins, and resins. They are more costly to produce and show little increase in response to grazing (Briske 1991).

Generally plants with biochemical resistance to grazing have slower growth rates and are less effective competitors than those without these mechanisms (Dirzo and Harper 1982; Coley 1986). The costs to the plant of morphological and physiological grazing resistance mechanisms are less clear. However, there has been a concern that placing excessive emphasis on species with these grazing resistance characteristics may decrease productivity (Hyder 1972). The reader is referred to Briske (1991), Molyneaux and Ralphs (1992), Bryant et al. (1992), Briske and Richards (1994), and Briske and Richards (1995) for more detailed discussions of grazing resistance in plants.

GRAZING OPTIMIZATION THEORIES

The statement by Caldwell (1984) that "removal of foliage from a plant must to some degree reduce the potential of plants to compete and retain their status in the community" would lead one to believe that any grazing reduces productive capacities of plants. Yet it has been known for some time that certain degrees of defoliation can increase plant productivity. Removal of apical dominance by grazing or browsing has been long understood as one means of increasing productivity. However, recently there has been increased interest in understanding the relationship between herbivores and plants. From his studies of game animals grazing on the Serengeti Plains of East Africa, McNaughton (1979, 1983, 1984) developed what he called the "grazing optimization" hypothesis. McNaughton believed that herbivores and grasses developed together on the Serengeti and that some degree of grazing benefits the vegetation. Other authors have supported these ideas and have extended the concept (Owen and Wiegert 1976; Hilbert et al. 1981; Dyer et al. 1982). The first curve presented by McNaughton (1979) showed an initial increase in plant growth as grazing intensity increased up to an optimum level followed by decline until plant death occurred (Figure 5.9, top curve C). Later, other alternatives were incorporated as it became clear that plant responses among species vary considerably. Some species are extremely susceptible to grazing and might be injured by even light levels of grazing. These are represented by curve A in Figure 5.9. Others may not be influenced until grazing has reached a given level and they are affected detrimentally. These are represented by curve B in Figure 5.9. McNaughton (1983) regarded these as three alternative hypotheses, but it is likely that examples of all three situations can be found on most rangelands. Recently, Belsky (1986) has argued against the grazing optimization theory, while Paige and Whitham (1987) presented evidence that grazing can increase the reproductive potential of some plants.

The herbivory process is extremely complex and difficult to study and understand. Figure 5.10 is an attempt to depict some of the possible responses of plants to grazing. This equation shows that plant performance (fitness, productivity, reproduction, etc.) is

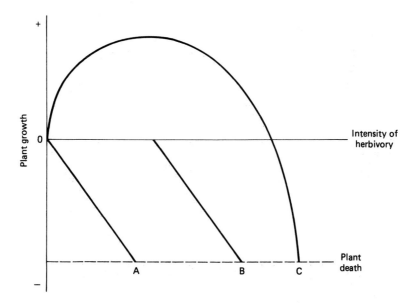

***Figure* 5.9** Three possible outcomes of changes in level of herbivory for individual grazed plants. (From McNaughton 1983.)

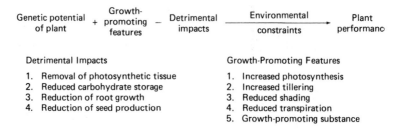

Figure **5.10** Equation showing grazing impacts on perennial grass plants.

TABLE 5.2   Negative Effects of Heavy Grazing Versus the Possible Positive Effects of Light to Moderate Grazing on Range Plant Physiology

HEAVY GRAZING	LIGHT TO MODERATE GRAZING
Decreased photosynthesis.	Increased photosynthesis.
Reduced carbohydrate storage.	Increased tillering.
Reduced root growth.	Reduced shading.
Reduced seed production.	Reduced transpiration losses.
Reduced ability to compete with ungrazed plants.	Inoculation of plant parts with growth-promoting substances.
Reduced mulch accumulation. This decreases soil water infiltration and retention. Mulch is also necessary to prevent soil erosion.	Reduction of excessive mulch accumulations that may physically and chemically inhibit vegetative growth. Excessive mulch can provide habitat for pathogens and insects that can damage forage plants.

Source: Holechek 1981; McNaughton 1983.

controlled by the genetic potential of the plant plus growth-promoting features minus detrimental impacts of grazing operating under a given set of environmental constraints. Detrimental impacts of grazing have been demonstrated by several studies, and Holechek (1981) and McNaughton (1983) have discussed growth-promoting features of grazing (Table 5.2). Positive effects of controlled grazing compared to no grazing are most likely in areas receiving over 400 mm of average annual precipitation. Below this level of precipitation, excessive accumulations of vegetation usually do not occur, due to aridity.

Better understanding of plant physiology and morphology will enable us to manage our rangelands more efficiently. Ample opportunity exists for innovative researchers to expand our horizons greatly in this area.

RANGE MANAGEMENT PRINCIPLES

■ Excessive removal of plant leaves destroys photosynthetic capability and ultimately kills the plant. Excessive accumulation of plant tissue also lowers the photosynthetic capability of the plant. The primary goal in grazing management is to maximize photosynthetic activity with controlled defoliation.

■ Vegetation residue plays an important part in sustaining plant welfare throughout the year. Although more herbage can be removed during dormancy, adequate residues must still be maintained to protect the plant crown and soil.

■ Plants that are highly resistant to grazing are generally less productive and palatable than those with low grazing resistance. This is because photosynthetic products used for physical and chemical protection from grazing could otherwise be used for plant growth and reproduction.

Literature Cited

Archer, S., and J. K. Detling. 1984. The effects of defoliation and competition on regrowth of tillers of two North American mixed-grass prairie graminoids. *Oikos* 43:351–357.

Bedunah, D. J., and R. E. Sosebee. 1995. *Wildland plants: Physiology, ecology, and developmental morphology.* Society for Range Management, Denver, CO.

Belsky, A. J. 1986. Does herbivory benefit plants? A review of the evidence. *Am. Nat.* 127:870–892.

Branson, F. A. 1953. Two factors affecting resistance of grasses to grazing. *J. Range Manage.* 6:165–171.

Briske, D. 1991. Developmental morphology and physiology of grasses. In R. K. Heitschmidt and J. W. Stuth (Eds.). *Grazing management: An ecological perspective.* Timber Press, Portland, OR.

Briske, D., and J. Richards. 1994. Physiological responses of individual plants to grazing: current status and ecological significance. In *Ecological implications of livestock herbivory in the West.* Society for Range Management, Denver, CO.

Briske, D. D., and J. H. Richards. 1995. Plant responses to defoliation: A physiologic, morphologic and demographic evaluation. In D. J. Bedunah and R. E. Sosebee (Eds.). *Wildland plants: Physiology, ecology, and developmental morphology.* Society for Range Management, Denver, CO.

Brown, R. W. 1977. Water relations of range plants. In R. E. Sosebee (Ed.). *Rangeland plant physiology. Range Science Series No. 4.* Society for Range Management, Denver, CO.

Brown, R. W. 1995. The water relations of range plants: Adaptations to water deficits. In D. J. Bedunah and R. E. Sosebee (Eds.). *Wildland plants: Physiology, ecology, and developmental morphology.* Society for Range Management, Denver, CO.

Bryant, J., P. Reichardt, and T. P. Clausen. 1992. Chemically mitigated interactions between woody plants and browsing mammals. *J. Range Manage.* 45:18–25.

Buwai, M., and M. J. Trlica. 1977. Multiple defoliation effects on herbage yield, vigor, and nonstructural carbohydrates of five range species. *J. Range Manage.* 30:164–171.

Caldwell, M. M. 1984. Plant requirements for prudent grazing. In National Research Council/National Academy of Sciences (Eds.). *Developing strategies for rangeland management.* Westview Press, Inc., Boulder, CO.

Caldwell, M. M., J. H. Richards, D. A. Johnson, R. S. Nowak, and R. S. Dzurec. 1981. Coping with herbivory: Photosynthetic capacity and resource allocation in two semiarid *Agropyron* bunchgrasses. *Oecologia* 50:14–24.

Cline, M. G. 1991. Apical dominance. *Bot. Rev.* 57:318–358.

Coley, P. D. 1986. Cost and benefits of defense by tannins in a neotropical tree. *Oecologia* 70:238–241.

Colinvaux, P. 1986. *Ecology.* John Wiley & Sons, Inc., New York.

Cook C. W. 1966. Carbohydrate reserves in plants. *Utah Agric. Exp. Stn. Res. Ser.* 31.

Cook, C. W. 1971. Effects of season and intensity of use on desert vegetation. *Utah Agr. Exp. Sta. Bull.* 483.

Coyne, P. I., M. J. Trlica, and C. E. Owensby. 1995. Carbon and nitrogen dynamics in range plants. In D. J. Bedunah and R. E. Sosebee (Eds.). *Wildland plants: Physiology, ecology, and developmental morphology.* Society for Range Management, Denver, CO.

Dahl, B. E. 1995. Development morphology of plants. In D. J. Bedunah and R. E. Sosebee (Eds.). *Wildland plants: Physiology, ecology, and developmental morphology.* Society for Range Management, Denver, CO.

Dahl, B. E., and D. N. Hyder. 1977. Developmental morphology and management implications. In R. E. Sosebee (Ed.). *Rangeland plant physiology. Range Science Series No. 4.* Society for Range Management, Denver, CO.

Dirzo, R., and J. L. Harper. 1982. Experimental studies on slug-plant interaction. IV. The performance of cyanogenic and acyanogenic morphs of trifolium reopens in the field. *J. Ecol.* 70:119–138.

Dyer, M. I., J. Detling, D. C. Coleman, and D. W. Hilbert. 1982. The role of herbivores in grasslands. In J. R. Estes, R. J. Tyrl, and J. N. Brunken (Eds.). *Grasses and grasslands: Systematics and ecology.* University of Oklahoma Press, Norman, OK.

Hilbert, D. W., D. M. Swift, J. K. Detling, and M. I. Dyer. 1981. Relative growth rates and the grazing optimization hypothesis. *Oecologia* 51:14–18.

Holechek, J. L. 1981. Livestock grazing impacts on public lands: A viewpoint. *J. Range Manage.* 34:251–254.

Hyder, D. N. 1972. Defoliation in relation to vegetative growth. In V. B. Younger and C. M. McKell (Eds.). *Biology and utilization of grasses.* Academic Press, Inc., New York.

Hyder, D. N. 1974. Morphogenesis and management of perennial grasses in the United States. In K. W. Kreitlow (Ed.). *Plant morphogenesis as the basis for scientific management of range resources.* U.S. Dep. Agric. Misc. Publ. 1271.

Kramer, P. J. 1969. *Plant and soil water relationships: A modern synthesis.* McGraw-Hill Book Company, New York.

Launchbaugh, J. L. 1957. The effect of stocking rate on cattle gains and on native shortgrass vegetation in west-central Kansas. *Kans. Agric. Exp. Stn. Bull.* 394.

McCarty, E. C. 1935. Seasonal march of carbohydrates in *Elymus ambiguus* and *Muhlenbergia gracilis* and their reaction under moderate grazing use. *Plant Physiol.* 10:727–738.

McCarty, E. C., and R. Price. 1942. Growth and carbohydrate content of important mountain forage plants in central Utah as affected by clipping and grazing. *U.S. Dep. Agric. Tech. Bull.* 818.

McNaughton, S. J. 1979. Grazing as an optimization process: Grass-ungulate relationships in the Serengeti. *Am. Nat.* 113:691–703.

McNaughton, S. J. 1983. Compensatory plant growth as a response to herbivory. *Oikos* 40:329–336.

McNaughton, S. J. 1984. Grazing lawns: Animals in herds, plant form, and coevolution. *Am. Nat.* 124:863–886.

Menke, J. W., and M. J. Trlica. 1981. Carbohydrate reserve, phenology, and growth cycles of nine Colorado range species. *J. Range Manage.* 34:269–277.

Menke, J. W., and M. J. Trlica. 1983. Effects of single and sequential defoliations on the carbohydrate reserves of four range grasses. *J. Range Manage.* 36:70–75.

Molyneaux, R. J., and M. H. Ralphs. 1992. Plant toxins and palatability to herbivores. *J. Range Manage.* 45:133–136.

Mueggler, W. F. 1967. Response of mountain grassland vegetation to clipping in southwestern Montana. *Ecology* 48:942–949.

Mueggler, W. F. 1972. Influence of competition on the response of bluebunch wheatgrass to clipping. *J. Range Manage.* 25:88–92.

Murphy, J. S., and D. D. Briske. 1992. Regulation of tillering by apical dominance: Chronology, interpretive value, and current perspectives. *J. Range Manage.* 45:419–429.

Nsinamwa, M. 1993. *Effects of grazing intensity and season of grazing on cow diets and plant responses in northern Chihuahuan desert.* Masters Thesis. New Mexico State University, Las Cruces, NM.

Owen, D. F., and R. G. Wiegert. 1976. Do consumers maximize plant fitness? *Oikos* 27:488–492.

Paige, K. N., and T. G. Whitham. 1987. Overcompensation in response to mammalian herbivory: The advantage of being eaten. *Am. Nat.* 129:407–416.

Painter, E. L., and J. K. Detling. 1981. Effects of defoliation of net photosynthesis and regrowth of western wheatgrass. *J. Range Manage.* 33:68–71.

Parker, K. G. 1969. The nature and use of Utah range. *Utah State Univ. Ext. Circ.* 359.

Rhodes, D. F. 1979. Evolution of plant chemical defense against herbivores, pp. 3–54. In G. A. Rosenthal and D. H. Janzen (Eds.). *Herbivores: Their interaction with secondary plant metabolites.* Academic Press, Inc., New York.

Rhodes, D. F. 1985. Offensive-defensive interactions between herbivores and plants: Their relevance in herbivore population dynamics and ecological theory. *Am. Nat.* 125:205–238.

Sampson, A. W., and E. C. McCarty. 1930. The carbohydrate metabolism of *Stipa pulchra. Hilgardia* 5:61–100.

Sauer, R. H. 1978. Effect of removal of standing dead material on growth of agropyron spicatum. *J. Range Manage.* 31:121–122.

Slatyer, R. O. 1967. *Plant-water relationships.* Academic Press, Inc., New York.

Sneva, F. A. 1980. Crown temperature of Whitmar Wheatgrass as influenced by standing dead material. *J. Range Manage.* 33:314–315.

Sosebee, R. E. (Ed.). 1977. *Rangeland plant physiology. Range Science Series No. 4.* Society for Range Management, Denver, CO.

Stubbendieck, J., S. L. Hatch, and K. J. Kjar. 1986. *North American range plants.* 3d ed. University of Nebraska Press, Lincoln, NE.

Trlica, M. J. 1977. Distribution and utilization of carbohydrate reserves in range plants. In R. E. Sosebee (Ed.). *Rangeland plant physiology. Range Science Series No. 4.* Society for Range Management, Denver, CO.

White, L. M. 1973. Carbohydrate reserves of grasses: A review. *J. Range Manage.* 26:13–18.

CHAPTER 6

RANGE ECOLOGY

ECOLOGY DEFINED

Ecology involves the study of the interrelationships between organisms and their environment. Range management is applied ecology because it deals with manipulation of organisms and sometimes their environment with the goal of increasing output usable to man. A fundamental concept in range management is that the welfare of plants and animals depends on each other. Inputs by man associated with regulating animals (control of grazing animal numbers, timing of grazing, frequency of grazing, etc.) are generally much lower than those for directly regulating plants (fertilization, cultivation, seeding, irrigation), and rangelands generally have low vegetation productivity compared to farmlands. Therefore, range management has focused on manipulating vegetation and soil by control of the grazing animal. The living and nonliving elements comprising a piece of rangeland on which man has placed boundaries for management purposes are referred to as a *rangeland ecosystem.*

RANGELAND ECOSYSTEM COMPONENTS AND FUNCTIONS

An ecosystem is a "functional unit consisting of organisms (including man) and environmental variables of a specific area" (Van Dyne 1966). It is important to realize that an ecosystem contains living and nonliving elements and that there is an exchange of energy and matter among these elements or components (Lewis 1969). These components are the abiotic (nonliving) factors, primary producers, consumers, and decomposers (Figure 6.1).

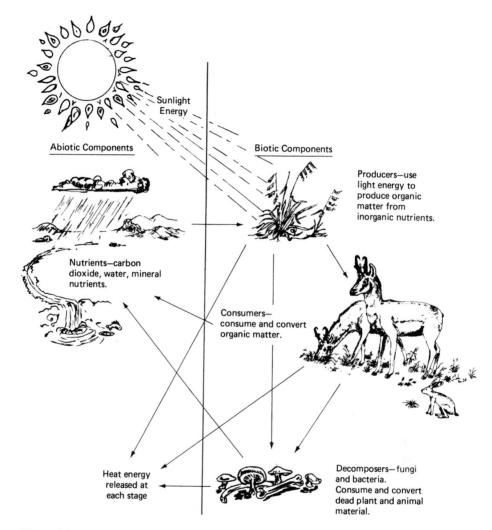

Figure **6.**1 General diagram showing interactions among ecosystem components. (Adapted from Nebel 1981 by John N. Smith.)

Ecosystem Components

One of the essential features of ecosystems is that the components are related functionally. A change in one ecosystem component influences all others. Traditionally, we have used a one-factor approach to study range relationships: the influence of grazing on individual plants, the influence of fertilization on vegetation, and so on. However, it is implicit that the grazing process influences many other components (Crisp 1964). With the advent of high-speed computers, it became possible to study many possible relationships and to model ecosystems or ecosystem processes (see Chapter 17).

Abiotic Components. The abiotic components consist mainly of the soil and climatic factors and are not usually manipulated by the range manager. The functions of the soil are for anchorage of plants and as a reservoir of water and nutrients for plants. The soil also serves as habitat for soil animals and microorganisms. Soil manipulation generally involves fertilization and mechanical treatments to increase infiltration, water storage, and so on.

Climatic factors impinge directly on the biotic components and indirectly on consumer and decomposer groups through their influence on the plants. Major features of the climate include temperature and precipitation. Climate patterns for major rangelands were discussed in Chapter 4.

Biotic Components. Biotic components are often divided into primary producers, consumers, and decomposers. The ***primary producers*** are plants with the pigment chlorophyll, which is responsible for converting solar energy to chemical energy that can then be used by the plant itself as well as by animals that consume the plants. The primary producers support, directly or indirectly, all other groups of organisms. ***Primary productivity*** has been defined as "the rate of organic matter storage by photosynthetic and chemosynthetic activity of producer organisms in the form of organic substances which can be used as food materials" (Odum 1959).

Biomass and *standing crop* have been used nearly synonymously to refer to the weight of organisms at a given time. Standing crop also has broader connotations, such as "the standing crop of nitrogen" or some other component. However, the essential distinction between productivity and biomass or standing crop is that productivity is a rate process with a specified time interval, while biomass and standing crop refer to quantities at a particular point in time.

Herbage is a term often used by range workers and refers to the biomass of all herbaceous vegetation at one point in time (Pieper 1978). Not all the herbage is usually eaten by livestock or other herbivores, since some may be unavailable (out of reach or protected by a shrub or spiny plant) or not readily acceptable at conservative stocking rates. *Forage,* although defined in various ways by different authors, generally refers to the herbage available and acceptable to grazing animals (Pieper 1978). Thus forage is always less than herbage. *Browse* has been defined as "that part of leaf and current twig growth of shrubs, woody vines, and trees available for animal consumption" (Duvall and Blair 1963). Thus browse is comparable in some ways to forage, but some authors include browse as part of forage. Because of these inconsistencies in definitions, it is sometimes in order to define the terms before using them.

Herbage or browse biomass is usually expressed in terms of dry weight per unit area. In the English system, units are generally pounds per acre, while in the metric system the units are kilograms per hectare or grams per square meter (g/m^2 can be converted to kg/ha by multiplying by 10). One pound per acre equals 1.121 kilograms per hectare.

The quality of herbage and browse is never stable on rangelands within or between years. Most rangelands are characterized by a single growing season when soil water and temperatures are suitable to support plant growth. Early in the growing season plant growth is slow, then reaches a peak during mid-growing season, and finally slows and then ceases during the dormant season. Often it is assumed that herbage standing crop is nearly stable during the dormant season. However, many processes, such as respiration, translocation, shattering, herbivory, and so on, continue during the dormant season and contribute to the decline in herbage biomass (Pieper et al. 1974b). Thus, herbivores face a declining food supply as the dormant season proceeds even without considering their own consumption.

Seasonal changes in plant roots are not well understood because of the difficulty of studying root systems. However, apparently root biomass varies considerably on different ranges depending partly on plant species involved and soil characteristics (Table 6.1). Root/shoot ratios also vary considerably and are influenced by grazing pressure.

Traditionally, range workers have considered livestock, big game animals, and destructive smaller animals such as grasshoppers and rodents as the main primary consumers (or herbivores) in range ecosystems. Herbivores can be looked upon as ecosystem regulators that have a direct impact on the vegetation and are a source of food for predators.

Decomposers. Decomposers and microconsumers are critical but often overlooked components of range ecosystems. They function primarily in the decomposition process and are responsible for preventing accumulations of organic matter. Without decomposers, ecosystem functioning would not be possible because there could be no nutrient cycling, and elements would eventually be tied up in undecomposed organic material. The decomposer microorganisms are generally bacteria and fungi as well as actinomycetes, algae, and lichens, which possess enzyme systems necessary to break down resistant organic materials. Several other groups of organisms also function as microconsumers and are active in detrital food chains (Paris 1969). Some of the most obvious microconsumers in rangelands are ants, termites, and nematodes. Nematodes are recognized as being very important as the result of International Biological Program studies.

Manipulators: The Human Component. One critical influence on all ecosystems remaining to be discussed is the human factor. Humans are often referred to as manipulators because they are the only species of animal deliberately rearranging the components of the

TABLE 6.1   Average Belowground Biomass for Grassland Sites in the United States (g/m^2)

SITE	UNGRAZED	GRAZED
Jornada (desert grassland)	157	142
Osage (tallgrass)	924	868
Pawnee (shortgrass)	1,119	1,317
Cottonwood (northern mixed prairie)	1,312	2,139

Source: Sims et al. 1978.

ecosystems for their own benefit. Human activities affect all trophic levels, and have had both positive and negative influence on natural processes and functions. In the past, intentional manipulations of ecosystems have been largely directed toward increasing the amount of wood, meat, minerals, and fiber with little regard for other animals such as wild ungulates, rabbits, rodents, reptiles, birds, insects, and other invertebrates. More recently managers have recognized that all organisms are intertwined in the pyramid of life. Therefore, any management practice affecting one animal species will in turn affect all others. This has led to the present policy of multiple use management on public lands in the United States.

Multiple use involves managing wild lands for livestock, lumber, wildlife, minerals, water, and recreation. Obviously, no acre of land can be used for maximum production of all six products. However, most land units will provide two or more products, and regionally all can be produced. The product or products emphasized will depend primarily on the characteristics of the ecosystem involved and demands by society.

In the future, the demands placed on the earth's ecosystems for all products will accelerate because of a rapidly expanding world population. This means management must be much more intensive than in the past. Each ecosystem responds differently to management. However, the same basic ecological principles apply to all ecosystems.

Ecosystem Function

Range ecosystem function can be viewed mainly from two standpoints: energy flows and chemical cycles. These really represent physiological processes within the ecosystem. Energy flows throughout the ecosystem operate under the first law of thermodynamics: Energy can be neither created nor destroyed, only changed in form.

Energy Flow. Figure 6.2 is a simplified version showing transformation of energy among compartments of a range ecosystem. Solar energy is received by grasses, forbs, and shrubs and transferred by the process of photosynthesis into stored chemical energy (see Chapter 5 for a discussion of the process of photosynthesis) in green plant tissues. When herbivores eat plant tissue, they gain energy stored in plant tissues through the process of digestion (Figures 6.3 and 6.4). Carnivores, in turn, eat other animals and derive their energy requirements from their food. However, energy is dissipated at each step in the food chain through respiration.

In addition, the organisms at each step in the food chain are not completely efficient at harvesting all available food resources, so energy transfer diminishes considerably at each stage. Once energy is dissipated in the form of heat, it can never be recovered and reused; thus energy flow is a one-way path and must be continually refueled by energy from the sun.

A few studies have been conducted to estimate magnitudes of energy flow in range ecosystems. These studies illustrate that less than 1% of the usable solar radiation received by plants in range ecosystems is utilized in photosynthesis (Sims et al. 1978). In addition, only a relatively small portion of the aboveground primary productivity is harvested by herbivores, including livestock (Lewis 1971; Pieper 1983).

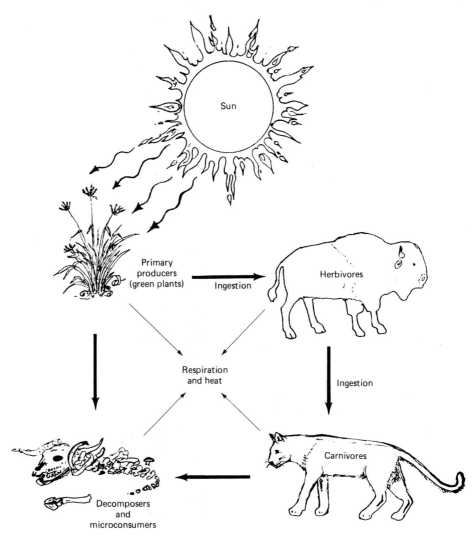

Figure **6.2** Generalized diagram showing energy flow through a range ecosystem. (Drawing by John N. Smith.)

Chemical Cycling. The second basic functional process of range ecosystems is chemical cycling. Unlike energy, chemical elements cycle through the various compartments and can be reused (Figure 6.5). The source of many elements, except for nitrogen, is the soil parent material. In many cases the soil acts as a sink or reservoir for chemical elements. For nitrogen, there are large quantities in the atmosphere, but in the gaseous state it is unavailable for plants. It must be "fixed" or transformed by "free-living" micoorganisms, or those living in symbiotic relationships with certain plants in nodules on their roots. These organisms convert atmospheric nitrogen into forms that can be used by the plants.

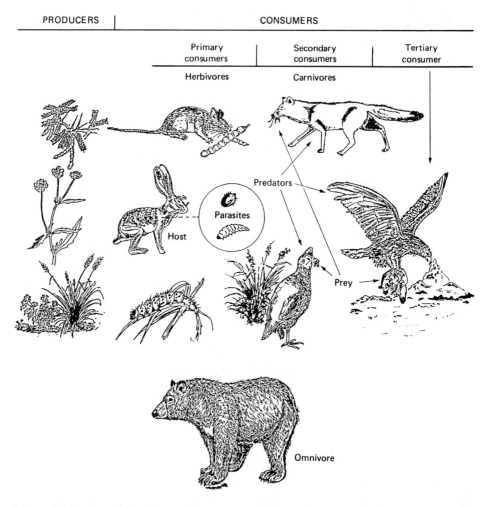

PRODUCERS	CONSUMERS		
	Primary consumers	Secondary consumers	Tertiary consumer
	Herbivores	Carnivores	

Predators

Parasites

Host

Prey

Omnivore

***Figure* 6.3** Feeding relationships among plants and animals. (Redrawn from Nebel 1981 by John N. Smith.)

Individual elements are absorbed by plant roots and function in many important roles within the plant. Herbivores consume plant tissue and the elements contained there (Figure 6.5). Some plant material is not consumed by herbivores and is instead broken down by decomposers and microconsumers. The chemical elements are then returned to the soil. For herbaceous plants these turnover rates can be rather rapid, but for woody plants, elements may be tied up in plant material for some time. Feces and urine of all consumers are deposited on rangeland and eventually returned to the soil, where they may be taken up by the plants. Grazing by domestic livestock results in the removal of some elements when livestock are removed and changes in distribution of others as feces accumulate around water and in other areas of livestock concentration.

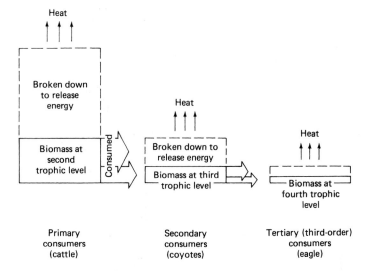

Figure 6.4 The loss of energy as it moves through trophic levels. (From Bernard J. Nebel, *Environmental Science: The Way the World Works,* copyright © 1981, p. 33. Reprinted by permission of Prentice-Hall, Inc., Englewood Cliffs, NJ.)

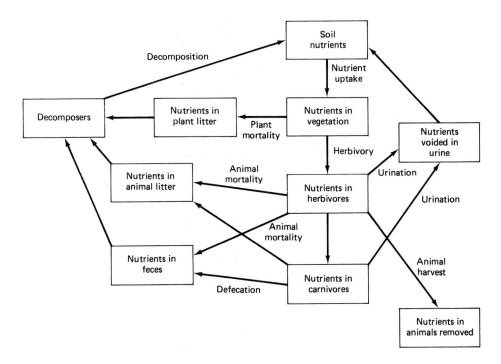

Figure 6.5 Generalized nutrient cycle diagram for range ecosystems. (From Pieper 1977.)

SUCCESSION AND CLIMAX

Range ecosystems are dynamic and changing continuously. It is important for the range manager to understand these changes and which ones influence management decisions.

Successional changes are among the most important to the range manager. Ideas concerning successional changes and stability have been quite controversial and subject to different interpretations. In the classical sense, plant succession involves the replacement of one plant community by another until the final community is reached. This final, somewhat stable community, is often called the *climax*.

Primary Succession

Primary successions are those starting from bare ground and open water (primary areas) (Figure 6.6). On bare ground, there may be microsites where lichens, algae, or moss are supported. As the rock is weathered, and water and organic matter are added from the lichens or algae, a rudimentary soil is formed. Seeds from nearby plants may be available to germinate and to support vascular plants. These are often annuals that can survive under rather harsh conditions. With further weathering and soil formation, some perennial plants may become established. These are generally herbaceous plants, but eventually if the climate will support them, woody plants will become established. Each assemblage of plants influences soil and microclimate, sometimes making it more suitable for plants that have higher

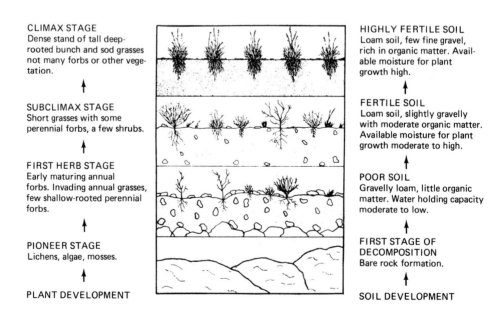

CLIMAX STAGE
Dense stand of tall deep-rooted bunch and sod grasses not many forbs or other vegetation.

↑

SUBCLIMAX STAGE
Short grasses with some perennial forbs, a few shrubs.

↑

FIRST HERB STAGE
Early maturing annual forbs. Invading annual grasses, few shallow-rooted perennial forbs.

↑

PIONEER STAGE
Lichens, algae, mosses.

↑

PLANT DEVELOPMENT

HIGHLY FERTILE SOIL
Loam soil, few fine gravel, rich in organic matter. Available moisture for plant growth high.

↑

FERTILE SOIL
Loam soil, slightly gravelly with moderate organic matter. Available moisture for plant growth moderate to high.

↑

POOR SOIL
Gravelly loam, little organic matter. Water holding capacity moderate to low.

↑

FIRST STAGE OF DECOMPOSITION
Bare rock formation.

↑

SOIL DEVELOPMENT

Figure **6.6** Primary succession on grassland rangelands. (From Gay 1965.)

requirements for water, nutrients, and so on. Thus some plant species alter the environment such that it is no longer suitable for them and they are replaced by other plants. Total biomass (plants and animals), total energy storage, diversity, and rate of mineral cycling increase as succession proceeds (Lewis 1969). The processes of primary succession are summarized as follows:

1. The development of soil from parent materials.
2. Increasing longevity with successional advance.
3. Replacement of species with broad ecological requirements by those occupying narrow niches complementary with other species.
4. Greater accumulation of living tissue and litter per unit area with successional advance.
5. Modification of microenvironmental extremes.
6. Change in size of plants from small to large.
7. Increase in the number of pathways of energy flow.
8. More nutrients tied up in living and dead organic matter.
9. Greater resistance to fluctuation in the controlling factors.

Secondary Succession

Secondary successions are those that occur following some type of disturbance, such as fire or destructive grazing. Range managers routinely deal with secondary succession, but rarely with primary succession; however, sometimes erosion does change the initial soil surface conditions. Generally, we are concerned mainly with vegetational changes in secondary succession and how these changes influence habitats for other organisms.

Secondary succession generally occurs much faster than primary succession, and generally in a more predictable fashion. The variability in secondary succession is reduced as the climax is approached (Huschle and Hironaka 1980).

Figure 6.7 depicts secondary successional stages following fire in piñon-juniper woodlands in three different states. These diagrams support the idea that succession is somewhat predictable since the changes are very similar. There is a skeleton of dead trees following a fire, but these areas are soon occupied by annuals, then some perennial grasses and forbs. The herbaceous vegetation is invaded by some shrubs and eventually the climax piñon-juniper woodland. Two of the diagrams show what happens when fire occurs during one of the seral stages. The intensity of the fire, characteristics of the fuel, time of year, and so on, all influence the damage which the fire does.

Continual heavy grazing also influences vegetation. The process of change away from climax is called induced retrogression. Usually, it decreases grazing values for domestic livestock and results in lower watershed values. If the grazing pressure is relaxed, secondary succession back to the climax can resume.

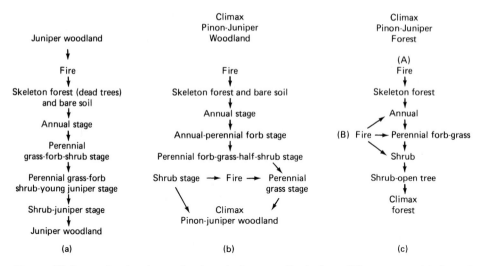

***Figure* 6.7** Successional pathways in piñon-juniper woodlands from different areas: (a) Central Utah (Barney and Frischknecht 1974), (b) Arizona (Arnold et al. 1964), and (c) Colorado (Erdman 1970).

Climax Theory

The final stage of succession is the climax. The climax has been viewed in different ways by different authors. Some have considered the climax to be "stable," others that it is in "dynamic equilibrium" with the environment. Clements (1916) viewed the climax as controlled primarily by the macroclimate. This he referred to as the "climate climax." He viewed large areas of landscape as having the same climax governed by the climate. Development of climax vegetation was considered a very slow process on the same time scale as geologic changes. Variations in the climate climax were viewed as deviations and were handled with modifications, such as *pre*climax, which would represent an earlier stage than climax. *Dis*climax would represent retrogressive successions due to some type of disturbance. Clements developed a separate terminology to cover all these departures from climax. Other authors considered other types of climax in addition to climatic climax, such as edaphic climaxes, where soil characteristics may have an influential role in the type of vegetation that develops. This view was sometimes called the "polyclimax" viewpoint. Others considered vegetational development more a matter of chance, with discrete communities difficult to discern. This "individualistic" concept later gave way to the idea of the "continuum," which stated that the distribution of each species was independent of that of other species and that these distributions overlapped with each other (Curtis 1959). Whittaker (1967) believed that species were distributed along environmental gradients and that individual communities were only delineated somewhat arbitrarily. Although it is obvious that environmental gradients do occur on rangelands, it is also obvious that some ecotones are quite sharp between adjacent plant communities.

Different Theories on Succession

In some cases succession may be looked at simply as a rearrangement of species that were present during initial stages, perhaps only as seeds or other propagules, and that the proportion of the various species changes during succession. However, different groups of species may dominate different stages for different time periods. These shifts or rearrangement of different species over time emphasize the *initial floristic composition* theory of succession (Egler 1954). Studies in southern Idaho showed substantial change in sagebrush grass vegetation following protection from grazing (Anderson and Holte 1981). Secondary succession on these sites was characterized by changes in species composition, but no loss or gain of species. Another type of succession involves immigration of new species to the site from other sites with time progression. Different groups of species dominate the site for various periods. Egler (1954) used the term "relay floristics" to describe this type of successional change.

Primary succession starts from bare areas and proceeds to the development of a somewhat stable climax vegetation. Such changes require extremely long periods, on the scale of hundreds or even thousands of years. Consequently, primary successions may be of interest, but they play a small role in range management. It is important to recognize that other ecosystem components undergo succession as well as vegetation. Jenny (1980) found that nitrogen accumulated in response to soil and vegetational development. It is also evident that animal populations undergo successional changes, although these have not been emphasized. In reality, the entire ecosystem undergoes successional changes that act in concert and not independently. The overall macroclimate does not undergo succession, but the microclimate does undergo change as the soil and vegetation develop. Jenny (1961) has recognized that ecosystem properties are a function of the same soil forming factors which he formulated 20 years earlier (Jenny 1941). These factors are climate, organisms, topography, parent material, and time. Jenny stated that any ecosystem property is determined by these five factors.

There have been many problems with the various successional theories. Some authors doubt that succession occurs or at least that it is so orderly as depicted by some authors, especially in desert ecosystems. There are many kinds of changes that occur on rangelands (Hanson and Churchill 1961; Heady 1973). Some of these changes are cyclic in nature, some merely fluctuations, and so on, but all these other changes tend to mask successional change. Succession does not necessarily proceed in a regular, smooth pattern as some diagrams would indicate. The rate of succession generally slows down during a series of years with low precipitation and speeds up during years of favorable growing conditions. Figure 6.8 depicts some of these fluctuations that occur during succession. While a general trend occurs toward the climax, there can be much variation and fluctuation associated with this trend.

Recently, additional successional models have been described for rangeland situations. Figure 6.9a shows the traditional successional model and Figure 6.9b the model with climatic variability included (Westoby et al. 1989). Others have described a *state and transition* or *multiple steady state* situation where different fire, climatic, or grazing regimes result in different, relatively stable, states (Friedel 1991; George et al. 1992; Laycock 1991;

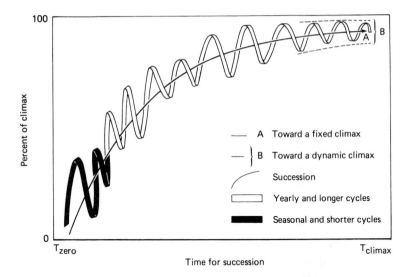

***Figure* 6.8** Stylized succession asymptotically approaching a dynamic climax (b) that includes variations associated with daily, seasonal, and yearling phenomena. The absolute climax might be defined as the midpoint of these variations at (a). (From Heady 1973.)

Westoby et al. 1989). These modifications have certain similarities with the earlier polyclimax hypothesis.

It is apparent that many arid and semiarid ecosystems depart from classical Clementsian successional theory and do not return to some previous state following disturbance (Ellis and Swift 1988; Behnke et al. 1993). These discontinuous, irreversible changes may not be predictable. This has practical significance related to range condition assessment that is discussed in Chapter 7.

Woody plants have increased in density on many western ranges in the last 150 years (Buffington and Herbel 1965; Branson 1985; Johnson and Mayeux 1992). Figure 6.10 shows that increase in cholla cactus (*Opuntia imbricata*) in New Mexico over a 40-year period. However, apparently, cholla cactus populations are somewhat cyclic in nature. The increase in woody plant abundance and distribution has been well documented on the desert grasslands (Buffington and Herbel 1965; Hastings and Turner 1965) and the piñon-juniper woodland in the southwestern United States (Johnsen 1962; Springfield 1976). These changes are complex, but several factors have often been given as causal mechanisms as follows:

1. Heavy grazing by domestic livestock.

2. Spread of seed by livestock.

3. Improved fire control methods, which reduced influence of wildfire.

4. Activities of rodents and rabbits.

(a) GENERAL SCHEME OF THE RANGE SUCCESSION MODEL

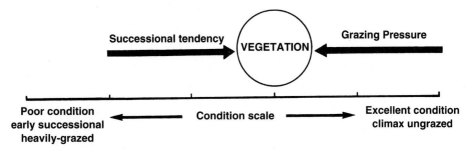

(b) INCORPORATION OF RAINFALL VARIABILITY IN THE RANGE SUCCESSION MODEL

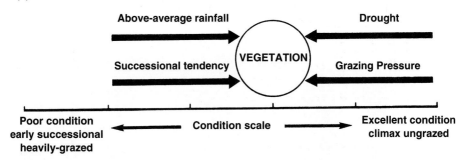

***Figure* 6.9** (a) General range succession model with grazing pressure the main factor operating against successional tendency in a linear model. (b) General model with rainfall variability and grazing pressure included. (From Westoby et al. 1989. Reprinted with permission.)

5. Climatic shifts.

6. Atmospheric CO_2 enrichment.

In most cases only the change has been documented, not the causal mechanism. It is likely that a combination of factors is responsible for these dramatic shifts. The reader is referred to Archer (1994), Lauenroth et al. (1994), and Miller et al. (1994) for detailed discussions of causes of vegetation changes on specific rangelands in the western United States.

Succession and Grazing

Under sustained heavy grazing the more palatable plants that generally dominate grassland ranges near the climax are replaced by a succession of plants that tend to be increasingly lower in palatability, lower in productivity, and more poisonous (Ellison 1960). This process is referred to as retrogression. When defoliation is reduced to a moderate or light

(a)

(b)

Figure **6.10** Shortgrass vegetation in southcentral New Mexico showing increase in cholla cactus (*Opuntia imbricata*) from 1937 (a) to 1977 (b).

rate, the palatable plants again have the competitive advantage and succession occurs back to the original or climax vegetation.

Under moderate or light grazing levels the poisonous, unpalatable plants are at a competitive disadvantage because they invest part of their products from photosynthesis in poisonous compounds (alkaloids, oxalates, glycosides, etc.) and appendages (spines, thorns, stickers, etc.) that discourage defoliation rather than contribute to growth (Cronin et al. 1978; Laycock 1978; Molyneaux and Ralphs 1992). In contrast the palatable plants use their photosynthetic products mainly for growth in the form of roots, leaves, stems, rhizomes, stolons, seeds, and so forth. Under heavy defoliation levels the photosynthetic capacity of the palatable plants is reduced to the point that they are unable to produce enough carbon compounds for maintaining root systems, regeneration of leaves, respiration and reproduction. Over time, they shrink and die, and gradually are replaced with the unpalatable plants that are able to defend themselves against defoliation.

The driving forces in succession are moisture (rainfall) and temperature. In wet humid range types such as the southern pine forest in the southeastern United States or the tall grass prairie in the eastern Great Plains, recovery after retrogression is both rapid and predictable. The climax plants will usually again dominate the site within 5 years if severe erosion has not occurred. In the drier range types such as the Chihuahuan desert, the palatable plants tend to be less resistant to grazing. Here retrogression can occur within just a few years under heavy grazing but recovery is a slow process often requiring 20 or more years. On some sites with serious soil erosion, only minor improvement has been observed after 20 or more years even under complete elimination of grazing (Grover and Musick 1990). The reader is referred to Archer and Smeins (1991) and Pieper (1994) for more detailed discussions on plant succession and grazing.

Retrogression

Progressive succession, or progression, refers to vegetation changes that lead to more diverse communities with higher productivity. In contrast retrogression involves vegetation changes away from the climax vegetation. It is usually, but not always, caused by some type of disturbance such as logging, fire, overgrazing, cultivation, and so forth.

Allogenic retrogressions involve plant community changes that are brought about by forces outside the community such as fire set by man or heavy grazing by livestock. However some retrogressions can be autogenic (caused by the community itself). In the tallgrass prairie of the central Great Plains several studies have shown that the productivity of the climax grasses such as big bluestem is reduced when there is no fire or grazing (Dyksterhuis and Schmutz 1947; Weaver and Rowland 1952; Duvall and Linnartz 1967). These same studies also showed composition shifts away from the climax grasses toward earlier seral grasses such as Kentucky bluegrass (*Poa pratensis*). Excessive mulch accumulation was the cause of this retrogression. Ehrenreich and Aikman (1963) concluded that when quantity of mulch exceeds annual yields of herbage, herbage yields will be depressed. High levels of mulch tie up excessive amounts of nutrients, physically inhibit plant growth in the spring by depressing soil temperatures, and provide habitat for pathogens and insects that can be harmful to the climax plants. Even in the Chihuahuan desert of New Mexico there is evidence of autogenic retrogression in the absence of fire and grazing (Paulsen and Ares 1962). Barbour et al. (1987) provide examples of autogenic retrogression in forest types of California and Alaska.

Although fire and grazing are often considered forces that cause retrogression, there is also much evidence they play a crucial role in maintaining the climax vegetation in many forest and grassland ecosystems. The timing, intensity, and frequency of both fire and grazing are critical factors in determining whether they result in retrogression or progression.

Practical Application of Successional Theory

On grassland and semidesert range types, retrogression in ecological condition from the climax has been well associated with decreased financial returns from livestock production (Johnson 1953; Klipple and Costello 1960; Shoop and McIlvain 1971; Holechek 1994; Workman 1995; Holechek 1996b). This was recognized by Dyksterhuis (1949) when he put Clementsian succession and climax theories into on applied framework for evaluating range condition and trend. Under the Dyksterhuis approach, rangeland in a near climax (76–100% remaining climax) condition was classified excellent, rangeland in a late successional stage (51–75% remaining climax) was classified good, rangeland in a midsuccessional stage (26–50% remaining climax) was classified fair, and rangeland in an early seral stage (0–25% remaining climax) was classified poor. The basic theory is that with overgrazing, the palatable productive perennial grasses (called decreasers) associated with the climax are gradually replaced by plants of increasingly lower palatability and productivity (called increasers) until the site is ultimately occupied by unpalatable shrubs and annual plants (invaders) that in many cases are poisonous.

The model described by Dyksterhuis (1949) has been and continues to be heavily used by government agencies (USDA-Natural Resources Conservation Service, USDI-Bureau of Land Management, and USDA-Forest Service) in characterizing rangeland condition and trend. Generally it works fairly well for grassland and semidesert areas but has some limitations for woodland, annual grassland, and seeded pasture range types (see Chapter 7). In recent years there has been a tendency to drop the terms excellent, good, fair, and poor and use climax, late seral, mid seral, and early seral for characterizations of range ecological condition. Modern range managers now refer to vegetation successional characterizations under the Dyksterhuis approach as range ecological condition rather than range condition. This is because they want to clarify that successional status of an area (range ecological condition) and how well the existing vegetation is suited for the intended uses, are separate issues.

In arid and semiarid areas range ecological condition under the Dyksterhuis approach has been strongly linked with financial return from livestock production (Workman 1995; Holechek 1996b). This relationship still holds, but to a lesser extent, for the more humid prairie rangelands. This is because many palatable annuals and short statured palatable perennial grasses occur on degraded prairie ranges while poisonous forbs, unpalatable shrubs, annual grasses, and bare soil predominate on degraded arid areas.

In southern New Mexico Holechek (1996b) showed that cattle grazing was quite profitable on Chihuahuan desert rangelands in near climax ecological condition but financial losses were probable for those in the early seral stage (Table 6.2). In the shortgrass prairie, changes in ecological condition from late to mid seral appear to reduce financial returns from cattle operations about 20% to 25% (Klipple and Costello 1960; Holechek 1994). Reductions in financial returns of 30% to 50% appear likely for cattle operations in the midgrass and tallgrass prairies when ecological condition shifts from a late to early seral stage (Shoop and McIlvain 1971).

On areas where the climax vegetation is forest or woodland, early and mid seral stages caused by logging or burning will give higher financial returns from livestock production

TABLE 6.2   Grazable Perennial Forage Production and Financial Returns ($/acre) for Chihuahuan Desert Cattle Ranches in New Mexico, in Different Ecological Condition Classes (Holechek 1996b)

	RANGE ECOLOGICAL CONDITION			
	CLIMAX (EXCELLENT)	LATE SERAL (GOOD)	MID SERAL (FAIR)	EARLY SERAL (POOR)
Forage production (lbs/ac/yr)[a]	764	501	210	55
Financial returns ($/ac/yr)[b]	+3.49	+2.14	+0.67	−0.16

[a] Forage production in year of average precipitation conditions.
[b] Average returns per year for 1986–1993 period.

than the climax (Pearson and Whitaker 1974; Skovlin et al. 1976; Quigley et al. 1984). This is because trees increasingly shade-out the grasses with progression from early seral to climax conditions. Annual grasslands and rangelands seeded to introduced forage plants are other situations where the Dyksterhuis condition and trend procedures have little practical application. This is because these areas no longer support native vegetation but are productive for some uses.

A critical part of range management is the decision on what successional stage will best meet management objectives and how to achieve it. On most grassland and semidesert rangelands a late seral ecological condition (51–75% of climax) will provide a good balance between forage for livestock and habitat for wildlife. This level of climax vegetation provides adequate vegetation cover for soil stability and will in many cases maximize plant and animal diversity (Smith et al. 1996). A wide variety of studies discussed later in this book show that a late seral condition can be attained and sustained on most rangelands through use of scientifically developed range management practices.

Climate and Plant Succession

When plant successional changes are evaluated, climate is generally assumed to be a constant. However climate is a dynamic, changing entity. Figure 6.11 (top) shows long-term temperature world fluctuations. Weather records taken over the last 100 years by the United States Weather Bureau show statistically different though minor changes. In the last 20 years, New Mexico has had a series of wet years. Precipitation in this period has been about 20% above the 100-year average (Betancourt 1996; Holechek 1996a,c). This has caused forage production on many of its rangelands to increase by 50% or more (Holechek et al. 1994; Herbel and Gibbens 1996). In contrast, New Mexico experienced extended drought periods in the 1930s and 1950s that caused several grassland communities to change to shrublands even where livestock grazing did not occur (Herbel et al. 1972). Changes back

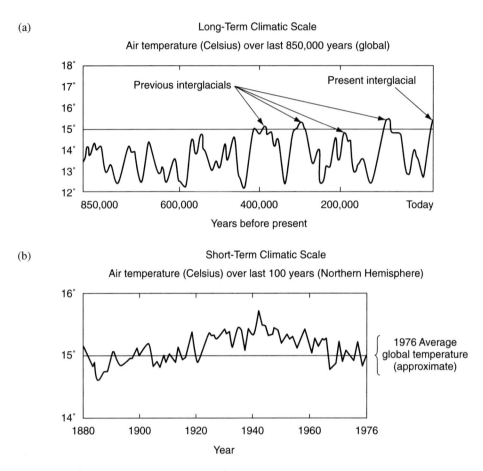

(a)

Long-Term Climatic Scale

Air temperature (Celsius) over last 850,000 years (global)

(b)

Short-Term Climatic Scale

Air temperature (Celsius) over last 100 years (Northern Hemisphere)

Figure **6.11** These two graphs of global temperature fluctuations show that the earth's climate is variable on both long (a) and comparatively short (b) time scales. (National Center for Atmospheric Research, from National Academy of Sciences and other sources.) (From Roberts and Lansford 1979.)

to grassland have been erratic under the wetter conditions of the last 20 years (Betancourt 1996; Herbel and Gibbens 1996).

The concept of climax vegetation for an area implies long-term (500 years) stability. In reality vegetation is much more stable under some climatic types than others. Wet areas that support forests and extremely dry desert areas with scattered shrubs tend to have a more stable climax than grasslands or grassland transition areas. This is because the climax plants that occur in grassland areas tend to be shorter lived and have shallower root systems (grasses and herbs) than those of deserts (shrubs) and forests (trees).

Johnson and Mayeux (1992) provide a critique of the Clementsian climax vegetation theory in the context of climatic shift. They use brush increase on southwestern U.S. rangelands to demonstrate their viewpoint. They challenge the conventional view that increased abundance of shrubs and decline of grassland has been a result of introduced livestock upsetting a natural balance. They present an alternative view that a balance never existed in the first place. Under the alternative viewpoint, the introduction of livestock only added another environmental factor that tended to modify the rate rather than the direction of changes already in progress. Johnson and Mayeux (1992) imply gradual changes in climatic elements such as timing and amount of precipitation, temperature, and level of atmospheric CO_2 are the primary factors causing the shift from grasses to shrubs; while introduction of livestock has been a more minor, secondary factor. They do point out that activities by man, other than livestock grazing, may be a cause of climatic shift. The final point of their discussion is that vegetation dynamics are much more complex than indicated by Clementsian climax theory. They expressed concern that the Clementsian model may cause natural resource managers to have unrealistic goals and expectations. Our interpretation of this point is that in the western United States many shrublands that were former grasslands will not return back to grasslands from grazing management alone because of ongoing climatic shift. Conversion of these areas back to grassland using fire, herbicides, or mechanical means will probably be only temporary because climatic and other environmental forces now favor shrubland.

The periodic droughts in the central Great Plains such as in the 1930s and 1950s caused massive shifts in land use between farming and ranching. The 1930s "Dust Bowl" in the Great Plains has provided a classic example of what can happen when climatic variability is ignored. Both short- and long-term climatic fluctuations are still often overlooked in natural resource management and planning. However, preparing for climatic uncertainty is one of the most crucial factors in successful range management.

Fire and Plant Succession

Fire has been a dominant force in the evolution of most of the world's grasslands and forests, as well as some desert shrublands. Lightning, sparks from falling rocks, volcanic activity, spontaneous combustion, and human activity are the primary causes of fires (Barbour et al.1987).

Grassland vegetation and grassland climates favor fire (Barbour et al. 1987). These areas typically go through wet periods where fuel accumulates followed by a dry period with high temperatures. Grasses on rangelands in the southwestern United States are favored by periodic fires (every 3–10 years) that suppress shrubs such as mesquite (Figure 6.12).

Huge areas of continuous, flammable grassland vegetation occurred in the central Great Plains of the United States before the advent of settlement in the mid-1800s (Barbour et al. 1987). Early accounts describe how summer thunderstorms moving across the prairie would rapidly ignite a series of fires in their path. Until the advent of fences, roads, farming, and livestock grazing, these fires could spread long distances until they burned out due to lack of fuel or a weather change.

Native Americans burned the grasslands of the Great Plains to improve conditions for game animals. The nutritious lush regrowth that occurs shortly after a fire was a strong attractant to bison, elk, deer, pronghorn, and a wide variety of other wildlife (Riggs et al. 1996).

Fire has a number of benefits to grassland vegetation (Wright and Bailey 1982; Barbour et al. 1987). It reduces competition from shrubs, speeds the cycling of nutrients

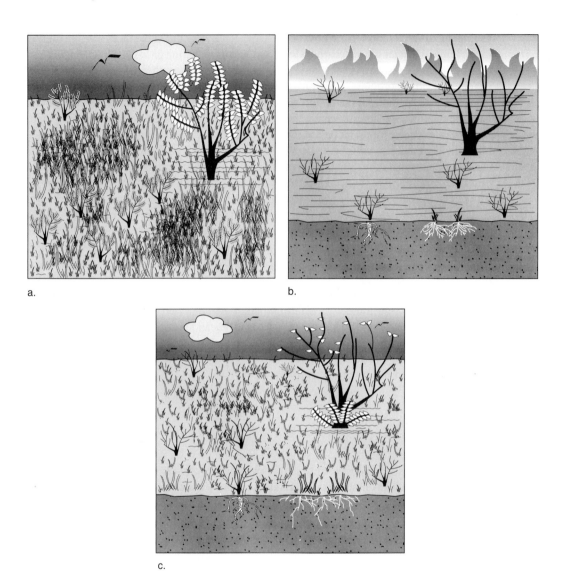

a.

b.

c.

***Figure* 6.12** Desert grasslands of the southwestern U.S. were maintained by fires that periodically reduced mesquite and other shrubs. (a) The build up in vegetation creates a fuel load that will carry fire. (b) After the fire. (c) The perennial grasses gain a competitive advantage over the mesquite because they have dense fibrous root systems and they regenerate more quickly then shrubs.

tied up in dead vegetation, and reduces allelopathic and pathogenic substances associated with old decaying vegetation. Most grasslands appear dependent on fire for recycling nutrients from vegetation to soil to new biomass (Mutch 1970).

Many forests are adapted to fire. Ponderosa pine, the primary timber tree in the western United States, has heavy, fire resistant bark. Periodic fires cleared out the small trees and shrubs maintaining most ponderosa pine areas as an open parkland. In the southeastern

United States longleaf pine, another important timber tree, depends on fire for its maintenance. Fire is necessary to reduce understory competition, prepare a mineral seedbed, reduce fungal disease, and open the forest canopy (Barbour et al. 1987).

Fire does have some negative effects particularly if it occurs too frequently or excessive amounts of fuel accumulate between fires (Wright and Bailey 1982). Negative effects of fire can include loss of some soil nutrients to volatilization, reduced soil water-holding capacity due to less soil organic matter and bacteria, and increased soil erosion. The temporary loss of food and cover from fire can adversely affect many wildlife species, particularly, if the fire covers a large area. In some areas fires can cause long-term changes in the types of wildlife species occupying an area. This is the case when mature woodland areas burn and regenerate as grasslands. We refer the reader to Riggs et al. (1996) for a detailed discussion of prescribed fire influences on rangeland wildlife. Wright and Bailey (1982) provide a thorough discussion of fire ecology on rangelands.

Ecosystem Stability and Grazing

Ecosystems may develop some degree of stability over time with certain levels of herbivores present. These herbivore levels may fluctuate and at times may be destructive of the vegetation (e.g., jackrabbit populations in the western United States, which cycle with about a 10-year periodicity, and lemmings in the arctic tundra, where they undergo a 3- to 4-year cycle). Hence climax equilibria must encompass considerable variation in producer and consumer organisms. Many ecologists believed that range ecosystems in the western United States represented climax conditions prior to the introduction of livestock from Europe. In some cases, sufficient time may not have elapsed since the last Ice Age for stability to develop. At any rate, introduction of livestock onto western ranges drastically changed herbivore pressures. In some areas native herbivores were replaced (e.g., bison in the central United States) and in others herbivore densities increased. In many cases livestock numbers were in excess of the capacity of the resources to support them. In these cases induced regression (changes away from climax caused by fire, grazing, drought, etc.) occurred that resulted in less primary production, accelerated erosion, and so on. When range people began to realize the limits of arid and semiarid rangelands to support livestock, management of these ranges often improved and secondary succession (induced progression) proceeded toward a different equilibrium position.

Even under protection from large herbivores, vegetation is dynamic and fluctuates in response to other controlling factors, especially climate. Figure 6.13 shows fluctuations in black grama (*Bouteloua eriopoda*) over time on the Jornada Experimental Range in southern New Mexico for nearly a 40-year period, under protection from grazing, and three intensities of grazing. In this study heavy grazing represented utilization of more than 55% for black grama; intermediate grazing, 40% to 55% utilization; and conservative grazing, less than 40% utilization. Even without grazing, black grama cover was extremely low during the mid-1920s. Black grama cover apparently was highest when the plants were grazed conservatively, a phenomenon supporting the grazing optimization theory discussed in Chapter 5.

In other situations shifts in species composition accompany livestock grazing. For example, in Alberta, Canada, blue grama (*Bouteloua gracilis*) cover was 50% higher on grazed

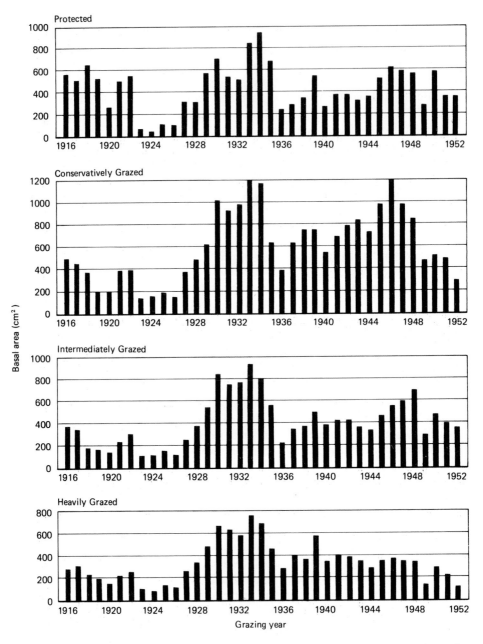

Figure 6.13 Basal area of black grama on meter-square quadrants protected from grazing and at three intensities of grazing on the Jornada Experimental Range, southern New Mexico, 1916–1953. (From Paulsen and Ares 1962.)

compared to protected areas (Smoliak 1965). On the other hand, a cool-season species, needleandthread (*Stipa comata*), responded to protection from grazing. Lacey and Van Poollen (1981) compared 11 studies throughout the West and found that protected areas produced an average of 68% more herbage than comparable areas grazed at a "moderate" rate (Table 6.3). However, the response varied from -12% (grazed areas produced more than protected areas) on the Edwards Plateau in Texas to $+438\%$ on a mountain range in Colorado.

Thus we conclude that the impacts of grazing by domestic livestock are varied. It is extremely difficult to generalize because of differences in climate, resistance of different species to grazing, stocking levels, composition of vegetation, grazing season, and many other factors. In some cases shifts in species composition may be minor, while in other cases they may represent changes in life-forms. Despite the development of sophisticated analytical techniques, separation of grazing impacts from climatic impacts remains difficult.

TABLE 6.3 Herbage Production (kg/ha) under No Grazing and Moderate Grazing, and Mean Difference in Production for 20 Observations of Different Rangelands in the United States

REFERENCE AND STUDY INFORMATION	PRODUCTION		PERCENTAGE DIFFERENCE
	MODERATE USE	NO GRAZING	
Albertson et al. (1953)			
Kansas (5 years protection)	1,232	2,016	64
Evanko and Peterson (1955)			
Montana (18 years protection)			
Festuca idahoensis	329	778	136
Agropyron spicatum	46	65	41
Johnson (1956)			
Colorado (10 years protection)	1,052	2,218	111
Larson and Whitman (1942)			
South Dakota (described as a relict)	1,838	2,650	44
Pieper (1968)			
New Mexico (12 years protection)			
Stony hills	526	627	19
Loamy upland	616	728	18
Loamy bottomland	330	683	107
Reardon and Merrill (1976)			
Texas (20 years protection)	1,331	1,166	-12

(Continued)

TABLE 6.3   Herbage Production (kg/ha) under No Grazing and Moderate Grazing, and Mean Difference in Production for 20 Observations of Different Rangelands in the United States *(Continued)*

REFERENCE AND STUDY INFORMATION	PRODUCTION		PERCENTAGE DIFFERENCE
	MODERATE USE	NO GRAZING	
Riegel et al. (1963)			
Kansas (20 years protection)			
Andropogon spp. (mixture)	2,432	6,188	154
Bouteloua gracilis	2,374	2,475	4
Agropyron sp-*Bouteloua* sp.	3,358	3,852	15
Schwan et al. (1949)			
Colorado (7 years protection)	1,318	1,817	38
Sims et al. (1978)			
Montana (relatively undisturbed)	899	1,519	69
Kansas (60 years protection)	1,109	1,159	5
Oklahoma (15 years protection)	2,399	2,698	13
Colorado (31 years protection)	899	1,049	17
Smith (1967)			
Colorado (17 years protection)			
Open grass	336	1,810	438
Open timber	224	371	66
Vogel and Van Dyne (1966)			
Montana (4 years protection)	661	739	12

Source: Lacey and Van Pollen 1981.

DROUGHT

Droughts are common phenomena on rangelands, but are sometimes confused with aridity. *Aridity* is a permanent condition of a general lack of water. *Drought,* on the other hand, is a period of low precipitation in relation to a longer-term average. There have been several definitions of drought, but it is difficult to define quantitatively. Drought has been defined as a period when precipitation is less than 75% of the average amount (Society for Range Management 1989). Others have been less specific and list interference with plant growth processes as the major criterion (Klages 1942).

Droughts can be relatively short term or long term, lasting several years. Droughts can reduce the vigor of plants and ultimately result in plant mortality. The effects of short-term drought during the dormant and early growing season in 1970–1971 at the Fort Stanton Experimental Range in southcentral New Mexico were quite dramatic (Pieper and Donart 1973). On some sites dominate blue grama was replaced by other grasses, such as sand dropseed (*Sporobolus cryptandrus*), Hall's panic (*Panicum hallii*), and wolftail (*Lycurus phleoides*). Forbs such as lemonweed (*Pectis papposa*) and showy goldeneye (*Viguira multiflora*) also increased drastically following the drought. These changes reduced grazing value of the range because these forbs are relatively low in palatability. Grass production was reduced by about 66% the year following the drought.

Other droughts have been relatively long term, such as the one in South Texas during the period 1961–1963 (Chamrad and Box 1965). This drought resulted in considerable mortality of perennial grass species. However, mortality differed on different soil types.

One of the most severe droughts in the Southwest occurred during the early 1950s. This drought had profound influence on the vegetation (Herbel et al. 1972). Cover and yield of the dominant black grama was reduced to nearly 25% of predrought figures (Table 6.4). Total herbage production was also reduced drastically (Table 6.5). Recovery has been relatively rapid on some sites (e.g., shallow sandy) and virtually nil on others (e.g., deep sandy) (Herbel and Gibbens 1996). Apparently, soil water differences account for some of these dissimilarities.

TABLE 6.4   Precipitation, Yield, and Cover of Black Grama During Drought on Several Sites on the Jornada Experimental Range in Southern New Mexico

SITE	PRECIPITATION INDEX (1951–1956 ÷ 1941–1950)	*BOUTELOUA ERIOPODA*[a]	
		YIELD (100)	COVER
Deep sandy	54	18	13
Sandy flats	57	29	32
Shallow sandy	55	37	51
Low hummocky	52	14	8
Flood plains	56	20	18
Heavy sandy flats	54	32	35
Slopes	58	32	46
Averages	55	26	29

Source: Herbel et al. 1972.

[a] Yield and cover during drought years as a percentage of predrought years.

TABLE 6.5   Herbage Production on Jornada Experimental Range Before and During the Drought of the 1950s

	HERBAGE PRODUCTION (KG/HA)	
SITE	PREDROUGHT	DROUGHT
Deep sandy	527	126
Sandy flats	730	224
Shallow sandy	794	265
Low hummocky	596	78

Source: Herbel et al. 1972.

COMPETITION

Competition is a very important concept in plant ecology, but has been defined in many ways by different ecologists (Risser 1969). From the standpoint of plant ecology, competition has been defined as "a process which occurs when two or more organisms are making a common endeavor to gain one or more requisites in excess of the immediate supply" (Heady et al. 1962). As such, competition may be viewed as a process (and not a result) between or among organisms. Competition can occur only when some required element is in short supply.

Generally, the factors for which competition occurs among plants are water, nutrients, light, oxygen, and carbon dioxide (Risser 1969). In some cases competition may occur for pollinating agents, suitable germination sites, and so on, but these processes are often more subtle than those for water, nutrients, and so on.

Results of competition may be manifested on individual plants or at the population level. Individual plants may show the effects of competition by reduced growth rate and reduced vigor, or in final stages, mortality. Plant community response to competition may be demonstrated by reduced density and occurrence of individuals of a species or by changing patterns of two competing species (Risser 1969). Plants have evolved many mechanisms to adapt to competition. Some plants have a rapid growth rate, whereas others have an extensive root system. Simply because two plants are growing in close proximity does not indicate that competition is occurring. For example, blue grama and walkingstick cholla (*Opuntia imbricata*) apparently compete very little even though they grow in close proximity. The two species grow at different times of the year and have different root distributions (Pieper et al. 1974a). In other cases, competition may be very severe as exemplified by broom snakeweed (*Gutierrezia sarothrae*), which severely restricts production of associated grasses (Ueckert 1979; McDaniel et al. 1982). Competition may occur between or among species (interspecific) or among individuals of the same species (intraspecific).

In many cases it is difficult to separate competitive from allelopathic impacts. Allelopathy is the production of toxic substances that may be detrimental to other plants. Species with allelopathic properties tend to delay or halt successional advance. Shrub species in the genera of *Juniperus, Artemisia, Larrea,* and *Adenostoma* have shown allelopathic effects. When these types of shrubs dominate a rangeland as a result of overgrazing, recovery is very slow when grazing pressure is removed. Improvement in productivity of palatable forage species can best be accomplished by reducing the influence of these shrubs through burning, herbicides, or mechanical control (see Chapter 15).

PLANT SUCCESSION AND RANGE MANAGEMENT: A CONCLUSION

One of the most controversial questions in rangeland management in recent years has been whether or not the more arid ecosystems can recover part of their original species composition after lengthy human induced disturbances such as heavy livestock grazing. We have previously reviewed conventional successional theory developed by Clements (1916) and put into practical application by Dyksterhuis (1949). Under this theory progression back to the climax would be expected when grazing pressure is reduced even when woody plants dominate the area. Range managers both in the past and present have relied heavily on Clements's theory, but there have been important alternative viewpoints (Westoby et al. 1989; Laycock 1991). Recently long term (30–40 years) studies across broad, arid areas in Utah and New Mexico have permitted a better understanding of the validity of the ideas advanced by Clements and Dyksterhuis. In the salt desert of Utah (Yorks et al. 1992), on piñon-juniper woodland in Utah (Yorks et al. 1994), and in the Chihuahuan desert of New Mexico (McCormick and Galt 1993), increases occurred in range forage plants over a 30- to 40-year period under conditions of reduced livestock grazing pressure. These studies and others (Potter and Krenetsky 1967; Smith and Schmutz 1975; Holechek et al. 1994) have shown that forage plants can increase on many arid shrubland and woodland areas in response to livestock grazing management alone, if soil degradation has not been too severe and palatable perennial forage plants still remain. In grassland rangelands the Clements-Dyksterhuis successional theory has been fairly well supported by a wide range of long-term stocking rate studies. We have previously discussed some of these studies, and they are reviewed in more detail in Chapters 8 and 9.

In conclusion, we consider the basic successional model of Clements and Dyksterhuis to provide a useful description of vegetation changes in response to grazing from a general standpoint. However, we emphasize range managers need to understand that many exceptions to this model do exist. Under conditions where arid rangelands are severely degraded and heavily dominated by unpalatable shrubs, the state and transition models (Westoby et al. 1989; Laycock 1991) have much practical application. In these situations the use of herbicides, mechanical treatment, and burning and/or seeding will probably be needed to obtain a more desirable plant community within a reasonable time period.

RANGE MANAGEMENT PRINCIPLES

- The driving forces in succession are precipitation and temperature. As precipitation and temperatures increase, ranges become more resistant to retrogression from heavy grazing and recover more quickly when grazing pressure is reduced from heavy to moderate or light levels.

- Succession after retrogression is less linear and predictable in the arid compared to humid range types. Because retrogression is more difficult to reverse in the arid types, they require lighter grazing intensities than used in the humid types.

- Climatic fluctuation, grazing, and fire are major factors causing vegetation changes on rangelands. Separating the influences of these factors is a critical part of range management.

- On most grassland and shrubland ranges a late seral stage (51–75% remaining climax vegetation) will maximize plant and wildlife diversity. This level of climax vegetation will also ensure soil stability and give high financial returns from livestock production.

- The conventional Clements-Dyksterhuis successional model provides a good general description of plant responses to grazing. However important exceptions do exist particularly on severely degraded arid shrubland ranges. Here vegetation manipulation by man will often be needed for meaningful improvement.

Literature Cited

Albertson, F. W., A. Riegel, and J. C. Launchbaugh Jr. 1953. Effects of different intensities of clipping on short grass in west-central Kansas. *Ecology* 34:1–20.

Anderson, J. E., and K. E. Holte. 1981. Vegetation development over 25 years without grazing on sagebrush-dominated rangeland in southeastern Idaho. *J. Range Manage.* 34:25–29.

Archer, S. 1994. Woody plant encroachment into southwestern grasslands and savannas: Rates, patterns, and proximate causes. In *Ecological implications of livestock herbivory in the West.* Society for Range Management, Denver, CO.

Archer, S., and F. E. Smeins. 1991. Ecosystem-level processes. In R. Heitschmidt and J. Stuth (Eds.). *Grazing management: An ecological perspective.* Timber Press, Inc., Portland, OR.

Arnold, J. F., D. A. Jameson, and E. H. Reid. 1964. The piñon-juniper type of Arizona: Effects of grazing, fire, and tree control. *U.S. Dep. Agric. Prod. Res. Rep.* 84.

Barbour, M. G., J. H. Burk, and W. D. Pitts. 1987. *Terrestrial plant ecology.* (2d ed.). Benjamin-Cummings Publishing Co., Menlo Park, CA.

Barney, M., and N. C. Frischknecht. 1974. Vegetation changes following fire in the piñon-juniper type of west-central Utah. *J. Range Manage.* 27:91–96.

Behnke, R. H. Jr., I. Scoones, and C. Kerven. 1993. *Range ecology at disequilibrium. New models of natural variability and pastoral adaptation in African savannas.* Overseas Dev. Inst. London.

Betancourt, J. L. 1996. Long- and short-term climatic influences on southwestern shrublands, pp. 5–9. Proceedings: Shrubland ecosystem dynamics in changing environment. *U.S. Dept. Agr. For. Serv. Gen. Tech. Rep.* INT-GTR-338. Ogden, UT.

Branson, F. A. 1985. *Vegetation changes on western rangelands. Range Monograph No. 2.* Society for Range Management, Denver, CO.

Briske, D. 1991. Developmental morphology and physiology of grasses. In R. Heitschmidt and J. Stuth (Eds.). *Grazing management: An ecological perspective.* Timber Press, Inc., Portland, OR.

Buffington, L. C., and C. H. Herbel. 1965. Vegetational changes on a semi-desert grassland range 1858 to 1963. *Ecol. Monogr.* 35:139–161.

Chamrad, A. D., and T. W. Box. 1965. Drought-associated mortality of range grasses in south Texas. *Ecology* 46:780–785.

Clements, F. E. 1916. Plant succession: An analysis of the development of vegetation. *Carnegie Inst. Wash. Pub.* 242.

Cole, D. R., and H. C. Monger. 1994. Influence of atmospheric CO_2 on the decline of C_4 plants during the last deglaciation. *Nature* 368:533–536.

Cooper, C. F. 1961. The ecology of fire. *Scientific Amer.* 204:150–160.

Crisp, D. J. (Ed.). 1964. Grazing in terrestrial and marine environments. In *British Ecological Society Symposium Number 4.* Blackwell Scientific Publications Ltd., Oxford.

Cronin, E. H., P. Ogden, J. Young, and W. Laycock. 1978. The ecological niches of poisonous plants in ranch communities. *J. Range Manage.* 31:328–334.

Curtis, J. T. 1959. *The vegetation of Wisconsin.* University of Wisconsin Press, Madison, WI.

Duvall, V. L., and R. M. Blair. 1963. Terminology and definitions. In *Range research methods.* U.S. Dept. Agric. For. Serv. Misc. Publ. 94.

Duvall, V. L., and N. E. Linnartz. 1967. Influences of grazing and fire on vegetation and soil of longleaf pine-bluestem range. *J. Range Manage.* 20:241–247.

Dyksterhuis, E. J. 1949. Condition and management of rangeland based on quantitative ecology. *J. Range Manage.* 2:104–115.

Dyksterhuis, E. J., and E. M. Schmutz. 1947. Natural mulches or "litter" of grasslands: With kinds and amounts of southern pine. *Ecology* 28:163–179.

Egler, F. E. 1954. Vegetation science concepts. 1. Initial floristic composition—a factor in old field vegetation development. *Vegetation* 4:412–417.

Ellis, J. E., and D. M. Swift. 1988. Stability of African pastoral ecosystems: Alternate paradigms and implications for development. *J. Range Manage.* 41:450–459.

Ellison, L. 1960. Influence of grazing on plant succession of rangeland. *Bot. Rev.* 26:1–78.

Ehrenreich, J. L., and J. M. Aikman. 1963. An ecological study of the effect of certain management practices on native prairie of Iowa. *Ecol. Monogr.* 33:113–130.

Erdman, J. A. 1970. Piñon-juniper succession after natural fires on residual soils of Mesa Vedge, Colorado. *Brigham Young Univ. Sci. Bull. Biol. Ser.* 11:1–24.

Evanko, A. B. and R. A. Peterson. 1955. Comparisons of protected and grazed mountain rangelands in southwestern Montana. *Ecology* 36:71–82.

Friedel, M. H. 1991. Range condition assessment and the concept of thresholds. *J. Range Manage.* 44:422–427.

Gay, C. 1965. Range management. How and why. *New Mexico State Univ. Coop. Ext. Circ. 376,* Las Cruces, NM.

George, M. R., J. R. Brown, and W. J. Clawson. 1992. Application of nonequilibrium ecology to management of Mediterranean grasslands. *J. Range Manage.* 45:436–440.

Grover, H. D., and H. B. Musick. 1990. Shrub land encroachment in southern New Mexico, USA: An analysis of desertification processes in the American Southwest. *Climate Change* 17:305–330.

Hanson, H. C., and E. D. Churchill. 1961. *The plant community.* Reinhold Publishing Corporation, New York.

Hastings, J. R., and R. M. Turner. 1965. *The changing mile.* The University of Arizona Press, Tucson, AZ.

Heady, H. F. 1973. Structure and function of climax. In *Arid shrublands.* Proc. 3rd Workshop, U.S./Australia Rangeland Panel. Society for Range Management, Denver, CO.

Heady, H. F. and other seminar members. 1962. *Biological competition: Definition and comments.* Mimeo report. Range Management 201B. University of California, Berkeley, CA.

Herbel, C. H., F. N. Ares, and R. A. Wright. 1972. Drought effects on a semidesert grassland range. *Ecology* 53:1084–1093.

Herbel, C. H., and R. P. Gibbens. 1996. Post-drought vegetation dynamics on arid rangelands of New Mexico. *New Mexico Agr. Exp. Bull.* 776.

Holechek, J. L. 1994. Financial returns from different grazing management systems in New Mexico. *Rangelands* 16:237–240.

Holechek, J. L. 1996a. Drought and low cattle prices: Hardship for New Mexico ranchers. *Rangelands* 18:11–13.

Holechek, J. L. 1996b. Financial returns and range condition on southern New Mexico ranches. *Rangelands* 18:52–56.

Holechek, J. L. 1996c. Drought in New Mexico: Prospects and management. *Rangelands* 18:225–227.

Holechek, J. L., A. Tembo, A. Daniel, M. Fusco, and M. Cardenas. 1994. Long term grazing influences on Chihuahuan Desert rangeland. *Southw. Nat.* 39:342–349.

Huschle, G., and M. Hironaka. 1980. Classification and ordination of plant communities. *J. Range Manage.* 33:179–182.

Jenny, H. 1941. *Factors of soil formation.* McGraw-Hill Book Company, New York.

Jenny, H. 1961. Derivation of the state factors equations of soils and ecosystems. *Proc. Soil Sci. Soc. Am.* 25:385–388.

Jenny, H. 1980. *The soil resource orgin and behavior.* Springer-Verlag Inc., New York.

Johnsen, T. N. 1962. One-seed juniper invasion of northern Arizona grasslands. *Ecol. Monogr.* 32:187–207.

Johnson, H. B., and H. S. Mayeux. 1992. Viewpoint: A view on species additions and deletions and the balance of nature. *J. Range Manage.* 45:322–334.

Johnson, W. M. 1953. *Effect of grazing intensity upon vegetation and cattle gains on ponderosa pine-bunchgrass ranges of the front range of Colorado.* U.S. Dep. Agric. Circ. 929.

Johnson, W. M. 1956. The effect of grazing intensity on plant composition, vigor, and growth of pine-bunchgrass ranges in central Colorado. *Ecology* 37:790–798.

Klages, K. H. W. 1942. *Ecological crop geography.* The Macmillan Company, New York.

Klipple, G. E., and D. F. Costello 1960. Vegetation and cattle responses to different intensities of grazing on shortgrass ranges of the central Great Plains. *U.S. Dept. Agric. Circ.* 929.

Lacey, J. R., and W. Van Poollen. 1981. Comparison of herbage production on moderately grazed and ungrazed western ranges. *J. Range Manage.* 34:210–212.

Larson, F., and W. A. Whitman. 1942. A comparison of used and unused grassland mesas in the badlands of South Dakota. *Ecology* 23:438–445.

Lauenroth, W. E., D. E. Milchunas, J. Dodd, R. Hart, R. Heitschmidt, and L. Rittenhouse. 1994. Effects of grazing on ecosystems of the Great Plains. In *Ecological implications of livestock herbivory in the West.* Society for Range Management, Denver, CO.

Laycock, W. A. 1978. Coevolution of poisonous plants and large herbivores on rangelands. *J. Range Manage.* 31:335–343.

Laycock, W. A. 1991. Stable states and thresholds of range condition on North American rangelands: A viewpoint: *J. Range Manage.* 44:427–434.

Lewis, J. K. 1969. Range Management viewed in the ecosystem framework. In G. M. Van Dyne (Ed.). *The ecosystem concept in natural resource management.* Academic Press, Inc., New York.

Lewis, J. K. 1971. The grassland biome: A synthesis of structure and function, 1970. In N. R. French (Ed.). *Preliminary analysis of structure and function in grasslands. Range Science Series No. 10.* Colorado State University, Fort Collins, CO.

McCormick, J., and H. Galt. 1993. Forty years of vegetation trend in southwestern New Mexico, pp. 68–79. In *Vegetation management of hot desert rangeland ecosystems symposium.* Univ. Arizona, Tucson, AZ.

McDaniel, K. C., R. D. Pieper, and G. B. Donart. 1982. Grass response following thinning of broom snakeweed. *J. Range Manage.* 35:219–222.

Miller, R., T. Svejcar, and N. West. 1994. Implications of livestock grazing in the intermountain sagebrush region: Plant composition. In *Ecological implications of livestock herbivory in the West.* Society for Range Management. Denver, CO.

Molyneaux, R. J., and M. Ralphs. 1992. Plant toxins and palatability to herbivores. *J. Range Manage.* 45:13–18.

Mutch, R. E. 1970. Wildland fires and ecosystems—a hypothesis. *Ecology* 51:1046–1051.

Nebel, B. J. 1981. *Environmental science: The way the world works.* Prentice-Hall, Inc., Englewood Cliffs, NJ.

Odum, E. P. 1959. *Fundamentals of Ecology.* 2d ed. W. B. Saunder Company, Philadelphia.

Paris, O. H. 1969. The function of soil fauna in grassland ecosystems. In R. L. Dix and R. G. Beidleman (Eds.). *The grassland ecosystems: A pro synthesis. Range Science Series No. 2.* Colorado State University, Ft. Collins, CO.

Paulsen, H. A., and F. N Ares. 1962. Grazing values and management of black grama and tobosa grasslands and associated shrub ranges of the southwest. *U.S. Dept. Agric. Tech. Bull.* 1270.

Pearson, H. A., and L. B. Whitaker. 1974. Forage and cattle responses to different grazing intensities on southern pine range. *J. Range Manage.* 27:444–446.

Pieper, R. D. 1968. Comparison of vegetation on grazed and ungrazed piñon-juniper grassland sites in southcentral New Mexico. *J. Range Manage.* 51–53.

Pieper, R. D. 1977. Effects of herbivores on nutrient cycling and distribution. *Proc. 2nd U.S./Australia Rangeland Panel.* Australian Rangeland Society, Western Australia.

Pieper, R. D. 1978. *Measurement techniques for herbaceous and shrubby vegetation.* New Mexico State University, Las Cruces, NM.

Pieper, R. D. 1983. Consumption rates of desert grassland herbivores. In J. A. Smith and V. W. Hays (Eds.). *Proc. 14th International Grassland.* Westview Press, Inc., Boulder, CO.

Pieper, R. D. 1994. Ecological implications of livestock grazing. In *Ecological implications of livestock herbivory in the West.* Society for Range Management, Denver, CO.

Pieper, R. D., and G. B. Donart. 1973. *Drought effects of blue grama rangeland.* N. Mex. State. Univ. Livest. Feeders Rep.

Pieper, R. D., K. G. Rea, and J. G. Fraser. 1974a. Ecological characteristics of walkingstick cholla. *N. Mex. Agric. Exp. Stn. Bull.* 623.

Pieper, R. D., D. D. Dwyer, and R. C. Banner. 1974b. Primary production of blue grama grassland in southcentral New Mexico under two soil nitrogen levels. *Southwest. Nat.* 20:293–302.

Potter, L. D., and J. C. Krenetzky. 1967. Plant succession with release from grazing on New Mexico Rangelands. *J. Range Manage.* 20:145–151.

Quigley, T. M., J. M. Skovlin, and J. P. Workman. 1984. An economic analysis on two systems and three levels of grazing on ponderosa pine-bunchgrass range. *J. Range Manage.* 37:309–312.

Reardon, P. O., and L. B. Merrill. 1976. Vegetative response under various grazing management systems on the Edwards Plateau of Texas. *J. Range Manage.* 29:195–198.

Riegel, D. A., F. W. Albertson, G. W. Tomanek, and F. Kinsinger. 1963. Effects of grazing and protection on a twenty-year-old seeding. *J. Range Manage.* 16:60–63.

Riggs, R. A., S. C. Bunting, and S. E. Daniels. 1996. Prescribed fire, pp. 295–321. In P. R. Krausman, (Ed.). *Rangeland wildlife.* The Society for Range Management, Denver, CO.

Risser, P. G. 1969. Competitive relationships among herbaceous grassland plants. *Bot. Rev.* 35:251–284.

Roberts, W. O., and H. Lansford. 1979. *The climate mandate.* W. H. Freeman and Company, San Francisco, CA.

Schwan, H. E., D. L. Hodges, and C. N. Weaver. 1949. Influence of mulch and grazing on forage growth. *J. Range Manage.* 2:142–148.

Shoop, M. C., and E. H. McIlvain. 1971. Why some cattlemen overgraze and some don't. *J. Range Manage.* 24:252–257.

Sims, P. L., J. S. Singh, and W. K. Lauenroth. 1978. The structure and function of ten western North American grasslands. I. Abiotic and vegetational characteristics. *J. Ecol.* 66:251–258.

Skovlin, J. M., R. W. Harris, G. S. Strickler, and G. A. Garrison. 1976. Effects of cattle grazing methods on ponderosa pine-bunchgrass range in the Pacific northwest. *U.S. Dep. Agric. Tech. Bull.* 1531.

Smith, D. A., and E. M. Schmutz. 1975. Vegetative changes on protected versus grazed desert grassland ranges in Arizona. *J. Range Manage.* 28:453–458.

Smith, D. R. 1967. Effects of cattle grazing on a ponderosa pine-bunchgrass range in Colorado. *U.S. Dep. Agric. For. Serv. Tech. Bull.* 1371.

Smith, G., J. L. Holechek, and M. Cardenas. 1996. Wildlife numbers on excellent and good condition Chihuahuan Desert rangelands: An observation. *J. Range Manage.* 49:489–493.

Smoliak, S. 1965. A comparison of ungrazed and lightly grazed *Stipa-Bouteloua* prairie in southeastern Alberta. *Can. J. Plant Sci.* 45:270–275.

Society for Range Management. 1989. *A glossary of terms used in range management.* 3d ed. Society for Range Management, Denver, CO.

Springfield, H. W. 1976. Characteristics and management of southwestern piñon-juniper ranges: The status of our knowledge. *U.S. Dep. Agric. For. Serv. Res. Pap. RM-160.*

Ueckert, D. N. 1979. Broom snakeweed: Effect on shortgrass forage production and soil water depletion. *J. Range Manage.* 32:216–219.

Van Dyne, G. M. 1966. *Ecosystems, systems ecology, and systems ecologists.* ORNL-3957. Oak Ridge National Laboratory, Oak Ridge, TN.

Vogel, W. G., and G. M. Van Dyne. 1966. Vegetation responses to grazing management on a foothill range. *J. Range Manage.* 19:80–85.

Weaver, J. E., and N. W. Rowland. 1952. Effects of excessive natural mulch on development, yield, and structure of native grassland. *Botan. Gazette* 114:1–19.

Westoby, M., B. Walker, and I. Noy-Meir. 1989. Opportunistic management for rangelands not at equilibrium. *J. Range Manage.* 42:266–274.

Whittaker, R. H. 1967. Gradient analysis of vegetation. *Biol. Rev.* 42:207–264.

Workman, J. P. 1995. The value of increased forage from improved rangeland condition. *Rangelands* 17:46–48.

Wright, H. A., and A. W. Bailey. 1982. *Fire ecology.* John Wiley and Sons, New York.

Yorks, T. P., N. B. West, and K. M. Capels. 1992. Vegetation differences in desert shrublands of western Utah's Pine Valley between 1933 and 1989. *J. Range Manage.* 45:569–578.

Yorks, T. P., N. E. West, and K. M. Capels. 1994. Changes in piñon-juniper woodlands in western Utah's pine valley. *J. Range Manage.* 47:359–365.

CHAPTER 7

RANGE INVENTORY AND MONITORING

Inventory and monitoring activities are essential features of a range management plan. They can be as detailed as necessary to meet the objectives of the plan. Inventories usually involve an assessment of vegetation resources or physical features at one point in time. As such, inventories serve as baseline data to aid in the development of a range management plan. Inventories include such features as vegetation type, topography, soils, streams, stock water developments, fences, and so on. The primary purpose of an inventory is to provide as accurate representation of existing conditions.

Monitoring, on the other hand, is an evaluation process usually conducted to determine the response to some management program. Hence monitoring is usually conducted several times over a fairly long time span. One might, for example, develop a monitoring scheme to evaluate a particular grazing system or to determine the efficacy of a herbicide treatment.

In this chapter we provide an overview of basic techniques used in rangeland vegetation inventories, grazing surveys, and monitoring range management effectiveness.

We refer the reader to Cook and Stubbendieck (1986), Tueller (1988), and Bonham (1989) for more detailed discussion of these subjects.

VEGETATIONAL MAPPING

The first step in an inventory program involves development of a vegetational map. In some cases a good map will serve as the basic inventory, while in other cases additional information is needed (Figure 7.1).

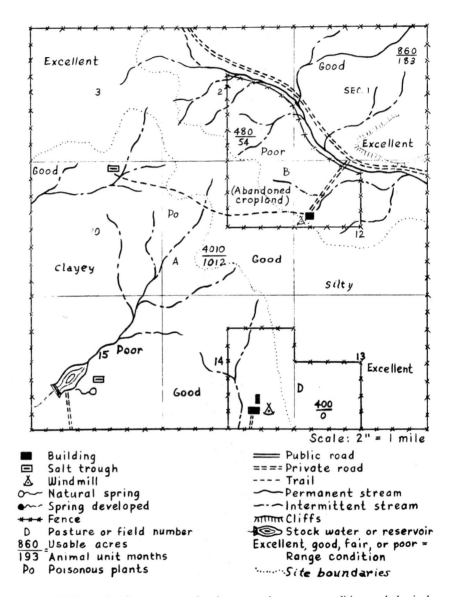

Figure **7.1** Example of a range map showing range sites, range condition, and physical features. (From Parker 1969.)

Vegetational mapping has been greatly expedited with the advent of aerial photography and recent advances in remote-sensing techniques. Kuchler (1967) one of the world's foremost vegetational mappers, has outlined four major approaches to vegetational mapping. If one simply desires to map broad vegetational categories, Kuchler's (1955) comprehensive approach or the California vegetation soil survey (Jensen 1947) might be useful. Gaussen's (Kuchler 1967) ecological approach provides for additional information on the environment. The table method of Braun-Blanquet et al. (1947) is a procedure for quantitative evaluation of the vegetation.

Aerial-photo coverage is currently available for most areas in the United States as well as other parts of the world (Figure 7.2). Often, coverage is available at different scales and at different times of the year and different years. Aerial photos can be used to differentiate among broad vegetation types, such as grassland, shrubland, savanna, and forests. Forest rangelands can often be subdivided according to dominant tree species using aerial photos

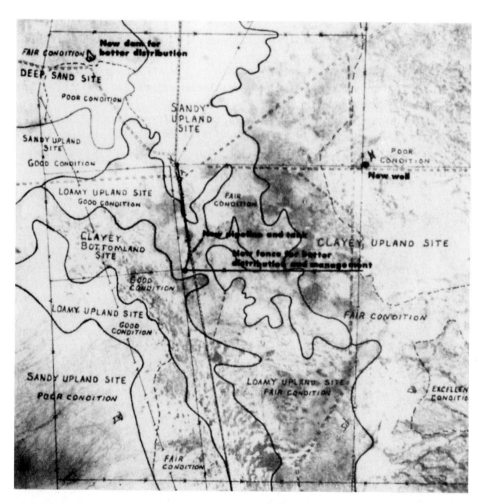

Figure **7.2** This aerial photograph shows range sites, range condition, present land use, and practices already applied. With this as a base, a complete inventory of the range can be made. Proposed changes are shown in black. Fencing is planned to take better advantage of the site, and additional water will be developed to improve livestock distribution. (From Gay 1965.)

or other remote-sensing techniques (Colwell 1968; Avery 1986). However, delineation of different grassland types is nearly impossible from small-scale black-and-white aerial photographs. Large-scale (70 mm) photos can be used to differentiate grassland types (Poulton 1970; Reppert and Driscoll 1970).

The development of satellite imagery has opened up an entirely new field of remote-sensing opportunities in the natural resource field. The large number of satellites launched in the past 15 years has resulted in large numbers of satellite images available. Initially, techniques were not sensitive enough to provide details except at the largest resolution. Refinement of techniques and development of computer-assisted analysis provides mechanisms for mapping range sites as well as other attributes (Maxwell 1976; McGraw and Tueller 1983) such as green biomass. Although these techniques depend on very sophisticated equipment, there are several labs where such analysis can be done.

For any remote-sensing operation to be successful, the observer needs to be thoroughly familiar with the ecology of the area. For specific attributes of the area, ground truth stations where actual sampling is conducted are necessary for correlation with the remote-sensing techniques. Although aerial photographs and remote-sensing techniques for range inventory have been primarily used to delineate range types, they also offer opportunities for monitoring with ERTS, Landsat, or other satellite imagery. The reader is referred to Tueller (1988) for more detailed discussion on the use of remote sensing in range management.

DETERMINATION OF VEGETATIONAL ATTRIBUTES

In many cases, delineation of vegetational types may not be sufficient for either monitoring or survey objectives. It may be necessary to sample the vegetation directly to determine the characteristics of interest. There are many methods available for determining vegetational characteristics (Brown 1954; 't Mannetje 1978; Pieper 1978; Risser 1984). Methods that exhibit high accuracy are the most useful since the purpose of an inventory is usually to depict conditions as they actually exist. Monitoring, on the other hand, is usually conducted to determine changes in vegetation over time. In this case, sampling is often done by different observers. Here methods that offer consistency among observers should have high priority.

Three quantifiable attributes of vegetation important in monitoring and inventory are weight, area, and number. Each of these may be sampled for different objectives in a range management plan.

Weight or Biomass

Biomass or standing crop usually refers to the weight of organisms present at one time (Pieper 1978; Society for Range Management 1989). Increases in biomass through the growth process of photosynthesis over time are generally considered productivity estimates that include a time dimension. In the English system, biomass is generally expressed in pounds per acre or in the metric system, kilograms per hectare. Productivity estimates include a unit of time (e.g., annual, per day, week, month, etc.).

Most estimates of plant biomass or standing crop include only that above the soil surface. This material is commonly available to larger herbivores. Belowground biomass is very important for plant functions, but is difficult to measure and generally not included in inventory or monitoring procedures.

Direct harvesting is considered the most reliable method of determining aboveground biomass. However, this method is too time-consuming to be of practical value for inventory or monitoring of extensive range areas. Several weight estimate techniques have been developed for rapid and fairly reliable determination of herbage weight (Pechanec and Pickford 1937; Shoop and McIlvain 1963). These procedures involve estimating herbage weight by species from small quadrats in the field. Training of observers in the field is necessary. This can be done easily by checking the estimates with clipped quadrats (Goebel 1955). The method is considered reliable enough to be used on detailed research studies (Shoop and McIlvain 1963).

Weight estimates can be adjusted by clipping a portion of the quadrats that have been estimated. Double-sampling procedures involving regression adjustments have been outlined by several workers. These studies are discussed in detail by Cook and Stubbendieck (1986) and Tueller (1988).

Area or Cover

Aerial or canopy cover refers to the area covered by the vertical projection of the crown of plants onto the soil surface (Brown 1954). Basal cover or area refers to the area occupied at the intersection of the plant and soil surface. Woody plant cover is often expressed in terms of canopy cover since the basal area of trees and shrubs is very small in relation to the role of these plants in the plant community. Basal area is most often used for herbaceous plants since it was assumed that basal area was not influenced greatly by seasonal precipitation and temperature. However, these assumptions concerning the relative stability of basal area of grasses may be misleading (Young 1980).

Cover determination is often conducted for inventory and monitoring purposes. Three methods that appear to meet time requirements for inventory and monitoring procedures are estimation (Stewart and Hutchings 1936; Daubenmire 1958), the step point method (Evans and Love 1957), and line intercept procedures (Canfield 1941). Estimating procedures usually involve estimating cover by species in relatively small plots. Often, cover classes are used instead of whole percent unit estimates (Table 7.1). In this case it is only necessary to estimate cover in the nearest cover class and then to use midpoints for data summarization.

The point-step method was developed as a rapid, objective method of determining cover and species composition of large range areas (Evans and Love 1957). The method involves cutting a notch or marking a spot on the observer's boot. The observer paces across the range area, recording whatever is directly beneath the notch or mark of his or her boot. Individual

TABLE 7.1   Cover Classes Rated According to
Percentage of Ground Surface Covered by Vegetation

| CLASS | PERCENTAGE COVERED | |
	RANGE	MIDPOINT
1	0–5	2.5
2	5–25	15.0
3	25–50	37.5
4	50–75	62.5
5	75–95	85.0
6	95–100	97.5

Source: Daubenmire 1958.

species, litter, bare ground, rock, and so on, can be recorded. Other devices, such as a fine rod or tripod, can be used to make placement of the point more objective (Owensby 1973). Care must be taken to make the point as small as possible, to avoid overestimation of cover.

Measures of cover by intercept length along a line is called the "line-transect method" (Bonham 1989). In contrast a system of cross-hairs, grid points, or dot matrix are grouped under the heading of "point-intercept methods." Line-intercept and point intercept methods are two of the most popular methods used to estimate cover (Bonham 1989). The original line intercept techniques of Canfield (1941) has been modified by Holechek and Stephenson (1983) so micro-lines of 1-M in length are evaluated using step-point methods of Evans and Love (1957). The reader is referred to Bonham (1989) for more detailed information on evaluating plant cover.

Density and Frequency

Density is defined as the number of individual plants per area (Cooper 1959). In some cases it is difficult to identify an individual plant for sod-forming species (Dix 1961). In these situations it may be necessary to use plant units such as an individual shoot. Density can be determined by counting the number of plants in quadrats, but quadrat size is critical. Large quadrats serve well for vegetation with low density but may be too time consuming for areas with high density.

Distance measures (Cottam and Curtis 1956) have also been used for determining density. The wandering quarter method (or zigzag method, Catana 1963) has been used by the Soil Conservation Service for determining tree density in woodland surveys. However, the method does entail measurement of distances between trees and may be somewhat time-consuming for large areas.

Frequency sampling is fast and easy to conduct in the field. If one determines density from quadrats, frequency can be calculated from the same data since frequency represents the percentage of the quadrats in which the species occurs. Quadrat size is critical with frequency sampling. If the quadrat is too large, many species will have high frequencies, while if the quadrat is too small, frequencies will be too small, especially for the less abundant species. Hyder et al. (1963) provide guidelines for frequency sampling.

GRAZING SURVEYS

Grazing surveys involve determination of grazing capacity, forage utilization, ecological condition, and trend. This information is used in monitoring the effectiveness of past range management practices and formulating new management plans.

Agencies managing public rangelands, primarily the Forest Service and Bureau of Land Management, are now being required to collect more quantitative biological data for their decision-making processes than in previous years. At the same time their budgets and personnel resources are being squeezed due to congressional budget cutting. Controversies surrounding grazing decisions on public lands are on the rise.

At the center of many of these controversies is the reliability of the grazing capacity, condition, trend, and forage utilization data on which decisions regarding livestock grazing adjustments are being based. Ranchers have often resisted grazing changes on their federal land allotments because they lacked confidence in the validity of the vegetation surveys. On the other hand, environmental groups have tended to believe more severe adjustments were needed than those proposed by federal rangeland managers.

We will discuss the primary approaches used in monitoring range management effectiveness and also explore changes in present approaches that might improve their reliability. Our focus will be on grazing capacity, forage utilization, ecological condition, and trend surveys.

Determining Grazing Capacity

Determination of grazing capacity remains as one of the biggest problems in managing rangelands. Climatic fluctuation, increases in wildlife grazing (particularly elk), shrub and tree invasion, changes in grazing methods, and ranch ownership transfers contribute to the need to periodically reevaluate the numbers of livestock and big game animals an area will sustain. Because both climate and vegetation are dynamic and always changing, any determination will only be an estimate of a moving target. However, any estimate should be one that will perpetuate and promote a healthy resource.

An important problem has been that grazing capacity survey techniques have varied widely within and among government agencies. Guidelines used on animal intake rates, forage use, correction for slopes, and correction for distance from water have often been inconsistent within and among agencies. This has created confusion about technique reliability and fairness among various groups that have an interest in the management outcome. A grazing capacity technique described by Holechek (1988) and tested by Holechek and Pieper (1992), attempts to consolidate research on the various components of grazing capacity determination into a unified procedure that can be applied consistently across western range types. This and other grazing capacity procedures are discussed in detail in Chapter 8.

The real problems in grazing capacity determination involve precipitation variability over years and vegetation variability among land units. Any grazing capacity determination has to take into account precipitation conditions in prior years as well as the present year. For this reason, ideally the vegetation sampling should be averaged over a 3-year period. However, in reality, short-term needs and cost often necessitate use of data from a single year. Under this condition there should be some downward or upward adjustment in forage production if growing season precipitation is 20% above or below average. USDA Natural Resources Conservation Service site guidelines and long-term grazing studies (reviewed in Chapter 8) can be helpful in these adjustments. If growing season precipitation departs more than 30% from the long-term average, any assessment of grazing capacity will probably be unreliable due to erratic forage production.

If they are carefully selected, about one key site per 4,000 acres has been an acceptable minimum on the Bureau of Land Management (BLM) and Forest Service lands in New Mexico. Generally key sites should be about 0.75 to 1.00 mile from water and representative of average grazing pressure and vegetation composition on primary sites within a pasture.

DETERMINING UTILIZATION

Utilization has been defined as the percentage of the current year's herbage production consumed or destroyed by herbivores (Society for Range Management 1989). Utilization surveys can provide important information severity of defoliation correctness of stocking rate, livestock distribution patterns, cover and food available for wildlife, soil cover, and esthetic quality. Utilization estimates can be used to check stocking at the end of the grazing season to see if actual use agrees with desired use. Utilization estimates made during the grazing season can also be made to adjust stocking before the end of the grazing season.

Several methods have been developed for estimating utilization, but all have drawbacks (Risser 1984; Cook and Stubbendieck 1986). Utilization can be estimated either on a quadrat or an individual plant basis (Pieper 1978; Risser 1984). However, the skill of individual estimators varies considerably. In addition, variations in shape and form of individual species make estimates difficult.

Direct estimates of utilization have been made using paired plots or quadrats. In one case, one quadrat of each pair is clipped prior to grazing and the other after grazing, and the difference is considered the amount utilized (Pieper 1978; Cook and Stubbendieck 1986). The method can be applied only when plants are not growing or over short time spans. The other approach is to protect one quadrat in each pair with a cage and to clip both quadrats at the end of the grazing season. Again, the difference in average weight between the clipped and protected quadrats is considered the amount utilized. This approach has promise during the dormant season, but has limitations during the growing season because protected plants grow at a different rate than grazed plants (Pieper 1978).

Considerable work has been devoted to the development of relationships between the percentage of plants grazed and percent utilization. This method is simple to use in the field once the basic relationship is determined, since all that is required is to classify each plant as grazed or ungrazed. The method appears suitable for some species, but not for others at higher utilization levels (Pieper 1978; Cook and Stubbendieck 1986). For some species, 100% of the plants have been grazed when utilization by weight is only 35% to 40%. Consequently, utilization levels higher than 40% could not be detected by this method.

Height-weight methods have been used for some time to estimate utilization. These methods involve developing tables or graphs showing how weight is distributed by height for each plant species. Grazed and protected plant heights are measured to determine height reduction by grazing. Height reduction is converted to weight reduction using the corresponding table or graph for the species (Risser 1984; Cook and Stubbendieck 1986). There is considerable variation in height-weight relations for each species among sites, years, and so on. Consequently, it is recommended that separate data be developed and used for plants of different height classes. Work with sedges in California (McDougald and Platt 1976) and fireweed in Washington (Harshman and Forsman 1978) illustrates the utility of the method.

Utilization estimates have been used to determine if actual utilization matched "proper utilization" guidelines established for different species on each allotment or range site. Utilization estimates are used to make adjustments in stocking and to determine if grazing distribution meets expectations. However, care must be used in interpreting utilization data. Grazing is often spotty and some plants may be used excessively while others are not used (Frost et al. 1994). In some situations season of use may be more important than actual utilization (Frost et al. 1994; Sharp et al. 1994). Sharp et al. (1994) questioned the validity of using utilization guidelines for making management decisions. They recommended using photos taken at appropriate times of the year instead of utilization estimates for determining if management objectives are being met.

Another approach is to estimate residue or stubble left on the range at the end of the grazing season rather than the amount removed. Bement (1969) illustrated this approach for shortgrass range in Colorado, considering both livestock performance and vegetational composition. Stubble height guidelines are discussed in Chapter 8. The reader is referred to Jasmer and Holechek (1984), Cook and Stubbendieck (1986) and Tueller (1988) for more detailed discussion on how to measure utilization and other vegetational attributes.

Utilization surveys by both the Bureau of Land Management and Forest Service have involved qualitative techniques. As an example the Bureau of Land Management in southern New Mexico uses categories of unused (0–5%), lightly used (6–40%), moderately used (41–60%), heavily used (61–80%), or severely used (81–100%) based primarily on visual appearance. Both Forest Service and Bureau of Land Management personnel often map out utilization zones for allotments they survey. The primary concern regarding these surveys has been they are subjective and their reliability cannot be readily quantified with standard statistical procedures.

Grass stubble height measurements can provide a useful quantitive cross check on qualitative utilization surveys. While stubble heights do not necessarily reflect percentage use by weight of forage plants, they do reflect remaining vegetation residue to protect the soil, provide food and cover for wildlife, and provide forage for livestock. Grass stubble heights have been well related to severity of grazing (Johnson 1953; Valentine 1970). Minimum stubble height guidelines have been developed for various range grasses (see Chapter 8).

RANGE CONDITION

Some type of range condition classification is often included in a range inventory. Changes in range condition scores over time are usually the basis for monitoring management effectiveness. Range condition classification provides an indication of management inputs necessary. If ranges are in good or excellent condition, maintaining them in a stable condition may be the best management strategy. However, if they are in poor or fair condition, management that is aimed at "improvement" may be indicated.

Generally, four or five condition classes are recognized: excellent, good, fair, and poor. Sometimes a fifth category is added. Differences between condition classes are somewhat arbitrary since they really form a continuum from badly depleted ranges to those with maximum cover and productivity. Differences in range condition are often indicated by differences in species composition, but range condition is generally defined as departures from some conceived potential for a particular site. It is important to recognize differences in time on one site from differences from site to site at the same time.

A *range site* is defined as "a distinctive kind of rangeland, which in the absence of abnormal disturbance and physical site deterioration, has the potential to support a native plant community typified by an association of species different from that of other sites. This differentiation is based upon significant differences in kind or proportion of species, or total productivity" (Society for Range Management 1989). The range site concept was fostered and developed by the Soil Conservation Service as the basic management unit. They have published descriptions of the major range sites by states and range condition guidelines.

A similar idea is the habitat-type approach developed by Daubenmire (1968) and used by the U.S. Forest Service (e.g., Moir and Hentzel 1983). He defined a *habitat type* as a "term for all parts of the earth's surface which support or is capable of supporting the same kind of plant association, i.e., the same climax" (Daubenmire 1984). Although some authors insist that the two concepts are dissimilar (Anderson 1983; Dyksterhuis 1983), they have many similarities and can be used for planning and management.

Many approaches have been used to determine range condition on different range sites or habitat types. Perhaps the most familiar method is the one developed by Dyksterhuis (1949, 1958). This approach is ecological, in that range condition is measured in degrees of departure from climax. The approach assumes that climax can be determined for each range site. Excellent condition class would represent climax, and poor condition, the most removed from climax. The following ratings were used to determine condition:

RANGE CONDITION	PERCENT OF CLIMAX
Excellent	76–100
Good	50–75
Fair	26–50
Poor	0–25

Originally, species occurring on each site were classified, by their reaction to grazing, as decreasers, increasers, or invaders. Recently, the SCS has changed terminology slightly. Decreasers are highly palatable plants that decline in abundance with grazing pressure (Figure 7.3). Plants classified as increaser I types are moderately palatable and serve as secondary forage plants. They may increase slightly or remain stable under moderate grazing. As grazing pressure increases or as range condition reaches fair condition, these species also decline. Other plant species present in the climax vegetation but that are unpalatable may increase under grazing pressure or as site deterioration occurs. These species are classified as increaser II plants (Figure 7.4). Invaders are species that encroach onto the site from adjacent sites in later stages of deterioration. Type I invaders may eventually decrease if forced utilization occurs at later stages of deterioration; Type II invaders are generally unpalatable and increase through final stages of deterioration. Figure 7.5 shows an example of a range site in different condition classes as determined by the climax approach.

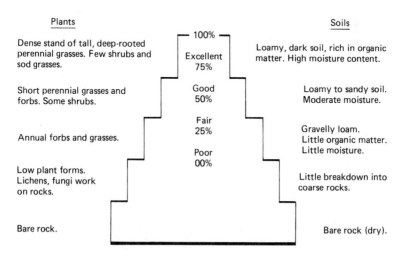

Figure **7.3** Example of vegetation and soil characteristics associated with the Soil Conservation Service approach to range condition for prairie ranges in the central Great Plains. (From Parker 1969.)

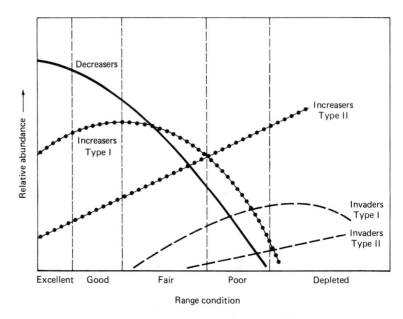

Figure 7.4 Changes in abundance of species groups for different range condition classes of the Soil Conservation Service.

Table 7.2 provides an example of how range ecological condition was calculated for a silty site in a 15–19 inch precipitation zone in western Montana using USDA Soil Conservation Service (now Natural Resources Conservation Service) 1971 guidelines. Under this approach decreasers were considered 100% allowable but none of the invaders were allowable toward the climax vegetation. It should be noted that USDA Natural Resources Conservation Service site guidelines differ substantially among states. In some states low amounts of certain invaders are now counted towards the condition score and there can be restrictions on how much of a decreaser can be used. Rather than using terms of excellent, good, fair, and poor to describe range condition with the Dyksterhuis approach, many range managers now use categories of climax, late seral, mid-seral, and early seral. This distinguishes ecological condition from how well the existing vegetation may be suited for the intended uses.

Although the Dyksterhuis method has been widely applied, it has numerous weaknesses in practice (Smith 1979). These are listed as follows:

1. Excellent, good, fair, and poor may not have relevance with respect to management objectives. In some cases livestock grazing or wildlife objectives may be maximized at fair or good condition classes. Pieper (1982) suggested using the ecological terms *climax, high seral, mid-seral,* and *low seral* to replace the terms *excellent, good, fair,* and *poor.* Ecological condition represented by these stages in secondary succession can then be related to specific uses or outputs, such as forage for livestock, habitat for game, infiltration or erosion, and so on, to establish management goals.

(a)

(b)

***Figure* 7.5** Range condition classes on a sandy upland range site in southern New Mexico: (a) excellent; (b) good. (*Continued*)

(c)

(d)

***Figure* 7.5** *(Continued)* Range condition classes on a sandy upland range site in southern New Mexico: (c) fair; (d) poor.

TABLE 7.2   Calculation of Range Ecological Condition for a Silty Site in Western Montana Using the Dyksterhuis Approach

LIST OF PLANT SPECIES	WEIGHT (G)	ESTIMATED PERCENT COMPOSITION	MAXIMUM ALLOWABLE PERCENT COMPOSITION[a]	PERCENT USED
Rough fescue (decreaser)	10	13	100	13
Idaho fescue (increaser)	33	44	25	25
Sandberg bluegrass (increaser)	12	16	5	5
Big sagebrush (increaser)	7	9	5	5
Cheatgrass (invader)	5	7	0	0
Goatweed (invader)	8	11	0	0
TOTAL	75	100		48[b]

[a]Based on USDA-Soil Conservation Service 1971 guidelines.
[b]48% remaining climax vegetation = fair or mid-seral ecological condition.

2. Determining climax is very difficult. Most rangelands in the United States have been disturbed to some degree, and finding climax conditions representatives for all range sites is difficult.

3. The method does not allow for a realistic evaluation of rangelands invaded by exotic species such as the California annual grasslands or crested wheatgrass seedings in the intermountain region. In some cases grazing values of seedings are improved over climax conditions.

4. The method is not well suited for forest or wooded rangeland, where some clearing might actually improve grazing conditions.

In addition, there is some concern about the basic assumptions concerning the predictability of the secondary succession-to-climax model for western rangelands. The cone model of Huschle and Hironaka (1980) indicates that there may be considerable variation in vegetation of seral stages. Nevertheless, the climax approach does provide a basis for range condition classification which is difficult to obtain with other approaches.

The National Research Council (NRC) (1994) recommends that the term "health" be used to describe the status of rangeland. This report considers ranges in one of three categories: healthy, at risk, and unhealthy (Figure 7.6). Rangeland health is defined as the sustainability of basic soil and ecological processes. Ranges classified as healthy require no change in management, but those classed as at risk may require a change in management to restore them to healthy condition. The early warning line in Figure 7.6 showing the change from healthy to at risk represents changes that are reversible while the change from the at risk category to unhealthy cannot be reversed without expenditure of energy (e.g., brush control). These concepts are compatible with the "state and transition model"

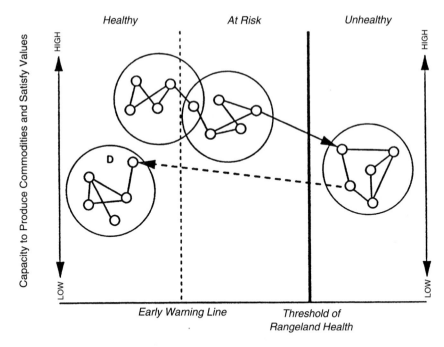

Figure **7.6** Diagram showing pathways from rangelands considered healthy, to at risk, to unhealthy. Several states can exist on similar range sites within each category. (Reprinted with permission from *Rangeland Health: New Methods to Classify Inventory and Monitor Rangelands.* (Copyright 1994 by the National Academy of Sciences. Courtesy of the National Academy Press, Washington, D.C.)

discussed in Chapter 6. A committee of the Society for Range Management (Task Group on Unit in Concepts in Terminology [UCT]) developed terminology and methods of range condition classification with some similarities to those of the NRC. Instead of the term healthy, UTC (1995a, 1995b) used the terms sustainable, and unsustainable. The main criterion for range condition classification is soil protection from accelerated erosion. The point at which accelerated erosion appreciably accelerates from management influences is the Site Conservation Threshold (SCT). The SC denotes those rangelands that could not be returned to the sustainable category without intrusive management (West et al. 1994; UTC 1995b). Rangelands listed as unhealthy or unsustainable represent conditions of accelerated soil erosion and degraded plant communities.

Basing range condition standards on ecological criteria does not imply suitability for a particular use. Several workers have recommended resource value ratings applied to a particular ecological setting as a valuable tool for management decisions (Hann 1986; Kindschy 1986; Pieper and Beck 1990). Thus a mixed grassland-forb community might be more suitable as pronghorn antelope habitat than a grassland with few forbs (Howard et al. 1990).

Techniques for rangeland inventory, monitoring and condition classification are developing rapidly. Various approaches were compared by West et al. (1994). Table 7.3 shows the main characteristics of these different monitoring systems developed by different agencies. Apparently there is not much coordination among these agencies since the systems were developed with different goals and objectives. Different statistical methods for range condition analysis have been developed and tested (Wilson and Tupper 1982; Hacker 1983; Foran et al. 1986; Pamo et al. 1991; Uresk 1990). However, it is likely that some of the basic ecological concepts underlying the idea of range condition will be debated for some time to come.

DETERMINING TREND

Trend has been defined as the "direction of change in range condition" (Society for Range Management 1989). Generally, trend is considered upward (or improving) or downward (declining) or stable. Trend ratings were initially used to indicate conditions for livestock grazing as indicated by increasing productivity, cover, and succession toward climax conditions. Generally, these changes also improved conditions for other uses, such as watershed values, but in some cases, a downward trend for some uses might be an upward trend for others. In the southwestern United States, pronghorn antelope prefer forbs in their diet over grasses. Disturbed sites at some stage below climax are usually best suited to produce pronghorn food. Forested areas in Idaho and western Montana provided excellent elk habitat in subclimax conditions following extensive wild fires in 1910. Now these subclimax conditions must be maintained artificially to maintain the elk populations.

Thus, to say whether a trend is upward or downward, one must specify the use or criteria used. If the trend is used to correspond to successional stages, upward trend would be toward climax and downward trend would be away from climax.

Exclosures are an important tool when trend is measured. They are necessary to separate climatic influences from those caused by grazing. If range improvement occurs during periods of average or near-average precipitation, it is probably due to grazing management. However, a downward trend in drought years or an upward trend in above-average years may be more the result of climatic conditions than of grazing management. Ungrazed areas (exclosures) can be valuable in separating these influences.

An important concern with trend analysis has been that intervals between sampling are often 20 to 30 years. Usually only 1 year of information has been collected. To accurately identify trend, consecutive years of information are needed to smooth out variation from annual climatic fluctuations. Collection of range condition data at intervals of no more than 5 years is necessary to effectively monitor trend, in our opinion.

Various qualitative and quantitive approaches can be used to determine if a significant trend in range condition has occurred over some specific time period. From a qualitative standpoint, many range scientists and managers might agree a definite trend would involve a 5% or more change in average condition score on a pasture or allotment with half or more of the key areas showing this change. However, standard statistical

TABLE 7.3 Characteristics of Various Monitoring Systems in Use in the United States

	RANGE CONDITION (SCS)	NATIONAL RESOURCE INVENTORY (NRI)	SVIM (BLM)	LCTA (DOD)	PARKER 3-STEP (USFS)	EMAP (EPA)	UCT (SRM)	RANGELAND HEALTH (NAS)
Purpose	conserve soil while grazing livestock	set priorities for action and budgeting	to allocate forage for all animals on rangelands	conserve natural resources while continuing military training	to sustain land for multiple uses	develop a national report card of environmental "health"	to inventory land for sustainability	to assess rangeland ecosystems
Scope	individual ranches, data collected as funding is available	entire nation and states, about every 5–10 years since 1967	one point in time survey on allotments	military installations nationwide	livestock grazing allotments	national survey taken systematically with each point visited once each 4–5 years	individual ranches and national survey collected periodically	national survey with no time frame defined
Implementation	bottom-up	top-down	top-down at first, but bottom-up modifications have emerged over time	top-down	bottom-up	top-down	top-down	top-down
Conceptual underpinnings	succession to climax	succession to climax for vegetation	succession to climax	??	succession to desired plant community and protection of soil	??	soil stability	soil stability

Indicators	plants, ranked by response to livestock grazing, range condition estimated	soil surface characters and plant community composition	range condition and ecological status	modeling of soil erosion and vegetation change	plants ranked by response to grazing, soil surface condition	many being tested	? plant, soil, and others being tested	? plant, soil, nutrient cycles and others
Trend	apparent trend based on comparison of condition collected over time	determined from successive 10-year data bases of condition	apparent trend based on plant species and surface soil factors	real trend based on baseline data	? apparent trend	real trend based on baseline data	? real trend based on baseline data	? real trend based on baseline data
Benchmarks	relict areas and intuition	comparison with earlier surveys	begins with first inventory	initial baseline	? relict areas and intuition	initial baseline	? site potential	??
Ecological integrity	condition: excellent, good, fair, poor	condition: excellent, good, fair, poor	condition: excellent, good, fair, poor	??	high seral, mid seral, low seral	nominal or sub-nominal	sustainable, marginal, nonsustainable	healthy, at risk, unhealthy

Source: West et al. 1994.

TABLE 7.4   Average Botanical Composition on Heavily Grazed Chihuahuan Desert
Rangeland in New Mexico in Fall 1965 and Fall 1995

	COVER		RELATIVE COVER	
SPECIES	FALL 1965	FALL 1995	FALL 1965	FALL 1995
Grasses				
Black grama	20[a]	10[b]	53[a]	26[b]
Dropseed	t[a]	8[b]	t[a]	21[b]
Threeawn	2	4	5[a]	10[b]
Other grasses	14[a]	1[b]	37[a]	2[b]
Total grasses	36[a]	23[b]	95[a]	59[b]
Forbs				
Croton	t	t	t	t
Globemallow	t	t	t	t
Other forbs	2	t	5	2
Total forbs	2	1	5	2
Shrubs				
Broom snakeweed	t[a]	6[b]	t[a]	15[b]
Honey mesquite	t[a]	7[b]	t[a]	18[b]
Yucca	t	2	t	5
Other shrubs	t	t	t	t
Total shrubs	1[a]	15[b]	1[a]	38[b]

Source: Holechek and Galt 1997.
NOTE: Overall plant community composition in fall 1965 is different from fall 1995 (p < 0.05)
using multivariate analysis variance (MANOVA).
[a,b]Row means with different letters differ (p < 0.05).
t = trace.

procedures such as t-tests and analysis of variance (ANOVA) provide the most reliable
assessment of vegetation composition changes in a pasture or allotment. Multivariate
analysis of variance (MANOVA) is a practical tool for determining if overall plant
species composition has changed over some time period. Various statistical procedures
for detecting change in plant composition are discussed in detail by Holechek et al.
(1984) and Bonham (1989).

   A practical example of how to present data on plant composition changes through time
is given in Table 7.4. In this case the ecological condition score of the pasture changed from
69% to 57% remaining climax vegetation using the Dyksterhuis (1949) method.
Multivariate analysis of variance showed both the ecological condition score and black

grama, the key forage species, had significantly ($p < 0.05$) declined between the two periods. Therefore it is concluded a definite downward trend occurred between 1965 and 1995. A combination of reduced stocking and mesquite control would be appropriate management practices for this pasture, since the goal is to increase forage for livestock.

Interpretation of Grazing Surveys

In our opinion, a combination of grazing capacity, utilization, condition, and trend data are needed for sound range management decisions. Grazing capacity is dynamic and can show great fluctuations with climatic trends. Although grazing capacity provides information on how many animals can graze an area at a particular time, it tells little about recent management effectiveness. Utilization surveys reflect grazing severity on forage plants for a particular year. They also provide an indication of soil cover, food available to livestock and wildlife on an area, and esthetic appearance. On the other hand, one year of overuse does not necessarily mean management is unsound or that degradation is occurring. Experienced ranchers and range personnel know well that having forage resources and level of use always synchronized is nearly impossible due to the vagaries of climate. However, a repetition of overuse for a 2- to 3-year period would indicate a need for reduction in animal numbers and/or improved livestock distribution practices, particularly if vegetation trend was in the opposite direction of management goals.

Range ecological condition information, if properly interpreted, can provide important insight into how well the existing vegetation is suited to intended uses. Generally late seral condition range (51–75% of climax), will provide adequate vegetation cover for soil stability, maximize plant and wildlife diversity, and give moderate returns from livestock production. If the goals are to maximize soil stability and returns from livestock production, then excellent condition range (76–100% of climax) would often be best on arid and semi arid rangelands.

When range ecological condition drops into a mid-seral category, usually the opportunity for meaningful recovery still exists because residual, productive decreaser grasses are still present. However, at this stage continued overuse can completely eliminate residual climax grasses. Once ecological condition drops into the early seral category, natural recovery through grazing management alone becomes less likely. Costly vegetation manipulation procedures such as herbicidal and mechanical control in conjunction with seeding will usually be needed. The point we wish to make is that stocking rate and other grazing management adjustments are more justified on rangelands at threshold levels (30–40% of climax) than for rangelands in an acceptable condition (over 50% of climax) or where degradation is so far advanced that natural recovery is unlikely. This is not to say we condone the practice of heavy grazing on rangelands in poor condition.

Many managers would agree a definite trend away from the desired ecological condition would necessitate a major management change such as a reduction in livestock numbers. However, major downward shifts in vegetation due to drought and/or overgrazing usually require a minimum of 3 to 5 years. Most managers would prefer to avoid a definite downward trend by attempting to keep stocking rate and utilization patterns at a sustainable level on an annual basis.

The point is that when reliable information is available on grazing capacity, utilization, ecological condition, and trend, managers can make more informed decisions than if they have only part of these elements. This is particularly true for situations where only 1 year of condition and trend information is collected at widely spaced intervals (10 to 30 years). The case for a grazing permit reduction on public lands is much more compelling if utilization is above the desired level, the allotment shows a significant trend away from the management goal, and ecological condition is below what is desired than if based merely on a 1-year grazing capacity survey that indicates a forage deficit. By the same token, grazing capacity determinations may indicate adequate or surplus forage for present animal numbers. However, distribution problems may be causing a downward trend on important key areas and range ecological condition may be at a threshold level necessitating management changes. Here again all four elements in combination permit much better decision making than reliance on part of these elements.

Generally a reduction in livestock numbers would be most justified where 50% or more of the key areas showed a definite trend away from the desired vegetation composition. In cases where less than half of the key areas showed a downward trend, management practices such as rotational grazing or rotation of access to water might be more justified than a stocking rate reduction.

Presentation of Information

Consolidation of biological and climatic data into easily understood tables is an important part of the monitoring process. Table 7.5 shows an approach developed by Holechek and Hess (1995) for displaying grazing survey information for management decisions. The average condition score of the last 3 years is used to separate the short-term from the long-term trend. Basically data in Table 7.5 show a rise and then a decline in grazing capacity and range condition due to changing precipitation conditions. The goal on this allotment is to increase the remaining climax vegetation. There is no definite trend, based on comparing long-term and short-term ecological condition scores. The utilization scores are based on ocular observations. We suggest grazing pressure be further evaluated through stubble height measurement or some other quantitative technique. Because the ecological condition is near the threshold level (30–40% of climax), failure to partially destock could cause lasting harm if drought should persist. In 1995 the permittee wisely reduced cattle numbers in accordance with lower rainfall.

Grazing Survey Costs

Vegetation monitoring is now mandated on most federal rangelands. Increased scrutiny by ranchers and environmental groups and budget reductions have caused the Bureau of Land Management and Forest Service to look for more reliable and cost effective ways to do various grazing surveys. This is leading to more standardization in procedures and to sub-contracting certain components of these surveys. The use of a team consisting of government agency personnel and certified range consultants may be an efficient low-cost way to do grazing surveys. Table 7.6 provides an estimation of cost on a per acre ba-

TABLE 7.5 Livestock Use, Range Ecological Condition, Forage Production, and Precipitation Data for the Afton Allotment (Bureau of Land Management) in Southcentral New Mexico in the 1986–1995 Period

	1986	1987	1988	1989	1990	1991	1992	1993	1994	1995	LAST 10-YEAR AVERAGE	LAST 3-YEAR AVERAGE
Actual cattle animal unit years (AUY)	150	150	150	168	179	185	190	202	202	170	174	191
Stocking rate (acre/AUY)	173	173	173	156	153	141	138	128	128	153	152	136
Precipitation Afton Allotment (in./year)[1]	11.9	NA	NA	13.8	8.3	12.4	8.5	7.1	6.5	8.0	9.6	7.2
Forage utilization (%)[2]	20	38	31	29	26	30	23	32	23	NA	27	26
Fall vegetation standing crop (lb/acre)	551	349	566	446	495	650	568	306	159	276	437	247
Fall perennial forage standing crop (lb/acre)	292	160	337	325	338	486	481	288	130	224	306	214
Fall black grama standing crop (lb/acre)[3]	88	57	38	91	72	107	134	113	19	16	74	49
Ecological range condition score	42	43	41	48	52	54	53	48	43	35	46	42

Source: Holechek and Hess 1995.

[1]Long-term average precipitation is 8.47 inches.

[2]Utilization data were collected in late spring and reflect use of forage produced in the previous year. Utilization data are ocular (nonquantitative) and therefore subjective.

[3]Black grama is a perennial grass that is considered to be a key forage for cattle and a dominant component of the climax vegetation. NA Not available.

TABLE 7.6 Cost Estimates for Different Types of Grazing Surveys Using a Team of Agency Range Conservationists and Range Consultants

GRAZING ASSESSMENT TYPE[1]	COST ($/ACRE)
Grazing capacity (rugged terrain)	0.25
Grazing capacity (flat terrain)	0.17
Ecological condition	0.07
Ecological condition and trend	0.15
Utilization (stubble height)	0.05
Utilization (ocular estimate)	0.03
Grazing capacity, condition, trend, and utilization	0.50–0.80

Source: Holechek and Galt 1997.
[1]Based on a 25,000 acre land unit with six key sites.

sis for different types of grazing surveys on forest service lands in southwestern New Mexico. Ideally, grazing surveys should be averaged across a 3 year period to reduce the effects of climatic variation and other possible aberrations (grasshoppers, abnormal wildlife use, trespassing livestock, etc.).

RANGE MANAGEMENT PRINCIPLES

■ Historically the amount of climax vegetation remaining on a range site has been used as an important measure of range condition. Although natural grassland ranges at or near the climax generally produce maximum forage for livestock use, they often have lower plant and wildlife diversity than those in a lower successional stage. Soil stability and suitability for particular uses as well as ecological position should be used when range condition ratings are assigned by managers.

■ In most situations, the trend in vegetation change will be the most important element in evaluating range management effectiveness. Exclosures can be important in separating changes caused by climatic fluctuations from those brought about by management.

■ A combination of grazing capacity, utilization, ecological condition, and trend information is needed for sound range management decisions.

Literature Cited

Anderson, E. W. 1983. Ecological site/range site/habitat type-a viewpoint. *Rangelands* 5:187–188.
Avery, T. E. 1986. *Interpretation of aerial photographs,* 2d ed. Burgess Publishing Company, Minneapolis, MN.
Bement, R. E. 1969. A stocking rate guide for beef production on blue grama range. *J. Range Manage.* 22:83–86.
Bonham, C. D. 1989. *Measurements for terre trial vegetation.* John Wiley and Sons. New York.

Braun-Blanquet, J., L. Emberger, and R. Molinier. 1947. *Instructions pour l'établissement de la carte de groupéments végétaux.* Centre National de la Recherche Scientifique, Paris.

Brown, D. 1954. *Methods of measuring vegetation.* Commonwealth Agricultural Bureau, Farnham Royal, Bucks, England.

Canfield, R. H. 1941. Application of the line-interception method in sampling range vegetation. *J. Forest.* 39:388-394.

Catana, A. J., Jr. 1963. The wandering quarter method of estimating population destiny. *Ecology* 44:349–360.

Colwell, R. N. 1968. Remote sensing of natural resources. *Sci. Am.* 218:54–69.

Committee on Rangeland Classification. 1994. *Rangeland health.* National Academy Press, Washington, DC.

Cook, C. W., and J. Stubbendieck (Eds.). 1986. *Range research: Basic problems and techniques.* Society for Range Management, Denver, CO.

Cooper, C. F. 1959. Cover vs. density. *J. Range Manage.* 12:215.

Cottam, G., and J. T. Curtis. 1956. The use of distance measures in phytosociological sampling. *Ecology* 37:451–460.

Daubenmire, R. F. 1958. A canopy-coverage method of vegetational analysis. *Northwest Sci.* 53:43–64.

Daubenmire, R. F. 1968. *Plant communities: A textbook of plant synecology.* Harper & Row Publishers, Inc., New York.

Daubenmire, R. F. 1984. Viewpoint: Ecological site/range site/habitat type. *Rangelands* 6:263–264.

Dix, R. L. 1961. An application of the point-centered quarter method to the sampling of grassland vegetation. *J. Range Manage.* 14:63–69.

Dyksterhuis, E. J. 1949. Condition and management of rangeland based on quantitative ecology. *J. Range Manage.* 2:104–115.

Dyksterhuis, E. J. 1958. Ecological principles in range evaluation. *Bot. Rev.* 24:253–272.

Dyksterhuis, E. J. 1983. Habitat-type: A review. *Rangelands* 5:270–271.

Evans, R. A., and R. M. Love. 1957. The step-point method of sampling—A practical tool in range research. *J. Range Manage.* 10:208–212.

Foran, B. D. 1986. Range assessment and monitoring in arid lands: The use of classification and ordination in range survey. *J. Environ. Manage.* 22:67–84.

Frost, W. E., E. L. Smith, and P. R. Ogden. 1994. Utilization standards. *Rangelands* 16:256–259.

Gay, C. 1965. Range management: How and why. *New Mexico State Univ. Coop. Ext. Circ.* 376, Las Cruces, NM.

Goebel, C. J. 1955. The weight-estimate method at work in southeastern Oregon. *J. Range Manage.* 8:212–213.

Hacker, R. B. 1983. Use of reciprocal averaging ordination for the study of range condition gradients. *J. Range Manage.* 36:25–30.

Hann, W. J. 1986. Evaluation of resource values in the northern region of the Forest Service. *Rangelands* 8:159–161.

Harshman, E. P., and R. Forsman. 1978. Measuring fireweed utilization. *J. Range Manage.* 31:383–396.

Holechek, J. L. 1988. An approach for setting the stocking rate. *Rangelands* 10:10–14.

Holechek, J. L. and D. Galt. 1997. Grazing assessments on public rangelands: Problems and opportunities. *A report to the Gila Natural Forest.* Mimeo. New Mexico State University, Las Cruces.

Holechek, J. L. and K. Hess, Jr. 1995. Government policy influences on rangeland conditions in the United States: A case example. *Environmental Monitoring and Assessment* 37:179–187.

Holechek, J. L. and R. D. Pieper. 1992. Estimation of stocking rate on New Mexico rangeland. *J. Soil and Water Conserv.* 47:116–119.

Holechek, J. L. and T. Stephenson. 1983. Comparison of big sage brush in north central New Mexico under moderately grazed and grazing excluded conditions. *J. Range Manage.* 36:455-456.

Holechek, J. L., M. Vavra, and R. D. Pieper. 1984. Methods for determining the botanical composition, similarity and overlap of range herbivore diets. In G. M. Van Dyne, J. Brotnov, B. Burch, S. Fairfax, and B. Huey eds. *Developing Strategies for Rangeland Management.* Westview Press, Boulder, CO.

Howard, V. W. Jr., J. L. Holechek, R. D. Pieper, K. Green-Hammond, M. Cardenas, and S. L. Beasom. 1990. Habitat requirements for pronghorn on rangeland impacted by livestock and net wire in eastcentral New Mexico. *New Mex. State Univ. Agr. Exp. Stn. Bull.* 750, Las Cruces, NM.

Huschle, G., and M. Hironaka. 1980. Classification and ordination of plant communities. *J. Range Manage.* 33:179–182.

Hyder, D. N., C. E. Conrad, P. T. Tuller, L. D. Calvin, C. E. Poulton, and F. A. Sneva. 1963. Frequency sampling in sagebrush-bunchgrass vegetation. *Ecol.* 44:740–746.

Jasmer, G. E., and J. L. Holechek. 1984. Determining grazing intensity on rangeland. *J. Soil Water Conserv.* 39:32–35.

Jensen, H. A. 1947. A system for classifying vegetation in California. *Calif. Fish Game* 34:199–266.

Johnson, W. M. 1953. Effect of grazing intensity upon vegetation and cattle gains on ponderosa pine-bunchgrass ranges of the front range of Colorado. *U.S. Dept. Agric. Circ.* 929.

Kindschy, R. R. 1986. Rangeland vegetative succession—Implications to wildlife. *Rangelands* 8:157–159.

Kuchler, A. W. 1955. A comprehensive method of mapping vegetation. *Ann. Assoc. Am. Geogr.* 155:404–415.

Kuchler, A. W. 1967. *Vegetation mapping.* The Ronald Press Company, New York.

't Mannetje, L. 1978. *Measurement of grassland vegetation and animal production.* Commonwealth Agricultural Bureau, Farnham Royal, Bucks, England.

Maxwell, E. L. 1976. A remote rangeland analysis system. *J. Range Manage.* 29:66–73.

McDougald, N. E., and R. C. Platt. 1976. A method of determining utilization for wet meadows on the summit allotment, Sequoia National Forest, California. *J. Range Manage.* 29:497–501.

McGraw, J. F., and P. T. Tueller. 1983. Landsat computer-aided analysis techniques for range vegetation mapping. *J. Range Manage.* 36:627–631.

Moir, W. H., and L. Hentzel (Tech. Coords.). 1983. *Proc. workshop on southwestern habitat types.* U.S. Department of Agriculture Forest Service, Albuquerque, NM.

National Research Council. 1994. *Rangeland Health: New methods to classify inventory and monitor rangelands.* Natural Academy Press, Washington, D.C.

Owensby, C. E. 1973. Modified step-point system for botanical composition and basal cover estimates. *J. Range Manage.* 26:302–303.

Pamo, E. T., R. D. Pieper, and R. F. Beck. 1991. Range condition analysis: Comparison of 2 methods in southern New Mexico. *J. Range Manage.* 44:374–378.

Parker, K. G. 1969. The nature and use of Utah range. *Utah State Univ. Ext. Circ.* 359, Logan, UT.

Pechanec, J. F., and G. D. Pickford. 1937. A weight-estimate method for determination of range or pasture production. *J. Am. Soc. Agron.* 29:894–904.

Pieper, R. D. 1978. *Measurement techniques for herbaceous and shrubby vegetation.* New Mexico State University, Las Cruces, NM.

Pieper, R. D., and R. F. Beck. 1990. Range condition from an ecological perspective: Modification to recognize multiple use objectives. *J. Range Manage.* 43:550–552.

Poulton, C. E. 1970. Practical applications of remote sensing in range resources development and management. In *Range and wildlife habitat evaluation—A research symposium.* U.S. Dep. Agric. Misc. Pub. 1147.

Reppert, J. N., and R. S. Driscoll. 1970. 70-mm. Aerial photography—A remote sensing tool for wild land research and management. In *Range and wildlife habitat evaluation—A research symposium.* U.S. Dep. Agric. Misc. Publ. 1147.

Risser, P. G. 1984. Methods for inventory and monitoring of vegetation, litter, and soil surface condition. In National Research Council/National Academy of Sciences (Eds.). *Developing strategies for rangeland management.* Westview Press, Inc., Boulder, CO.

Sharp, L., K. Sanders, and N. Rimbey. 1994. Management decisions based on utilization—Is it really management? *Rangelands* 16:38–40.

Shoop, M. C., and E. H. McIlvain. 1963. The micro-unit forage inventory method. *J. Range Manage.* 16:172–179.

Smith, E. L. 1979. Evaluation of the range condition concept. *Rangelands* 1:52–54.

Society for Range Management. 1989. *A glossary of terms used in range management.* 3d ed. Society for Range Management, Denver, CO.

Society for Range Management. 1989. *Report of unity in concepts and terminology committees.* Society for Range Management, Denver, CO.

Stewart, G., and S. S. Hutchings. 1936. The point-observation-plot (square-foot-density) method of vegetation survey. *J. Am. Soc. Agron.* 28:714–722.

Tueller, P. T. (Ed.). 1988. *Vegetation science applications for rangeland analysis and management.* Kluver Academic Publ., Boston.

Unity in Concepts and Terminology Committee. 1995a. New concepts for assessment of rangeland condition. *J. Range Manage.* 48:271–282.

Unity in Concepts and Terminology Committee. 1995b. Evaluating rangeland sustainability: The evolving technology. *Rangelands* 17:85–92.

Uresk, D. W. 1990. Using multivariate techniques to quantitatively estimate ecological stages in a mixed grass prairie. *J. Range Manage.* 43:282–285.

Valentine, K. A. 1970. Influence of grazing intensity on improvement of deteriorated black grama range. *N. Mex. Agric. Exp. Stn. Bull.* 553.

West, N. E., K. C. McDaniel, E. L. Smith, P. T. Tueller, and S. Leonard. 1994. *Monitoring and interpreting ecological integrity on arid and semi-arid lands of the western United States.* New Mex. Range Improve. Task Force. New Mex. State. Univ., Las Cruces, NM.

Westoby, M., B. Walker, and I. Noy-Meir. 1989. Opportunistic management for rangelands not at equilibrium. *J. Range Manage.* 42:166–174.

Wilson, A. D., and G. J. Tupper. 1982. Concepts and factors applicable to the measurement of range condition. *J. Range Manage.* 35:684–689.

Young, S. A. 1980. *Phenological development and impact of season and intensity of defoliation on Sporobolus flexuosus (Thurb.) and Bouteloua eriopoda (Torr.).* Ph.D. dissertation. New Mexico State University, Las Cruces, NM.

CHAPTER 8

CONSIDERATIONS CONCERNING STOCKING RATE

IMPORTANCE OF CORRECT STOCKING RATE

Selection of the correct stocking rate is the most important of all grazing management decisions from the standpoint of vegetation, livestock, wildlife, and economic return. Although this has been the most basic problem confronting ranchers and range managers since the initiation of scientific range management early in the twentieth century, specific approaches to this problem are still generally unavailable. It is agreed that there is no substitute for experience in stocking-rate decisions on specific ranges. However, experience is still lacking for some ranges in the United States and many ranges in other parts of the world.

Stocking rate is defined by the Society for Range Management (1989) as the "amount of land allocated to each animal unit for the grazable period of the year." In the southwestern United States, stocking rate is typically expressed as animal units per section of land. In the intermountain region of the United States, stocking is usually expressed as animal unit months per acre of land. The Society for Range Management (1989) defines an animal unit as one mature [1,000 lb (455 kg)] cow either dry or with a calf up to 6 months of age. Based on the most recent research, this animal would be expected to consume 20 lb (9.1 kg) of forage per day, 600 lb (273 kg) per month, and 7,300 lb (3,318 kg) per year. An animal unit month is the amount of feed or forage [600 lb (273 kg)] required by one animal unit for one month.

Carrying or *grazing capacity* are terms commonly used when discussing stocking rate. These terms refer to the maximum stocking rate possible year after year without causing damage to vegetation or related resources. Although actual stocking rates may vary considerably between years due to fluctuating forage conditions, grazing capacity is generally considered to be the average number of animals that a particular range will sustain over time. In most cases, ranches are bought and sold on the basis of their grazing capacity.

INFLUENCE OF STOCKING RATE ON FORAGE PRODUCTION

Stocking rate has more influence on vegetation productivity than any other grazing factor. When all North American studies were averaged, annual herbage production increased by 13% when specialized grazing systems were implemented at a moderate stocking rate (Table 8.1). The average increase was much larger (35%) when continuous livestock use was reduced from heavy to moderate (Table 8.2). An average increase in forage production of 28% resulted from switching from moderate to light. On some ranges forage production was actually less under light grazing than under moderate grazing. Herbage production on most ranges can be substantially increased by switching from heavy to moderate or light grazing intensities. This is particularly true for grassland ranges. After careful consideration of grazing studies in the central Great Plains, Klipple and Bement (1961) concluded that most of the improvement in forage production from light grazing occurs during the first 5 to 7 years following application and that there is little added improvement after 7 years. They found light grazing to be an economically effective means of improving shortgrass prairie ranges, with a poor cover of desirable forages but little competition from undesirable shrub species.

In the more arid shrubland ranges of the Southwest and intermountain regions, light grazing can be a useful means of improving forage production during the early stages of range deterioration when the desirable forages are still present but in low vigor. Valentine (1970) found that a 32% utilization level gave a high rate of recovery for deteriorated black grama (*Bouteloua eriopoda*) ranges in southcentral New Mexico. On the Santa Rita Range in Arizona, Martin and Cable (1974) found that an average utilization level of 40% maintained the perennial grasses over a 10-year period. Areas that received lighter use than average tended to have the highest perennial grass production.

Light grazing has the lowest potential for recovery of highly deteriorated, brush-infested ranges (Figure 8.1). Holechek and Stephenson (1983) found that 20 years of complete rest had almost no influence on recovery of desirable forages compared to moderate grazing (30% to 40% use) on ranges heavily infested with big sagebrush (*Artemisia tridentata*) in northcentral New Mexico. They considered control of big sagebrush the only feasible means to improve forage production. Hughes (1980) found that 25 years of protection from grazing resulted in a 30% to 40% density increase in big sagebrush, with a similar decrease in grasses in northwestern Arizona. In southeastern Idaho, Anderson and Holte (1980) reported

TABLE 8.1 Herbage Production (kg/ha) under Grazing Systems and Continuous Use for Different Range Types in North America

RANGE TYPE	STATE	GRAZING SYSTEM	CONTINUOUS USE	PERCENT DIFFERENCE	REFERENCE
Tallgrass	Kansas	1,976	1,627	+ 21	Herbel and Anderson 1959
Tallgrass	Kansas	4,627	4,022	+ 15	Owensby et al. 1973
Tallgrass	Kansas	3,090	2,196	+ 41	Smith and Owensby 1978
Southern mixed prairie	Texas	1,428	1,036	+ 38	Kothmann et al. 1975
Southern mixed prairie	Texas	2,700	2,500	+ 8	Heitschmidt et al. 1990
Northern mixed prairie	Alberta, Canada	548	578	– 5	Smoliak 1960
Northern mixed prairie	Saskatchewan, Canada	1,512	1,302	+ 16	Campbell 1961
Coastal prairie	Texas	6,372	5,695	+ 12	Drawe 1988
Shortgrass	New Mexico	1,214	1,199	+ 1	White et al. 1991
Shortgrass	New Mexico	910	821	+ 11	Pieper et al. 1991
Palouse prairie	Oregon	398	397	0	Skovlin et al. 1976
Oak savannah	Texas	1,319	534	+ 147	Reardon and Merrill 1976
Coniferous forest	Colorado	1,240	1,242	0	Currie 1976
Coniferous forest	Oregon	132	128	+ 3	Skovlin 1976
Coniferous forest	Utah	997	836	+ 19	Laycock and Conrad 1981
Desert grassland	Arizona	246	616	– 60	Martin 1970
Desert grassland	Arizona	100	72	+ 39	Martin 1973
Desert grassland	Arizona	412	337	+ 22	Martin and Ward 1976
Desert grassland	Arizona	405	439	– 8	Martin and Severson 1988
Desert grassland	Arizona	161	168	– 4	Beck and McNeely 1993
Average across all studies		—	—	+ 16	

Source: Updated from Van Poollen and Lacey 1979.

TABLE 8.2 Long-Term Influence of Grazing Intensity on Forage Production

ANIMAL	LOCATION	GRAZING INTENSITY (% USE)	FORAGE PRODUCTION (KG/HA/YEAR)	REFERENCE
Cattle	Starkey, Oregon (forest)	Heavy (34%)	105	Skovlin et al. 1976
		Moderate (25%)	124	
		Light (18%)	158	
	Manitou, Colorado (forest)	Heavy (58%)	1,411	Johnson 1953
		Moderate (33%)	1,758	
		Light (16%)	2,289	
	Southern Pine Range, Louisiana (forest)	Heavy (57%)	2,350	Pearson and Whitaker 1974
		Moderate (49%)	2,057	
		Light (35%)	1,927	
	Flint Hills, Kansas (tallgrass)	Heavy (60%)	1,475	Herbel and Anderson 1959
		Moderate (50%)	1,956	
		Light (35%)	2,326	
	Central Great Plains, Experimental Range, Colorado (shortgrass)	Heavy (54%)	536	Klipple and Costello 1960
		Moderate (37%)	689	
		Light (21%)	735	
	Scotts Bluff Experimental Range, Nebraska (shortgrass)	Heavy (74%)	1,377	Burzlaff and Harris 1969
		Moderate (58%)	1,418	
		Light (53%)	1,558	
	Fort Stanton, New Mexico (shortgrass)	Heavy (63%)	674	Pieper et al. 1991
		Moderate (43%)	821	
	Eastern Colorado Range Station, Colorado (shortgrass)	Heavy (64%)	1,299	Sims et al. 1976
		Moderate (44%)	1,472	
		Light (30%)	1,713	

(Continued)

TABLE 8.2 Long-Term Influence of Grazing Intensity on Forage Production (*Continued*)

	Location	Grazing intensity	Forage	Reference
	Rolling Plains, Texas (southern mixed prairie)[a]	Heavy (50%)	2,592	Heitschmidt et al. 1985
		Moderate (40%)	2,852	
		No use	3,418	
	Lethbridge, Alberta (northern mixed prairie)	Very Heavy (81%)	1,170	Willms et al. 1986
		Heavy (47%)	1,865	
		Moderate (36%)	2,171	
		Light (26%)	2,199	
	Starkey, Oregon (bunchgrass)	Heavy (53%)	361	Skovlin et al. 1976
		Moderate (35%)	423	
		Light (20%)	405	
		No use	344	
	Southcentral New Mexico (desert grassland)	Heavy (60%)	85	Valentine 1970
		Moderate (35%)	202	
		Light (26%)	208	
	Benmore, Utah (crested wheatgrass)	Heavy (80%)	364	Frischknecht and Harris 1968
		Moderate (65%)	485	
		Light (55%)	495	
Sheep	Manyberries, Alberta (midgrass prairie)	Heavy (68%)	346	Smoliak 1974
		Moderate (53%)	417	
		Light (45%)	474	
	Desert Experimental Ranch, Utah (salt desert)	Heavy (50%)	134	Hutchings and Stewart 1953
		Moderate (34%)	353	
		No use	398	
	Hopland Field Station (California annual grassland)	Heavy (75%)	3,627	Rosiere 1987
		Moderate (55%)	5,041	
		Light (41%)	4,619	

[a] Percent utilization for moderate grazing was not given but is estimated at 40% based on personal communication with R. Heitschmidt. Percent utilization for the heavy grazing treatment is estimated to be 50%.

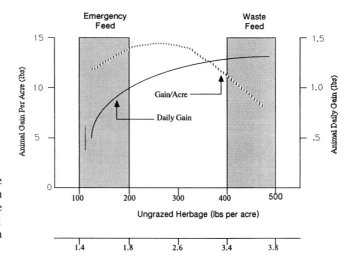

Figure **8.1** Stocking-rate guide for beef production on upland blue grama range in Colorado, grazed May 1 through October 31. (From Bement 1969.)

that canopy cover of big sagebrush increased 54% and minimal recovery of understory grasses occurred under 28 years of complete protection from grazing. On ponderosa pine bunchgrass range in Colorado, heavily grazed for over 23 years, protection from grazing did not promote recovery (Currie 1976). A combination of fertilizer herbicide treatment greatly increased herbage yield under a variety of grazing treatments. On desert grassland in central Arizona, a range protected for 30 years and a heavily grazed range had both rapid and nearly equal rates of mesquite invasion (Smith and Schmutz 1975). The protected range did have substantially more perennial grass cover than the heavily grazed range.

It is important to recognize that precipitation drives plant succession. If soil erosion has not been severe, recovery from severe overgrazing requires less than 10 years in the prairie country of the central and eastern United States, where precipitation averages over 500 mm per year. In the big sagebrush and creosotebush (*Larrea tridentata*) areas of the West with less than 250 mm average precipitation per year, range recovery after severe degradation and brush invasion is slow to nonexistent (Holechek and Stephenson 1983; Beck and Tober 1985).

INFLUENCE OF STOCKING RATE ON RANGE LIVESTOCK PRODUCTIVITY

A number of studies have evaluated the influence of grazing intensity on animal productivity (weight gains, weaning weights, calf or lamb crops, and wool yield) per unit area and per animal unit. Generally, as stocking rate is increased, productivity per animal declines (Table 8.3). Differences in animal productivity between light and moderate stocking rates are much less than between moderate and heavy stocking rates. Although productivity per animal unit declines as stocking rate increases, productivity per unit area increases up to a

TABLE 8.3 Livestock Weight Gains per Unit Area and per Animal Unit from Various Studies Involving Grazing Intensity

ANIMAL	LOCATION	GRAZING INTENSITY	LEVEL OF USE	WEIGHT GAIN PER UNIT AREA (KG/HA)	WEIGHT GAIN PER ANIMAL (KG)	REFERENCE
Cattle	Starkey, Oregon (forest)	Heavy	34%	8.9	74	Skovlin et al. 1976
		Moderate	28%	8.1	95	
		Light	17%	6.9	105	
	Manitou, Colorado (forest)	Heavy	58%	41.4	82	Johnson 1953
		Moderate	33%	43.7	100	
		Light	16%	23.2	107	
	Throckmorton, Texas (midgrass prairie)	Heavy	—	37.1	199	Kothmann et al. 1971
		Moderate	—	23.7	200	
		Light	—	18.1	210	
	Eastern Colorado Range Station, Colorado (midgrass prairie)	Heavy	64%	64.4	91	Sims et al. 1976
		Moderate	44%	48.8	105	
		Light	30%	26.6	110	
	Lethbridge, Alberta (midgrass prairie)	Very heavy	81%	132.[b]	164[b]	Willms et al. 1986
		Heavy	47%	82[b]	205[b]	
		Moderate	36%	62[b]	230[b]	
		Light	26%	45[b]	224[b]	
	Fort Stanton, New Mexico (shortgrass)	Heavy	63	7.8	193	Pieper et al. 1991
		Moderate	43	6.9	198	
	Central Plains Experimental Range, Colorado (shortgrass)	Heavy	(166 kg/ha residue)	15.5	57	Bement 1969
		Moderate	(333 kg/ha residue)	16.5	106	
		Light	(550 kg/ha residue)	12.9	118	
	Central Plains Experimental Range, Colorado (shortgrass)	Heavy	54%	24.6	99	Klipple and Costello 1960
		Moderate	37%	37.9	122	
		Light	21%	11.2	129	

ANIMAL	LOCATION	GRAZING INTENSITY	LEVEL OF USE	WEIGHT GAIN PER UNIT AREA (KG/HA)	WEIGHT GAIN PER ANIMAL (KG)	REFERENCE
	Scotts Bluff Experimental Range, Nebraska (shortgrass)	Heavy	74%	53.3	0.74[a]	Burzlaff and Harris 1969
		Moderate	58%	40.7	0.74[a]	
		Light	53%	24.6	0.74[a]	
	Benmore, Utah (crested wheatgrass)	Heavy	80%	41.4	0.96[a]	Frischknecht and Harris 1968
		Moderate	65%	48.6	1.20[a]	
		Light	53%	41.3	1.29[a]	
	Saylor Creek, Idaho (cheatgrass)	Heavy	—	23.7	0.66[a]	Murray and Klemmedson 1968
		Moderate	—	15.6	0.66[a]	
Sheep	Manyberries, Alberta (midgrass prairie)	Heavy	68%	18.2	58	Smoliak 1974
		Moderate	53%	15.2	60	
		Light	45%	12.3	62	
	Archer, Wyoming (shortgrass)	Heavy	—	52[c]	28[c]	Lang et al. 1956
		Moderate	—	42[c]	33[c]	
		Light	—	27[c]	35[c]	
	Hopland Field Station (California annual grassland)	Heavy	75	209	39	Rosiere et al. 1996
		Moderate	55	160	35	
		Light	41	124	36	

[a] Average daily gain.
[b] Cow plus calf.
[c] Ewe plus lamb.

TABLE 8.4   Influence of Grazing Intensity on Range Livestock Nutritional Characteristics

STUDY	GRAZING INTENSITY (% USE)	FORAGE INTAKE (KG/DAY)	DIET CRUDE PROTEIN (%)	DIET DIGESTIBILITY (%)
Pieper et al. 1959,	Moderate (37%)	1.9	6.0	48
sheep, Utah	Heavy (75%)	1.6	5.8	38
Vavra et al. 1973,	Light (20%)	4.4	11.3	57
cattle, Colorado	Heavy (60%)	3.7	11.3	52

point. It then decreases as scarcity of forage reduces nutrient intake by livestock (see Figure 8.1). This is why heavy grazing is often tempting to ranchers.

Maximum gains per animal and per unit area are not possible concurrently. Further, there is little leeway between maximum gains per unit area and no gains per unit area under heavy stocking rates. This is because animals will cease to gain weight as forage becomes increasingly scarce and often lower in nutritive quality. During drought periods, heavy stocking can be economically disastrous because the complete lack of forage will necessitate that all livestock be removed from the range and fed hay or sold at low prices. Ranchers using moderate to light stocking rates have much higher levels of forage standing crop throughout the year and generally more vigorous plants than do those using heavy stocking rates. This forage reserve permits much less adjustment in animal numbers than would be necessary under heavy stocking rates.

The decline in livestock performance per animal unit as grazing intensity increases is explained by reduced forage intake and diet quality (Table 8.4). Decreased forage availability reduces animal selectivity and forces them to consume diets lower in quality. It also forces animals to spend more energy on foraging activity that could otherwise go into production.

INFLUENCE OF STOCKING RATE ON ECONOMIC RETURNS

Heavy grazing generally maximizes gross economic returns, but net economic returns are maximized by moderate grazing (Tables 8.5 and 8.6). Death losses and supplemental feed costs are higher for heavy grazing compared with moderate grazing (Klipple and Costello 1960; Shoop and McIlvain 1971; Abdalla 1980). On cow-calf or ewe-lamb operations, weaning percentages (percentage of female animals in the herd producing a marketable offspring) are lower for heavy compared with moderate grazing. Shoop and McIlvain (1971) found that calf weaning percentages were 78% and 89% for heavily grazed and moderately grazed ranges, respectively, during drought years. In nondrought years, calf weaning percentages were 82% and 94% for heavily and moderately grazed ranges, respectively.

TABLE 8.5   Net Economic Returns from Different Grazing Intensities

RANGE TYPE/ LOCATION	LIVESTOCK TYPE	GRAZING INTENSITY	FORAGE (% USE)	NET RETURN (ACRE $)	REFERENCE
Desert:					
Salt desert (Utah)	Ewe-lamb	Heavy	68	+1.69[a]	Hutchings and Stewart 1953
		Moderate	35	+3.45[a]	
Chihuahuan desert (New Mexico)	Cattle and sheep	Heavy	45–60	+0.11	Abdalla 1980
		Moderate	30–45	+0.17	
Chihuahuan desert (New Mexico)	Cow-calf	Heavy	50–60	+0.32	Holechek 1991
		Moderate	30–35	+0.75	
Mulga (Australia)	Sheep-wool	Heavy	80%	+0.81	Beale et al. 1986
		Heavy-Moderate	50%	+0.94	
		Moderate	30%	+1.01	
		Light-Moderate	20%	+0.97	
		Light	10%	+0.84	
Coniferous Forest:					
Colorado	Cattle-yearlings	Heavy	58%	+0.74	Johnson 1953
		Moderate	33%	+1.34	
		Light	16%	+0.98	
Oregon	Cow-calf	Heavy	34%	+2.55	Quigley et al. 1984
		Moderate	28%	+2.69	
		Light	17%	+2.15	
Arizona	Cattle-yearlings	30–38% use gave highest returns			Pearson 1973
Piñon-Juniper/Shortgrass:					
New Mexico	Cow-calf	Heavy	60–65	2.02	Holechek 1994
		Moderate	40–45	2.34	

(Continued)

TABLE 8.5   Net Economic Returns from Different Grazing Intensities   *(Continued)*

RANGE TYPE/ LOCATION	LIVESTOCK TYPE	GRAZING INTENSITY	FORAGE (% USE)	NET RETURN (ACRE $)	REFERENCE
Shortgrass Prairie:					
Colorado	Cattle- yearlings[b]	Heavy Moderate Light	54% 37% 21%	+1.54 +1.93 +1.41	Klipple and Costello 1960
Colorado	Cattle- yearlings[c]	Heavy Moderate Light	55% 35% 20%	+1.55 +1.94 +1.06	Bement 1969
Wyoming	Cattle- yearlings	45% use gave highest returns		Hart et al. 1988	
Mid-Grass Prairie:					
Oklahoma	Cattle- yearlings Cow-calf	Heavy Moderate Heavy Moderate	60–70 50–60 62 44	+1.38 +1.01 +0.70 +1.88	Shoop and McIlvain 1971
Texas (rolling plains)	Cow-calf	Heavy Moderate	50–55 40–45	+5.25 +4.26	Heitschmidt et al. 1990
Eastern Colorado	Cattle- yearlings	45% use gave highest returns		Torell et al. 1991	
Southern Pine Forest:					
Louisiana	Cow-calf	Heavy Moderate Light	57% 49% 35%	+5.55 +3.73 +3.44	Pearson and Whitaker 1974

[a]Net returns per animal.
[b]1940–1953
[c]1954–1963

TABLE 8.6 Influence of Grazing Intensity on Winter Sheep Production at the Desert Experimental Range in Utah, and Cattle Production at the Southern Great Plains Experimental Range in Oklahoma

	HEAVY GRAZING	MODERATE GRAZING
DESERT EXPERIMENTAL RANGE		
Percent utilization of forage	68	35
Ewe weight change (fall to spring) (lb.)	+ 1.1	+9.3
Average fleece weight (lb.)	9.68	10.63
Lamb crop (percent)	79	88
Death loss (percent)	8.1	3.1
Lamb weaned per ewe (lb.)	67.0	77.0
Net income (3,000-herd flock) (dollars)	5,072	10,390
Net income per ewe (dollars)	1.69	3.45
SOUTHERN GREAT PLAINS EXPERIMENTAL RANGE		
Acres per cow	12	17
Estimated percent utilization of forage	62	44
Calf crop weaned (percent)	81	92
Calf weaning weight per cow (lb.)	314	424
Calf weaning weight (lb.)	388	461
Net returns per cow (dollars)	9.00	29.44
Net returns per acre (dollars)	0.70	1.88

Sources: Hutchings and Stewart 1953; Shoop and McIlvain 1971.

Livestock can make high gains on heavily grazed range for a few years particularly if they are given supplemental feed and precipitation is average or above. However, in drought years, the reduction in livestock productivity both per animal unit and per unit area is far more severe than on moderately grazed ranges.

In the Rolling Plains of northcentral Texas, a 26-year study from 1961 to 1987 showed that heavy grazing actually gave higher net returns than did moderate grazing (Whitson et al. 1982; Heitschmidt et al. 1990) (Table 8.5). However, it was concluded that continually stocking at heavy rates increased financial risk because of the need to periodically destock or provide a substitute feed due to insufficient amounts of forage in drought years. Forage use levels on the heavily grazed pasture were estimated to be near 50–55% compared to 40–45% for the pasture moderately grazed. There is some doubt as to whether 50–55% use of key forage species really represents a heavy or unsustainable rate in the southern mixed prairie.

Death losses from poisonous plants are much higher on heavily grazed ranges because the nonpoisonous, palatable species are less available. Average livestock death losses from poisonous plants were about 50% higher on heavily grazed compared to moderately grazed ranges in southcentral Texas (Taylor and Ralphs 1992). On shortgrass prairie in Colorado, heavily grazed ranges had nearly three times the cattle death loss of moderately grazed ranges (Klipple and Costello 1960).

Continued heavy grazing results in gradual degradation of soil and vegetation resources that often goes unnoticed by ranchers because forage yield is usually not measured (Figure 8.2). Finally, ranchers are forced out of business unless they can subsidize their ranches with outside income.

Today, few range livestock producers remain who severely overgraze the range. Most of the overgrazing that occurs at present is on leased rangeland that the operator can readily abandon after exploitation. These operators usually graze yearling cattle because they can yield high profits under heavy grazing (particularly in nondrought years) and require less overhead than do other types of livestock.

The relatively low price of livestock and the high price of supplemental feed in recent years have made overgrazing an economically unattractive proposition for ranch owners. It is more recognizable now than 20 years ago that the extra supplemental feed costs required to replace range forage divert money and labor that could be used more profitably in some other manner. Today's ranchers generally wish to leave their ranches to the next generation, in better condition than when they acquired them. They recognize that this cannot be done by heavy grazing.

Stocking Rate and Risk

Ranchers in the past have often equated livestock numbers with their wealth and income level. However, in recent years, it has become increasingly recognized that livestock management practices play as big a role in determining net income as forage harvest efficiency. This is particularly true for the arid and semiarid rangelands. Pratchett and Gardiner (1991)

Figure 8.2 This fenceline contrast in the Chihuahuan desert near Las Cruces, New Mexico, shows the difference in forage production between a long-term heavily stocked ranch (foreground) and conservatively stocked ranch (background).

addressed the issue of how a large reduction in stocking rate would impact net ranch income in the semiarid region of western Australia. They found that reducing average stocking rates by as much as 50% could actually increase net income if accompanied by improvements in ranch infrastructure (fencing and water development), a calf weaning program, and if brahman bloodlines were incorporated into the herd. Higher calving percentages, lower livestock mortality rates, and lower expenses explained the net income increase under the destocking and intensive herd management strategy. Most importantly in western Australia reduced stocking was considered essential in reversing the degradation of soil and vegetation resources that was occurring on many ranches.

Boykin et al. (1962) evaluated grazing and herd management practices of financial survivors of the 1950s drought in the southern Great Plains of the United States. All the ranchers studied believed that use of conservative stocking was the key element in their survival. During drought, reducing livestock numbers to levels supportable by range forage resources was financially more effective than holding livestock and providing them with harvested feed. Holechek (1996a,b) made the same observation during the mid-1990s drought in southern New Mexico.

Conservative stocking involves using about 35% of forage resources on arid and semiarid rangelands. There appears to be little biological benefit from lighter use levels. On the other hand, increased financial returns from heavier use levels are doubtful, particularly when risk is taken into account (Martin 1975).

In the humid types such as the tallgrass prairie and southern pine forest, the reward relative to risk from maximizing forage harvest efficiency is more favorable than in the arid types (Shoop and McIlvain 1971; Pearson and Whitaker 1974; Whitson et al. 1982; Heitschmidt et al. 1990). Playing the cattle price cycle can be potentially profitable for some ranchers (particularly yearling cattle operators) in humid areas. The cattle price cycle has historically involved about 5 to 7 years of improving prices and 3 to 5 years of price declines (Holechek et al. 1994; Holechek 1996a). A rancher playing the cattle price cycle would begin building up his herd after 3 or 4 years of low cattle prices. The plan would be to have the ranch stocked to capacity just prior to the cattle cycle peak and sell heavily as prices top and begin to decline. Any degradation of rangeland that occurs when the ranch is fully stocked should be overcome by the 3 to 4 year period of conservative stocking when cattle prices are in a downturn. This is because vegetation recovers quickly and degrades slowly in the humid types in contrast to desert areas. Extended drought is the primary risk associated with this strategy. Drought is less of a risk in the humid than arid types, but it can still severely impact forage resources and financial returns (Shoop and McIlvain 1971; Whitson et al. 1982). It is essential for ranchers who play the cattle price cycle to have a plan to deal with climatic adversity. Whitson et al. (1982) and Carande et al. (1995) provide risk management approaches for ranchers interested in playing the cattle price cycle.

FLEXIBLE VERSUS FIXED STOCKING RATES

One of the most basic questions regarding stocking rate concerns whether to use a flexible or a constant stocking rate. Basically, livestock producers strive to maintain a stable enterprise.

However, the forage crop can fluctuate more than 100% from one year to the next. Producers also wish to supply most of the forage needs of their livestock from the range. Basically, producers can select from four strategies (Pieper 1981):

1. Stock conservatively so that the range is not overstocked, even in fairly severe droughts.
2. Stock at a high rate commensurate with forage available during favorable years.
3. Stock at the average forage supply.
4. Vary stocking to meet forage supplies for the year.

Flexible stocking to keep a relatively constant ratio between forage and livestock would be a rangeman's dream. Unfortunately, it is difficult to follow in practice (Schmutz 1977; La Baume 1981). Guidelines such as heavy culling of cows during a drought and maintaining part of the herd as young animals that can be sold when forage conditions warrant have been proposed.

In a southcentral Arizona study, Martin (1975) found net returns obtained by increasing stocking rate 120, 130, or 140% of the average in the best years were only $1 to $2 greater per animal unit than for constant stocking at 90% of the level required for proper stocking in an average year. In actual practice he speculated that the small monetary advantage would probably be offset by the disadvantages of the flexible system. The disadvantages of flexible stocking include:

1. The sheer difficulty of estimating forage crops.
2. The administrative costs of buying extra animals to stock the range in good years.
3. The possibility of introducing parasites or disease with livestock purchased from other areas.
4. The natural reluctance to cull as heavily as necessary for the good of the range in drought years.

In comparison with flexible stocking, Martin (1975) reported the following advantages for constant stocking at 90% of the average proper stocking rate:

1. High economic stability to the ranching operation.
2. Relatively the same income as flexible stocking.
3. Moderate to low risk of damage to the range.
4. Moderate to low risk of financial crisis during drought.

Martin (1975) concluded that the best approach would be stocking at or not to exceed 90% of average proper stocking, but with some reductions during prolonged severe drought.

Yearling cattle are better suited to flexible stocking than are mother-young operations. However, yearlings are usually grazed only during the growing season, and the forage crop is unknown when the yearlings are bought and placed on the range. There is a general reluctance to sell part of the yearlings midway during the growing season if precipitation is below average. Conversely, it is often difficult to acquire more animals in the middle of the growing season when forage production is above average.

Retention of calves as yearlings has been advocated as a practical means of incorporating flexible stocking into cow-calf operations (Paulsen and Ares 1961). Martin (1975) pointed out some problems with calf retention in the fall during years of high forage production in the southwestern United States. In most years, fall weaner calves gain little weight during the winter dormant season unless supplemented. Therefore, these animals must be retained through the next growing season (summer) in order for their value to increase appreciably if only range feed is provided. This will result in overgrazing unless forage growth is above average.

In high-forage-production years, the profitability of retaining weaner calves in the fall and selling them prior to the next growing season depends on the cost of supplemental feeds, the selling price of calves in the fall, and the selling price for yearlings in the following spring. When feed grains are cheap, cattle feeders will pay more per unit weight for calves than for yearlings because they can add 1 kg of grain cheaper than they can buy it on the animal. Conversely, high feed grain prices reverse the situation. If the prices and weights of calves and yearlings are known, the relative advantages of cow-calf and cow-yearling operations can be estimated.

Martin (1975) pointed out that a major shift from cow-calf to cow-calf-yearling operations would help reduce the oversupply of beef and the amount of grain fed to beef. Cow-calf-yearling operations do offer flexibility for reducing stocking rate after dry summers. However, the Martin study shows that net returns are highest when the breeding herd is maximized and calves are not retained over winter in years following summers of high forage production.

GRAZING INTENSITY CONSIDERATIONS

After the question of flexible versus constant stocking is resolved, the next issues regarding stocking rate on a range involve determining the average forage production (kg/ha/year) and the average level of use that the principal forage species can withstand. The subjects of measuring forage production and utilization are addressed in Chapter 7. The reader is referred to Cook and Stubbendieck (1986) and Bonham (1989) for detailed information on measuring herbage production and utilization. From the standpoint of management decisions by ranchers, standing crop (dry-matter basis) measurements are the most useful. Most managers can be trained to estimate standing dry matter ocularly to within 25 kg/ha (Smith 1944; Valentine 1970). On yearlong ranges most decisions regarding adjustment in stocking rates are made at the end of the growing season in the fall. After the standing crop is estimated, animal numbers can be adjusted so that a minimum residue of dry matter remains just prior to the average time when growth is initiated the following year. The premise here is that a certain minimum level of dry matter should always be present on a particular range to maintain the soil, forage plant vigor, livestock diet quality, and wildlife habitat. The importance of maintaining minimum residues for soil, vegetation, and animal resources has been well established. Bement (1969) and Hooper and Heady (1970) have demonstrated the economic effectiveness of basing stocking rates on minimum residues.

Critical dry-matter residue levels have been derived from some range types in the United States. Enough information is available so that they can be deduced for others. In the shortgrass prairie country of eastern Colorado, 335 kg/ha will give maximum economic returns and maintain forage production (see Figure 8.1, Bement 1969). On southeastern Oregon big sagebrush ranges, grass residues of 180 kg/ha will maintain or improve range condition on most sites (Hyder 1953). In the California annual grassland type, from 300 kg/ha to 1,200 kg/ha of minimum residue is needed, depending on the site (Hooper and Heady 1970; Bartolome et al. 1980).

Most information regarding critical grazing intensities on rangeland involves utilization data. Utilization data can readily be used in stocking-rate decisions. If a reasonable estimate of the average forage production (kg/ha/year, dry-matter basis) is available for a particular range, this can be combined with the level of utilization that will maintain soil and vegetation resources to derive the critical minimum residue. As an example, Martin and Cable (1974) reported that perennial grass vigor started to decline on southcentral Arizona ranges when the average utilization level of the perennial grasses for a 10-year period exceeded 40%. The 10-year average perennial grass dry matter production on the ranges studied was 59 kg/ha. Therefore, the critical dry-matter residue is 35 kg/ha (59 kg/ha $\times$ 0.60 = 35 kg/ha). On southern pine forest ranges in Louisiana, an average utilization level of 57% over a 10-year period did not affect forage yields (Pearson and Whitaker 1974). A minimum residue of 960 kg/ha would appear to be adequate for most of this range type.

We have developed a simple utilization guide for determining critical minimum residues for various rangelands in the United States based on careful analysis of available research (Table 8.7). Our analysis of the literature shows that 35% to 45% use of grazable forage will generally maintain forage production on semiarid (shortgrass) grassland ranges where brush encroachment is not a problem. In the more arid regions (under 300 mm of mean annual precipitation) of the Southwest and intermountain areas, utilization levels between 25% and 40% are recommended. In the humid tallgrass and southern pine regions, utilization levels of 45% to 60% will maintain forage productivity. Research from California shows that annual grasslands can generally withstand higher grazing intensities (50% to 60%) than can perennial grasslands (Rosiere 1987). Generally, as average annual precipitation increases, utilization can be increased, with some exceptions. The coniferous forest range type in the West is easily damaged by grazing because of rugged terrain that causes livestock, particularly cattle, to concentrate in the flatter, more convenient areas. Both Johnson (1953) in Colorado and Skovlin et al. (1976) in Oregon reported that, to prevent degradation, utilization levels of the primary forage grasses must be kept around 35%. The palouse prairie is much less grazing resistant than is the shortgrass prairie, although average annual precipitation is similar. This is due in part to the fact that the palouse prairie receives most of its precipitation in the winter, compared to the summer for the shortgrass prairie. The longer growing season, coupled with the higher grazing resistance of its grasses, explains the higher recommended utilization level for the shortgrass compared with the palouse prairie.

TABLE 8.7 Utilization Guidelines for Different Range Types in the United States

RANGE TYPES	AVERAGE ANNUAL PRECIPITATION		PERCENT USE OF KEY SPECIES FOR MODERATE GRAZING[a]	REFERENCES
	CM	IN		
Salt desert shrubland	10–13	4–8	25–35	Hutchings and Stewart 1953; Cook and Child 1971
True desert (Mojave)	10–13	4–8	25–35	Hughes 1982
Semidesert grass and shrubland	13–30	8–12	30–40	Valentine 1970; Paulsen and Ares 1961; Martin and Cable 1974; Holechek 1991
Sagebrush grassland	13–30	8–12	30–40	Pechanec and Stewart 1949; Laycock and Conrad 1981
Palouse prairie (bunchgrass)	30–50	12–20	30–40	Pickford and Reid 1948; Skovlin et al. 1976
Shortgrass prairie	25–40	10–16	40–50	Klipple and Costello 1960; Burzlaff and Harris 1969; Hart et al. 1988; Sims et al. 1976; Pieper et al. 1991
California annual grassland	25–100	10–40	50–60	Hooper and Heady 1970; Bartolome et al. 1980; Rosiere 1987
Northern mixed prairie	40–65	16–25	40–50	Lewis et al. 1956; Houston and Woodward 1966; Smoliak 1974; Willms et al. 1986
Southern mixed prairie	40–65	15–25	40–50	McIlvain and Shoop 1965; Kothmann et al. 1975; Heitschmidt et al. 1987
Coniferous forest	40–130	16–50	30–40	Pickford and Reid 1948; Johnson 1953; Skovlin et al 1976
Mountain shrubland	40–130	16–50	30–40	Pickford and Reid 1948; Skovlin et al. 1976
Oak woodland	40–130	16–50	30–40	Pieper 1970
Piñon-juniper woodland	25–40	9–16	30–40	Pieper 1970
Alpine tundra	40–130	16–50	20–30	Thilenius 1979
Tallgrass prairie	65–100	25–40	45–55	Herbel and Anderson 1959; Drawe 1988
Southern pine forest	100–175	40–70	50–60	Pearson and Whitaker 1974
Eastern deciduous forest	100–175	40–70	50–60	Research not available

Source: J. L. Holechek, 1988. An approach for setting the stocking rate. *Rangelands* 10:10–14, Table 1.

[a]Ranges in good condition and/or grazed during the dormant season can withstand the higher utilization level. Those in poor condition or grazed during active growth should receive the lower utilization level.

A 10% range of utilization is given rather than a single value. Several factors determine whether a low, medium, or high value should be selected from within the range. These include range condition, season of use, distribution of water, type of livestock, wildlife, and site characteristics (soils, precipitation, and topography). On ranges in high condition with flat terrain and good water distribution, the upper utilization value can be applied if the goal is to maximize profits from livestock. Low range condition, thin soils, rough topography, and poor water distribution all necessitate some lower utilization level than the maximum, depending on their severity.

Grazing Intensity and Timing

Clipping studies have indicated that both timing and intensity of grazing can greatly affect plant productivity and vigor (Cook and Child 1971; Trlica et al. 1977; Miller and Donart 1981). It is well established that forage plants can withstand greater use during certain times of the year as compared to others. Most plants can withstand higher intensities of grazing during dormancy than active growth. Defoliation when plants initiate growth is generally less harmful than during the latter half of the growth cycle when plants are completing reproduction and storing carbohydrates (Stoddart 1946; Cook and Child 1971; Miller and Donart 1979). Damage to plants from early spring grazing on temperate mountain rangelands may be avoided if the plants are given grazing relief during culm elongation and reproduction.

A commonly held view has been that grazing intensity is of minor importance if grazing is properly timed. This concept appears to have more validity for the humid rangelands and possibly riparian areas than the semiarid and arid rangelands (Cook and Child 1971; Miller and Donart 1981). However, even if heavy grazing use is timed so damage to key forage plants is avoided, there are likely to be adverse impacts on livestock performance and financial returns (Johnson 1953; Klipple and Costello 1960; Shoop and McIlvain 1971). Another problem is that lack of vegetation residue may result in severe soil erosion (Dunford 1949; McCalla et al. 1984) and be harmful to desirable wildlife and fish species. Therefore we advocate that managers try to stay with the forage use guidelines we provide in Table 8.7, but acknowledge that with careful timing, heavier use may be possible in certain situations.

Range Readiness and Timing

Many rangelands in the United States and other parts of the world are grazed only part of the year. This is usually due to adverse weather conditions such as winter snowfall that make much of the forage unavailable. In the intermountain region of the western United States, severe damage to rangeland vegetation resulted from intense early spring grazing of new growth after snow melt. An important part of early range management was development of range readiness guidelines for the forested rangelands in the western states. Range readiness is defined as "that point in the plant growth cycle at which grazing may begin without permanent damage to vegetation and soil" (Heady and Child 1994). Range readiness guidelines are generally based on delay of grazing until primary forage grasses have reached a certain height or have a certain number of leaves and the soil is dry enough that large animals will not make deep tracks. Generally it is recommended that the shortgrasses such as sandberg bluegrass and blue grama be allowed to obtain a 2-inch height while midgrasses such as bluebunch wheatgrass, prairie junegrass, Idaho fescue, and Arizona fescue be allowed to reach a 4-inch height. More detailed guidelines on range readiness are provided by Heady and Child (1994).

ADJUSTMENT FOR DISTANCE FROM WATER

Failure to adjust stocking rates for travel distance to water has resulted in considerable range degradation, particularly in the hot, arid rangelands of the southwestern United States, parts of Australia, and in the Sahel region of Africa. On the cold desert ranges of the intermountain United States, snow reduces water availability problems in winter. Unlike cattle, sheep and goats do not require water every day. Therefore they will readily use areas that are 3.2 km (2 miles) or more from water (McDaniel and Tiedeman 1981).

In contrast, several studies show that cattle make little use of areas farther than 3.2 km (2 miles) from water (Valentine 1947; Martin and Ward 1970, 1973). Table 8.8 provides guidelines on adjustments in cattle stocking rates as distance from water increases.

ADJUSTMENT FOR SLOPE

On mountainous rangeland, overgrazing and deterioration often occur on the flatter, more convenient sites even though the total forage supply is adequate. In rough, rugged terrain, cattle congregate on the more convenient flat areas, such as valley bottoms, riparian zones, and ridgetops. Forage on the steeper slopes (over 60%) receives little to no use by cattle (Mueggler 1965; Cook 1966), and these areas must be deleted from the grazable land area. Table 8.9 gives guidelines on grazing-capacity adjustments for slope.

TABLE 8.8   Suggested Reductions in Cattle Grazing Capacity with Distance from Water

DISTANCE FROM WATER		
MILES	KM	PERCENT REDUCTION IN GRAZING CAPACITY[a]
0–1	0–1.6	None
1–2	1.6–3.2	50
2	Over 3.2	100 (consider this area ungrazable)

Source: J. L. Holechek, 1988. An approach for setting the stocking rate. *Rangelands* 10:10–14, Table 4.

[a]Supporting literature includes Valentine (1947), Martin and Ward (1973), Sneva et al. (1973), Squires (1973), Beck (1978), Pinchak et al. (1991), and Hart et al. (1993).

TABLE 8.9   Suggested Reductions in Cattle Grazing Capacity for Different Percentages of Slope

PERCENT SLOPE	PERCENT REDUCTION IN GRAZING CAPACITY[a]
0–10	None
11–30	30
31–60	60
Over 60	100 (consider these slopes ungrazable)

Source: J. L. Holechek, 1988. An approach for setting the stocking rate. *Rangelands* 10:10–14, Table 3.

[a]Supporting literature includes Glendening (1944), Mueggler (1965), Cook (1966), Gillen et al. (1984), Ganskopp and Vavra (1987), and Pinchak et al. (1991).

Sheep and goats make much better use of rugged terrain than do cattle. Because of smaller size, more surefootedness, and a stronger climbing instinct, they naturally use steep areas much more than do cattle. In most cases, sheep are under the control of a herder and can readily be forced to use the steeper hillsides, minimizing overuse of the valley bottoms. McDaniel and Tiedeman (1981) found that sheep on winter range in New Mexico uniformly used slopes of less than 45%. However, utilization was sharply reduced when slopes exceeded 45%. Based on their study, slopes greater than 45% should be considered unusable by sheep, but little or no adjustment appears necessary for slopes under 45%.

FORAGE DEMAND BY GRAZING ANIMALS

Forage demand is a function of the number of animals and the number of days they will occupy a particular range. We believe that the best way to derive daily forage demand (dry-matter basis) of ruminant animals is to multiply their body weight by 2%. In Table 11.2 we will review a wide range of studies that were consistent in showing that range ruminants consume 2% of body weight per day in dry matter when forage availability is not restricted. Intake may go as high as 2.6% body weight for short periods when forage quality is high, and it may drop to 1.5% or lower when quality and/or quantity is low. However, the yearly averages given for cattle, sheep, goats, deer, elk, pronghorn, moose, and so on, are all about 2%. Forage intake by horses and donkeys averages about 50% higher than that for ruminants (see Chapter 11). Daily forage intake by various range animals is shown in Table 8.10.

CALCULATION OF STOCKING RATE

Once the average forage production and the minimum residue required to maintain the site are determined, the initial stocking rate can be set. It is important to recognize that this rate

TABLE 8.10 Daily Dry-matter Consumption by Various Range Animals Based on Their Body Weight

ANIMAL	ANIMAL WEIGHT[a]		DAILY DRY-MATTER INTAKE (% BODY WEIGHT)	DAILY DRY-MATTER INTAKE		ANIMAL UNIT EQUIVALENTS (AU₁)
	LB	KG		LB	KG	
Cattle (mature)	1,000	455	2	20.0	9.1	1.00
Cattle (yearlings)	750	318	2	15.0	6.8	0.75
Sheep	150	68	2	3.0	1.4	0.15
Goat	100	45	2	2.0	0.9	0.10
Horse	1,200	545	3	36.0	10.9	1.80
Donkey	700	318	3	21.0	6.4	1.05
Bison	1,800	818	2	36.0	16.4	1.80
Elk	700	318	2	14.0	6.4	0.70
Moose	1,200	545	2	24.0	10.9	1.20
Bighorn sheep	180	82	2	3.6	1.6	0.18
Mule deer	150	68	2	3.0	1.4	0.15
White-tailed deer	100	45	2	2.0	0.9	0.10
Pronghorn antelope	120	55	2	2.4	1.1	0.12
Caribou	400	182	2	8.0	3.6	0.40

Source: J. L. Holechek, 1988. An approach for setting the stocking rate. *Rangelands* 10:10–14, Table 5.
[a]Average weight of mature male or female animal.

will often need to be modified as experience is gained for the particular range. The stocking rate is determined by dividing the total usable forage per unit area by the total forage demand of the grazing animals for the grazing period.

We are now ready to solve some hypothetical stocking-rate problems using the procedures developed by Holechek (1988) and validated by Holechek and Pieper (1992). Three cases will be used as examples.

Case 1

You are contemplating buying a ranch on shortgrass prairie range in eastern Colorado. You have determined that range condition is good. The range is flat and well-watered (no part of the pasture is over 2.4 km from water). Based on information from the Natural Resources Conservation Service and your own ocular estimates, production of key forage species averages about 700 kg/ha of dry matter per year. The ranch is 2,000 ha in size and you are planning a cow-calf operation.

Question: How many 400-kg cows can you have in your base herd?

Calculation of total usable forage:

Forage production (kg/ha) × percent allowable use
$$× \text{ area (ha)}$$
$$= \text{total forage (kg) available for grazing}$$

$$700 × 0.50 × 2,000 = 700,000 \text{ kg}$$

Calculation of forage demand:

Weight of cows (kg) × daily dry-matter intake (2% body weight)
$$× \text{ number of days pasture will be grazed (365)}$$
$$= \text{forage demand per cow per year}$$

$$400 × 0.02 × 365 = 2,920 \text{ kg of forage/cow/year}$$

Calculation of stocking rate:

Total usable forage (kg) ÷ forage/cow/year
$$= \text{number of cows pasture will carry}$$

$$700,000 ÷ 2,920 = 240 \text{ total cattle}$$

One bull is recommended per 20 cows. Therefore, this range would support a base herd of about 228 cows and 12 bulls.

Question: If sheep (ewes) were substituted for cattle, the number of sheep in the base herd (assume that sheep weigh 65 kg) would be calculated as follows:

$$240 ÷ \frac{65 \text{ kg (weight per sheep)}}{400 \text{ kg (weight per cow)}} = 1,477 \text{ sheep}$$

If this range were used for only 9 months, the total number of cattle would be calculated as follows:

$$\frac{12 \text{ months}}{9 \text{ months}} × 240 \text{ cattle} = 320 \text{ cattle}$$

At the end of the dormant season (mid-April), 350 kg/ha should remain to protect the site.

Case 2

You have summer range in the mountains of northeastern Oregon. Condition of the range is poor. Although the terrain is rugged, water is well distributed. You graze this

We recommend keeping the base herd at 90% of grazing capacity on this range to maximize stability during drought. This would result in grazing capacity of 83 total cattle (79 cows + 4 bulls).

Question: How many cows and how many yearlings should you have in your herd in an average forage production year if 30% of your grazing capacity is used for 275-kg yearlings?

92 cows $\times$ 0.7 (% cows in base herd) = 64 cows

92 cows in base herd (unadjusted for yearlings)
$$- \ 64 \text{ cows (adjusted for yearlings)}$$
$$= 28 \text{ cows that can be converted to yearlings}$$

$$28 \text{ cows} \times \frac{400 \text{ kg (average weight/cow)}}{275 \text{ kg (average weight/yearling)}} = 41 \text{ yearlings}$$

In an average year, the base herd would be comprised of 61 cows, 3 bulls, and 41 yearlings.

Question: During a drought year when forage production is only 150 kg/ha, how should cattle numbers be adjusted in mid-October after the growing season?

On this range 210 kg/ha of residue is required for protection (300 kg/ha forage production in average years $\times$ 0.70). Theoretically, based on the current year's forage production, nearly all cattle must be removed to protect this range. However, this could be financially disastrous to the rancher and probably is not necessary to maintain the health of the range. In this situation, empirical judgment on the part of the rancher would be of critical importance. If the drought followed 2 or more years of average or above-average forage production, sufficient carryover residue from previous years should maintain site stability. Grazing on perennial grasses would not become heavy until after the growing season. In some years, winter precipitation in southcentral New Mexico results in substantial growth of palatable forbs in late winter and early spring. These forbs take much of the pressure off perennial grasses. Areas long distances from permanent water with large forage supplies can serve as a forage reserve in drought. Utilization is possible by hauling water to these areas.

The best plan would be to sell all yearlings in mid-October and any dry or otherwise undesirable cows. If there was little fall-winter precipitation and forage was showing signs of depletion, the remaining cow herd could be brought into a drylot and fed harvested forage until initiation of forage growth on the range in the spring or summer. Herbel et al. (1984) provides guidelines for feeding confined cattle and marketing of calves on desert ranges during drought. Their data show that a part-year confinement of the cow herd (spring), coupled with early weaning of calves in late spring or summer rather than in October, can be economically advantageous over yearlong grazing during periods of drought in southcentral New Mexico. Good ranchers plan for drought by having reserves of range forage and/or harvested forage. They cull heavily and reduce herd size after 3 to 4 wet years when the probability of drought becomes high. Consecutive droughts lasting 2 or more years are the ones most damaging to good ranchers and the range. Under these conditions, the most effective strategy financially has been to sell livestock down to levels supportable by range forage resources (Boykin et al. 1962; Holechek 1996b).

Troxel and White (1989) have developed a simpler, more conservative procedure than Holechek (1988) that allocates 25% of current year forage production to livestock, another 25% to natural disappearance (insects, wildlife, weathering), and 50% is left for site protection. The approach developed by Holechek (1988) is based on maximizing forage use by livestock while that of Troxel and White (1988) works well for range betterment and minimization of risk. On most western ranges, partial or complete destocking would be necessary in only about 3 to 4 years out of 20 with the Troxel and White (1989) procedure.

KEY-PLANT AND KEY-AREA PRINCIPLES

The key-plant and key-area concepts have proven highly useful to managers in evaluating grazing effects on range vegetation (Holechek 1988). A key species is defined as "a forage species whose use serves as an indicator to the degree of use of associated species, and because of its importance, must be considered in any management program" (Society for Range Management 1989). Key management species are those on which management of grazing on a specific range is based. The key species and key area serve as indicators of management effectiveness. Generally, when the key species and key area are considered properly used, the entire pasture is considered correctly used.

In most cases one to three plant species are used as key species. These plants should be abundant, productive, and palatable. They should provide the bulk of the forage for grazing animals within the pasture. The "ice-cream" plants are not used because of their scarcity and low resistance to grazing. Key species are usually decreaser plants that are an important part of the climax vegetation. If the range has been heavily grazed, decreasers may be in short supply but they have the potential to become abundant if grazing pressure is reduced. Conditions do exist where the climax plants are not the most desirable or in which a reduction in stocking rate will not restore the climax plants within a reasonable period (5 to 15 years). In these cases a palatable increaser plant may be selected as a key species. It is important to recognize that key species for one type of animal may be different than for another type due to differences in food habits. As an example, bitterbrush (*Purshia tridentata*) is the key species for mule deer on many eastern Oregon ranges, but the key species for cattle on these same ranges is bluebunch wheatgrass (*Agropyron spicatum*). The key species for elk will be Idaho fescue (*Festuca idahoensis*) in most of this country.

Under the key-species approach, secondary forage species [i.e., sandberg bluegrass (*Poa sandbergii*) in eastern Oregon] will receive light use (10% to 25%), key species (bluebunch wheatgrass) will receive moderate use (30% to 40%), and the ice-cream plants [arrowleaf balsamroot (*Balsamorhiza sagittata*)] may be used excessively (over 40%).

The key area is a portion of range which, because of its location, grazing or browsing value, and/or use serves as an indicative sample of range conditions, trend, or degree of seasonal use (Society for Range Management 1989). The key area guides the general management of the entire area of which it is part. Successful range management practices within a pasture are usually judged by the response of the key plant species on the key area.

The key-area concept is based on the premise that no range of appreciable size will be utilized uniformly. Even under light grazing intensities, areas around watering points, salt grounds, valley bottoms, and driveways will often be heavily used. These preferred areas are referred to as sacrifice areas because setting stocking rates for proper use of these areas will result in under use of the bulk of the pasture. A major objective of specialized grazing systems is to minimize the size of sacrifice areas and provide them with periodic opportunity for recovery. These strategies are discussed in detail in Chapter 9.

When selecting the key area, parts of the pasture remote from water, on steep slopes, or with poor accessibility due to physical barriers should be disregarded. Proper use of these areas will generally result in destructive grazing on most of the pasture. These areas should be omitted when carrying capacity is estimated.

A number of qualitative guidelines have been developed for judging intensity of grazing on a range. We have found that a simple categorization into heavy, moderate, and light use is most practical using the following criteria:

Heavy use. Range has a "clipped" or mowed appearance. Over half of the fair and poor forage-value plants are used. All accessible parts of the range show use, and key areas are closely cropped. They may appear stripped if grazing is very severe. There is evidence of livestock trailing to forage.

Moderate use (proper use). About one-half of the good and fair forage-value plants are used. There is little evidence of livestock trailing. Most of the accessible range shows some use.

Light use. Only choice plants and areas are used. There is no use of poor forage plants. The range appears practically undisturbed.

On key areas average stubble heights of 30 cm to 35 cm (12 in. to 14 in.) for tallgrasses, 15 cm to 20 cm (6 in. to 8 in.) for midgrasses, and 5 cm to 8 cm (2 in. to 3 in.) for shortgrasses are recommended minimums for proper use.

Guidelines for minimum stubble heights under proper use for selected grass species are provided in Table 8.11. Considerable research exists on minimum stubble height guidelines for some grass species such as Kentucky bluegrass, blue grama, and black grama while for other plants such as big bluestem and sideoats grama the main basis for our guidelines is practical experience by range professionals. We freely acknowledge that situations exist where these guidelines may be conservative. However in nearly all situations their application should ensure protection of soil and vegetation resources as well as maintaining livestock performance and wildlife habitat.

TABLE 8.11   Minimum Recommended Stubble Heights for Selected Grass Species Under Proper Grazing Use

GRASS SPECIES	MINIMUM STUBBLE HEIGHT (IN.)[a]	RANGE TYPE	AUTHORITY
Shortgrasses			
Blue grama	1½–2	Shortgrass	Crafts and Glendening 1942
Buffalo grass	1–2	Shortgrass	Costello and Turner 1944
Curly mesquite	1½	Chihuahuan desert	Parker and Glendening 1942
Black grama	3	Chihuahuan desert	Paulsen and Ares 1962; Valentine 1971
Sandberg bluegrass	3–4	Sagebrush-palouse	Practical experience
Mountain muhly	4	Coniferous forest	Johnson 1953
Kentucky bluegrass	3–5	Mountain meadows	Clary 1995; Hall and Bryant 1995
Sedges	3–5	Mountain meadows	Clary 1995
Midgrasses			
Arizona fescue	6–7	Coniferous forest	Johnson 1953
Idaho fescue	5–6	Coniferous forest-palouse	Practical experience
Bluebunch wheatgrass	6	Sagebrush-palouse	Anderson 1969
Little bluestem	6–8	Tallgrass-mixed prairie	Practical experience
Sand dropseed	6–8	Mixed prairie-Chihuahuan desert	Practical experience
Sideoats grama	6	Mixed prairie-Chihuahuan desert	Practical experience
Green needlegrass	6	Northern mixed prairie	Practical experience
Western wheatgrass	3–4	Shortgrass-mixed prairie	Holscher and Woolfolk 1953
Crested wheatgrass	3–4½	Sagebrush	Frischknecht and Harris 1968
Threeawns	3–5	Mixed prairie-Chihuahuan desert	Practical experience
Tallgrasses			
Big bluestem	12–14	Tallgrass	Practical experience
Indian grass	12–14	Tallgrass	Practical experience
Switchgrass	12–14	Tallgrass	Practical experience
Giant sacaton	12–14	Chihuahuan desert	Practical experience
Basin wildrye	12–14	Sagebrush	Practical experience

TABLE 8.11 Minimum Recommended Stubble Heights for Selected Grass Species under Proper Grazing Use

GRASS SPECIES	MINIMUM STUBBLE HEIGHT (IN.)[a]	RANGE TYPE	AUTHORITY
Riparian grasses	3–7	Coniferous forest	Clary and Webster 1990; Clary 1995; Hall and Bryant 1995; Clary et al. 1996

[a]Recommended stubble height minimums should maintain or improve soil, vegetation and wildlife resources, and provide adequate plant material to meet livestock nutritional needs. We recognize in some cases our guidelines may be conservative if the only goal is maintenance of key forage plants.

Stubble height is one of the few measurements of range use that is highly repeatable and can be collected quickly. We have found that measurement of 40 randomly selected plants of each key forage species in key areas will usually give a reliable estimate of grazing use. Long-term studies by Johnson (1953), Paulsen and Ares (1962), and Valentine (1970) have shown grass heights to be well related to grazing intensity and forage productivity. Readers are referred to Clary and Webster (1990), Clary (1995), and Hall and Bryant (1995) for detailed stubble height guidelines on riparian zones.

FORAGE ALLOCATION TO MORE THAN ONE ANIMAL SPECIES

Many ranges are grazed by a combination of animals rather than by a single species. The grazing of two or more animals on the same range to obtain more efficient use is referred to as *common use.* It is well recognized that forage species selection varies considerably among different animal species on the same range. Mule deer on northwestern Colorado ranges heavily use big sagebrush but make little use of needlegrass (*Stipa* sp.) (Hansen et al. 1977). Conversely, on these same ranges, needlegrass is an important component of cattle diets, but cattle will not consume big sagebrush. This range can be used more efficiently by a combination of cattle and deer than by deer or cattle alone. The important questions relate to how much grazing capacity can be increased by the use of both animals, and what amount of the grazing capacity on these ranges should be allocated to deer and to cattle.

In the case above, little dietary overlap (less than 5%) occurs between the two animals, and grazing capacity is therefore additive when both animals area grazed. Because the key species are different for the two animals, no adjustment in cattle or deer numbers is necessary to compensate for forage consumed by the other animal.

On low-elevation winter range in northcentral New Mexico, cattle and sheep use the same ranges and have high dietary overlap (over 80%) (Holechek et al. 1986). On these ranges, common winterfat (*Ceratoides lanata*) and western wheatgrass (*Agropyron smithii*) are key species for both animals. Here, grazing with cattle and sheep in combination is nonadditive and animal unit equivalents of one animal can be substituted directly for the other animal. Grazing both animals in combination gives little improvement in efficiency of use of the forage resource.

However, on many ranges, cattle and sheep have moderate dietary overlaps (30% to 60%). This is also often true of cattle and elk. Here, allocation of forage is more complicated.

Controversy exists over how grazing capacity should be evaluated when common use is involved. Scarnecchia (1985, 1986) argues that grazing capacity should be based on animal-related factors because dietary overlaps between different animal species vary with terrain, season of use, grazing system, stocking rate, and year-to-year weather fluctuations that affect forage production and species composition.

In contrast, Hobbs and Carpenter (1986) advocate that animal unit equivalents should be weighted relative to degree of dietary overlap. They base their argument on the fact that different herbivores have different impacts on the range due to their consumption of different forages. Several computer models have been developed to resolve this problem (Cooperrider and Bailey 1984; Jensen 1984; Nelson 1984; Van Dyne et al. 1984). These models consider herbage yields of the various forage species consumed by different animals on the range, utilization levels these species will tolerate, and degree of dietary overlap between animals using the range.

Our own position is that when partially additive grazing capacities are involved (dietary overlap of 30% to 70%) and animals using the range share one or more of the same key forages, grazing capacity is, in most cases, nonadditive. Our position is supported by a New Mexico study of common use by cattle, sheep, and pronghorn (Howard et al. 1990). During this 4-year study involving several moderately stocked pastures with different combinations of cattle, sheep, and pronghorn, dietary overlaps between sheep and pronghorn averaged about 45%. However, during periods of drought, overlaps increased to 60%. In most of these pastures, pronghorn suffered heavy to complete mortality in drought periods and their populations generally declined during the study.

Cattle and pronghorn in this study had an average dietary overlap of about 18%. During drought periods this increased to 30%. On cattle-pronghorn pastures, pronghorn survived well in drought periods and maintained or increased their populations through the study.

Cattle and sheep diet overlap on this range averaged about 55%. Unlike pronghorn, both animals easily shifted their diets according to forage availability (see Chapter 11). Both animals can survive on diets of much lower quality than that required by pronghorn. Key species in this study were:

Cattle	*Sheep*	*Pronghorn*
Blue grama	Blue grama	Buckwheat
Sideoats grama	Globemallow	Globemallow
Threeawn	Bladderpod	Bladderpod

During both drought and nondrought periods, cattle consumed primarily grasses. Sheep preferred forbs, but after they were depleted during drought periods, they shifted to grasses. Pronghorn required a high forb and/or shrub diet throughout the year due to a small digestive system relative to body weight (see Chapter 11). After depletion of forbs by sheep, pronghorn perished, as shrubs were in low supply on the ranges studied.

On juniper (*Juniperus* sp.), oak (*Quercus* sp.), and sagebrush (*Artemisia* sp.) ranges, pronghorn and deer are able to survive drought periods when grazed with sheep because they will readily use the shrubs listed above. These shrubs are little used by sheep because of their high content of volatile oils and/or tannins (see Chapter 11).

Goats are generally additive when stocked with cattle because they will use many shrubs receiving little or no use by cattle. Conversely, goats are generally nonadditive when stocked with white-tailed deer or domestic sheep. During drought periods, all three animals heavily use many of the same shrubs (McMahan 1964; Bryant et al. 1979; Pfister and Malechek 1986).

CONCLUSION

The procedures we have discussed previously provide some guidelines for establishing an initial stocking rate for a particular range that can be adjusted as experience is gained. It is important to recognize that there is no substitute for experience. Local ranchers, state extension personnel, and Natural Resources Conservation Service personnel can provide useful advice on setting initial stocking rates to new ranch owners.

Downward trends in range condition are not always due to overgrazing. A few small exclosures (1 ha to 3 ha) on key grazing areas on a ranch can be useful in separating climatic from grazing influences. Although we advocate maintaining critical residues, we recognize that this may be impractical or impossible with extended droughts lasting 2 or more years. The utilization guidelines we have developed in Table 8.8 are based on long-term studies involving 5 or more years. Data from several studies shows that underuse in wet years will compensate for some overuse in dry years even on desert ranges. For most mother-young operations in the more arid areas, a stocking rate at 90% of the carrying capacity, with some adjustment in drought periods, will provide relatively high sustained ranch income and maintain or improve range condition. This strategy appears more practical than use of variable stocking rates and cow-calf-yearling operations on the more arid ranges.

Light stocking rates can be a useful tool to improve overgrazed ranges not heavily infested with unpalatable shrubs. However, once these shrubs become established, light grazing will generally give little improvement in forage production or composition. These ranges will require brush control for improvement.

RANGE MANAGEMENT PRINCIPLES

- The four basic components of grazing management are proper stocking rate, proper timing of use, proper distribution, and proper grazing system. Proper stocking is the most important part of successful range management.

■ Lighter grazing intensities are necessary for sustainability in the arid compared to humid range types. Areas with long growing seasons, high amounts of growing season precipitation, deep soils, and flat terrain can withstand heavier grazing intensities than where the opposite conditions prevail.

■ Stocking rates that sustain desirable vegetation will in most cases give the highest financial returns on both a short- and long-term basis.

■ Conservative stocking is a low cost and low risk approach with proven effectiveness in improving forage production on most degraded ranges. However, on some ranges heavily dominated by brush, other management tools such as fire, herbicides, mechanical manipulation, or biological manipulation may be required for meaningful improvement.

Literature Cited

Abdalla, S. H. 1980. *Application of simulation techniques to evaluate grazing management policies in the semidesert grasslands of southern New Mexico.* Ph.D. Thesis. New Mexico State University, Las Cruces, NM.

Anderson, J. A., and K. E. Holte. 1980. Vegetation development over 25 years without grazing on sagebrush dominated rangeland in southeastern Idaho. *J. Range Manage.* 34:25–29.

Anderson, E. W. 1969. Why proper grazing use? *J. Range Manage.* 22:361–363.

Bartolome, J. W., M. C. Stroud, and H. F. Heady. 1980. Influence of natural mulch on forage production on differing California annual range sites. *J. Range Manage.* 33:4–8.

Beale, F. F., D. M. Orr, W. E. Holmes, N. Palmer, C. J. Erenson, and P. S. Bowly. 1986. The effect of forage utilization levels on sheep production in the semiarid southwest of Queensland. *Proc. Intn't Rangel. Cong.* 2:30.

Beck, R. F. 1978. A grazing system for semiarid lands. *Proc. Int. Rangel. Congr.* 1:569–572.

Beck, R. F., and D. A. Tober. 1985. Vegetational changes on creosotebush sites after removal of shrubs, cattle, and rabbits. *New Mexico State Univ. Agr. Exp. Stn. Bull.* 717.

Beck, R. F., and R. P. McNeely. 1993. *Twenty-five year summary of year-long and seasonal grazing on the College Ranch.* Livestock Research Briefs and Cattle Growers Short Course. New Mexico State University, Las Cruces, NM.

Bement, R. E. 1969. A stocking rate guide for beef production on blue grama range. *J. Range Manage.* 22:83–86.

Bonham, C. D. 1989. *Measurements for terrestrial vegetation.* John Wiley and Sons. New York.

Boykin, C. C., J. R. Gray, and D. P. Caton. 1962. Ranch production adjustments to drought in eastern New Mexico. *New Mexico Agr. Expt. Sta. Bull.* 470.

Bryant, F. C., M. M. Kothmann, and L. B. Merrill. 1979. Diets of sheep, angora goats, Spanish goats, and white-tailed deer under excellent range conditions. *J. Range Manage.* 23:412–418.

Burzlaff, D. E., and L. Harris. 1969. Yearling steer gains and vegetation changes of western Nebraska rangelands under three rates of stocking. *Nebr. Agric. Exp. Stn. Bull.* 505.

Campbell, J. B. 1961. Continuous versus repeated-seasonal grazing of grass-alfalfa mixtures at Swift Current, Saskatchewan. *J. Range Manage.* 14:72–77.

Carande, V. G., E. T. Bartlett, and P. H. Guitierrez. 1995. Optimization strategies under rainfall and price risks. *J. Range. Manage.* 46:68–72.

Clary, W. P. 1995. Vegetation and soil responses to grazing simulation on riparian meadows. *J. Range Manage.* 48:18–26.

Clary, W. P., and B. F. Webster. 1990. Riparian grazing guidelines for the Intermountain region. *Rangelands* 12:209–213.

Clary, W. P., C. I. Thornton, and S. R. Abt. 1996. Riparian stubble height and recovery of degraded streambanks. *Rangelands* 18:137–141.

Cook, C. W. 1966. Factors affecting utilization of mountain slopes by cattle. *J. Range Manage.* 19:200–204.

Cook, C. W., and R. D. Child. 1971. Recovery of desert plants in various states of vigor. *J. Range Manage.* 22:339–343.

Cook, C. W., and J. Stubbendieck (Eds.). 1986. *Range research: Basic problems and techniques.* Society for Range Management, Denver, CO.

Cooperrider, A. Y., and J. A. Bailey. 1984. A simulation approach to forage allocation. In National Research Council/National Academy of Sciences (Eds.). *Developing strategies for rangeland management.* Westview Press, Inc., Boulder, CO.

Costello, D. F., and G. T. Turner. 1944. Judging condition and utilization of short-grass ranges on the central Great Plains. *USDA Farmers' Bull.* 1949.

Crafts, E. C., and G. E. Glendening. 1942. How to graze blue grama on southwestern ranges. *U.S. Dept. Agr. Leaflet* 215:1–8.

Currie, P. O. 1976. Recovery of ponderosa bunchgrass ranges through grazing and herbicide or fertilizer treatments. *J. Range Manage.* 29:444–448.

Drawe, D. L. 1988. Effects of three grazing treatments on vegetation, cattle production, and wildlife on the Welder Wildlife Foundation Refuge, 1974–1982. *Welder Wildlife Foundation Contrib.* B-8, Sinton, TX.

Dunford, E. G. 1949. Relation of grazing to runoff and erosion on bunchgrass ranges. *U.S. Dep. Agric. For. Serv. Note* RM-7.

Frischknecht, N. C., and L. E. Harris. 1968. Grazing intensities and systems on crested wheatgrass in central Utah: Response of vegetation and cattle. *U.S. Dep. Agric. Tech. Bull.* 1338.

Ganskopp, D., and M. Vavra. 1987. Slope use by cattle, feral horses, deer, and bighorn sheep. *Northwest Sci.* 61:74–81.

Gillen, R. F., W. C. Krueger, and R. F. Miller. 1984. Cattle distribution on mountain rangeland in northeastern Oregon. *J. Range Manage.* 37:549–553.

Glendening, G. E. 1944. Some factors affecting cattle use of northern Arizona pine-bunchgrass ranges. *U.S. Dept. Agr. For. Serv. Southwest For. Range Exp. Stn. Res. Rep. 6.*

Hall, F. C., and L. Bryant. 1995. Herbaceous stubble height as a warning of impending cattle grazing damage to riparian areas. *Gen. Tech. Rep.* PNW-GAR-362, Portland, OR. U.S. Department of Agriculture, Forest Service, Pacific Northwest Research Station. 9 p.

Hansen, R., R. C. Clark, and W. Hawhorn. 1977. Food of wild horses, deer, and cattle in the Douglas Mountain area, Colorado. *J. Range Manage.* 30:116–119.

Hart, R. H., J. Bissio, M. J. Samuel, and J. W. Waggoner, Jr. 1993. Grazing systems, pasture size, and cattle grazing behavior, distribution, and gains. *J. Range Manage.* 46:81–88.

Hart, R. H., M. J. Samuel, P. S. Test, and M. A. Smith. 1988. Cattle vegetation, and economic responses to grazing systems and grazing pressure. *J. Range Manage.* 41:282–286.

Heady, H. F., and R. D. Child. 1994. *Rangeland ecology and management.* Westview Press, San Francisco, CA.

Heitschmidt, R. K. 1986. Short duration grazing at the Texas Experimental Ranch. In *Proc. Short-Duration Grazing and Current Issues in Grazing Management Shortcourse.* Washington State University Cooperative Extension Service, Pullman, WA.

Heitschmidt, R. K., S. L. Dowhower, R. A. Gordon, and D. L. Price. 1985. Response of vegetation to livestock grazing at the Texas Experimental Ranch. *Tex. A&M Univ. Agric. Exp. Stn. Bull.* 1515.

Heitschmidt, R. K., S. L. Dowhower, and J. W. Walker. 1987. 14 vs 42-paddock rotational grazing: Aboveground biomass dynamics, forage production, and harvest efficiency. *J. Range Manage.* 40:216–224.

Heitschmidt, R. K., J. R. Conner, S. K. Canon, W. E. Pinchak, J. W. Walker, and S. L. Dowhower. 1990. Cow/calf production and economic returns from yearlong continuous deferred rotation and rotational grazing treatments. *J. Agric. Prod.* 3:92–99.

Herbel, C. H., and K. L. Anderson. 1959. Response of true prairie vegetation on major Flint Hills range sites to grazing treatment. *Ecol. Monogr.* 29:171–198.

Herbel, C. H., J. D. Wallace, M. D. Finkner, and C. C. Yarbrough. 1984. Early weaning and part-time confinement of cattle on arid rangelands of the southwest. *J. Range Manage.* 37:127–130.

Hobbs, N. T., and L. H. Carpenter. 1986. Viewpoint: Animal-unit equivalents should be weighted by dietary differences. *J. Range Manage.* 39:470–471.

Holechek, J. L. 1988. An approach for setting the stocking rate. *Rangelands* 10:10–14.

Holechek, J. L. 1991. Chihuahuan desert rangeland, livestock grazing and sustainability. *Rangelands* 13:115–120.

Holechek, J. L., and R. D. Pieper. 1992. Estimation of stocking rate on New Mexico rangeland. *J. Soil and Water Conser.* 47:116–119.

Holechek, J. L. 1996a. Drought and low cattle prices: Hardship for New Mexico ranchers. *Rangelands* 18:11–13.

Holechek, J. L. 1996b. Drought in New Mexico: Prospects and management. *Rangelands* 18:225–228.

Holechek, J. L., and T. Stephenson. 1983. Comparison of big sagebrush vegetation in northcentral New Mexico under moderately grazed and grazing excluded conditions. *J. Range Manage.* 36:455–457.

Holechek, J. L., J. Jeffers, T. Stephenson, C. B. Kuykendall, and S. A. Butler-Lance. 1986. Cattle and sheep diets on low elevation winter range in northcentral New Mexico. *Proc. West. Sec. Am. Soc. Anim. Sci.* 37:243–248.

Holechek, J. L., J. Hawkes, and T. Darden. 1994. Macro-economics and cattle ranching. *Rangelands* 16:118–123.

Holscher, C. E., and E. J. Woolfolk. 1953. Forage utilization by cattle on northern Great Plains ranges. *USDA Circ.* 918.

Hooper, J. F., and H. F. Heady. 1970. An economic analysis of optimum rates of grazing in the California annual type ranges. *J. Range Manage.* 23:307–311.

Houston, W. R., and R. R. Woodward. 1966. Effects of stocking rates on range vegetation and beef cattle production in the northern Great Plains. *U.S. Dep. Agric. Tech. Bull.* 1357.

Howard, V. W., J. L. Holechek, R. D. Pieper, K. Green-Hammond, M. Cardenas, and S. L. Beasom. 1990. Habitat requirements for pronghorn on rangeland impacted by livestock and net wire in eastcentral New Mexico. *New Mexico Agr. Exp. Stn. Bull.* 750.

Hughes, L. E. 1980. Six enclosures with a message. *Rangelands* 2:17–18.

Hughes, L. E. 1982. A grazing system in the Mojave desert. *Rangelands* 4:256–258.

Hutchings, S. S., and G. Stewart. 1953. Increasing forage yields and sheep production on intermountain winter ranges. *U.S. Dep. Agric. Circ.* 925.

Hyder, D. N. 1953. Grazing capacity as related to range condition. *J. For.* 51:206.

Jensen, J. C. 1984. Perspectives on BLM forage allocation: Calculations with special reference to the limiting factor approach. In National Research Council/National Academy of Sciences (Eds.). *Developing strategies for rangeland management.* Westview Press, Inc., Boulder, CO.

Johnson, W. M. 1953. Effect of grazing intensity upon vegetation and cattle gains on ponderosa pine-bunchgrass ranges of the front range of Colorado. *U.S. Dep. Agric. Circ.* 929.

Klipple, G. E., and D. F. Costello. 1960. Vegetation and cattle responses to different intensities of grazing on shortgrass ranges of the central Great Plains. *U.S. Dep. Agric. Tech. Bull.* 1216.

Klipple, G. E., and R. E. Bement. 1961. Light grazing—Is it economically feasible as a range improvement practice? *J. Range Manage.* 14:57–62.

Kothmann, M. M., G. W. Mathis, and W. J. Waldrip. 1971. Cow-calf response to stocking rates and grazing systems on native range. *J. Range Manage.* 24:100–105.

Kothmann, M. M., W. S. Rawlins, and J. Bluntzer. 1975. Vegetation and livestock responses to grazing management on the Texas Experimental Ranch. *Tex. Agric. Exp. Stn.* PR-3310.

La Baume, J. T. 1981. "Accounting" for adjustable stocking rates on public lands. *Rangelands* 3:145–146.

Lang, R. L., O. K. Barnes, and F. Rauzi. 1956. Shortgrass range: Grazing effects on vegetation and sheep gains. *Wyo. Agric. Exp. Stn. Bull.* 343.

Laycock, W. A., and P. W. Conrad. 1981. Responses of vegetation and cattle to various systems of grazing on seeded and native mountain rangelands in eastern Utah. *J. Range Manage.* 34:52–58.

Lewis, J. K., G. M. Van Dyne, L. R. Albee, and F. W. Whetzal. 1956. Intensity of grazing: Its effect on livestock and forage production. *S. Dak. Agric. Exp. Stn. Bull.* 459.

Lodge, R. W. 1970. Complementary grazing systems for the northern Great Plains. *J. Range Manage.* 23:268–271.

Martin, S. C. 1970. Vegetation changes on semi-desert ranges during 10 years of summer, winter, and year-long grazing by cattle. *Proc. Int. Grassl. Congr.* 11:23–26.

Martin, S. C. 1973. Responses of semi-desert grasses to seasonal rest. *J. Range Manage.* 26:165–170.

Martin, S. C. 1975. Stocking strategies and net cattle sales on semi-desert range. *U.S. Dep. Agric. For. Serv. Res. Pap.* RM-146.

Martin, S. C., and D. R. Cable. 1974. Managing semidesert grass-shrub ranges: Vegetation responses to precipitation, grazing, soil texture, and mesquite control. *U.S. Dep. Agric. Tech. Bull.* 1480.

Martin, S. C., and D. E. Ward. 1970. Rotating access to water to improve semidesert cattle range near water. *J. Range Manage.* 23:22–26.

Martin, S. C., and D. E. Ward. 1973. Salt and meal-salt help distribute cattle use on semidesert range. *J. Range Manage.* 26:94–97.

Martin, S. C., and D. E. Ward. 1976. Perennial grasses respond inconsistently to alternate year seasonal rest. *J. Range Manage.* 29:346.

Martin, S. C., and K. E. Severson. 1988. Vegetation response to the Santa Rita grazing system. *J. Range Manage.* 41:291–296.

McCalla, G. R. II, W. H. Blackburn, and L. B. Merrill. 1984. Effects of livestock grazing on sediment production, Edwards Plateau of Texas. *J. Range Manage.* 37:211–295.

McDaniel, K. C., and J. A. Tiedeman. 1981. Sheep use on mountain winter range in New Mexico. *J. Range Manage.* 34:102–105.

McIlvain, E. H., and M. C. Shoop. 1965. Forage, cattle, and soil responses to stocking rates and grazing systems on sandy rangeland in the Southern Plains. *Abstr. Annu. Meet. Soc. Range Manage.* 18:31–34.

McMahan, C. A. 1964. Comparative food habits of deer and three classes of livestock. *J. Wildl. Manage.* 28:798–808.

Miller, R. F., and G. B. Donart. 1979. Response of *Bouteloua eriopoda* (Torr.) and *Sporobolus flexuosus* (Thrub.) Rybd. to season of defoliation. *J. Range Manage.* 32:63–67.

Miller, R. F., and G. B. Donart. 1981. Response of *Muhlenbergia porteri* Schibn. to season of defoliation. *J. Range Manage.* 34:91–94.

Mueggler, W. F. 1965. Cattle distribution on steep slopes. *J. Range Manage.* 18:255–257.

Murray, R. B., and J. O. Klemmedson. 1968. Cheatgrass range in southern Idaho: Seasonal cattle gains and grazing capacities. *J. Range Manage.* 21:308–313.

Nelson, J. R. 1984. A modeling approach to large herbivore competition. In National Research Council/National Academy of Sciences (Eds.). *Developing strategies for rangeland management.* Westview Press, Inc., Boulder, CO.

Owensby, C. E., E. F. Smith, and K. L. Anderson. 1973. Deferred rotation grazing with steers in the Kansas Flint Hills. *J. Range Manage.* 26:393–395.

Parker, K. W., and G. E. Glendening. 1942. General guide to satisfactory utilization of the principal southwestern range grasses. *S.W. For. and Range Expt. Sta. Research Note* 104.

Paulsen, H. A., Jr., and F. N. Ares. 1961. Trends in carrying capacity and vegetation on an arid southwestern range. *J. Range Manage.* 14:78–83.

Paulsen, H. A., and F. N. Ares. 1962. Grazing values and management of black grama and tobosa grasslands and associated shrub ranges of the southwest. *U.S. Dept. Agric. Tech. Bull.* 1270.

Pearson, H. A. 1973. Calculating grazing intensity for maximizing profit on ponderosa pine range in northern Arizona. *J. Range Manage.* 26:277–278.

Pearson, H. A., and L. B. Whitaker. 1974. Forage and cattle responses to different grazing intensities on southern pine range. *J. Range Manage.* 27:444–446.

Pechanec, J. F., and G. Stewart. 1949. Grazing spring-fall sheep ranges in southern Idaho. *U.S. Dep. Agric. Circ.* 808.

Pfister, J. A., and J. C. Malechek. 1986. Dietary selection by goats and sheep in deciduous woodland of northeastern Brazil. *J. Range Manage.* 39:24–29.

Pickford, G. D., and E. H. Reid. 1948. Forage utilization on summer cattle ranges in eastern Oregon. *U.S. Dep. Agric. Circ.* 796.

Pieper, R. D. 1970. Species utilization and botanical composition of cattle diets on piñon-juniper grassland. *New Mexico Agr. Exp. Sta. Bul.* 566.

Pieper, R. D. 1981. The stocking rate decision. *Proc. Int. Rancher's Roundup* 1:199–204.

Pieper, R. D., L. W. Cook, and L. E. Harris. 1959. Effect of intensity of grazing upon nutrient content of the diet. *J. Anim. Sci.* 18:1031–1037.

Pieper, R. D., G. B. Donart, E. E. Parker, and J. D. Wallace. 1978. Livestock and vegetational response to continuous and four-pasture one-herd grazing systems in New Mexico. *Proc. Int. Rangel. Congr.* 1:560–562.

Pieper, R. D., E. E. Parker, G. B. Donart, and J. D. Wright. 1991. Cattle and vegetational response to four-pasture and continuous grazing systems. *New Mexico Agr. Exp. Sta. Bul.* 576.

Pinchak, W. E., M. A. Smith, R. H. Hart, and J. W. Wagoner. 1991. Beef cattle distribution patterns on foothill ranges. *J. Range Manage.* 44:267–276.

Pratchett, D., and G. Gardiner. 1991. Does reducing stocking rate necessarily mean reducing income? *Intnl. Rangel. Cong.* 4:714–716.

Quigley, T. M., J. M. Skovlin, and J. P. Workman. 1984. An economic analysis of two systems and three levels of grazing on ponderosa pine-bunchgrass range. *J. Range Manage.* 37:309–312.

Reardon, P. O., and L. B. Merrill. 1976. Vegetative response under various grazing management systems in the Edward Plateau of Texas. *J. Range Manage.* 29:195–198.

Rosiere, R. 1987. An evaluation of grazing intensity influences on California annual range. *J. Range Manage.* 40:160–165.

Rosiere, R. E., and D.T. Torell. 1996 Performance of sheep grazing California annual range. *Sheep and Goat Res. J.* 12:49–58.

Scarnecchia, D. L. 1985. The animal-unit and animal-unit equivalent concepts in range science. *J. Range Manage.* 39:346–349.

Scarnecchia, D. L. 1986. Viewpoint: Animal-unit equivalents cannot be meaningfully weighted by indices of dietary overlap. *J. Range Manage.* 39:471.

Schmutz, E. E. 1977. Adjustable vs fixed stocking rates on public lands. *Rangeman's J.* 4:178.

Shoop, M. C., and E. H. McIlvain. 1971. Why some cattlemen overgraze and some don't. *J. Range Manage.* 24:252–257.

Sims, P. L., B. E. Dahl, and A. H. Denham. 1976. Vegetation and livestock response at three grazing intensities on sandhill rangeland in eastern Colorado. *Colorado State University Exp. Sta. Tech. Bull.* 130.

Skovlin, J. M., R. W. Harris, G. S. Strickler, and G. A. Garrison. 1976. Effects of cattle grazing methods on ponderosa pine-bunchgrass range in the Pacific northwest. *U.S. Dep. Agric. Tech. Bull.* 1531.

Smith, A. D. 1944. A study of reliability of range vegetation estimates. *Ecology* 25:441–448.

Smith, D. A., and E. M. Schmutz. 1975. Vegetative changes on protected versus grazed desert grassland ranges in Arizona. *J. Range Manage.* 28:453–458.

Smith, E. F., and C. E. Owensby. 1978. Intensive-early stocking and season-long stocking of Kansas Flint Hills range. *J. Range Manage.* 31:14–17.

Smoliak, S. 1960. Effect of deferred rotation and continuous grazing on yearling steer gains and shortgrass prairie vegetation of southeastern Alberta. *J. Range Manage.* 13:239–243.

Smoliak, S. 1974. Range vegetation and sheep production at three stocking rates on *Stipa-Bouteloua* prairie. *J. Range Manage.* 27:23–26.

Sneva, F. A., L. R. Rittenhouse, and L. Foster. 1973. Stockwater restriction and trailing effects on animal gain, water drunk, and mineral consumption. *Water-Animal Relations Symposium Proc.,* pp. 34–48.

Society for Range Management. 1989. *A glossary of terms used in range management.* 3d ed. Society for Range Management, Denver, CO.

Squires, V. R. 1973. Distance to water as a factor in performance of livestock on arid and semiarid rangelands. *Water-Animal Relations Symposium Proc.,* pp. 28–33.

Stoddart, L. A. 1946. Some physical and chemical responses of *Agropyron spicatium* to herbage removal at various seasons. *Utah Agric. Exp. Sta. Bull.* 324.

Taylor, C. A., and M. H. Ralphs. 1992. Reducing livestock losses from poisonous plants through grazing management. *J. Range Manage.* 45:9–12.

Thilenius, J. F. 1979. Range management in the alpine zone. In D. A. Johnson (Ed.). *Special management needs of alpine ecosystems. Range Science Series No. 5,* pp. 43–65. Society for Range Management, Denver, CO.

Torell, L. A., K. S. Lyon, and E. B. Godfry. 1991. Long-run versus short-run planning horizons and rangeland stocking rate decision. *Amer. J. Agr. Econ.* 73:795–807.

Trlica, M. Bawai, and J. W. Meake. 1977. Effects of rest following defoliations on the recovery of several range species. *J. Range Manage.* 30:21–27.

Troxel, T. R. and L. D. White. 1989. Balancing forage demand with forage supply. *Texas A&M Univ. Ext. Serv. Publ.* B–1606.

Valentine, K. A. 1947. Distance to water as a factor in grazing capacity of rangeland. *J. For.* 45:749–754.

Valentine, K. A. 1970. Influence of grazing intensity on improvement of deteriorated black grama range. *N. Mex. Agric. Exp. Stn. Bull.* 553.

Van Dyne, G. M., P. T. Kortopates, and F. M. Smith. 1984. Quantitative frameworks for forage allocation. In National Research Council/National Academy of Sciences (Eds.). *Developing strategies for rangeland management.* Westview Press, Inc., Boulder, CO.

Van Poollen, H. W., and J. R. Lacey. 1979. Herbage response to grazing systems and stocking intensities. *J. Range Manage.* 32:250–253.

Vavra, M., R. W. Rice, and R. E. Bement. 1973. Chemical composition of the diet, intake, and gain of yearling cattle under different grazing intensities. *J. Anim. Sci.* 36:411–414.

White, M. R., R. D. Pieper, G. B. Donart, and L. White-Trifaro. 1991. Vegetational response to short-duration and continuous grazing in southcentral New Mexico. *J. Range Manage.* 44:399–403.

Whitson, R. E., R. K. Heitschmidt, M. M. Kothmann, and G. K. Lundgren. 1982. The impact of grazing systems on the magnitude and stability of ranch income in the Rolling Plains of Texas. *J. Range Manage.* 35:526–533.

Willms, W. D., S. Smoliak, and G. B. Schaalje. 1986. Cattle weight gains in relation to stocking rate on rough fescue grassland. *J. Range Manage.* 39:182–187.

CHAPTER 9

SELECTION OF GRAZING METHODS

Specialized grazing systems have been a major focus of range researchers and managers since the 1950s. During the 1950s and 1960s, deferred-rotation systems received considerable attention. Rest-rotation grazing was heavily applied on public lands in the intermountain West during the 1970s. Both deferred- and rest-rotation systems are still being utilized, particularly in mountainous areas. In the 1980s, short-duration (cell) grazing became the newest fad in grazing systems. Through the years, impressive claims have been made for each new system regarding increased stocking rates and livestock production. However, actual research has shown that specialized grazing systems generally give either modest (10% to 30%) or no increase in grazing capacity over season-long or continuous systems (see Table 8.1). In this chapter, we discuss the major considerations involved in the selection of grazing methods.

DEFINITION OF GRAZING SYSTEM TERMS

Deferment, rest, and *rotation* are terms that receive constant use when grazing systems are discussed. *Deferment* involves delay of grazing in a pasture until the seed maturity of the key forage species. This permits the better forage plants to gain vigor and reproduce. *Rest* is distinguished from deferment in that the range receives nonuse for a full year rather than just during the growth period. This gives plants a longer period to recover from past grazing influences and provides wildlife with a pasture free from livestock use during the critical dormant period. A disadvantage of both deferment and rest is that the grazing load on other pastures must be increased during the critical growth period. It is questionable if periodic nonuse during critical periods compensates for periodic heavy use. However, deferment and rest do provide plants on sacrifice areas with some opportunity for recovery. On mountainous areas, season-long or continuous grazing generally results in degradation of areas convenient to livestock even under light grazing. *Rotation* involves the movement of livestock from one pasture to another on a scheduled basis. It is the critical feature of all specialized grazing systems. The main advantage of rotation is that key forage plants are provided with periodic nonuse during the critical growing season. Systems with deferment and rest typically involve livestock rotations. Short-duration (rapid-rotation, cell, and time-control) grazing is distinguished from other specialized systems in that a pasture (paddock) will receive several periods of nonuse and grazing during the growing season.

CONSIDERATIONS IN GRAZING SYSTEM SELECTION

Grazing systems commonly used in the United States and other parts of the world include continuous, deferred-rotation, rest-rotation, short-duration, Merrill three-herd/four-pasture, high intensity-low frequency, best-pasture, and seasonal-suitability. Climate, topography, vegetation, kind or kinds of livestock to be grazed, wildlife needs, watershed protection, labor requirements, and developments (fence and water) are important considerations involved in grazing system selection. Specialized systems have been most useful where:

1. Terrain is rugged.
2. Wildlife are an important consideration.
3. Water distribution is poor.
4. Poor distribution of precipitation over the range occurs within years (e.g., the southwestern United States).
5. Carefully timed grazing is necessary to prevent tree damage.
6. The vegetation has low grazing resistance.

We will discuss the conditions where the various grazing systems give best results, updating the discussion of Holechek (1983).

Continuous Grazing

Although it has been speculated that desirable plants, particularly grasses, will be grazed excessively under continuous grazing, actual research does not support this speculation. Studies from the various range types in the United States show that during the critical growing season both cattle and sheep select a variety of plants. During this period many forbs are highly preferred and their use greatly reduces grazing pressure on the grasses. Rotation systems that restrict livestock to part of the range during the growing season can result in waste of much of the forb crop. This is because many forbs complete their life cycle quickly and are generally low in palatability after they mature.

Another often overlooked advantage of continuous grazing is that actual grazing pressure during the critical growing season is relatively light (10% to 20%) since adequate forage must be left to carry animals through the dormant season. Numerous studies throughout the United States and other parts of the world show that rangeland productivity and condition can be maintained under moderate continuous grazing (see Table 8.1).

In terms of vegetation productivity, continuous grazing at a moderate stocking rate has been superior to various rotation systems in the flat shortgrass and northern mixed prairie country of the Great Plains, where watering points are seldom farther than 3.2 km apart (Black et al. 1937; Smoliak 1960; Klipple 1964) (Figure 9.1). The grasses in this area evolved with heavy bison grazing and are quite grazing resistant. Precipitation occurs as several light rains throughout the summer months. Therefore, considerable opportunity exists for regrowth after defoliation. The flat nature of the terrain and the close proximity of watering points minimize the tendency of livestock to congregate and linger in the most convenient areas. Sacrifice areas can be improved by temporary fencing to allow vegetation recovery. Since most of the sacrifice areas occur around water, control of access to watering points can be used to provide sacrifice areas with periodic nonuse.

The real problem with continuous grazing is that livestock have preferred areas for grazing. These areas generally occur where water, forage, and cover are in close proximity and are often the most productive parts of the pasture. Even under light stocking rates, these areas will often receive excessive use.

However, in relatively flat terrain this problem can be overcome by rotation of livestock access to watering points. Martin and Ward (1970) found that regulating access to watering points could be effective in controlling where cattle grazed on Arizona semidesert ranges, with little extra cost for labor or fence. Within a year, cattle had learned to move

Figure **9.1** Continuous grazing has maintained range vegetation productivity in the shortgrass prairie; where terrain is flat, water points are in close proximity, and most plants have some grazing value.

and use the range where watering points were open over the 1,279-ha study area. Rotation of access to watering points nearly doubled the yield of perennial grasses compared to continuous grazing during the 8-year study. Most of the vegetation improvement occurred within 300 m of the watering points. Rotation of watering points was most effective if the closed period included the summer growing season.

On shortgrass range in southeastern Wyoming, Hart et al. (1988) reported that deferred-rotation, short-duration, and continuous grazing systems at the same stocking rates did not differ in vegetation or livestock production over a 6-year period. This study was unique in that both moderate and heavy stocking rates were used for each system. There were no vegetation composition differences between the grazing systems. This study indicates short-duration and deferred-rotation grazing have no advantages over continuous grazing on shortgrass prairie range.

Continuous grazing has given comparable or superior vegetation productivity compared to specialized systems in the California annual grassland type when use was moderate and practices such as salting, fencing, and water development were used to obtain proper distribution (Heady 1961). Annual grasses need only to set seed year after year to maintain themselves, unlike perennial grasses, which must store carbohydrates.

In both the Great Plains prairie and California annual grassland types, individual livestock performance has generally been better under continuous grazing than specialized systems (Table 9.1). This is explained by the fact that continuous grazing allows livestock to exhibit maximum forage selectivity and minimizes livestock disturbance due to gathering, trailing, and quick change in forage quality.

Results from various studies comparing livestock production per unit area among continuous and rotation grazing systems are shown in Table 9.2. Generally, livestock production per unit area has been greater under continuous grazing, but there are exceptions (Table 9.2). Recent short-term studies with short-duration grazing consistently show higher livestock production per unit area than under continuous grazing (Daugherty et al. 1982; Sharrow 1983; Jung et al. 1985; Heitschmidt et al. 1990) (Tables 9.3 and 9.4). Based on our analysis of the literature, it appears that rotation grazing systems are most likely to increase livestock production per unit area in the humid grassland range types, where average annual precipitation exceeds 500 mm or on seeded ranges such as crested wheatgrass (*Agropyron cristatum*) pastures (Table 9.4). We found no studies involving native rangeland showing higher livestock production from rotation systems than for continuous grazing in areas receiving less than 300 mm (12 in.) of average annual precipitation.

Season-long grazing is distinguished from continuous grazing in that animals are grazed on a particular pasture for only part of the year. Season-long grazing is commonly used in the intermountain portions of the western United States, where cold weather conditions make grazing feasible only during part of the year, such as on alpine or forested ranges. Low-elevation ranges are usually saved for winter use if the operator has both types of rangeland.

A variation of season-long grazing called intensive-early stocking has proven superior to regular season-long summer grazing for steer production on tallgrass ranges of the Kansas Flint Hills (Smith and Owensby 1978) and eastern Oklahoma (McCollum et al. 1990). This grazing strategy involves doubling the stocking rate compared to normal season-long grazing, but shortening the grazing period from 154 days to 75 days. Under intensive-early stocking, steers are grazed from May 2 to July 15 compared with May 2 to

TABLE 9.1 Individual Livestock Weight Gains from Several Studies Comparing Rotation and Continuous Grazing

LOCATION	YEARS OF STUDY	CLASS OF LIVESTOCK	CHARACTERISTIC	WEIGHT GAIN (KG)		REFERENCE
				CONTINUOUS	ROTATION	
			ADVANTAGE CONTINUOUS GRAZING			
Tallgrass						
Woodward, Oklahoma	1	Steers	Seasonal gain	118	95	McIlvain and Savage 1951
Sinton, Texas	5	Calves	Weaning weight	233	221	Drawe 1988
Flint Hills, Kansas	17	Steers	Average daily gain	0.66	0.50	Owensby et al. 1973
Midgrass prairie						
Mandan, North Dakota	8	Yearling steers	Seasonal gain	119	110	Rogler 1951
Mayberries, Canada	6	Calves	Weaning weight	160	150	Hubbard 1951
Mayberries, Canada	9	Calves	Average daily gain	0.76	0.72	Smoliak 1960
Shortgrass						
Fort Stanton, New Mexico	10	Calves	Weaning weight	193	185	Pieper et al. 1990
Desert grassland						
Santa Rita, Arizona	2	Cows	End-of-season weight	429	396	Ward 1975
College Ranch, New Mexico	25	Calves	Weaning weight	221	213	Beck and McNeely 1993
Bunchgrass						
Harvey Valley, California	4	Calves	Average daily gain	66	62	Ratliff et al. 1972
Starkey, Oregon	10	Calves	Seasonal gain	83	81	Skovlin et al. 1976
Annual grassland						
Hopland, California	5	Lambs	120-day gain	35	32	Heady 1961
San Joaquin Range, California	8	Calves	Seasonal gain	229	204	Ratliff 1986
Sagebrush grassland						
Squaw Butte, Oregon	5	Cows	Seasonal gain	57	53	Hyder and Sawyer 1951
Saylor Creek, Idaho	3	Yearling steers	Average daily gain	0.66	0.63	Murray and Klemmedson 1968
Crested wheatgrass						
Benmore, Utah	13	Calves	Average daily gain	0.79	0.77	Frischknecht and Harris 1968

ADVANTAGE ROTATION GRAZING

Location	n	Animal	Measure			Reference
Tallgrass						
Flint Hills, Kansas	—	Steers	Average daily gain	0.53	0.60	Anderson 1940
Midgrass prairie						
Throckmorton, Texas	8	Calves	Weaning weight	228.	237	Kothmann et al. 1971
Throckmorton, Texas	2	Calves	Weaning weight	181	185	Mathis and Kothmann 1971
Throckmorton, Texas	18	Calves	Weaning weight	212	215	Heitschmidt et al. 1982c
Throckmorton, Texas	4	Calves	Weaning weight	262	270	Heitschmidt et al. 1990
Foothills						
Soap Creek, Oregon	3	Lambs	Seasonal gain	26	28	Sharrow and Krueger 1979
Africa						
Zimbabwe	2	Calves	Weaning weight	230	250	Savory 1978
			NO DIFFERENCE			
Tallgrass						
Woodward, Oklahoma	7	Steers	Seasonal gain	139	134	McIlvain and Savage 1951
Midgrass prairie						
Spur, Texas	8	Steers	Average daily gain	0.40	0.41	Fisher and Marion 1951
Shortgrass						
Panhandle Experimental Range, Nebraska	15	Steers	Average daily gain	0.95	0.92	Reece 1986
Forest						
Manitou, Colorado	6	Calves	Seasonal gain	50	49	Currie 1978
Bighorn Mountains, Wyoming	3	Steers	Average daily gain	0.95	1.00	Smith et al. 1967
Starkey, Oregon	5	Yearling heifers	Average daily gain	0.58	0.56	Holechek et al. 1987

Source: Revised from Pieper 1980.

TABLE 9.2 Weight Gains per Unit Area from Various Studies Involving Grazing Systems

LOCATION	GRAZING SYSTEM	WEIGHT GAIN (KG/HA)	REFERENCE
Tallgrass			
Flint Hills, Kansas	Season-long	44.8	Anderson 1940
	Deferred-rotation	73.0	
Welder Wildlife Refuge, Texas	Continuous	47.7	Drawe 1988
	Merrill three-herd/four-pasture	50.0	
	High-intensity/low-frequency	31.1	
Midgrass			
Throckmorton, Texas	Heavy continuous	38.6	Kothmann et al. 1971
	Moderate continuous	23.7	
	Light continuous	18.4	
	Merrill three-herd/four-pasture	27.6	
	Heavy continuous	32.0	Heitschmidt et al. 1982c
	Moderate continuous	24.9	
	Merrill three-herd/four-pasture	27.4	
	Heavy continuous	41.5	Heitschmidt et al. 1990
	Moderate continuous	35.8	
	Merrill three-herd/four-pasture	36.9	
	Short-duration	49.2	
Barnhart, Texas	Continuous	7.2	Taylor and Garza 1986
	South African switchback	7.2	
	High intensity/low frequency	6.4	
	Merrill three-herd/four-pasture	6.6	

Shortgrass			
Fort Stanton, New Mexico	Heavy continuous	7.8	Pieper et al. 1991
	Moderate continuous	6.9	
	Four-pasture rotation	7.1	
Desert grassland			
College Ranch, New Mexico	Moderate continuous	3.5	Beck and McNeely 1993
	Best-pasture rotation	3.5	
Sagebrush grassland			
Saylor Creek, Idaho	Heavy continuous	23.7	Murray and Klemmedson 1968
	Moderate continuous	15.6	
	Rotation	13.0	
Forest			
Bighorn Mountains, Wyoming	Moderate season-long	63.9	Smith et al. 1967
	Moderate rotation	61.7	
	Heavy rotation	72.9	
Starkey, Oregon	Season-long	7.8	Skovlin et al. 1976
	Deferred-rotation	6.8	
Crested wheatgrass			
Benmore, Utah	Continuous	43.7	Frischknecht and Harris 1968
	Rotation	48.0	

Source: Revised from Pieper 1980.

TABLE 9.3 Average Herbage Standing Crop and Live Weight Changes of Ewes and Lambs from Pastures under Season-long and Rapid-rotation (Short-duration) Grazing Systems at the Same Stocking Rate in Western Oregon[a]

	SEASON-LONG	RAPID-ROTATION
Average herbage standing crop (kg/ha)	2,140	2,503
Ewe live weight change (g/ewe/day)	−45	−53
Lamb live weight change (g/lamb/day)	172	205

Source: Sharrow 1983.

[a]Average annual precipitation, 1,020 mm; pastures were grazed in spring and summer; data were averaged over 2 years; the same stocking rate was used for each system.

TABLE 9.4 Effects of Grazing Treatment on Yearling Steer Performance on Crested Wheatgrass Pasture in Southeastern Oregon[a]

	SEASON-LONG	SHORT-DURATION
Number of Steers	30	28[b]
Grazing period (days)	100	100
Total gain per steer (kg)	100	109
Total gain per system (kg)	3,000	3,052

Source: Daugherty et al. 1982.

[a]Stocking rates were the same for season-long and short-duration treatments.

[b]Two steers died of unknown causes on the short-duration pastures.

October 3 for normal season-long grazing. Total steer production per hectare is substantially increased under intensive-early stocking compared to regular season-long grazing (Table 9.5). Intensive-early stocking is advantageous to important forage species such as big bluestem (*Andropogon gerardii*), and more uniform grazing occurs on all sites than under regular season-long grazing (Smith and Owensby 1978). Removal of cattle by mid-July under intensive-early stocking gives the desirable warm-season grasses such as big bluestem a long growing period (July 15 to October 15) without defoliation. On shortgrass range in western Kansas, intensive-early stocking and season-long stocking gave equal vegetation and steer production responses (Olson et al. 1993).

Deferred-Rotation Grazing

The first specialized grazing system developed in the United States was deferred-rotation. Initial research on this system was conducted by Arthur Sampson in the Blue Mountains of

TABLE 9.5 Steer Gains on Intensive-early and Season-long Stocked Pastures

	INTENSIVE-EARLY STOCKED MAY 2–JULY 15, 75 DAYS	SEASON-LONG STOCKED		
		MAY 2–JULY 15, 75 DAYS	JULY 15–OCTOBER 3, 79 DAYS	MAY 2–OCTOBER 3, 154 DAYS
Hectares per steer	0.69	1.38	1.38	1.38
Gain per steer (kg), average	64	90	36	95
Daily gain per steer (kg), average	0.85	0.80	0.45	0.62
Gain per hectare (kg), average	93	43	26	70

Source: Smith and Owensby 1978.

northeastern Oregon in the early twentieth century (Sampson 1913). The system involved dividing the range into two pastures. Each pasture received deferred grazing every other year (Figure 9.2).

Several modifications of this system have been used involving more than two pastures. However, the key feature remains that periodically (every 2 to 4 years) each pasture will receive deferment. Vegetation response under this system has been slightly to moderately better than continuous or season-long grazing on palouse bunchgrass ranges (Hubbard 1951; Dillion 1958; Skovlin et al. 1976) and mountain coniferous forest ranges (Johnson 1965; Skovlin et al. 1976). It has given best results on tallgrass prairie ranges (Herbel and Anderson 1959; Owensby et al. 1973). However, on flat sagebrush grassland (Hyder and Sawyer 1951) and shortgrass (Hart et al. 1988) rangelands it has shown no vegetation benefits over continuous or season-long grazing. Deferred-rotation grazing provides a better opportunity for preferred plants and areas to maintain and gain vigor than does continuous grazing. It works best where considerable differences exist between palatability of plants and convenience of areas for grazing. On mountain ranges, stringer meadows and riparian zones will often receive excessive use by cattle even under light grazing intensities, while surrounding uplands will receive little or no use (Roath and Krueger 1982a; Gillen et al. 1985). The deferred-rotation system provides forage species on the lowland sacrifice areas with the opportunity to store carbohydrates and set seed every other year. On palouse bunchgrass ranges and coniferous forest ranges, individual cattle weight gains have shown little difference between deferred-rotation and season-long systems under moderate stocking rates (Skovlin et al. 1976; Holechek et al. 1987) (Table 9.6). In tallgrass prairie, deferred-rotation has reduced individual cattle weight gains compared to continuous grazing (Herbel and Anderson 1959; Owensby et al. 1973). However, the increased stocking rates possible under the deferred-rotation system more than compensate for the lower animal performance. On flat arid and semiarid rangelands, deferred-rotation grazing has shown no advantages over continuous grazing.

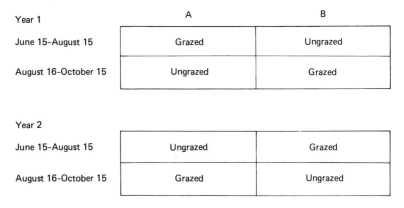

***Figure* 9.2** A one-herd, two-pasture deferred-rotation grazing plan used on the Starkey Range in northeastern Oregon.

TABLE 9.6   Average Cattle Gains and Diet Quality over a 5-year Period for Season-long, Rest-Rotation, and Deferred-Rotation Grazing Systems at the Starkey Range in Northeastern Oregon[a]

	SEASON-LONG	REST-ROTATION	DEFERRED-ROTATION
Average daily gain (kg)	0.58	0.58	0.59
Average diet crude protein (%)	11.3	11.6	10.6
Average diet digestibility (%)	57.3	56.5	57.6

Source: Holechek et al. 1987.

[a]The same stocking rate was used for each system.

The Merrill Three-Herd/Four-Pasture System

Merrill (1954), in south-central Texas, developed a grazing system involving three herds and four pastures. With this system, each pasture is grazed continuously for a year and then given a 4-month nonuse period (Figure 9.3). The nonuse period has occurred during all times of the year by the end of a 4-year cycle. This system gives good results where effective precipitation and plant growth can occur at any time during the year. It also works well where common use of the range by more than one grazing animal is practiced. In Texas, some combination of cattle, sheep, goats, and white-tailed deer graze many ranches. Each grazed pasture is assigned a different type of livestock and every 4 months the types of livestock are interchanged among pastures. White-tailed deer prefer the pasture receiving nonuse by livestock (Reardon et al. 1978). The Merrill system has been the best studied of all the specialized grazing systems. In Texas it is superior to continuous grazing from the standpoint of sustained livestock, forage, and wildlife production (Kothmann et al. 1971; Reardon and Merrill 1976; Wood and Blackburn 1984; Heitschmidt et al. 1990). On the Edwards Plateau rangeland in south Texas, the Merrill system has given higher financial returns than moderate continuous, heavy continuous, and short-duration grazing (Figure 9.4). The Merrill system can be recommended in areas where forage plants have the potential for growth throughout the year and common use grazing is practiced.

The Merrill system can be modified for areas with seasonal precipitation by dividing the ranch into four pastures and providing each pasture with growing season nonuse once every 4 years. During the dormant season stocking rates are adjusted so all pastures will receive near equal use. Although this strategy has not been experimentally evaluated, it is theoretically sound for all parts of the world that will support yearlong grazing and have annual patterns of seasonal wetness and dryness.

Seasonal-Suitability Grazing

This system involves partitioning a range into pastures based on vegetation types. These vegetation types are then fenced and integrated into a grazing strategy based on vegetation and

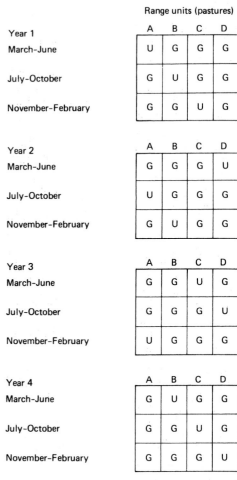

Range units (pastures)

Year 1

	A	B	C	D
March–June	U	G	G	G
July–October	G	U	G	G
November–February	G	G	U	G

Year 2

	A	B	C	D
March–June	G	G	G	U
July–October	U	G	G	G
November–February	G	U	G	G

Year 3

	A	B	C	D
March–June	G	G	U	G
July–October	G	G	G	U
November–February	U	G	G	G

Year 4

	A	B	C	D
March–June	G	U	G	G
July–October	G	G	U	G
November–February	G	G	G	U

G = grazed, U - ungrazed.

***Figure* 9.3** The Merrill three-herd/four-pasture grazing plan used in Texas.

livestock requirements. These fenced units may or may not be contiguous, depending on the location, size, and specific ownership of the land parcels involved. In some cases control of livestock access to water rather than fencing can be used to manipulate where livestock graze. Seeded pastures are often but not always an important part of the grazing system. Terrain, ranch operation requirements, range condition classes, and range site differences are all considered when pastures are declined and fenced. Seeded pastures are fenced separately from native range and are important tools for promoting native range improvement as well as increasing livestock production. Seeded pastures often provide highly nutritious forage earlier and later than native range, and can withstand more intensive grazing.

It is important to recognize that a seasonal-suitability system will probably not be superior to continuous or season-long grazing unless a ranch contains diverse forage re-

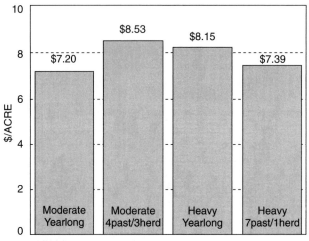

Figure **9.4** Net returns from four grazing systems at moderate and heavy stocking rates 1970–1982 on the Edwards Plateau of south Texas. (From Conner and Taylor 1988 and Taylor et al. 1993.)

sources. The productivity of individual forage types must be high enough that separate fencing or control of water to force livestock use is practical. We will discuss a few examples of seasonal-suitability systems.

Mountain ranges in the northwestern United States are typically comprised of forest (north-facing slopes), grassland (south-facing slopes), and meadow (riparian) vegetation types. Fencing off these vegetation types and using them separately is advantageous to vegetation, livestock, and wildlife. The mountain grassland forage species initiate growth earlier in the spring and mature earlier in the summer than do forested ranges. They are best used from June to mid-July, at which time cattle should be moved to forested areas (Holechek et al. 1981) (Figure 9.5). Mountain meadows will give cattle gains equal or superior to upland grassland or forested range in the latter part of the grazing season in September and October (Holechek et al. 1982). Late use (late August to mid-October) of mountain meadows results in concentration of livestock on low areas, where they are easily gathered. Separate fencing has the additional advantage of permitting spring-summer deferment of meadows. Unless use of mountain meadows is carefully controlled, excessive grazing occurs frequently because of the meadows' convenience to livestock.

Cook and Harris (1968) studied livestock performance and forage quality on desert, foothill, and mountain ranges in Utah. They reported that foothill ranges could be used most efficiently by livestock in the spring, while mountain ranges should be used during the summer because of delayed plant phenology and less susceptibility to grazing damage. Low-elevation desert ranges were found to be ideally suited to winter grazing because weather is less severe and the desert shrubs retain high nutritive value when dormant. The seeding of crested wheatgrass on foothill ranges can be an effective tool to increase forage quality and quantity in spring and fall.

Smoliak (1968), in Alberta, found that the integration of native range (summer), crested wheatgrass (spring), and Russian wildrye (*Elymus junceus*) (fall) greatly increased

Range units (pastures)

Period	A Grassland (south-facing slope)	B Forest (north-facing slope)	C Meadows (between north- and south- facing slopes)
June 15–July 15	Grazed	Ungrazed	Ungrazed
July 16–Sept 15	Ungrazed	Grazed	Ungrazed
Sept 16–Oct 15	Ungrazed	Ungrazed	Grazed

Figure **9.5** One-herd/three-pasture seasonal-suitability grazing plan used on mountain ranges in northeastern Oregon.

livestock production. The integrated use of the three forage types more than doubled grazing capacity.

The Best-Pasture System

Valentine (1967) first proposed the best-pasture system for use on semidesert ranges in southcentral New Mexico. This system is designed for areas with localized convectional rainstorms that cause forage production within years to vary greatly over short distances (less than 10 km). This system also incorporates elements of seasonal-suitability grazing. The basic strategy is to use mesa dropseed (*Sporobolus flexuosus*) grasslands in July and August when they are actively growing, and in the spring if moisture and temperature conditions permit growth. Black grama (*Bouteloua eriopoda*) areas are saved for winter, since black grama is more nutritious than other desert grasses at this time and is easily damaged by summer grazing. However, if local rains cause annual forbs to become available on parts of the range, livestock are moved to utilize these forages and then returned to their original pastures.

Beck and McNeely (1993) reported that the best-pasture system did not differ from continuous grazing in vegetation or livestock productivity over a 25-year period (Table 9.7). Average forage utilization levels for both systems were about 20%. The best-pasture system may have proven superior to the continuous system if the utilization was heavier.

On blue grama (*Bouteloua gracilis*) range in southcentral New Mexico, a one-herd/four-pasture variation of the best-pasture system has been superior to continuous grazing in terms of vegetation and livestock productivity (Pieper et al. 1991). This system involves grazing a pasture until moderate use is achieved (300 to 340 kg/ha dry-matter residue). Cattle are then moved to the pasture with the highest standing crop of forage unless this pasture was grazed the same season the previous year. If this is the case, cattle are moved to the pasture second highest in standing crop. The one-herd/four-pasture system was stocked at a 25% higher rate than moderately stocked, continuously grazed pasture. Over a 10-year period, forage production (kg/ha) averaged 11% higher and livestock pro-

TABLE 9.7   Rangeland and Cattle Production Characteristics for Different Grazing Management Strategies on the Fort Stanton Range in New Mexico

FORT STANTON EXPERIMENT RANGE—COW-CALF OPERATION			
	MODERATE CONTINUOUS	HEAVY CONTINUOUS	BEST-PASTURE ROTATION
Duration of study (yr.)	10	10	10
Average annual ppt (in.)	15	14.8	15.1
Average forage production (lb./acre)	740	607	819
Total forage production in 1974 drought, (lb./acre)	235	103	379
Range condition	good	fair	good
Acres/animal unit	67	54	54
Forage use (%)	40–45	60–65	60–65
Calf crop (%)	93	91	85
Calf weaning wt. (lb.)	435.0	425.0	406.0
Average calf wean wt./acre (lb.)	6.2	7.0	6.4
Death losses (%)	<2%	<2%	<2%
Total 10 year accumulated value ($)	388,575	358,744	382,380

Source: Adapted from Pieper et al. 1991 and Holechek 1994.

duction (kg/ha) was 3% higher for the one-herd/four pasture system compared to the moderately stocked, continuously grazed pasture system. However, total 10-year accumulated financial returns were slightly higher for the moderate continuous compared to the best-pasture system (Holechek 1994; Table 9.7) because of reduced cattle performance and financing costs associated with extra cattle and fence. On a long-term (11–30 years) basis, it was concluded the best-pasture system would be financially most effective because these costs would no longer occur and further increases in grazing capacity were likely.

Rest-Rotation Grazing

Gus Hormay of the Forest Service developed rest-rotation grazing in the 1950s and 1960s (Hormay and Evanko 1958). This system is unique in that one pasture receives 12 months of nonuse while the other pastures absorb the grazing load (Figure 9.6). Most rest-rotation

Pasture:

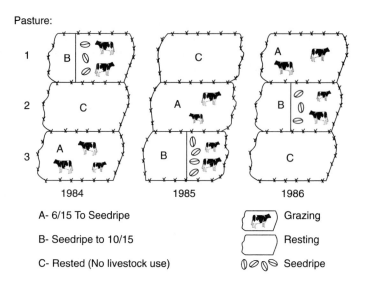

A- 6/15 To Seedripe
B- Seedripe to 10/15
C- Rested (No livestock use)

Grazing
Resting
Seedripe

Figure **9.6** A rest-rotation grazing system used to manage forage on elk summer range at the Mount Haggin wildlife area in Montana. (From Frisina 1992.)

Figure **9.7** Rest-rotation grazing has given good results in mountainous terrain where livestock distribution is a major problem.

schemes involve three or four pastures. Various sorts of rotation schemes are used on the three grazed pastures. The problem with rest-rotation grazing is that the benefits from rest may be nullified by the extra use that occurs on the grazed pastures.

Rest-rotation grazing has shown superiority to season-long grazing on mountain ranges where livestock distribution problems occur (Figure 9.7). In northeastern California, rest-rotation grazing resulted in higher vigor of the key forage plant, Idaho fescue (*Festuca idahoensis*), than that resulting from continuous grazing (Hormay 1970; Ratliff et al. 1972; Ratliff and Reppert 1974). On mountain range in Wyoming, range condition improved more under rest-rotation than under deferred-rotation grazing (Johnson 1965). In this study season-long grazing caused range condition to decline. On mountain rangelands in eastern Utah, rest-rotation and season-long grazing had similar influences on range condition and productivity (Laycock and Conrad 1981). The authors pointed out that rest-rotation grazing requires intensive management of water, salt, riding, and so on. These practices were applied in all pastures of their study to ensure uniform cattle distribution. Utilization levels were below 40% for all pastures in both the Johnson (1965) and Laycock and Conrad (1981) studies.

Rest-rotation grazing has been criticized for reducing livestock performance because of reduced selectivity and forced animal movement from pasture to pasture. However, actual research invalidates this criticism if the stocking rate is moderate. Cattle weight gains per hectare and per animal did not differ among rest-rotation, deferred-rotation, and season-long systems on mountain range in northeastern Oregon when the pastures were stocked so that forage utilization over a 5-year period averaged about 35% under each system (Holechek et al. 1987) (see Table 9.6).

On semidesert grassland range in southern Arizona, a three-pasture rest-rotation system has shown promise for improving the vigor and density of palatable perennial grasses (Martin and Ward 1976; Martin 1978). Under the Santa Rita system, as it is called, stocking rates on grazed pastures are kept at moderate levels (30% to 40% use). The rest-grazing sequence for one pasture is: (1) rest 12 months (November–October), (2) graze 4 months (November–February), (3) rest 12 months (March–February), (4) then graze 8 months (March–October) to complete the 3-year cycle. Each pasture is rested during both spring and summer growth periods for 2 out of 3 years, but each year's forage is utilized. A full year of rest before spring grazing provides old herbage that protects early growth from repeated close grazing. Benefits are greatest where animals congregate, such as around water. After 12 years of study, Martin and Severson (1988) concluded the Santa Rita system may accelerate recovery of hot desert ranges in poor condition, but it has little advantage over moderate continuous grazing for ranges in good condition.

Rest-rotation grazing has a number of multiple-use advantages. The benefits of rest-rotation grazing to wildlife are discussed in Chapter 13. From an esthetic standpoint, the public prefers to see a certain amount of the range ungrazed. On mountain riparian meadows in northeastern Oregon, rest-rotation grazing resulted in better soil properties (higher infiltration, reduced bulk density, and reduced sediment) than deferred-rotation and season-long grazing (Bohn and Buckhouse 1985).

Most of the failures associated with rest-rotation grazing are related to heavy stocking rates. It is well established that for most plants in arid areas, 1 or more years of rest will not compensate for 1 year of severe defoliation during the growing season (Cook and Child 1971; Trlica et al. 1977). Research by Holechek et al. (1987) showed that rest-rotation grazing did not cause cattle to more heavily use forage grasses of low palatability [Sandberg bluegrass (*Poa sandbergii*)] compared to season-long or deferred-rotation systems. This study shows that any benefit of rest-rotation grazing due to better use of secondary forage species is highly questionable if moderate stocking rates are used. A 10-year Nevada study showed that rest-rotation did not overcome the effects of heavy use (65%) during the growing season (Eckert and Spencer 1986, 1987). This study showed that heavy stocking rates will prevent range improvement under an otherwise appropriate grazing strategy. Many managers believed erroneously that much heavier stocking rates were possible under rest-rotation than under season-long grazing since they thought that the rest and deferment periods would more than offset the heavy use when a pasture was grazed. This proved disastrous for both range vegetation and livestock, particularly in desert areas. Under moderate stocking rates it appears that grazed rest-rotation pastures are used more evenly than those grazed season-long (Holechek et al. 1987). This compensates for the higher livestock numbers on the grazed rest-rotation pastures. In rugged terrain in the Northwest, where cattle heavily use riparian areas under all grazing strategies, the riparian vegetation [various sedges (*Carex* sp.), Kentucky bluegrass (*Poa pratensis*), timothy (*Phleum pratense*)] recovers well from heavy grazing if given periodic nonuse during the growing season. In contrast, the surrounding uplands support forages [Idaho fescue (*Festuca idahoensis*), and bluebunch wheatgrass (*Agropyron spicatum*)] easily damaged by heavy use. However, actual use is light under moderate grazing intensities, due to rugged terrain. Rest-rotation grazing increases utilization from 20% to 30% on upland areas and has a limited effect on utilization of riparian vegetation compared to deferred- or season-long systems.

We consider rest-rotation grazing if used with conservative stocking rates a good system for both vegetation and livestock in rugged, mountainous terrain. This system will, in most cases, improve grazing capacity over season-long grazing due to better livestock use of upland areas and improved vegetation vigor and composition in the more productive riparian zones. On flat desert or prairie ranges, we believe there are better systems than rest-rotation grazing when the goal is primarily livestock production. When multiple use is an important consideration, rest-rotation grazing has important advantages even on these ranges.

High Intensity-Low Frequency Grazing

The high intensity-low frequency (HILF) system of rotation grazing typically involves three or more pastures, with grazing periods longer than 2 weeks and nonuse periods over 60 days (Figure 9.8). A key feature of this system first reported by Acocks (1966) in South Africa is that livestock are forced to use the coarse and less palatable forage species as well as the more palatable plants. This supposedly reduces competition to the more palatable plants and prevents the development of tall, coarse, ungrazed "wolf" plants. The long period of nonuse is thought to more than offset the heavy use that occurs during grazing.

In the flat, humid grassland range types, such as the tallgrass prairie in North America or the mixed veld in South Africa, most of the plants have some grazing value, and recovery is relatively rapid after grazing. Continuous grazing at light to moderate stocking rates permits many plants to become tall, coarse, and unpalatable, while other plants are grazed repeatedly. Under these conditions, HILF grazing can promote range improvement and lead to higher carrying capacity (Acocks 1966; Merrill and Taylor 1975; Denny and Steyn 1977; Howell 1978).

Range units (pastures)

Year 1

	A	B	C	D
January–February	G	U	U	U
March–April	U	G	U	U
May–June	U	U	G	U
July–August	U	U	U	G
September–October	G	U	U	U
November–December	U	G	U	U

Year 2

	A	B	C	D
January–February	U	U	G	U
March–April	U	U	U	G
May–June	G	U	U	U
July–August	U	G	U	U
September–October	U	U	G	U
November–December	U	U	U	G

G = grazed, U - ungrazed

Figure **9.8** One-herd/four-pasture high intensity-low frequency grazing plan used in the southeastern United States.

The main problem with HILF grazing is that individual animal performance is adversely affected when livestock are forced to use old, coarse, unpalatable but grazable forage (Kothmann et al. 1975; Howell 1978; Corbett 1978; Taylor and Garza 1986; Drawe 1988) (Table 9.8). A 10-year Texas study showed that lamb production under HILF averaged 6.4 kg/ha compared to 7.1 kg/ha per year for continuous grazing (Taylor and Garza 1986). Taylor et al. (1980) reported cattle under HILF grazing had reduced-diet crude protein content and digestibility compared to those under Merrill and short-duration systems in south-central Texas. This was due to reduced animal selectivity and the mature condition of the forage under the HILF system. Because the HILF system has a negative impact on animal nutrition, it has been largely abandoned in favor of short-duration grazing, which allows greater animal selectivity, permits lighter levels of defoliation, and prevents more of the forage from maturing prior to grazing.

Although HILF grazing has shown advantages over continuous grazing from the standpoint of increased grazing capacity and range condition for flat, humid rangelands, it is theoretically unsound for the more arid, rugged rangelands in the intermountain United States and

TABLE 9.8 Livestock Production Characteristics and Financial Returns from Three Grazing Systems on the Welder Wildlife Foundation Refuge, 1977–1981

CHARACTERISTIC	MERRILL THREE-HERD/ FOUR-PASTURE	HIGH INTENSITY/ LOW FREQUENCY	CONTINUOUS
Calf crop (%)	91	89	89
Calf weaning weight (lb.)	498	476	512
Beef production/acre (lb.)	45	28	43
Beef produced/cow (lb.)	449	417	468
Net profit ($/acre)	2.87	1.52	3.65

Source: Drawe 1988.

other parts of the world. In these areas recovery from heavy grazing is slow to nonexistent. Many desirable forage species will not recover from heavy defoliation even after 2 or more years of complete rest (Cook and Child 1971; Trlica et al. 1977). Livestock will starve to death before consuming unpalatable plants such as big sagebrush, creosotebush, or juniper.

Short-Duration Grazing

Short-duration grazing (also called rapid-rotation, time-control, and cell grazing) was developed in Zimbabwe by Allan Savory in the 1960s and later introduced into the United States by Goodloe (1969). When Savory came to the United States in the late 1970s, he made further refinements, discussed by Savory and Parsons (1980), Savory (1983), and Savory (1988). Savory's modifications of short-duration grazing have been called the Savory grazing method and more recently, holistic resource management (Savory 1988). For a complete discussion of Allan Savory's views on grazing management the reader is referred to Savory (1988).

Short-duration grazing typically involves a wagon-wheel arrangement of fences, with water and livestock-handling facilities located in the center of the grazing area (cell). However, short-duration grazing can be applied successfully without the use of the wagon-wheel design (Figure 9.9). It is recommended that no fewer than eight pastures (paddocks) of equal grazing capacity be built that radiate as spokes from the central area where water is located. Ideally, the grazing period of each paddock should be 5 days or less followed by 4 or more weeks of nonuse (Figure 9.10). It is recommended that livestock be moved more quickly during periods of active forage growth than during forage dormancy. The high stock density (number of animals per unit area) is thought to

1. Improve water infiltration into the soil as a result of hoof action.

2. Increase mineral cycling.

3. Reduce selectivity so that more plants are grazed.

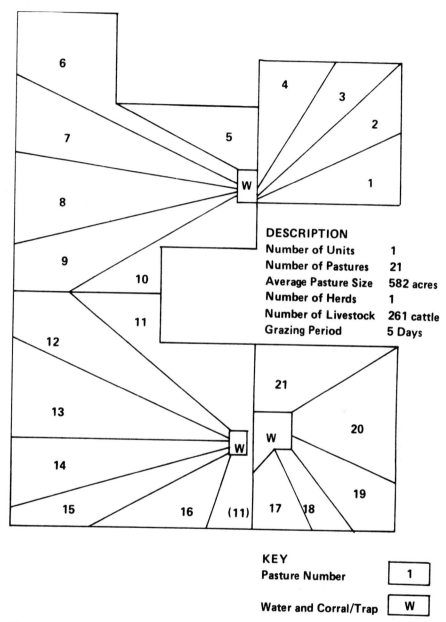

DESCRIPTION
Number of Units   1
Number of Pastures   21
Average Pasture Size   582 acres
Number of Herds   1
Number of Livestock   261 cattle
Grazing Period   5 Days

KEY
Pasture Number   [1]

Water and Corral/Trap   [W]

Figure **9.9** Short-duration grazing system on a small cow-calf ranch in southeastern New Mexico. (From Fowler and Gray 1986.)

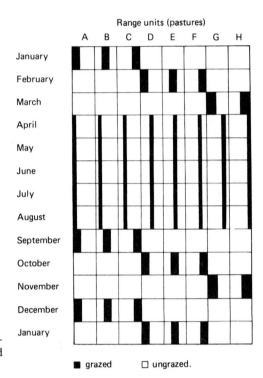

***Figure* 9.10** One-herd/eight-pasture short-duration grazing plan for ranges in Texas and the southeastern United States.

4. Improve the leaf area index.

5. Give more even use of the range.

6. Increase the period when green forage is available to livestock.

7. Reduce the percentage of ungrazed "wolf" plants.

Under good management it is claimed that a well-managed cell permits stocking rates to be increased substantially (doubled in many cases) compared to continuous and other grazing systems (Savory and Parsons 1980). Reduced labor costs, better individual animal performance, and rapid improvement in range condition are other benefits claimed. This system has been advocated for all rangeland types throughout the world. Presently, a lack of long-term research prevents drawing very many definite conclusions about the effectiveness of various short-duration grazing strategies.

Much of the increase in stocking rate claimed possible under short-duration grazing results from better livestock distribution (Dahl 1986). Confining a large number of animals to a small area for a short period improves uniformity of use and forces the use of areas and plants not used previously.

The claim that short-duration grazing improves water infiltration into the soil compared to continuous grazing does not appear valid based on studies from Texas and New Mexico (McCalla et al. 1984a; Thurow et al. 1986; Weltz and Wood 1986a; Pluhar et al.

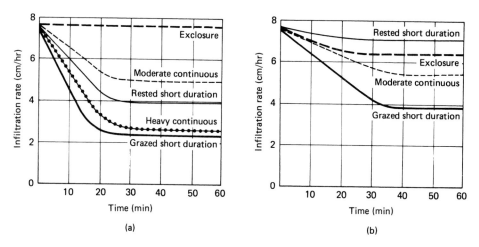

***Figure* 9.11** Mean infiltration rates for various grazing treatments at two locations in new Mexico: (a) Fort Stanton and (b) Fort Sumner. (From Weltz and Wood 1986a.)

1987). In all studies, infiltration rates were lower for short-duration grazing than for moderate continuous grazing (Figure 9.11). In the McCalla et al. (1984a) study, the short-duration system had double the stocking rate of the moderate continuous grazed treatment. Based on this study, any positive benefits through hoof action from having a large number of animals on a small area for a short time would appear questionable. Sediment losses from the short-duration treatments were higher than for the moderate continuous treatments (McCalla et al. 1984b; Thurow et al. 1986; Weltz and Wood 1986b; Pluhar et al. 1987).

The number and kinds of cattle trails can have a drastic impact on the relative amount of bare soil and subsequently on rate of soil erosion. In Texas, Walker and Heitschmidt (1986) found that there were no differences among heavy continuous, moderate continuous, and Merrill three-herd/four-pasture grazing systems in density of trails (Figure 9.12). Trail densities ranged from 9 per kilometer at the far end of the pasture to 14 per kilometer near the pasture center. However, a 14-paddock short-duration cell grazing system in this same study had trail densities ranging from 24 per kilometer at the far end of the paddock to 164 per kilometer near the cell center. The effect of increasing the number of paddocks from 14 to 42 was also studied for the cell system. Trail densities near the cell center were 32 per kilometer for 14 paddocks compared to 57 per kilometer for 42 paddocks with no increase at the far end of the paddocks. Walker and Heitschmidt (1986) concluded that cell-designed, short-duration grazing systems can increase the density and number of cattle trails over those under continuous grazing.

Claims have been made that increasing paddock numbers for short-duration grazing systems will increase forage production, improve livestock distribution, and improve watershed condition. A Texas study compared 14- and 42-paddock short-duration grazing treatments (Heitschmidt et al. 1987). There was no difference in forage production and

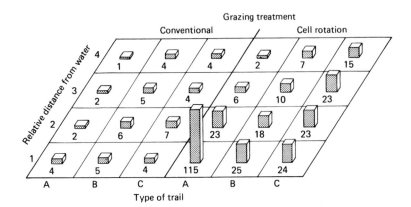

***Figure* 9.12** Cattle trail density (trails/km) in conventional and short-duration cell grazing treatments. Conventional grazing is the average for heavy continuous, moderate continuous, and Merrill three-herd/four-pasture treatments. ("Type of trail" refers to depth and vegetation characteristics of trail: A = deepest, least vegetated; B = intermediate; C = shallowest, most vegetated.) (From Walker and Heitschmidt 1986.)

harvest efficiency between the two treatments over a 4-year period. Increasing the number of paddocks did not increase water infiltration rates or reduce sediment production (Pluhar et al. 1987).

It has been speculated that short-duration grazing will improve livestock diet quality because forage maturation is delayed by repeated defoliation during the growing season. However, research by Heitschmidt (1986) indicates that differences between continuous and short-duration grazing are small during active forage growing (Table 9.9). During dormancy, diet quality was higher under moderate continuous grazing, probably due to higher-standing forage crop providing livestock with greater selectivity.

From a theoretical standpoint, short-duration grazing should work best in flat, humid grassland areas that have an extended period of plant growth (at least 3 months and over 500 mm of average annual precipitation). These conditions apply to the eastern portions of the Great Plains, southern pine forest, and eastern improved pastureland ranges in the United States. Here there is considerable opportunity for regrowth after the relatively short, intensive grazing periods associated with short-duration grazing. Droughts tend to be much less frequent and severe than in arid areas (under 300 mm of average annual precipitation). Removal of some herbage is most likely to enhance plant productivity because of considerable shading of young leaves due to the high volume of vegetation. Periodic grazing during periods of active growth will have the most tendency to delay forage maturation under these conditions. On these ranges, annual forage production averages over 2,200 kg/ha. Fencing requirements are less than one-tenth those of semidesert grassland or sagebrush grassland ranges, where annual forage production seldom averages over 300 kg/ha.

Short-term research in the northcentral Texas rolling plains, where the conditions above apply, indicates that short-duration grazing accelerates plant growth (Heitschmidt et

TABLE 9.9   Average Crude Protein and Digestible Organic Matter of Steer Diets in Short-Duration and Moderate Continuous Grazing Treatments at the Texas Experimental Ranch[a]

	JUNE 1983	JANUARY 1984	MAY 1984
Crude protein (%)			
Short-duration	10.8	7.9	9.9
Continuous	10.9	9.1	10.7
Digestibility (%)			
Short-duration	67	57.0	70.0
Continuous	64	60	63

Source: Heitschmidt 1986.

[a]The short-duration grazing system was stocked 40% higher than the continuous grazing treatment.

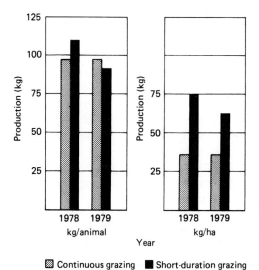

☒ Continuous grazing   ■ Short-duration grazing

Figure **9.13** Production per animal and production per hectare of growing heifers during 1978 and 1979 from the continuous and short-duration grazing treatments in the Texas Rolling Plains. (Adapted from Heitschmidt et al. 1982a.)

al. 1982d) improves forage quality (Heitschmidt et al. 1982b) and increases livestock production per unit area (Heitschmidt et al. 1982a) (Figure 9.13). These studies involved a 2-year period of near-average growing conditions in an area receiving 650 mm of average annual precipitation. The authors pointed out that several years of information would be needed before definite conclusions could be made.

Aside from the fencing cost, there are a number of other reasons why short-duration grazing may not be practical in arid areas. In many years, the period of active growth of the main forage grasses is less than 60 days. This minimizes the positive aspects of repeated periods of defoliation and nonuse. Precipitation that does occur is often from one or two intense thunderstorms. A concentration of animals therefore has the potential for severe soil compaction. One failure to move cattle at the correct time could severely damage grasses such as black grama or Arizona cottontop (*Digitaria californica*). Denying livestock access to a large portion of the range could lead to wastage of many palatable annual forbs and grasses that complete their life cycle in 10 to 21 days.

Lowlands that collect water runoff from uplands and have deep soils are much more productive than are surrounding uplands. On these ranges short-duration grazing may be more practical. It may also be useful on seedings of such plants as crested wheatgrass and Lehman lovegrass (*Eragrostis lehmanniana*), where grazing resistance and capacity are higher than for native range.

Scientific reevaluations of long-term vegetation response to short-duration grazing compared to less intensive management schemes are available from Zimbabwe, Africa (Gammon 1984). A number of ranchers were involved that had applied short-duration grazing for over 10 years. Areas under short-duration grazing were not markedly or consistently superior to adjacent, less intensively managed areas in terms of basal cover, litter cover, and species composition. Short-duration grazing was generally applied at higher stocking rates than normally recommended. These stocking rates were sustainable during the 1970s when precipitation was 50% above the average. However, during the early 1980s, precipitation returned to normal and stocking rates were lowered due to shortage of grass. On one ranch (Leibig's ranch) under short-duration grazing, the stocking rate had been doubled with good livestock performance for eight years (1972–1980). However, during the two following years of average rainfall (1980–1981 and 1981–1982), reductions in stocking rate were required, bringing the rate to 50% above average. At this level, paddocks became short of grass and by November 1982 the system was completely destocked. During the 1982–1983 drought, the area supported no livestock. Based on these results, a stocking rate even 50% above normal appeared too heavy for years of average or slightly below average precipitation.

With the exception of Gammon (1984), most of the research available on short-duration grazing involves short periods (less than 5 years) and is from the more humid grassland types. Short-term studies suggest that the stocking rate can be increased 20% to 30% under short-duration grazing compared to continuous grazing without greatly affecting individual animal performance (Daugherty et al. 1982; Heitschmidt et al. 1982a; Sharrow 1983; Jung et al. 1985; Dahl 1986; Skovlin 1987; Heitschmidt et al. 1990). This appears to be due more to improved livestock distribution than improved range condition. Short-duration grazing has had little effect on botanical composition of rangelands (Pitts and Bryant 1987; Hart et al. 1988; White et al. 1991). Stocking rates have greater potential than grazing systems for altering botanical composition of range plant communities (Hart et al. 1993).

Limited research is available comparing range condition trends under short-duration grazing with other rotation grazing schemes. A 4-year study at the Sonora Research Station in Texas indicated short-duration grazing did not promote secondary succession from shortgrasses to midgrasses as effectively as high intensity-low frequency grazing (Taylor et al. 1993).

Long term impacts of short-duration grazing on range vegetation and livestock productivity need to be determined. Considerable research is in progress throughout the United

States that should provide a better understanding of short-duration grazing influences on vegetation, soil, and livestock productivity.

One advantage of short-duration grazing using a cell is that much better livestock management is possible. The cell center provides a convenient area for provision of supplemental feed, vaccination, frequent examination of animals for health problems, and implementation of artificial insemination programs, since animals can be corraled and restrained when they come into the cell center for water. The cell center should be designed for adequate animal handling and watering facilities but should discourage animal use of the area for resting and bedding.

The economics of short-duration is a controversial issue. Short-duration grazing gave higher returns than those from moderate continuous grazing in northcentral Texas during a 6-year period (1982–1987) (Heitschmidt et al. 1990) (Table 9.10). However, the differences were not statistically significant. Further, it was concluded that the heavy stocking rate used with the short-duration system greatly increased financial risk. Economic returns should theoretically be best in high rainfall areas (over 650 mm per year), due to lower fence costs. Much of the inconsistency in reports from ranchers regarding economic returns results from variation in fencing, water development and extra cattle costs (Graham et al. 1992) (Table 9.11). The availability of fencing materials, labor, existing fence, and the number of paddocks used all heavily influence economic returns. Further, rancher ingenuity in designing and operating cell grazing systems varies tremendously. Therefore, we believe that the controversy surrounding short-duration grazing will persist for many years to come. Skovlin (1987) provides a comprehensive review of short-duration grazing in South Africa, where it has been practiced for many years. Based on his review, it appears that in some cases, short-duration grazing may permit modest increases (30% to 40%) in stocking rate over continuous grazing, due to better forage utilization. However, the doubling or tripling of stocking rates initially thought possible could not be sustained. Based on the Skovlin (1987) review, the

TABLE 9.10 Average Economic Returns from Heavy Continuous (HC), Moderate Continuous (MC), Merrill Three-herd/Four-pasture (DR), and Short-duration (SD) Grazing Treatments in Northcentral Texas over a 6-year period (1982–1987)[a]

	HC	MC	DR	SD
Per cow	$60.81	$69.57	$93.12	$62.72
Per hectare[b]	12.97	11.02	15.99	16.38
Per acre[b]	5.25	4.46	6.47	6.63

Source: Heitschmidt et al. 1990

[a] The short-duration grazing treatment was stocked 40% higher than the moderate continuous treatment. The stocking rate was the same for the moderate continuous and Merrill grazing treatments.

[b] There were no statistically significant differences in financial returns per unit area among treatments when averaged across years.

TABLE 9.11 Average Costs per Acre for Fence, Water Developments, and Extra cattle for Implementing Short-duration Grazing (Holistic Resource Management) at a 60% Higher Stocking Rate on Shortgrass Range in Northcentral New Mexico

	COST PER ACRE ($)
Fencing	2.96
Water development	1.81
Additional livestock	16.23
Total	21.00

Source: Graham et al. 1992.

cost effectiveness of short-duration grazing appears as controversial in South Africa as in the United States.

All too often managers on public and private rangelands have been led to believe that stocking rate can be ignored if some miracle specialized grazing system is applied. After careful consideration of available research on short-duration grazing, Pieper and Heitschmidt (1988) concluded *stocking rate is and always will be the major factor affecting degradation of rangeland resources.* They stated that no grazing system can counteract the negative impacts of long-term overstocking. Their conclusions are supported by Thurow et al. (1988) in Texas and Willms et al. (1990) in Alberta, Canada, who both found short-duration grazing at heavy stocking rates led to range deterioration.

GRAZING SYSTEMS FOR RIPARIAN ZONES

Continuous or season-long grazing is most damaging to streamside areas (riparian zones) and wetlands because livestock concentrate and linger on these areas due to the convenience of forage, water, and cover (Gunderson 1968; Evans and Krebs 1977; Severson and Boldt 1978). Riparian zones are the most important part of the range from the standpoint of wildlife, water quality, esthetics, and forage productivity (Figure 9.14). Because of their small size relative to the rest of a pasture, riparian zones in the past were considered sacrifice areas. However, in recent years, degradation of these areas, particularly on public lands, has become unacceptable to society. Methods available to rehabilitate riparian zones include complete livestock exclusion, rotation grazing schemes, changes in season of use, changes in the type or class of animal, and techniques that improve livestock distribution. Many managers and researchers have concluded that the only means of restoring and maintaining these valuable areas is complete livestock exclusion. However, this alternative is unacceptable to ranchers. Recent studies have shown that improvement of riparian zones may be possible without complete livestock exclusion (Skovlin 1984; Elmore and Kauffman 1994).

As discussed previously (see also Chapter 13), rest-rotation appears to be one of the most practical means of restoring and maintaining riparian zones. It is well established

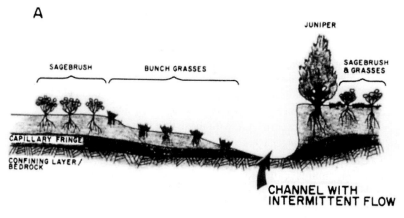

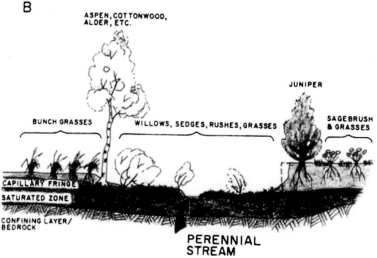

Figure **9.14** General characteristics and functions of riparian zones. (From Elmore and Beschta 1987.)

(A) Degraded riparian area
- little vegetation to protect and stabilize banks, little shading
- lowered saturated zone, reduced subsurface storage of water
- little or no summer streamflow
- warm water in summer and icing in winter
- poor habitat for fish and other aquatic organisms in summer or winter
- low forage production and quality
- low diversity of wildlife habitat

(B) Recovered riparian area
- vegetation and roots protect and stabilize banks, improve shading
- elevated saturated zone, increased subsurface storage of water
- increased summer streamflow
- cooler water in summer, reduced ice effects in winter
- improved habitat for fish and other aquatic organisms
- high forage production and quality
- high diversity of wildlife habitat

that under moderate stocking rates this system improves both streamside vegetation and physical characteristics (Hayes 1978; Davis 1982; Platts 1982; Bohn and Buckhouse 1985).

Replacing cattle with sheep that are herded is a workable solution in some areas where livestock operators graze both animals or can switch from one animal to another without economic hardship. Herding of sheep permits much better control of grazing timing, frequency, and intensity on riparian zones (Platts 1982). Selective culling of cattle that prefer riparian zones and replacing them with those that more readily use uplands also shows potential (Roath and Krueger 1982b; Howery et al. 1996).

Researchers in Oregon have found that fencing and delayed grazing of riparian zones on mountains can be beneficial to vegetation, stream banks, and livestock (Holechek et al. 1982; Kauffman et al. 1983a,b). Their scheme involves restriction of cattle to upland areas until late summer, when the gates are opened to the riparian zones and meadows. By this time nesting birds and small mammals have completed critical activities associated with reproduction. The growing season is over, so impacts on vegetation are minimal. The intensity of grazing can be controlled by having the rancher haul the livestock away as soon as the desired level of use is achieved. Livestock performance under this strategy has been found to equal or exceed season-long use of the riparian zone (Holechek et al. 1982). Problems associated with gathering cattle for removal in the fall are greatly minimized because cattle are concentrated on a small area of flat terrain with good visibility. The only drawback to this scheme is the cost of fencing. It has also been suggested that spring grazing may be beneficial under some conditions (Elmore and Beschta 1987). This allows riparian vegetation to recover in the summer and allows for deferment of upland areas. The effectiveness of different management systems for improving riparian zones based on qualitative riparian evaluations is provided in Table 9.12. The reader is referred to Skovlin (1984), Elmore and Kauffman (1994), and Ohmart (1996) for more detailed discussions of livestock grazing methods and impacts on riparian zones.

GRAZING SYSTEMS FOR DEVELOPING COUNTRIES

Livestock grazing in much of Africa is based on communal use of native forages, characterized by a rapid growth rate and a short period of high nutritive quality (1 to 2 months). This region has seasonal wet and dry periods averaging about 6 months each. In the traditional pastoral system, livestock are confined overnight near water and grazed during the day by a herder in various directions radiating away from the water. This type of strategy results in heavy use of forage around water during the critical growing period. During the dormant season livestock must be trailed increasing distances to find available forage. Both low-quality forage and long travel distance stress livestock heavily during the dry season.

Reserving forage near watering points for dry season use has the potential to improve livestock performance in central Africa (Niger). In Kenya, zebu cattle were watered every 2 days without greatly influencing their performance (Musimba et al. 1987). Therefore, it

TABLE 9.12 Qualitative Ratings for Improvement of Grazing Strategies of Stream-riparian Habitats

STRATEGY	LEVEL TO WHICH RIPARIAN VEGETATION IS COMMONLY USED	CONTROL OF ANIMAL DISTRIBUTION (ALLOTMENT)	STREAM-BANK STABILITY	BRUSHY SPECIES CONDITION	SEASONAL PLANT REGROWTH	RIPARIAN REHABILITATIVE POTENTIAL	RATING
Continuous season-long (cattle)	Heavy	Poor	Poor	Poor	Poor	Poor	1[a]
Holding (sheep or cattle)	Heavy	Excellent	Poor	Poor	Fair	Poor	1
Short-duration high intensity (cattle)	Heavy	Excellent	Poor	Poor	Poor	Poor	1
Three-herd/ four-pasture (cattle)	Heavy to moderate	Good	Poor	Poor	Poor	Poor	2
Holistic (cattle or sheep)	Heavy to light	Good	Poor to good	Poor	Good	Poor to excellent	2–9
Deferred (cattle)	Moderate to heavy	Fair	Poor	Poor	Fair	Fair	3
Seasonal-suitability (cattle)	Heavy	Good	Poor	Poor	Fair	Fair	3
Deferred-rotation (cattle)	Heavy to moderate	Good	Fair	Fair	Fair	Fair	4
Stuttered-deferred rotation (cattle)	Heavy to moderate	Good	Fair	Fair	Fair	Fair	4

(Continued)

TABLE 9.12 Qualitative Ratings for Improvement of Grazing Strategies of Stream-riparian Habitats (*Continued*)

STRATEGY	LEVEL TO WHICH RIPARIAN VEGETATION IS COMMONLY USED	CONTROL OF ANIMAL DISTRIBUTION (ALLOTMENT)	STREAM-BANK STABILITY	BRUSHY SPECIES CONDITION	SEASONAL PLANT REGROWTH	RIPARIAN REHABILITATIVE POTENTIAL	RATING
Winter (sheep or cattle)	Moderate to heavy	Fair	Good	Fair	Fair to good	Good	5
Rest-rotation (cattle)	Heavy to moderate	Good	Fair to good	Fair	Fair to good	Fair	5
Double rest-rotation (cattle)	Moderate	Good	Good	Fair	Good	Good	6
Seasonal riparian preference (cattle or sheep)	Moderate to light	Good	Good	Good	Fair	Fair	6
Riparian pasture (cattle or sheep)	As prescribed	Good	Good	Good	Good	Good	8
Corridor fencing (cattle or sheep)	None	Excellent	Good to Excellent	Excellent	Good to excellent	Excellent	9
Rest-rotation with seasonal preference (sheep)	Light	Good	Good to excellent	Good to excellent	Good	Excellent	9
Rest or closure (cattle or sheep)	None	Excellent	Excellent	Excellent	Excellent	Excellent	10

Source: Adapted from Platts and Nelson (1989) by Elmore and Kauffman (1994).

[a]Rating scale based on 1 (poorly compatible) to 10 (highly compatible with fishery needs).

appears that areas distant from water can be used during the growing season, saving areas close to water for dormant season use. Trailing routes can be established to minimize forage trampling. Use of a short-duration strategy during both the growing and dormant seasons is possible since the animals are herded. The use of short-duration grazing schemes in the tropical grasslands of Africa may be advantageous (Acocks 1966; Denny et al. 1977; Gammon 1978, 1984). The conservation of forage near watering points for dry season use in conjunction with short-duration grazing of areas distant from water in the wet season has been referred to as *centripetal grazing*. Although this system has not been tested, it is fundamentally sound. For additional discussion of centripetal grazing, see Chapter 16.

ECONOMIC ADVANTAGES OF SPECIALIZED GRAZING SYSTEMS

Information on cost-benefit analyses of specialized grazing systems is restricted. Much of the cost associated with specialized grazing systems involves additional fencing. Other costs are for water development and labor. From a rancher's point of view, there is little incentive to implement a specialized grazing system unless it can be shown that the additional livestock and/or wildlife production will more than repay the cost of installation. In New Mexico, the rate of abandonment of specialized grazing systems has been high (30%) (Fowler and Gray 1986). This is probably true in many other states as well.

In arid areas where average annual forage production is less than 500 kg/ha, systems that require considerable division of the range by additional fence may not be cost-effective. Here grazing strategies that control where livestock graze by regulating access to water and/or involve minimum fence construction should give the best economic returns. In shortgrass prairie, northern mixed prairie, and California annual grassland types, the lack of livestock and vegetation responses to rotation systems has made their implementation economically disadvantageous. Information is limited on the relatively new short-duration grazing system. This system does appear theoretically sound for prairie range types or for the humid ranges of the southeastern United States.

Although deferred-rotation had slight benefits from range condition and forage productivity standpoints in northeastern Oregon (Skovlin et al. 1976), season-long grazing was superior in economic returns (Quigley et al. 1984). On western Canada bunchgrass range, Hubbard (1951) reported that expected benefits of deferred-rotation grazing would not offset the costs of fencing and water developments. Cost-to-benefit ratios of grazing systems have been evaluated most thoroughly in Texas. Here long-term research has been somewhat inconsistent regarding the economic advantage of the Merrill system compared to continuous grazing (Stewart and Leinweber 1968; Huss and Allen 1969; Whitson et al. 1982; Conner and Taylor 1988; Drawe 1988; Heitschmidt et al. 1990). However in the Edwards Plateau region, the Merrill system has shown a definite financial advantage over continuous and other grazing strategies evaluated (Connor and Tayler 1988; Taylor et al. 1993).

Although economic documentation of specialized grazing systems is limited, much practical experience indicates that they have been economically successful in a variety of range types. Information collected by Gray and Fowler (1980) shows that well over 50% of the ranches in New Mexico use specialized systems.

Fowler and Gray (1986) reported a case study on economic impacts of specialized grazing systems on New Mexico ranches. Their sample size was reduced from 140 to 26 ranches, primarily because two criteria were not met: (1) the ranch was still using a system 3 years after installation, and (2) adequate records were kept to reflect grazing system economic impacts. With only a few exceptions, on the remaining 26 ranches, deferred-rotation, rest-rotation, and short-duration grazing systems all increased net annual monetary returns over those prior to system implementation. This was true in both drought and nondrought years. Calving rates and calf market weights almost always increased after specialized grazing systems were established.

Stocking rates were generally increased by about 4% after grazing systems were installed. Stocking rates declined 18% from nondrought to drought years, both with and without grazing systems. This study reflects that fact that management skills of the individual rancher have much to do with the success or failure of a specialized grazing system.

One benefit derived from specialized grazing systems that is difficult to quantify is improved livestock management. Systems such as short-duration and rest-rotation grazing that involve considerable concentration and handling of livestock can result in tamer animals and better health, breeding, and supplemental feeding programs. This increases pride in the operation and leads to a better-managed ranch. Pastures receiving rest are ideal for burning, seeding, and other forms of range improvement.

RANGE MANAGEMENT PRINCIPLES

- There is no one grazing system that will meet all management objectives on all types of rangeland.

- Continuous grazing with control of access to watering points works well biologically and financially in flat arid and semiarid rangeland types.

- Rotational grazing systems are most effective in rugged terrain, humid areas, and where heterogeneity exists in range plant communities.

- No specialized grazing will be biologically or financially effective if used with an excessive stocking rate.

Literature Cited

Acocks, J. P. H. 1966. Non-selective grazing as a means of veld reclamation. *Proc. Grassl. Soc. South. Afr.* 1:33–40.

Anderson, K. L. 1940. Deferred grazing of bluestem pasture. *Kans. Agric. Exp. Stn. Bull.* 291.

Beck, R. F., and R. P. McNeely. 1993. Twenty-five year summary of year-long and seasonal grazing on the College Ranch. *Livestock Research Briefs and Cattle Growers Short Course.* New Mexico State University, Las Cruces, NM.

Black, W. H., A. L. Baker, V. I. Clark, and D. R. Mathews. 1937. Effect of different methods of grazing on native vegetation and gains of steers in northern Great Plains. *U.S. Dep. Agric. Tech. Bull.* 547.

Bohn, C. C., and J. Buckhouse. 1985. Some responses of riparian soils to grazing management in northeastern Oregon. *J. Range Manage.* 38:378–382.

Conner, J. R., and C. A. Taylor, Jr. 1988. Economic analysis of grazing strategies in the Edwards Plateau. In *Abstracts. 41st Annual Meeting, Society for Range Management,* Corpus Christi, TX.

Cook, C. W., and L. E. Harris. 1968. Nutritive value of seasonal ranges. *Utah Agric. Exp. Stn. Bull.* 472.

Cook, C. W., and R. D. Child. 1971. Recovery of desert plains in various states of vigor. *J. Range Manage.* 24:339–343.

Corbett, Q. 1978. Short-duration grazing with steers—Texas style. *Rangeman's J.* 5:201–203.

Currie, P. O. 1978. Cattle weight gain comparisons under season-long and rotation grazing systems. *Proc. Int. Rangel. Congr.* 1:579–580.

Dahl, B. E. 1986. The west Texas experience in short-duration grazing. In *Proc. Short-Duration Grazing and Current Issues in Grazing Management Shortcourse.* Washington State University Cooperative Extension Service, Pullman, WA.

Daugherty, D. A., C. M. Britton, and H. A. Turner. 1982. Grazing management of crested wheatgrass range for yearling steers. *J. Range Manage.* 35:347–351.

Davis, J. W. 1982. Livestock vs riparian management—there are solutions. pp. 175–184. In *Wildlife Livestock Relationships Symposium: Proc. 10th University of Idaho Forest, Wildlife and Range Experiment Station,* Moscow, Idaho.

Denny, R. P., and J. S. H. Steyn. 1977. Trails of multi-paddock grazing systems on veld. 2. A comparison of a 16 paddock-to-one-herd system with a four paddock-to-one-herd system. *Rhod. J. Agric. Res.* 15:119–127.

Denny, R. P., D. L. Barnes, and T. C. D. Kennan. 1977. Trails of multi-paddock grazing systems on veld. An exploratory trial of systems involving 12 paddocks and one herd. *Rhod. J. Agric. Res.* 15:11–23.

Dillon, C. C. 1958. Benefits of rotation-deferred grazing on northwest ranges. *J. Range Manage.* 11:278–281.

Drawe, D. L. 1988. Effects of three grazing treatments on vegetation, cattle production, and wildlife on the Welder Wildlife Foundation Refuge 1974–1982. *Welder Wildlife Foundation Contribution* B-8, Sinton, TX.

Eckert, R. E. Jr., and J. S. Spencer. 1986. Vegetation response to rest-rotation grazing management. *J. Range Manage.* 39:166–174.

Eckert, R. E. Jr., and J. S. Spencer. 1987. Growth and reproduction of grasses heavily grazed under rest-rotation management. *J. Range Manage.* 40:156–159.

Elmore, W., and R. L. Beschta. 1987. Riparian areas: Perceptions in management. *Rangelands* 9:260–265.

Elmore, W., and Kauffman. 1994. Riparian and watershed systems, degradation and restoration. In *Ecological implications of livestock herbivory in the West.* Society for Range Management, Denver, CO.

Evans, K. E., and R. R. Krebs. 1977. Avian use of livestock watering ponds in western South Dakota. *U.S. Dep. Agric. For. Serv. Gen. Tech. Rep.* RM-35.

Fisher, C. E., and P. T. Marion. 1951. Continuous and rotation grazing on buffalo and tobosa grass. *J. Range Manage.* 4:48–51.

Fowler, J. M., and J. R. Gray. 1986. Economic impacts of grazing systems during drought and nondrought years on cattle and sheep ranches in New Mexico. *N. Mex. Agric. Exp. Stn. Bull.* 725.

Frischknecht, N. C., and L. E. Harris. 1968. Grazing intensities and systems on crested wheatgrass in central Utah: Response of vegetation and cattle. *U.S. Dep. Agric. Bull.* 1388.

Frisina, M. R. 1992. Elk habitat use within a rest-rotation grazing system. *Rangelands* 14:93–96.

Gammon, D. M. 1978. A review of experiments comparing systems on natural pastures. *Proc. Grassl. Soc. South Afr.* 13:75–82.

Gammon, D. M. 1984. An appraisal of short-duration grazing as a method of veld management. *Zimbabwe J. Agric. Res.* 84:59–64.

Gillen, R. L., W. C. Krueger, and R. F. Miller. 1985. Cattle use of riparian meadows in the Blue Mountains of northeastern Oregon. *J. Range Manage.* 38:205–210.

Goodloe, S. 1969. Short duration grazing in Rhodesia. *J. Range Manage.* 22:369–373.

Graham, K. T., L. A. Torell, and C. D. Allison. 1992. Costs and benefits of implementing holistic resource management on New Mexico ranches. *New Mexico Agr. Exp. Sta. Bull.* 762.

Gray, J. R., and J. M. Fowler. 1980. Requirements for economic analysis of grazing systems. In *Proc. Grazing Management Systems for Southwest Rangelands Symposium.* The Range Improvement Task Force, New Mexico State University, Las Cruces, NM.

Gunderson, D. R. 1968. Flood plain use related to stream morphology and fish populations. *J. Wildl. Manage.* 32:507–514.

Hart, R. H., S. Clapp, and P. S. Test. 1993. Grazing strategies, stocking rates and frequency and intensity of grazing on western wheatgrass and blue grama. *J. Range Manage.* 46:122–127.

Hart, R. H., M. J. Samuel, P. S. Test, and M. A. Smith. 1988. Cattle, vegetation, and economic responses to grazing systems and grazing pressure. *J. Range Manage.* 41:282–286.

Hayes, F. A. 1978. *Streambank and meadow condition in relation to livestock grazing in mountain meadows of central Idaho.* M.S. Thesis, University of Idaho, Moscow, ID.

Heady, H. F. 1961. Continuous vs specialized grazing systems: A review and application to the California annual type. *J. Range Manage.* 14:182–183.

Heitschmidt, R. K. 1986. Short-duration grazing at the Texas Experimental Ranch. In *Proc. Short-Duration Grazing and Current Issues in Grazing Management Shortcourse.* Washington State University Cooperative Extension Service, Pullman, WA.

Heitschmidt, R. K., J. R. Conner, S. K. Canon, W. E. Pinchak, J. W. Walker, and S. L. Dowhower. 1990. Cow/calf production and economic returns from year-long continuous deferred rotation and rotational grazing treatments. *J. Range Manage.* 3:92–99.

Heitschmidt, R. K., J. R. Frasure, D. L. Price, and L. R. Rittenhouse. 1982a. Short-duration grazing at the Texas Experimental Ranch: Weight gains of growing heifers. *J. Range Manage.* 35:375–379.

Heitschmidt, R. K., R. A. Gordon, and J. S. Bluntzer. 1982b. Short-duration grazing at the Texas Experimental Ranch: Effects on forage quality. *J. Range Manage.* 35:372–374.

Heitschmidt, R. K., M. M. Kothmann, and W. J. Rawlins. 1982c. Cow-calf response to stocking rates, grazing systems, and winter supplementation at the Texas Experimental Ranch. *J. Range Manage.* 35:204–211.

Heitschmidt, R. K., D. L. Price, R. A. Gordon, and J. R. Frasure. 1982d. Short-duration grazing at the Texas Experimental Ranch: Effects on aboveground net primary production and season growth dynamics. *J. Range Manage.* 35:367–372.

Heitschmidt, R. K., S. L. Dowhower, and J. W. Walker. 1987. 14- vs 42-paddock rotational grazing: Aboveground biomass dynamics, forage production and harvest efficiency. *J. Range Manage.* 40:216–224.

Herbel, C. H., and K. L. Anderson. 1959. Response of true prairie vegetation on major Flint Hills range sites to grazing treatment. *Ecol. Monogr.* 29:171–186.

Holechek, J. L. 1983. Considerations concerning grazing systems. *Rangelands* 5:308–311.

Holechek, J. L. 1994. Financial returns from different grazing management systems in New Mexico. *Rangelands* 16:237–240.

Holechek, J. L., M. Vavra, and J. Skovlin. 1981. Diet quality and performance of cattle on forest and grassland range. *J. Anim. Sci.* 53:291–299.

Holechek, J. L., M. Vavra, and J. Skovlin. 1982. Cattle diet and performance on mountain meadows in Oregon. *J. Range Manage.* 35:652–655.

Holechek, J. L., T. J. Berry, and M. Vavra. 1987. Grazing system influences on cattle diet and performance on mountain range. *J. Range Manage.* 40:55–60.

Hormay, A. L. 1970. Principles of rest-rotation grazing and multiple use land management. *U.S. Department of Agriculture Forest Service Training Text 4* (2200).

Hormay, A. L., and A. B. Evanko. 1958. Rest-rotation grazing: A management system for bunchgrass ranges. *U.S. Dep. Agric. Calif. For. Range Exp. Stn. Misc. Pap. 27.*

Howell, L. N. 1978. Development of multi-camp grazing systems in the Southern Orange Free State, Republic of South Africa. *J. Range Manage.* 31:459–465.

Howery, L. D., F. D. Provenza, R. E. Banner, and C. B. Scott. 1996. Differences in home range and habitat use among individuals in a cattle herd. *Appl. Anim. Behav. Sci.* 49:305–320.

Hubbard, W. A. 1951. Rotational grazing studies in Canada. *J. Range Manage.* 4:25–29.

Huss, D. L., and J. V. Allen. 1969. Livestock production and profitability comparisons of various grazing systems. Texas Range Station. *Tex. Agric. Exp. Stn. Bull.* 1089.

Hyder, D. N., and W. A. Sawyer. 1951. Rotation-deferred grazing as compared to season-long grazing on sagebrush-bunchgrass ranges in Oregon. *J. Range Manage.* 4:30–34.

Johnson, W. M. 1965. Rotation rest-rotation and season-long grazing on a mountain range in Wyoming. *U.S. Dep. Agric. For. Serv. Res. Pap.* RM-14.

Jung, H. G., R. W. Rice, and L. J. Koong. 1985. Comparison of heifer weight gains and forage quality for continuous and short-duration grazing systems. *J. Range Manage.* 38:144–149.

Kauffman, J. B., W. C. Krueger, and M. Vavra. 1983a. Effect of cattle grazing on riparian plant communities. *J. Range Manage.* 36:685–691.

Kauffman, J. B., W. C. Krueger, and M. Vavra. 1983b. Impacts of cattle grazing on streambanks in northeastern Oregon. *J. Range Manage.* 36:683–685.

Klipple, G. E. 1964. Early- and late-season grazing versus season-long grazing of shortgrass vegetation on the central Great Plains. *U.S. Dep. Agric. For. Serv. Res. Pap.* RM-11.

Kothmann, M. M., G. M. Mathis, and W. J. Waldrip. 1971. Cow-calf response to stocking rates and grazing systems on native range. *J. Range Manage.* 24:100–105.

Kothmann, M. M., W. S. Rawlins, and J. Bluntzer. 1975. Vegetation and livestock responses to grazing management on the Texas Experimental Ranch. *Tex. Agric. Exp. Stn.* RP-3310.

Laycock, W. A., and P. W. Conrad. 1981. Responses of vegetation and livestock to various systems of grazing on seeded and native mountain rangelands in eastern Utah. *J. Range Manage.* 34:52–58.

Martin, S. C. 1978. The Santa Rita grazing system. *Proc. Int. Rangel. Congr.* 1:573–575.

Martin, S. C., and D. E. Ward. 1970. Rotating access to water improvement semidesert cattle range near water. *J. Range Manage.* 23:22–26.

Martin, S. C., and D. E. Ward. 1976. Perennial grasses respond inconsistently to alternate year seasonal rest. *J. Range Manage.* 29:346–347.

Martin, S. C., and K. E. Severson. 1988. Vegetation response to the Santa Rita grazing system. *J. Range Manage.* 41:291–296.

Mathis, G. W., and M. M. Kothmann. 1971. Weaning weight and daily gain of calves on different grazing treatments on the Texas Experimental Ranch. *Tex. Agric. Exp. Stn.* 2963–2999.

McCalla, G. R. II, W. H. Blackburn, and L. B. Merrill. 1984a. Effects of livestock grazing on infiltration rates, Edwards Plateau of Texas. *J. Range Manage.* 37:265–269.

McCalla, G. R. II, W. H. Blackburn, and L. B. Merrill. 1984b. Effects of livestock grazing in sediment production, Edwards Plateau of Texas. *J. Range Manage.* 37:291–294.

McCollum, F. T., R. L. Gillen, D. M. Engle, and G. W. Horn. 1990. Stocker cattle performance and vegetation response to intensive-early stocking of Cross Timbers rangeland. *J. Range Manage.* 43:104–109.

McIlvain, E. H., and D. A. Savage. 1951. Eight-year comparisons of continuous and rotational grazing on the southern plains range. *J. Range Manage.* 4:42–47.

Merrill, L. B. 1954. A variation of deferred rotation grazing for use under southwest range conditions. *J. Range Manage.* 7:152–154.

Merrill, L. B., and C. A. Taylor. 1975. Advantages and disadvantages of intensive grazing management systems near Sonora and Barnhart. *Tex. Agric. Exp. Stn.* PR-3341.

Murray, R. B., and J. O. Klemmedson. 1968. Cheatgrass range in southern Idaho: Seasonal cattle gains and grazing capacities. *J. Range Manage.* 21:308–313.

Musimba, N. K. R., R. D. Pieper, J. D. Wallace, and M. L. Galyean. 1987. Influence of watering frequency on forage consumption and steer performance in southeastern Kenya. *J. Range Manage.* 40:412–415.

Ohmart, R. 1996. Historical and present impact of livestock grazing on fish and wildlife resources in western riparian habitats. In *Rangeland wildlife.* Society for Range Management, Denver, CO.

Olson, K. C., J. R. Brethour, and J. L. Launchbough. 1993. Shortgrass range vegetation and steer growth response to intensive-early stocking. *J. Range. Manage.* 46:127–132.

Owensby, C. E., E. F. Smith, and K. L. Anderson. 1973. Deferred-rotation grazing with steers in the Kansas Flint Hills. *J. Range Manage.* 26:393–395.

Pieper, R. D. 1980. Impacts of grazing systems on livestock. Proc. *Grazing Management Systems for Southwest Rangelands Symposium.* New Mexico State University, Las Cruces, NM.

Pieper, R. D., G. B. Donart, E. E. Parker, and J. D. Wallace. 1978. Livestock and vegetational response to continuous and four-pasture one-herd grazing systems in New Mexico. *Proc. Int. Rangel. Congr.* 1:560–562.

Pieper, R. D., and R. K. Heitschmidt. 1988. Is short-duration grazing the answer. *J. Soil and Water Cons.* 43:133–137.

Pieper, R. D., E. E. Parker, G. B. Donart, J. D. Wallace, and J. D. Wright. 1991. Cattle and vegetation in response to four-pasture rotation and continuous grazing. *New Mexico Agric. Exp. Sta. Bull.* 756.

Pitts, J. S., and F. C. Bryant. 1987. Steer and vegetation response to short-duration and continuous grazing. *J. Range Manage.* 40:386–390.

Platts, W. S. 1982. Sheep and cattle grazing strategies on riparian stream environments pp. 251–270. In *Wildlife Livestock Relationships Symposium. Proc.* 10th University of Idaho Forest, Wildlife and Range Experiment Station, Moscow, ID.

Platts, W. S., and R. L. Nelson. 1989. Characteristics of riparian plant communities with respect to livestock grazing. In *Practical approaches to riparian resource management.* United States Department of Interior-Bureau of Land Management, Billings, MT.

Pluhar, J. J., R. W. Knight, and R. K. Heitschmidt. 1987. Infiltration rates and sediment production as influenced by grazing systems in the Texas Rolling Plains. *J. Range Manage.* 40:240–244.

Quigley, T. M., J. M. Skovlin, and J. P. Workman. 1984. An economic analysis of two systems and three levels of grazing on ponderosa pine-bunchgrass range. *J. Range Manage.* 37:309–312.

Ratliff, R. D. 1986. Cattle responses to continuous and seasonal grazing of California annual grassland. *J. Range Manage.* 39:482–486.

Ratliff, R. D., J. N. Reppert, and R. J. McConnen. 1972. Rest-rotation at Harvey Valley: Range health, cattle gains, costs. *U.S. Dep. Agric. For. Serv. Res. Pap.* PSW-77.

Ratliff, R. D., and J. N. Reppert. 1974. Vigor of Idaho fescue grazed under rest-rotation and continuous grazing. *J. Range Manage.* 27:447–449.

Reardon, P. O., and L. B. Merrill. 1976. Vegetative response under various grazing management systems in the Edwards Plateau of Texas. *J. Range Manage.* 29:195–198.

Reardon, P. O., L. B. Merrill, and C. A. Taylor Jr. 1978. White-tailed deer preferences and hunter success under various grazing systems. *J. Range Manage.* 31:40–43.

Reece, P. A. 1986. Short-duration grazing research and case studies in Nebraska. In *Proc. Short-Duration Grazing and Current Issues in Grazing Management Shortcourse.* Washington State University Cooperative Extension Service, Pullman, WA.

Roath, L. R., and W. C. Krueger. 1982a. Cattle grazing influence on a mountain riparian zone. *J. Range Manage.* 35:100–104.

Roath, L. R., and W. C. Krueger. 1982b. Cattle grazing and behavior on a forested range. *J. Range Manage.* 35:332–338.

Rogler, G. A. 1951. A 25-year comparison of continuous and rotation grazing in the Northern Plains. *J. Range Manage.* 4:35–41.

Sampson, A. W. 1913. Range improvement by deferred and rotation grazing. *U.S. Dep. Agric. Bull.* 34.

Savory, A. 1978. A holistic approach to range management using short-duration grazing. *Proc. Int. Rangel. Congr.* 1:555–557.

Savory, A. 1983. The Savory grazing method or holistic resource management. *Rangelands* 5:155–159.

Savory, A. 1988. *Holistic resource management.* Island Press, Washington, DC.

Savory, A., and S. D. Parsons. 1980. The Savory grazing method. *Rangelands* 2:234–237.

Severson, K. E., and C. E. Boldt. 1978. Cattle, wildlife, and riparian habitat in the western Dakotas, pp. 94–103. In *Management and Use of Northern Plains Rangeland Symposium,* Bismark, ND.

Sharrow. S. H. 1983. Forage standing crop and animal diets under rotational vs continuous grazing. *J. Range Manage.* 36:447–450.

Sharrow, S. H., and W. C. Krueger. 1979. Performance of sheep under rotational and continuous grazing on hill pastures. *J. Anim. Sci.* 49:893–899.

Skovlin, J. M. 1984. Impacts of grazing on wetlands and riparian habitat. In National Research Council/National Academy of Sciences (Eds.). *Developing strategies for rangeland management.* Westview Press, Inc., Boulder, CO.

Skovlin, J. M. 1987. Southern Africa's experience with intensive short-duration grazing. *Rangelands* 9:162–168.

Skovlin, J. M., R. W. Harris, G. S. Strickler, and G. A. Garrison. 1976. Effects of cattle grazing methods on ponderosa pine-bunchgrass range in the Pacific northwest. *U.S. Dep. Agric. Tech. Bull.* 1531.

Smith, D. R., H. G. Fisser, N. Jefferies, and P. O. Stratton. 1967. Rotation grazing on Wyoming's Big Horn Mountains. *Wyo. Agric. Exp. Stn. Res. J.* 13.

Smith, E. F., and C. E. Owensby. 1978. Intensive-early stocking and season-long stocking of Kansas Flint Hills range. *J. Range Manage.* 31:14–17.

Smoliak, S. 1960. Effects of deferred-rotation and continuous grazing on yearling steer gains and shortgrass prairie vegetation of southeastern Alberta. *J. Range Manage.* 13:239–243.

Smoliak, S. 1968. Grazing studies on native range crested wheatgrass, and Russian wildrye pastures. *J. Range Manage.* 21:147–150.

Stewart, J. R., and C. L. Leinweber. 1968. An economic evaluation of range improvement and grazing practices, Texas Experimental Ranch. *Tex. Agric. Exp. Stn. Prog. Rep.* 2591.

Taylor, C. A. Jr., M. M. Kothmann, L. B. Merrill, and D. Elledge. 1980. Diet selection by cattle under high-intensity low-frequency, short duration, and Merrill grazing systems. *J. Range Manage.* 33:428–435.

Taylor, C. A. Jr., and N. E. Garza Jr. 1986. Rambouillet ewe response to grazing systems at the Texas Range Station. *Sheep Ind. Dev. Res. Dig.* 3:35–40.

Taylor, C. A. Jr., T. D. Brooks, and N. E. Garza Jr. 1993. Effects of short-duration and high-intensity, low-frequency grazing systems on forage production and composition. *J. Range Manage.* 46:118–122.

Taylor, C. A. Jr., N. E. Garza Jr., and T. D. Brooks. 1993. Grazing systems on the Edwards Plateau of Texas: Are they worth the trouble? II. *Rangelands* 15:57–61.

Thurow, T. L., W. H. Blackburn, and C. A. Taylor Jr. 1986. Hydrologic characteristics of vegetation types as affected by livestock grazing systems, Edwards Plateau, Texas. *J. Range Manage.* 39:505–509.

Thurow, T. L., W. A. Blackburn, and C. A. Taylor, Jr. 1988. Some vegetation responses to selected livestock grazing strategies, Edwards Plateau, Texas. *J. Range Manage.* 41:108–114.

Trlica, M. J., M. Buwai, and J. W. Menke. 1977. Effects of rest following defoliations on the recovery of several range species. *J. Range Manage.* 30:21–27.

Valentine, K. A. 1967. Seasonal suitability, a grazing system for ranges of diverse vegetation types and condition classes. *J. Range Manage.* 20:395–397.

Walker, J. W., and R. K. Heitschmidt. 1986. Effect of various grazing systems on type and density of cattle trails. *J. Range Manage.* 39:428–431.

Ward, D. E. 1975. Seasonal weight changes of cattle on semidesert grass-shrub ranges. *J. Range Manage.* 28:97–99.

Weltz, M., and M. K. Wood. 1986a. Short duration grazing in central New Mexico: Effects on infiltration rates. *J. Range Manage.* 39:365–368.

Weltz, M., and M. K. Wood. 1986b. Short-duration grazing in central New Mexico: Effects on sediment production. *J. Soil Water Conserv.* 41:262–266.

White, M. R., R. D. Pieper, G. B. Donart, and L. White-Trifaro. 1991. Vegetation response to short-duration and continuous grazing in southcentral New Mexico. *J. Range Manage.* 44:399–404.

Whitson, R. E., R. K. Heitschmidt, M. M. Kothmann, and G. K. Lundgren. 1982. The impact of grazing systems on the magnitude and stability of ranch income in the Rolling Plains of Texas. *J. Range Manage.* 35:526–533.

Willms, W. D., S. Smoliak, and J. F. Dormaar. 1990. Vegetation response to time-controlled grazing on mixed and fescue prairie. *J. Range Manage.* 43:513–518.

Wood, M. K., and W. H. Blackburn. 1984. Vegetation and soil response to cattle grazing systems in the Texas Rolling Plains. *J. Range Manage.* 37:303–308.

CHAPTER 10

METHODS OF IMPROVING LIVESTOCK DISTRIBUTION

FACTORS CAUSING POOR DISTRIBUTION

Uneven use of rangeland by livestock has been and continues to be a major problem confronting range managers. On many ranges, improvement will occur without reduction in livestock numbers if practices to secure more uniform utilization are implemented. Distribution problems are most severe in arid or desert areas, and in mountainous terrain. These conditions characterize most of the area west of the Rocky Mountains in the United States. Factors causing uneven use of rangeland include: distance from water, rugged topography, diverse vegetation, wrong type of livestock, pests, and weather. These factors are discussed together with management practices that can be used to improve livestock distribution. For more detailed discussion of spatial components of plant-animal relations the reader is referred to Coughenour (1991), Bailey et al. (1996), and a conceptional model of plant-animal spatial relations provided by Bailey and Rittenhouse (1989).

Distance from Water

Poor water distribution is the chief cause of poor livestock distribution on most ranges. In arid regions of the world, water is in short supply and poorly distributed. Where available watering points are infrequent, large sacrifice areas around watering points have often occurred. The heavy use of vegetation around watering points is well documented (Figure 10.1). A study on Montana mixed prairie range showed virtually 100% use of forage around water, with declining use as distance from water increased (Holscher and Woolfolk 1953). In southern Idaho, Mueggler (1965) reported 100% use of bunchgrasses around water. Under moderate grazing intensities, two long-term studies from southern New Mexico show that forage production is most severely reduced in the zone within 0.8 km (0.5 mi) of water (Valentine 1947; Fusco et al. 1995). On forested range in California, cattle concentrated in riparian habitats because of close proximity to water (Kie and Boroski 1996).

Although livestock will travel great distances to water, this is not in the best interest of the animal or the range resource. Travel increases energy expenditure by the animal that otherwise would go into production (weight gain, and milk production), and takes away from grazing and resting time.

If animals must travel large distances between water and available forage, a series of trails will be created that gradually become larger and more numerous. These trails become water channels that cause severe erosion. The proportion of land in livestock trails on a range is a good indicator of grazing severity.

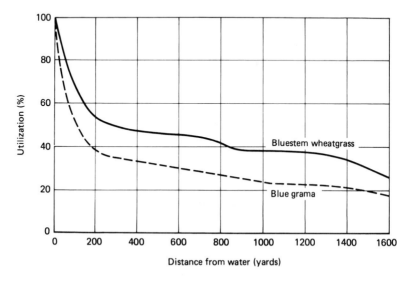

***Figure* 10.1** The relationship between percentage utilization of major forage species and distance from water on winter range in eastern Montana plains. (From Holscher and Woolfolk 1953; Stoddart et al. 1975.)

Recommendations on distances between watering points vary with terrain, type of animal, and breed of livestock. Herbel and Nelson (1966) found that the average daily travel distance of Hereford cows was 7.9 km (4.91 mi) compared to 12.6 km (7.83 mi) for Santa Gertrudis cows on desert grassland range in southcentral New Mexico. Studies with sheep have shown Dorset Horns will travel 9 km (5.59 mi) a day, Merinos 13.7 km (8.51 mi), and Border Leicesters 14 km (8.7 mi) (Squires et al. 1972). Barnes (1914) recommended that cattle in rough topography should not have to travel over 1.6 km (1 mi) to water. In flat country, he suggested travel distances should not exceed 3.2 km (2 mi). On mountain range in northeastern Oregon, Goebel (1956) believed .8 to 1.2 km (.5 to .75 mi) was the ideal distance between watering points. In his range management textbook, Bell's (1973) recommended travel distances to water (most practical from the standpoint of both livestock and the range) are as follows:

Rough country	0.8 km (0.5 mi)
Rolling, hilly country	1.6 km (1 mi)
Flat country	3.2 km (2 mi)
Smooth, sandy country	2.4 km (1.5 mi)
Undulating, sandy country (dunes)	1.6 km (1 mi)

We consider these recommendations to be sound based on our own experiences. A common rule of thumb is: no more than 50 cattle or 300 sheep per water facility.

Travel distance to water can influence livestock productivity. Animals forced to travel long distances between available forage and water have reduced productivity due to reductions in forage intake and increased energy expenditure in travel (Squires 1978). Sneva et al. (1973) studied the effects of trailing cattle to water on cold desert range in southeastern Oregon. When cows were trailed about 1.6 km (1 mi), their calves gained less weight than those of cows in pastures with water close by (Table 10.1). Squires, (1970) in Australia, found penned lambs gained 219 gm per day; while lambs forced to travel gradually-increasing distances, from 0.8 to 4.0 km per day, gained only 126 gm.

Topography

Rugged topography is the second most important cause of poor livestock distribution on rangelands. The reluctance of livestock to use steep slopes is not entirely undesirable, since these areas are often fragile and valley bottoms can better withstand grazing. However, in many cases slopes serve as barriers to the use of benches and ridgetops above valley bottoms.

Livestock vary considerably in their willingness to use steep terrain. Large, heavy animals such as mature cattle or horses have difficulty in traversing steep, rocky slopes. Therefore, cattle make little use of slopes over a 10% gradient (Table 10.2). Because of their smaller size, greater agility, and surefootedness, sheep and goats use these areas more readily. On winter range in New Mexico, McDaniel and Tiedeman (1981) found sheep uniformly used slopes of less than 45%, but utilization was sharply reduced on steeper slopes.

TABLE 10.1   Average Daily Gain (lb) of Cattle in Southeastern Oregon as Affected by Trailing One Mile to Water and No Trailing

	TREATMENT	
PERIOD	TRAILING 1 MILE	NO TRAILING
May 20–Aug 17, 1970		
Cows	1.11	1.32
Calves	1.68	2.05
Yearlings	1.10	1.37
April 16–Aug 3, 1971		
Cows	0.48	0.41
Calves	1.70	1.80
Yearlings	1.32	1.86

Source: Sneva et al. 1973.

TABLE 10.2   Preference Indices for Slope Gradient Classes During Three Grazing Periods as Determined by Direct Cattle Observations on Mountain Rangelands in Northeastern Oregon[a]

SLOPE GRADIENT (%)	SEASON-LONG GRAZING	EARLY SUMMER ONLY	LATE SUMMER ONLY
5 or less	3.0	3.7	4.0
6–10	5.3	2.9	4.8
11–15	0.8	1.6	1.6
16–20	0.8	1.1	0.6
21–30	0.4	0.5	0.5
31–45	0.3	0.4	0.4
Greater than 45	0.1	0.1	0.1

Source: Gillen et al. 1984.

[a]5.3 highest, 0.1 lowest, as determined by direct cattle observations.

Many rugged ranges can be better used by wild animals than by livestock. In Utah, slopes between 30% and 40% had the greatest use by mule deer; elk readily used slopes up to 30%; but cattle use was primarily on slopes of less than 10% (Julander and Jeffery 1964). In Oregon, bighorn sheep activities were not reduced by slopes up to 80% whereas cattle, horses, and mule deer had less activity as slope increased (Figure 10.2).

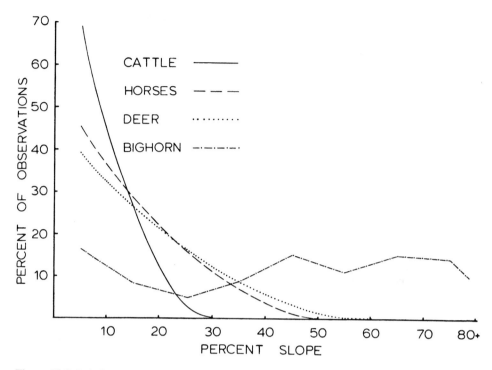

***Figure* 10.2** Relationship of slope gradient to the percentage of observations of cattle, feral horses, deer, and bighorn sheep. (From Ganskopp and Vavra 1987. Reprinted with permission.)

Vegetation Type

All animal species have higher preferences for certain vegetation types than for others. This causes different vegetation types within a pasture to receive different degrees of use. Assuming that other factors affecting distribution, such as water and terrain, are the same, livestock will prefer the vegetation type best meeting their nutritional needs. On shortgrass prairie in Colorado, cattle preferences for plant communities with seasonal advance were closely associated with the relative crude protein content of their standing crop (Senft et al. 1985).

Open grasslands are preferred by cattle over heavily forested areas (Pickford and Reid 1948). Separate fencing of grassland and forest vegetation types is necessary to obtain efficient use of both types in northeastern Oregon (Holechek et al. 1981). This is practical when both vegetation types occur as large discrete units.

Snow cover can influence livestock preference for vegetation types. Cattle preferred vegetation types dominated by western wheatgrass (*Agropyron smithii*) and blue grama (*Bouteloua gracilis*) during the spring and summer in northcentral New Mexico (Holechek et al. 1986). However, in winter, shrubland vegetation types dominated by fourwing saltbush (*Atriplex canescens*) and common winterfat (*Ceratoides lanata*) were preferred because snow covered up the grasses. These two shrubs were readily consumed by cattle because of their low content of volatile oils and tannins. On this same range, sheep preferred vegetation types with a high component of forbs in spring and summer, but like cattle, they used the shrubland vegetation types in winter.

Pests

Under some weather conditions and on some ranges, insects affect the grazing pattern. When potholes filled with rainwater become infested with mosquitoes, livestock will move to higher ground. Livestock will also avoid infestations of flies, as in parts of Africa where the tsetse fly curtails livestock production (see Chapter 13).

Weather

Distribution of grazing may be directly affected by weather. Temperature changes, snowfall, and excessive rainfall may limit grazing, while variations in day and night temperatures can cause livestock to alter their grazing locations. On mountain range in northeastern Oregon, cattle preferred open grasslands on south-facing slopes for morning grazing in summer due to warmer temperatures (Holechek and Vavra 1983). In afternoon, open forest vegetation types on north slopes were preferred because of shade. When it is hot, more use will be made of higher areas that are cooled by air currents. When it is cold, a reverse pattern will occur and animals will move to warmer lowland areas or areas having shrubs or trees (shade). When there is a cold wind, livestock will often travel in the same direction as the wind in an attempt to reduce the chilling effects of the wind.

Snowfall precludes the use of most mountainous rangelands in the intermountain United States in the winter. Too much rainfall is seldom a problem on rangeland. However, in some areas, high rainfall can result in marshlands where grazing is curtailed during wet periods. This problem occurs during the July through September period in parts of Africa (southwestern Sudan). Excessive rainfall, high temperatures, and insects on coastal ranges combine to affect the grazing pattern; and when these conditions prevail, livestock will move to the highest ground.

BETTER LIVESTOCK DISTRIBUTION METHODS

Proper grazing distribution within each range unit requires the scattering of animals within a grazed area to obtain uniform use of range forages. In the large pastures prevalent in most arid and semiarid areas, it is common to find serious overgrazing near watering points and little use of forage in portions of the grazing unit distant from water. Several practices can be used to improve livestock distribution. These include (1) increasing the number and changing location of watering points; (2) fencing; (3) increasing the number and changing the location of salt, minerals, and supplemental feeds; (4) predator control to permit various animal species to use areas avoided previously; (5) spot burning to improve palatability of areas lightly utilized; (6) fertilization to change species composition and improve the palatability of some forage plants; (7) mowing old growth to permit grazing of certain plants; (8) insect control to allow livestock to graze all parts of a range unit; (9) changing type or types of livestock; (10) use of specialized grazing systems; (11) constructing trails into inaccessible areas; (12) use of herding or drifting to move livestock to underutilized areas; and (13) construction of shade structures or planting shelter belts at strategic places where natural shade and protection from wind are unavailable.

Water

Stock water is the center of grazing activity. Additional watering locations will often improve both livestock distribution and their productivity. Development of new watering facilities can be accomplished by (1) drilling a well, installing a pumping unit, and constructing a storage tank and drinking troughs; (2) constructing an earthen tank to impound runoff water; and (3) piping water to a new location. Economics and location usually dictate the type of water development to be constructed.

Prospecting for well sites often considers slope position, fault lines, rock layering, and outcrops. If depth to the water strata is fairly uniform over the range, locating wells in an ungrazed forage supply can be considered. Pumping units in remote locations, such as jacks or submersible pumps, are sometimes powered by a solar source. Some of the considerations regarding selection of wind or solar energy to power a water well are given by Polk and Ervin (1996).

Reservoirs are constructed to hold as much water as needed by livestock in relation to the forage resource, yet with as little surface as possible, to reduce evaporation. Earthen reservoirs are often constructed with a trap above them to catch sediment carried by runoff water. Where feasible, earthen reservoirs are fenced and the water is piped to troughs.

Seeps and springs are often developed as a source of water. Sedges, rushes, and other water-loving plants reveal sites for developing surface water while spring water may be increased with horizontal drilling (Gartner 1986).

Plastic pipe is often used to get water from a permanent site to an area with additional forage. Water hauling may seem like the last alternative for improving distribution, yet it can be profitable where forage might not otherwise be consumed.

Changing access to watering points and hauling water are two means of avoiding the high cost of permanent watering facilities. On Arizona desert grassland range, rotation of cattle access to watering points improved the uniformity of pasture use and provided vegetation around watering points with the opportunity to recover from grazing (Table 10.3). This practice improved the use of forage distance from watering points and reduced the use of forage around watering points. Rotation of access to watering points is a very sound practice in arid and desert areas. This practice is often economically much more feasible than implementation of a specialized grazing system (high fencing costs).

Water is often hauled to sheep on salt desert ranges in Utah and Nevada. This practice has improved sheep gains and uniformity of range use compared to sheep herds that were trailed to water (Hutchings 1946).

TABLE 10.3   Herbage Yield (lb/acre) of Perennial Grasses from Part-time and Yearlong Access to Water on Southern Arizona Range

WATER AVAILABILITY	DISTANCE FROM WATER (YARDS)					
	100	200	300	400	500	AVERAGE
Part-time	14	105	167	66	101	91
Yearlong	1	22	37	61	112	47
Average	8	64	102	64	106	69

Source: Martin and Ward 1970.

Fencing

Fencing is used to subdivide large range units into small units. The location, size, and shape of a grazing unit, and the direction of livestock travel, are important considerations in fence placement. Uniform grazing is difficult when several range sites occur in the same grazing unit. On small grazing units, the forage selectivity is reduced and grazing is more uniform. However, fencing often is not possible because of the costs involved. Fencing can adversely affect wild ungulate populations by restricting movement between ranges. Passages that restrict livestock and permit wild ungulate crossing are discussed in Chapter 13 and by Kindschy (1996).

For distribution, fences serve to (1) control seasonal drift of livestock, (2) regulate use among forage types or protect choice grazing areas for special use, and (3) separate range units for special management. When fencing, consideration is given to range sites and potential production. If possible, fences are located so that permanent watering points will serve two or more range units. Electric fences may be used to subdivide large units and reduce livestock losses from predation. These fences may be electrified by a charger using solar panels as a source of energy. Some properties of barbed wire and fence posts are given in Tables 10.4 and 10.5. Fencing and herding costs can be reduced by audio-electric stimulation (Quigley et al. 1990; Rose 1991).

Drift Fences. Because plants develop more slowly at high elevations, animals habitually drift upward seeking more succulent forage. Drift fences are used to hold livestock on lower rangelands until higher range is ready for grazing.

Forage Types. Fencing is often used to separate high-value forage types, such as meadow or seeded pastures, from low-producing rangeland. Fencing is also used to separate forage types utilized in the summer-fall grazing season from those grazed in the winter-spring season (see Chapter 9).

Management Strategies. Animals distribute themselves more efficiently when large ranges have been divided into several range units by cross fencing. Greater use of less palat-

TABLE 10.4   Comparison of Selected Characteristics of Various Wire Types

TYPE OF WIRE	GAUGE	DIAMETER (MM)	BREAKING STRENGTH (KG)	YEARS UNTIL WIRE REACHES HALF-STRENGTH[a]
Two-strand barbed	12½	2.5	431	12
Standard high-tensile	12½	2.5	416	18

Source: Jepson et al. 1983.

[a]For coastal climates (dry climates will increase life threefold).

TABLE 10.5   Comparison of Properties of Common Western Fence Post Species[a]

SPECIES	HARDNESS	STRENGTH AS A POST	DECAY RESISTANCE OF HEARTWOOD
Aspen	C	C	C
Cottonwood	C	C	C
Douglas fir	B	A	B
Ponderosa pine	C	B	C
Western red cedar	C	B	A
Lodgepole	B	B	C

Source: Jepson et al. 1983.

[a]A, relatively high; B, intermediate; C, low.

able forage has been obtained by stocking each unit with a high livestock density for short periods (see Chapter 9). For example, forage utilization on shortgrass rangeland in Wyoming was not as uniform under continuous as under short-duration grazing (Hart et al. 1989).

Salt, Minerals, and Supplemental Feed

Proper salting, either alone or in a mixture of some other supplement, can be a great benefit in obtaining desired distribution of grazing animals. The preference of cattle for areas of varying distances from salt on mountain rangelands is shown in Table 10.6. Livestock usually go from water to grazing and then to salt; thus it is not necessary to place salt at watering points. By placing salt away from water, livestock can be enticed to use areas otherwise avoided (Figure 10.3). Salt increases the appetite of a grass-eating animal. General studies for amounts of salt needed are 1 kg to 2 kg (2.2 lb to 4.4 lb) per year per head for sheep and goats, 10 kg to 12 kg (22 lb to 26.4 lb) for mature cattle, and 14 kg to 18 kg (30.8 lb to 40 lb) for horses.

TABLE 10.6   Preference Indices for Distance Classes to Salt During Three Grazing Periods as Determined by Direct Cattle Observations on Mountain Rangeland in Northeastern Oregon[a]

DISTANCE (%)	SEASON-LONG GRAZING	EARLY SUMMER GRAZING ONLY	LATE SUMMER GRAZING ONLY
200 or less	1.2	1.2	1.2
201–400	0.8	1.4	1.1
401–600	0.6	1.2	1.7
601–800	1.5	0.6	0.4
Greater than 800	1.4	0.3	0.3

Source: Gillen et al. 1984.

[a]1.7 highest, 0.3 lowest. As determined by direct cattle observations.

Figure **10.3** Troughs with supplemental feed, 2 km from the nearest watering source, are used to improve the grazing distribution of this unit.

If the animals require phosphorus, dicalcium phosphate or other compounds containing phosphorus often are mixed with salt. A mixture might include 48% granulated salt, 48% dicalcium phosphate, and 4% cottonseed meal. The cottonseed meal is added to prevent hardening of the mixture due to salivation by the animal.

Intake of supplemental feed is often regulated by salt to prevent excess consumption. The proportion of salt can be reduced if a greater intake of meal is desired. Compared to feeding at water, placing salt or salt-meal 1.6 km to 4.0 km (1 mi to 2.5 mi) away from water improved the uniformity of perennial grass use (Martin and Ward 1973). In southern

New Mexico, ranges with salt-meal stations at water were compared to those with stations at least 1.3 km (0.8 mi) from water (Ares 1953). Feeding salt-meal away from water reduced the overgrazed area by 50%, the light or unused area by about 30%, and nearly doubled the zone of proper grazing over the unit.

Certain principles can be useful in the placement of salt on rangelands. Areas that provide easy access are recommended as salt grounds because they allow livestock to move freely and have sufficient forage to make increased livestock use advantageous. Desirable areas for the location of salt ground include little used ridges, knolls, benches, gentle slopes, and openings in forest or shrubland. On mountainous range, careful placement of salt can increase grazing capacity by as much as 20%.

Kinds of Livestock

An important consideration in determining stock adaptation is the vegetation. Grazing animals can be divided into three groups based on their vegetation preferences. These groups include the grazers (cattle and horses), which have a diet dominated by grasses, the browsers (goats), which consume primarily forbs and shrubs, and the intermediate feeders (sheep), which have no particular preference for grasses, forbs, or shrubs. These groups are discussed in Chapter 11. The suitability of livestock type or types for various rangelands is discussed in Chapter 12.

Within species, livestock differ in their use of rangeland. Yearling cattle make better use of rugged terrain than do cows with calves. Some cattle within breeds use rugged rangeland better than others (Roath and Krueger 1982). Skovlin (1957) reported that "cattle can be trained to use certain areas and will repeat the use year after year." Research by Roath and Krueger (1982) on mountain range in northeastern Oregon suggests that livestock operators could take cattle that have not grazed a range before and behaviorally bond these animals to an area previously underutilized, given that water, shade, and salt were available in that area. For this to be an effective management tool, livestock must be handled so that they disperse when released on the pasture, avoiding initial concentration and habituation on riparian bottoms. This study showed that cattle formed groups that occupied different parts of the same range. Cattle groups habitually using riparian bottoms could be culled and replaced by those trained to use uplands (Bailey and Rittenhouse 1989). In Idaho, selective culling changed cattle distribution and decreased the use of riparian areas (Howery et al. 1996).

On southern New Mexico range, Santa Gertrudis cattle traveled greater distances than did Hereford cattle (Herbel and Nelson 1966). The implications were that Santa Gertrudis cattle would use this range more evenly than would Herefords. In Australia, Squires and Wilson (1971) found that travel distance by range sheep differed among and within breeds when animals were in varying physiological status. Lactating females with young will travel less distance than animals without young.

The distribution of urine and feces over the range is an indicator of range use. It is also of concern because urine and feces are a source of nutrients returned to the soil. Areas with heavy concentrations of feces will receive light use because livestock will not graze vegetation fouled by feces. In contrast, both cattle and sheep preferentially graze herbage growing where urinations have occurred. A study in Australia showed that cattle had the most uniform distribution of feces over the range, followed by horses and then sheep (Arnold and Dudzinski 1978).

Herding

On some ranges herding of cattle can improve distribution. Riding to keep bulls and cows properly mixed has improved calf crops. In mountainous terrain, herding of cattle to areas with poor accessibility but adequate feed has improved uniformity of use (Skovlin 1957). Herding has been most effective where cattle were driven from lowland bottoms to upland benches and ridges that have available water in the form of springs and seeps. Once cattle become aware of these areas, they will be readily utilized even though steep terrain may hinder access.

With the exception of parts of Africa, cattle are not usually tended by a herder; while sheep, on the other hand, are usually herded. Methods of sheep herding can influence the uniformity of range use. As discussed previously (see Chapter 9), judiciously herded sheep can be a means of preventing degradation of mountain riparian zones. Open herding of sheep rather than close herding will minimize trampling of forage and compaction of soil (Heady et al. 1947). Return of sheep to a central camp for evening bedding causes excessive trampling of vegetation and the development of many trails around the bedding ground (Figure 10.4). The use of a one-night bedding system where sheep finish grazing for the day prevents the development of sacrifice areas associated with the bedding ground, and improves sheep performance since travel distance is reduced (Heady et al. 1947).

Figure **10.4** Herding of sheep improves uniformity of use of a grazing unit.

Range Fertilization

Most reports show the increase in forage production that may be obtained from soil amendments. However, some of the benefits of fertilization can be used to improve livestock distribution. For example, the longer green period obtained from range grasses fertilized with nitrogen can be used to improve utilization of parts of a pasture with light use. Furthermore, soil amendments sometimes change the plant species composition. If the new plant is palatable at a different time than the species on the other portions of the unit, the new plant will attract grazing animals to that area for the time it is available and more palatable. In some cases, nitrogen fertilizer makes unpalatable species attractive to grazing animals by increasing their protein content and succulence (Holt and Wilson 1961; Smith and Lang 1958).

Burning

Prescribed burning is the use of fire where fuel is adequate, where the areas to be burned are predetermined, and at times when the intensity of the fire can be controlled. Judicious use of fire can be used to improve the distribution of grazing animals. Removal of plant growth from previous years will improve accessibility of new plant parts and may permit earlier grazing because of an increase in soil temperatures. Buildup of litter in excess of 1,800 kg/ha lowers soil temperatures, which reduces bacterial activity, binds nutrients so that they are unavailable to growing plants, and slows the general nitrogen cycling process (Wright 1974). Therefore, burning in strips or in parts of the grazing unit with underuse usually improves the utilization of that area.

Grazing Systems

Generally, reducing the size of a grazing unit with fences improves grazing distribution. Rest-rotation and short-duration grazing systems have improved livestock distribution by concentrating animals on small range units for short periods. Grazing systems are discussed in Chapter 9.

Trail Building

Trails obtain better distribution by improving access to areas where natural barriers (escarpments or deep canyons) and dense stands of trees and shrubs prevent use of forage. Trails can also facilitate handling of livestock.

Shade

Provision of shade can entice livestock to use areas otherwise avoided in desert or prairie ranges, which are typically devoid of trees and tall shrubs. Shade increased summer-long live-weight gain of yearling steers by 8.6 kg (19 lb) on rangeland in the southern Great

Plains (McIlvain and Shoop 1971). Humidity above 45% and temperature above 30°C (86°F) depressed steer gains. Cattle eagerly sought shade during hot summer days and by manipulating shade structures, cattle were drawn to underutilized range areas.

Financial Considerations

One of the most important questions confronting ranchers centers around how much infrastructure (corrals, fences, and watering points) they need to efficiently use their forage resources. On the smaller ranches there has been a historic tendency to substitute watering points and fence for grass when the goal was increased grazing capacity. However, knowledgeable ranch buyers look for ranches with minimal infrastructure and high amounts of forage. They know it's usually much cheaper to create infrastructure than increase forage.

On most arid land ranches in the Chihuahuan desert or sagebrush type, infrastructure costs become excessive relative to potential earnings when permanent watering points exceed one per 4,000 acres (2 1/2 miles spacing) and the average pasture size is less than 2,000 acres (Holechek and Hawkes 1993). In the more productive prairie areas of the Great Plains less than 2,600 acres per watering point (2 miles spacing) and pasture sizes of less than a section (640 acres) would usually represent excessive capitalization. However, it is important to point out that more infrastructure is justified on ranches with high grazing capacity than those that are degraded, or have low forge production potential. Fence and watering points improve the efficiency of range forage use. As forage production per acre increases, there is more potential to increase financial returns from improvements in forage harvest efficiency with fence and water development (Holechek 1992). The key here is to know future value of the extra forage that can be used compared to the cost of the infrastructure.

A 3 mile spacing of permanent watering points may have some advantages for arid rangelands in poor or fair condition (Holechek 1996). This spacing of watering points is farther apart than the commonly recommended 2 miles. However, it minimizes the amount of sacrifice area that often results from continuous grazing at moderate or heavy stocking rates. It provides a forage reserve that can be beneficial to livestock and wildlife when drought occurs. Further, it reduces some of the depreciation, maintenance, and tax costs associated with infrastructure.

Annual costs to maintain watering points and fence can vary greatly with climatic conditions, terrain, big game use, and recreational activity. However some rough guidelines on annual maintenance costs are $200–$350 for wells and $5–$10 per mile of fence (Holechek and Hawkes 1993).

Rising production costs and low livestock prices in recent years have caused many ranchers to take a close look at their infrastructure needs. Not only are fences and watering points costly to build and maintain, but they are also depreciating taxable assets. Selecting the proper amount of fence and watering points to meet specific objectives is a critical part of range management. Approaches for evaluating costs and returns from watering point development and fencing are provided by Holechek (1992) and Williams and Lacey (1995).

Wildlife Considerations

Fencing and watering point development for livestock have had both positive and negative impacts on wildlife habitat and populations. Planning and foresight can greatly increase wildlife habitat values of fences, waterholes, and other range improvements (Kindschy 1996). Fencing and watering point designs and modifications to enhance wildlife habitat values are discussed in Chapter 13 and in detail by Kindschy (1996).

RANGE MANAGEMENT PRINCIPLES

■ Failure to correct grazing capacity estimates for distance from water has been an important cause of range degradation. Areas over two miles from water should generally be considered unusable by livestock.

■ Failure to correct grazing capacity estimates for slope has been an important cause of rangeland degradation in mountainous areas. Cattle use is greatly diminished for slopes with over a 10% gradient while sheep use is generally diminished for slopes over 45%. Stocking rate adjustments should be implemented when slopes exceed the above thresholds.

■ Selecting the optimal amount of fencing and watering points to achieve management objectives is a critical part of successful ranching. Generally, higher levels of fencing and watering point development are justified on the humid ranges with high forage production than on arid rangelands where forage production is low.

Literature Cited

Ares, F. N. 1953. Better cattle distribution through the use of meal-salt mix. *J. Range Manage.* 6:341–346.

Arnold, G. W., and M. L. Dudzinski. 1978. *Ethology of free ranging animals.* Elsevier Science Publishing Co., Inc. New York.

Bailey, D. W., and L. R. Rittenhouse. 1989. Management of cattle distribution. *Rangelands* 11:159–161.

Bailey, D. W., J. E. Gross, E. A. Laca, L. R. Rittenhouse, M. H. Coughenour, D. W. Swift, and P. L. Sims. 1996. Mechanisms that result in large herbivore distribution patterns. *J. Range Manage.* 49:386–401.

Barnes, W. C. 1914. Stock-watering places on western grazing lands. *U.S. Dep. Agric. Farmers Bull.* 582.

Bell, H. M. 1973. *Rangeland management for livestock production.* Univ. Oklahoma Press, Norman, OK.

Coughenour, M. B. 1991. Spatial components of plant-herbivore interactions in pastoral, ranching, and native ungulate ecosystems. *J. Range Manage.* 44:530–542.

Fusco, M., J. Holechek, A. Tembo, A. Daniel, and M. Cardenas. 1995. Grazing influences on watering point vegetation in the Chihuahuan desert. *J. Range Manage.* 48:32–38.

Ganskopp, D., and M. Vavra. 1987. Slope use by cattle, feral horses, deer, and bighorn sheep. *Northw. Sci.* 61:74–81.

Gartner, F. R. 1986. Horizontal wells-An economical water development option. *Rangelands* 8:8–11.

Gillen, R. F., W. C. Krueger, and R. F. Miller. 1984. Cattle distribution on mountain rangeland in northeastern Oregon. *J. Range Manage.* 37:549–553.

Goebel, C. J. 1956. Water development on the Starkey Experimental Forest and Range. *J. Range Manage.* 9:232–234.

Hart, R. H., M. J. Samuel, J. W. Waggoner Jr., and M. A. Smith. 1989. Comparisons of grazing systems in Wyoming. *J. Soil and Water Conserv.* 44:344–347.

Heady, H. F., R. T. Clark, and T. Lammasson. 1947. Range management and sheep production in the Bridger Mountains, Montana. *Montana Agric. Exp. Sta. Bull.* 444.

Herbel, C. H., and A. B. Nelson. 1966. Activities of Hereford and Santa Gertrudis cattle on a southern New Mexico range. *J. Range Manage.* 19:173–176.

Holechek, J. L. 1992. Financial benefits of range management practices in the Chihuahuan Desert. *Rangelands* 14:279–282.

Holechek, J. L. 1996. Financial returns and range condition on southern New Mexico ranches. *Rangelands* 18:52–56.

Holechek, J. L., and M. Vavra. 1983. Fistula sample numbers required to determine cattle diets on forest and grassland ranges. *J. Range Manage.* 36:323–326.

Holechek, J. L., and J. Hawkes. 1993. Desert and prairie ranching profitability. *Rangelands* 15:104–109.

Holechek, J. L., M. Vavra, and J. Skovlin. 1981. Diet quality and performance of cattle on forest and grassland range. *J. Anim. Sci.* 53:291–298.

Holechek, J. L., J. Jeffers, T. Stephenson, C. B. Kuykendall, and S. A. Butler-Nance. 1986. Cattle and sheep diets on low elevation winter range in northcentral New Mexico. *Proc. West. Sec. Am. Soc. Anim. Sci.* 37:243–248.

Holscher, C. E., and E. J. Woolfolk. 1953. Foraging utilization by cattle on northern Great Plains ranges. *U.S. Dep. Agric. Circ.* 918.

Holt, G. A., and D. G. Wilson. 1961. The effect of commercial fertilizers on forage production and utilization on a desert grassland site. *J. Range Manage.* 14:252–256.

Howery, L. D., F. D. Provenza, R. E. Banner, and C. B. Scott. 1996. Differences in home range and habitat use among individuals in a cattle herd. *Appl. Anim. Beh. Sci.* 49:305–320.

Hutchings, S. S. 1946. Drive the water to the sheep. *Natl. Wool Grow.* 36:10–11.

Jepson, R., R. G. Taylor, and D. W. McKenzie. 1983. Rangeland fencing systems, state-of-the-art review. *U.S. Dep. Agric. Forest Service Equipment Development Center Project Record* 8332-1201.

Julander, O., and D. E. Jeffery. 1964. Deer, elk, and cattle relations on summer range. *Trans. N. Am. Wildl. Nat. Resour. Conf.* 29:404–414.

Kie, J. G., and B. B. Boroski. 1996. Cattle distribution habitats, and diets in the Sierra Nevada of California. *J. Range Manage.* 49:482–488.

Kindschy, R. R. 1996. Fences, waterholes, and other range improvements. In P. R. Krausman, ed., *Rangeland wildlife,* pp. 369–381. The Society for Range Management, Denver, CO.

Martin, S. C., and D. E. Ward. 1970. Rotating access to water to improve semidesert cattle range near water. *J. Range Manage.* 23:22–26.

Martin, S. C., and D. E. Ward. 1973. Salt and meal-salt help distribute cattle use on semidesert range. *J. Range Manage.* 26:94–97.

McDaniel, K. C., and J. A. Tiedeman. 1981. Sheep use on mountain winter range in New Mexico. *J. Range Manage.* 34:102–105.

McIlvain, E. H., and M. C. Shoop. 1971. Shade for improving cattle gains and rangeland use. *J. Range Manage.* 24:181–184.

Mueggler, W. F. 1965. Cattle distribution on steep slopes. *J. Range Manage.* 18:255–257.

Pickford, G. D., and E. H. Reid. 1948. Forage utilization on summer cattle ranges in northeastern Oregon. *U.S. Dep. Agric. Circ.* 796.

Polk, M. W., and R. T. Ervin. 1996. Windmills or solar watering systems. *Rangelands.* 18:97–99.

Quigley, T. M., H. R. Sanderson, A. R. Tiedemann, and M. L. McInnis. 1990. Livestock control with electrical and audio stimulation. *Rangelands* 12:152–155.

Roath, L. R., and W. C. Krueger. 1982. Cattle grazing and behavior on a forest range. *J. Range Manage.* 35:332–339.

Rose, A. F. 1991. An alternative to fences. *Rangelands* 13:144–145.

Senft, R. L., L. R. Rittenhouse, and R. G. Woodmansee. 1985. Factors influencing patterns of cattle grazing behavior on shortgrass steppe. *J. Range Manage.* 38:82–88.

Skovlin, J. M. 1957. Range riding—the key to range management. *J. Range Manage.* 10:269–271.

Smith, D. R., and L. R. Lang. 1958. The effect of nitrogenous fertilizers on cattle distribution on mountain range. *J. Range Manage.* 11:248–249.

Sneva, F. A., L. R. Rittenhouse, and L. Foster. 1973. Stock water restriction and trailing effects on animal gain, water drunk, and mineral consumption. *Water-Animal Relations Symp. Proc.*, pp. 34–48.

Squires, V. R. 1970. Growth of lambs in a semiarid region as influenced by distance walked to water. *Proc. Aust. Soc. Anim. Prod.* 8:219–225.

Squires, V. R. 1978. Distance trailed to water and livestock response. *Proc. Intn'l Rangel. Cong.* 1:431–434.

Squires, V. R., and A. D. Wilson. 1971. Distance between food and water supply and its effect on drinking frequency and food and water intake of Merino and Border Leicester sheep. *Aust. J. Agric. Res.* 22:283–290.

Squires, V. R., A. D. Wilson, and G. T. Daws. 1972. Comparison of walking behavior of some Australian sheep. *Prog. Aust. Soc. Anim. Prod.* 9:376–380.

Stoddart, L. A., A. D. Smith, and T. W. Box. 1975. *Range management.* 3d ed. McGraw-Hill Book Company, New York.

Valentine, K. A. 1947. Distance from water as a factor in grazing capacity of rangeland. *J. For.* 10:749–754.

Williams, K., and J. Lacey. 1995. A guide for evaluating additional fencing and water development. *Rangelands* 17:7–12.

Wright, H. A. 1974. Range burning. *J. Range Manage.* 27:5–11.

CHAPTER 11

RANGE ANIMAL NUTRITION

Range nutrition differs from classical animal nutrition in that it involves land management rather than the formulation of rations. Grazing animal nutritional requirements are more difficult to define than those for confined animals because of the added energy required for travel and coping with environmental stresses, such as wind, heat, or severe cold.

Major factors influencing range animal nutritional status include stocking rate, grazing system, types of forage species, type of animal, and season for use. Burning, fertilization, use of seeded pastures, brush control, and supplemental feeding are other management inputs commonly used to manipulate range animal nutritional status.

NUTRITIONAL COMPONENTS OF GRAZING ANIMAL FOODS

All animals require food to maintain body structures, functions, and growth. Animal physiology and morphology largely determine the type of food selected. A basic knowledge of food nutritional components is essential in understanding range animal foraging habits and in developing effective range management strategies.

Carbohydrates

Carbohydrates, which are comprised of carbon, hydrogen, and oxygen, are the basic source of energy for range animals. Plants have two basic types of carbohydrates: those associated with the cell contents and those associated with the cell wall (Figure 11.1). Starches and sugars are found in the cell contents. They are easily broken down by the animal's digestive systems and are a readily available source of energy. Cellulose and hemicellulose cannot be broken down by enzymes in the animal's digestive tract. Only ruminant animals and animals with enlarged cecums (horses and rabbits) can efficiently use cellulose and hemicellulose because they have microorganisms capable of digesting these carbohydrates. Digestion of cellulose and hemicellulose is a much slower process than is that of starch and sugar. Lignin is the portion of the plant cell wall that cannot be utilized even by microorganisms. Lignin is higher in stems than in leaves; it increases as plants mature, and is considered the primary antiquality component in forages. Lignin is the principal constituent of woody material in trees and shrubs.

Fats

Fats and oils are distinguished from carbohydrates by having fewer oxygen atoms and more hydrogen atoms. Fats have about 2.25 times the energy content of carbohydrates; therefore fat deposits are the main source of stored energy in range animals. Vegetative parts of plants are typically very low in fats, but seeds of plants such as corn, peanut, and sunflower have high fat levels. Grazing and browsing animals generally do not have gastrointestinal systems suited for digesting fats. Bile, produced in the liver, is required for breaking down fats. Animals adapted to feeding on forages typically have low levels of bile secretion into the small intestine, where fat degradation takes place.

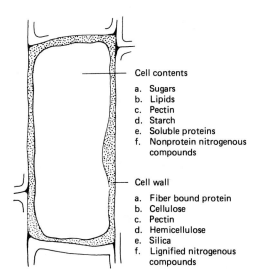

Cell contents

a. Sugars
b. Lipids
c. Pectin
d. Starch
e. Soluble proteins
f. Nonprotein nitrogenous
   compounds

Cell wall

a. Fiber bound protein
b. Cellulose
c. Pectin
d. Hemicellulose
e. Silica
f. Lignified nitrogenous
   compounds

Figure **11.1** The nutritional composition of a plant cell. (From Van Soest 1982.)

Proteins

Proteins are chemically distinguished from carbohydrates by having chains of amino acids that contain nitrogen as well as carbon, hydrogen, and oxygen. The unique chemical feature of protein is that it is comprised of amino acids linked together by nitrogen-carbon (peptide) bonds. Chains of amino acids (peptides) are linked into polypeptides via sulfur-to-sulfur bonds.

Proteins have many different functions in the animal's body. They are important as enzymes, hormones, and antibodies against disease; and as agents for transport and storage of nutrients within the body. Protein is the principal constituent of organs and skin of the animal's body. Unlike energy and most minerals, protein cannot be stored by the animal's body, so a continuous supply is required.

Actively growing plant parts have much higher protein levels than do those that are dormant. The leaves of grasses, forbs, and shrubs are much higher in protein than are the stems. However, leaves from forbs and shrubs are generally higher in protein than are grass leaves and stems at comparative stages of growth. Protein is often used as an indicator of forage quality because it is typically in short supply and is easy to measure.

Minerals

Macrominerals are distinguished from microminerals in that relatively large amounts are required by or represented in the animal's body. The macrominerals include calcium, phosphorus, sodium, potassium, magnesium, chlorine, and sulfur. Carbon, hydrogen, oxygen, and nitrogen do not fall into this group since they are considered organic constituents. The total mineral content of the animal's body is usually less than 5%.

Minerals serve many diverse but essential functions in the animal's body, such as regulation of muscle contraction, blood coagulation, nerve transmission, and osmotic balance. Mineral deficiencies and imbalances are reflected in poor animal condition, increased death losses, and poor productivity.

Total minerals as a group are determined in a feed by burning off the organic matter and weighing the remaining residue, which is called the ash. Ash analysis tells nothing about the amounts of specific minerals.

Phosphorus and calcium usually comprise over 70% of the ash. Phosphorus is the most limiting mineral to grazing animal productivity throughout the world. Calcium deficiency is a problem only in tropical areas with heavily leached soils (Oxisols). In arid and desert areas, range forages often contain high levels of calcium in relation to phosphorus. Forage calcium/phosphorus ratios from 1:1 to 2:1 have been considered optimal, although in many arid areas this ratio is much higher. Ruminants exhibit considerable tolerance to high calcium/phosphorus ratios (Underwood 1966). Cohen (1975) reviewed several studies showing that ratios from 1:1 to as high as 7:1 have given satisfactory and similar results in terms of nutrient conversion and performance in cattle and sheep. Because calcium/phosphorus ratios fall within the foregoing range for nearly all forages, they are not an important factor influencing grazing animal productivity.

Microminerals comprise minute quantities (less than 0.01%) of the animal's body. They include iron, iodine, copper, cobalt, fluorine, zinc, molybdenum, selenium, and manganese.

Recently, the trace minerals silicon, tin, vanadium, and nickel have been reported as being essential for animals.

Selenium is the micromineral that causes most problems in the United States. This micromineral is deficient in forages from the Pacific Northwest, the Northeast, and along the southeastern seaboard. Selenium deficiency causes white muscle disease in calves and lambs. In other parts of the western United States, selenium levels are so high in the soils and forages that they cause toxicity to grazing animals, manifested by hair loss, sloughing of the hooves, lameness, reduced food consumption, and liver injury.

Vitamins

Vitamins are organic compounds required in minute amounts for normal body function. Most of the vitamins function as coenzymes (metabolic catalysts), but some perform other essential functions. Vitamins are divided into the fat-soluble (A, D, E, K) and water-soluble (vitamin C and the B complex) on the basis of solubility properties. Fat-soluble vitamins are stored in the animal's body for use during periods of diet inadequacy. Conversely, water-soluble vitamins, with the exceptions of vitamin B_2 and B_{12}, are generally not stored and a constant supply is required. Bacteria associated with the digestive tract of grazing animals can synthesize the water-soluble vitamins and vitamin K. Because vitamin D is obtained from sunlight and vitamin E is high in plant leafy materials, these vitamins are seldom, if ever, deficient in grazing animal diets. Vitamin A is formed from carotene found in green plant material. When range animals consume dormant grasses for over 120 days, reserves of vitamin A stored in the body may be depleted and deficiency will be manifested. Vitamin A functions in vision, reproduction, bone growth, and epithelial cells. One of the first manifestations of vitamin A deficiency in range livestock is night blindness.

METHODS FOR DETERMINING THE NUTRITIONAL VALUE OF GRAZING ANIMAL DIETS

Due to space limitations, we refer the reader to reviews on the various aspects of this subject. Methods for sampling and evaluating a grazing animal's diet quality are comprehensively reviewed by Holechek et al. (1982c). Forage intake determination methods for grazing animals are covered by Cordova et al. (1978), Kartchner and Campbell (1979), and Van Dyne et al. (1980). Methods for determining botanical composition of the range herbivore diet are reviewed in detail by Holechek et al. (1982b).

DIET AND NUTRITIONAL QUALITY OF LIVESTOCK ON DIFFERENT RANGES

During the past 30 years, a tremendous amount of information has been collected on livestock nutritional status on various range types in the United States. An analysis of these studies shows that great differences within and between these range types occur with seasonal change (Table 11.1). This is because the timing and length of the growing season differ among range types because of climate. However, these studies are consistent in showing that cattle and sheep diets generally have crude protein levels of 10% to 12% and 12% to 16%, respectively, when forage is actively growing. During forage dormancy, crude protein levels drop to 4% to 7% for cattle and 6% to 12% for sheep. The higher values for sheep are explained by greater selectivity by sheep and the fact that sheep are generally grazed on ranges with a much higher forb and shrub component than that on cattle ranges.

Digestibility (percentage of the forage ingested) used by the animal's body values for cattle and sheep during active forage growth are generally over 50% (Table 11.1). However, during the dormant period, livestock diets on most ranges have digestibilities of less than 50%.

FORAGE INTAKE OF GRAZING ANIMALS

Forage intake of grazing animals varies with body weight as well as with forage quantity and availability. Presently, intake for grazing animals is most commonly expressed as the weight of the forage consumed as a percentage of the animal's body weight. Our review of the literature (Table 11.2) shows that dry-matter consumption by most range ruminants is about 2% of their body weight per day when values for different seasons are averaged across the year. However, considerable seasonal variation occurs, with values as low as 1% during periods when forage is low in quality and availability, to over 2.5% when quality and availability are high. Horses, which have enlarged cecums rather than rumens, consume about 60% to 70% more forage than ruminants when quality is comparable (Johnson et al. 1982). The comparative nutrition of grazing animals is discussed in more detail later in this chapter.

COMPARATIVE NUTRITIVE VALUE OF PLANT PARTS

Leaves of nearly all forages have higher crude protein, phosphorus, and cell soluble levels and lower fiber and lignin levels than those of stems (Table 11.3). Stems of woody plants are particularly low in quality because of their high level of lignification. Studies with grasses (Laredo and Minson 1973; Poppi et al. 1981) and legumes (Hendrickson et al. 1981) show that ruminants have substantially higher voluntary intakes of leaves than of stems.

Fruits and flowers from forbs and shrubs generally have much higher levels of cell solubles and protein than do leaves. Buds from shrubs also are particularly high in cell solubles and crude protein. Seeds from grasses are characterized by higher protein and cell solubles than those of grass leaves. However, as sources of protein and energy, seeds of most range grasses are inferior to fruits and flowers of forbs and shrubs. An exception is the Triticeae family (corn, wheat, barley, oats, etc.) which is characterized by large seeds high in cell solubles (energy).

TABLE 11.1 Percentage of Crude Protein and Digestibility of Cattle and Sheep Diets on Various Ranges

ANIMAL	LOCATION	SEASON	DIET CRUDE PROTEIN (%)	DIGESTIBILITY (%)	REFERENCE
Cattle	Desert, New Mexico	Winter	10.8	48.0	Rosiere et al. 1975b
	Prairie, Colorado	Winter	4.1	NA	Wallace et al. 1972
	Bunchgrass, Montana	Winter	4.3	NA	Van Dyne et al. 1964
	Prairie, Nebraska	Winter	4.4	47.8	Yates et al. 1982
	Shortgrass, New Mexico	Winter	11.1	45.0	Judkins et al. 1985a
	Prairie, Texas	Winter	7.0	54.0	Pinchak et al. 1990
	Southern pine forest, Florida	Winter	7.0	32.2	Long et al. 1986
	Shortgrass, Colorado	Summer	1.13	56.5	Vavra et al. 1973
	Shortgrass, New Mexico	Summer	16.2	60.4	McCollum et al. 1985
	Prairie, Texas	Summer	8.8	58.0	Pinchak et al. 1990
	Bunchgrass, Oregon	Summer	10.1	52.5	Holechek et al. 1981
	Forest, Oregon	Summer	11.6	57.1	Holechek et al. 1981
	Desert, New Mexico	Summer	10.1	54.0	Rosiere et al. 1975b
	Annual grassland, California	Summer	6.5	42.0[a]	Van Dyne and Heady 1965
	Prairie, Nebraska	Summer	9.8	61.0	Powell et al. 1982
	Southern pine forest, Florida	Summer	7.3	46.8	Long et al. 1986
Sheep	Annual grassland, California	Winter	10.7	54.1	Rosiere and Torell 1985
	Desert, Utah	Winter	6.0–12.6	NA	Pieper et al. 1959
	Annual grassland, California	Summer	7.2	51.9	Rosiere and Torell 1985
	Sagebrush grassland, Utah	Summer	13.8	NA	Cook et al. 1967
	Mountain range, Montana	Summer	10.5	NA	Buchanan et al. 1972

NA = not available.
[a]From Van Dyne and Lofgren 1964.

TABLE 11.2 Forage Intake by Range Ruminants

ANIMAL	LOCATION	SEASON	DRY-MATTER INTAKE (% BODY WEIGHT)	REFERENCE
Cattle	Annual grassland, California	Summer	1.5–1.8	Van Dyne and Meyer 1964
	Shortgrass, Wyoming	Summer	1.7–2.8	Jefferies and Rice 1969
	Bunchgrass, Oregon	Summer	1.5–2.5	Holechek and Vavra 1982
	Forest, Oregon	Summer	1.6–2.5[a]	Holechek and Vavra 1982
	Shortgrass, New Mexico	Summer	1.2–2.5[a]	Rosiere et al. 1980
	Prairie, Texas	Summer	2.45[b]	Pinchak et al. 1990
		Fall	1.95[b]	
		Winter	2.10[b]	
		Spring	2.42[b]	
	Desert grassland, New Mexico	Spring	1.7[b]	Hakkila et al. 1987
		Summer	1.8[b]	
		Autumn	1.5[b]	
		Winter	1.6[b]	
	Wiregrass, Georgia	Winter	1.4–2.1	Hale et al. 1962
	Veld, Zimbabwe	Yearlong	1.6–2.8	Elliott and Fokema 1961
	Harvested forages, Grasses	NA	0.9–2.6[bd]	Holechek et al. 1986
	Harvested forages, Nongrasses	NA	1.2–2.7[ab]	Holechek et al. 1986
	Harvested grasses	Summer	1.2–1.6	Johnson et al. 1982
	Harvested range forages	NA	1.76–2.09	Arthun et al. 1992
Sheep	Annual grassland, California	Summer	1.7–2.2	Van Dyne and Meyer 1964
	Mountain range, Utah	Summer	1.9–2.8	Cook et al. 1961
	Shortgrass, Colorado	Summer	1.9	Rice et al. 1974
	Desert, Utah	Winter	2.2–3.4	Cook and Harris 1951

Animal	Location/Feed	Season	Value	Reference
	Desert, Utah	Winter	2.0–3.3	Pieper et al. 1959
	Semiarid tropics, Brazil	Yearlong	1.2–2.8[b,e]	Pfister and Malechek 1986
	Harvested range forages	NA	1.86–2.41	Rafique et al. 1992
	Hyparrhenia pasture, Zambia	NA	1.3	Gihad 1976
Goat	*Hyparrhenia* pasture, Zambia	NA	1.8	Gihad 1976
	Semiarid tropics, Brazil	Yearlong	1.2–2.6[a]	Pfister and Malechek 1986
	Harvested range forages	NA	1.9–2.2[b]	Boutouba et al. 1990
Bison	Shortgrass, Colorado	Summer	1.7	Rice et al. 1974
	Harvested sedge hay	Summer	1.6	Hawley et al. 1981
		Winter	1.6	
Moose	Harvested feeds, Norway	Spring	NA	Hjeljord et al. 1982
	Grass-legume hay		1.6	
	Shrubs		1.4	
Mule deer	Mountain range, Colorado	Summer	2.1	Alldredge et al. 1974
		Winter	1.7	
	Harvested feeds	NA		Mubanga et al. 1985
	Alfalfa		2.0[b]	
	High shrub		1.5[b]	
	High forb		1.4[b]	
	High grass		1.2[b]	
Alpaca	Andes, Peru	Dry season	1.8	Reiner et al. 1987
		Wet season	1.6	

NA = not applicable.
[a] Average = 2.1.
[b] Organic matter basis.
[c] Average (cows) = 2.1, $\bar{x}$ (heifers) = 1.4.
[d] Average = 1.8.
[e] Average = 2.0.

TABLE 11.3   Crude Protein and Lignin Percentages of Three Forage
Classes in Northern Utah During Summer

	CRUDE PROTEIN (%)	LIGNIN (%)
Grass		
Stems	3.9	11.7
Leaves	12.3	8.9
Entire plant	6.2	11.0
Forbs		
Stems	4.5	12.4
Leaves	14.1	8.5
Entire plant	9.5	10.2
Shrubs		
Stems	6.3	20.6
Leaves	13.0	12.3
Current growth	11.6	15.0

Source: Cook and Harris 1950.

SEASONAL EFFECTS ON FORAGE NUTRITIONAL QUALITY

Forage nutritive quality on most ranges varies tremendously between seasons. Levels of cell
solubles, crude protein, and phosphorus are highest in actively growing forages and show
substantial declines as plants become dormant (Figure 11.2). These declines result from nutrient translocation from leaves and stems to crowns and roots with the onset of dormancy.
Leaching of soluble nutrients by rain and snow causes further losses during dormancy. In
the intermountain bunchgrass region of the western United States, the period when forage
is actively growing and nutritious is typically 60 to 90 days. On the shortgrass and tallgrass
prairie ranges of the Great Plains, this period is 140 to 180 days. The nutritive quality of the
prairie grasses during dormancy is higher than for the bunchgrasses, due, in part, to the
much lower winter precipitation (less leaching) of the shortgrass prairie. Annual grasses
show much greater declines in nutritive quality than do perennial grasses. Tallgrasses typically cure out with lower levels of nutrients than do shortgrasses. This is due to the fact that
tallgrasses have lower leaf/stem ratios than those of shortgrasses.

COMPARATIVE NUTRITIVE VALUE OF RANGE FORAGES

Considerable research has become available recently which adds greatly to our knowledge
of the nutritional response of grazing ruminants to grasses, forbs, and shrubs. Energy and

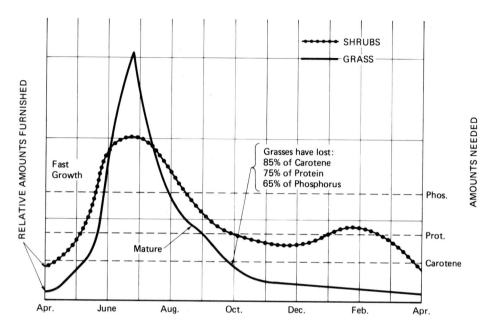

Figure **11.2** Seasonal trends in protein, phosphorous, and carotene content of range forage. (From Parker 1969.)

protein values of range forages will be emphasized because these two nutrients are usually the most limiting to range animal productivity and the most expensive to supply through supplementation programs.

Chemical Analyses

Reports of chemical analyses of various range forages are abundant. However, comprehensive studies comparing grasses, forbs, and shrubs are restricted to those summarized in Table 11.4. These investigations are consistent in showing actively growing material of forbs highest, shrubs intermediate, and grasses lowest in concentrations of crude protein, phosphorus, and cell solubles. When forage is dormant, evergreen shrubs are generally higher than grasses and forbs in concentrations of crude protein, phosphorus, and cell solubles. Grass leaves and stems are higher in cellulose and lower in lignin concentrations than are forbs and shrub twigs, fruits, and leaves at comparable stages of phenology (Cook and Harris 1968a; Nelson et al. 1970; Short et al. 1974).

Cool- and warm-season grasses and forb species often occupy the same range but grow at different times during the year. Comparisons regarding these groups are valid only if sample material represents the same plant part and stage of growth. Presently, this type of information is generally lacking for range plants, with the exception of limited data reported by Hart et al. (1983). Regardless of nutritional superiority, all four groups (warm season

TABLE 11.4 Summary of Studies Comparing the Nutritional Quality of Range Grasses, Forbs, and Shrubs

REFERENCE	SEASON	FORAGE CLASS	CRUDE PROTEIN (%)	FIBER (%)	PHOSPHORUS (%)	DIGESTIBILITY (%)
Cook et al. 1959 (Utah, desert shrub)	Winter	Grasses	4	NA	0.06	NA
		Forbs	NA	NA	NA	NA
		Shrubs	9	NA	0.09	NA
Nelson et al. 1970 (New Mexico, desert grass and shrubland)	Annual average	Grasses	7	49[a]	0.10	NA
		Forbs	13	38[a]	0.17	NA
		Shrubs	11	40[a]	0.16	NA
Pieper et al. 1978; Cordova and Wallace 1975 (New Mexico, shortgrass)	Annual average	Grasses	6	47[a]	0.16	50
		Forbs	13	31[a]	0.25	68
		Shrubs	11	267[a]	0.20	58
Short 1971 (Texas, east)	Annual average	Grasses	11	45[b]	NA	29
		Forbs	12	37[b]	NA	41
		Shrubs	10	38[b]	NA	29
Varner et al. 1979 (Texas, south)	Spring	Grasses	12	NA	0.22	53
		Forbs	17	NA	0.26	70
		Shrubs	22	NA	0.22	54
	Summer	Grasses	12	NA	0.27	50
		Forbs	14	NA	0.20	65
		Shrubs	18	NA	0.15	48
	Fall	Grasses	13	NA	0.23	44
		Forbs	16	NA	0.22	67
		Shrubs	18	NA	0.15	48
	Winter	Grasses	14	NA	0.21	51
		Forbs	21	NA	0.29	79
		Shrubs	17	NA	0.14	51
	Average	Grasses	13	NA	0.23	50
		Forbs	17	NA	0.26	70
		Shrubs	19	NA	0.17	50

Study	Season					
Huston et al. 1981 (Texas, Edwards Plateau)	Spring	Grasses	8	NA	0.13	44
		Forbs	19	NA	0.21	59
		Shrubs	16	NA	0.22	70
	Summer	Grasses	6	NA	0.11	43
		Forbs	11	NA	0.17	53
		Shrubs	14	NA	0.10	64
	Fall	Grasses	5	NA	0.08	34
		Forbs	11	NA	0.20	53
		Shrubs	9	NA	0.09	58
	Winter	Grasses	5	NA	0.06	31
		Forbs	NA	NA	NA	NA
		Shrubs	NA	NA	NA	NA
	Average	Grasses	7	66[b]	0.11	39
		Forbs	14	34[b]	0.18	54
		Shrubs	12	32[b]	0.15	64
Severson 1982 (South Dakota, Black Hills)	Summer	Grasses	7	44[a]	0.18	41
		Forbs	9	36[a]	0.22	59
		Shrubs	8	32[a]	0.27	53
Krysl et al. 1984b (Wyoming, shrubland)	Summer	Grasses	7	NA	0.12	53
		Forbs	10	1	0.14	58
		Shrubs	11	NA	0.16	55
	Winter	Grasses	5	NA	0.07	42
		Forbs	NA	NA	NA	NA
		Shrubs	7	NA	0.11	41
	Average	Grasses	6	NA	0.09	48
		Forbs	10	NA	0.14	58
		Shrubs	9	NA	0.14	49
Average, all studies		Grasses	8	50	0.13	43
		Forbs	13	35	0.20	58
		Shrubs	11	34	0.17	50

NA = Not available.
[a]Acid detergent fiber.
[b]Neutral detergent fiber.

grasses, cool season grasses, forbs, and shrubs) are considered desirable on most ranges because they provide high levels of nutrients at different times of the year.

Lignin, cutin, and silica are indigestible compounds associated with plant cell walls that reduce digestibility of cellulose and hemicellulose (Van Soest 1982). Forbs and shrubs have about twice the lignin content of grasses at the same digestibility (Short et al. 1974). Many grasses accumulate silica (Van Soest and Lovelace 1969; Smith et al. 1971a; Smith and Nelson 1975), which has been found to reduce digestibility about two to three percentage units for each percentage unit increase in the silica content of the forage (Smith et al. 1971a). Presently, little comparative data are available on the silica content of forages consumed by range ruminants.

Nitrogen-Complexing Compounds

Protein availability to ruminants varies considerably among grasses, forbs, and shrubs. Some forbs and shrubs are high in soluble phenolic and tannin compounds that cause elevated fecal nitrogen values in relation to the crude protein content of the diet (Mould and Robbins 1981; Nastis and Malechek 1981; Nunez-Hernandez et al. 1989). Soluble phenolic and tannin compounds can react directly with dietary proteins, forming complexes resistant to ruminal degradation (Chalupa 1975). However, protein protected by tannins and soluble phenolics appears to improve nitrogen retention and ruminant performance (Chalupa 1975) only if consumed at low levels. Nastis and Malechek (1981) fed goats diets containing various levels of Gambel oak (*Quercus gambelii*) and alfalfa (*Medicago sativa*). Tannins in the diets containing oak reduced dry-matter intake and digestibility of cellular contents, and elevated fecal nitrogen excretion. However, urinary nitrogen excretion was reduced, indicating a partial sparing effect of nitrogen. Holechek et al. (1990) found that forages high in soluble phenolics had reduced forage intake, digestibility, and nitrogen retention in goats compared to forages similar in crude protein content but low in soluble phenolics. More research is needed on the influence of forages containing nitrogen-complexing compounds on range ruminant nutritional status performance.

Digestibility

Generally, forbs and foliage from shrubs have higher in vitro digestibility values than do grasses at comparable stages of maturity (Table 11.5). Digestibility varies greatly among species, although leaves and fruits are consistently higher in digestibility than are stems and twigs (Short et al. 1974; Milchunas et al. 1978; Huston et al. 1981).

Forages with rapid rates of digestion generally have higher intakes than those with slower rates (Van Soest 1982). Leaves from forbs and shrubs reach their potential extent of digestion much more quickly than do those from grasses (Short et al. 1974; Milchunas et al. 1978; Wofford and Holechek 1982) (Table 11.5). This appears to be partly explained by higher levels of cell solubles and quicker microbial access to cell solubles in the forb and shrub leaves and stems. Studies by Smith et al. (1971b, 1972) show that the cell walls of legumes digest more quickly than do those of grasses. Smith et al. (1971a) found that range forages in New Mexico contained 5% to 10% of their dry matter as silica, and that silica contributed to indi-

TABLE 11.5   Comparison of 4- and 48-Hour In Vitro Digestibilities of Important Range Grasses, Forbs, and Shrubs on Southeastern New Mexico Range[a]

	PERCENTAGE 4-HOUR IN VITRO DIGESTIBILITY	PERCENTAGE 48-HOUR IN VITRO DIGESTIBILITY	$\dfrac{48 \times 100}{48\ HR}$
Grasses			
Sideoats grama	20	54	37
Black grama	18	52	35
California cottontop	22	62	35
Tobosa	17	40	43
Mesa dropseed	18	54	33
Average grasses	19	52	37
Forbs			
Scarlet globemallow	31	66	47
Leatherweed croton	28	44	64
Spectaclepod	31	58	53
Wright's knotweed	24	38	63
Verbena	37	68	54
Average forbs	30	55	56
Shrubs			
Fourwing saltbush	36	62	58
Mountain mahogany	25	54	46
Apache plume	20	36	56
Wright's silktassel	34	48	79
Gray oak	19	29	66
Average shrubs	27	46	59

Source: Adapted from Wofford and Holechek 1982.

[a]Collected at comparable stages of maturity.

gestibility as much as did lignin. Substances resistant to microbial digestion (lignin, cutin, and silica) are arranged differently in the cellulose/hemicellulose matrix in monocots, compared with dicots (Sinnott 1960; Van Soest 1982). One theory is that lignin, cutin, and silica appear to encrust the fibrils of cellulose and hemicellulose in grasses, whereas these substances appear to be evenly dispersed through the cellulose/hemicellulose matrix of forbs and shrubs. Encrusted hemicellulose and cellulose would have less surface for microbial attack.

Forages with thin cell walls and high levels of cell solubles, such as forb and shrub leaves, may have more rapid rates of digestion because the quick release of a high-quality food source (cell solubles) to rumen microbes allows rapid population expansion, and subsequently, faster cell wall degradation (Donefer 1969). Donefer (1969) theorized that rapid removal of solubles from the cell would provide maximum surface area for bacterial attachment within the cell, resulting sequentially in an increased rate of cell wall decomposition.

Essential oils contained in the leaves of some shrub species have been reported to inhibit digestion in the rumen (Nagy et al. 1964; Oh et al. 1967, 1968). However, recent studies by Welch and Pederson (1981) and Pederson and Welch (1982) show that certain essential oils, particularly monterpenoids, have a small effect on in vitro digestibility.

Tannins in oaks (*Quercus* sp.) suppress digestibility. Nastis and Malechek (1981) found that oak-containing diets fed to goats were less digestible and had lower intakes than those of a pure alfalfa diet, although the oak diets were higher in cell contents and had similar nitrogen concentrations. In another study with goats, oak-containing diets were lower in digestibility but had similar intakes to the alfalfa control (Nunez-Hernandez et al. 1989).

Intake

Considerable evidence from penned ruminants offered various forages in different physical forms, shows that intake is more affected by the particle passage rate through the digestive system than by palatability (Bines et al. 1969; Marten 1969). After a comprehensive review of forage palatability influences on intake and performance by ruminants, Marten (1969) concluded that forage palatability generally has a small impact on ruminant intake or performance, although there are exceptions, such as when the forage contains high levels of poisons or essential oils.

Several studies show higher intakes for legumes than for grasses (Ulyatt 1981). Research with penned cattle and sheep indicates that leafy materials from range shrubs are consumed at higher levels than grasses for a given level of digestibility if tannin and volatile oil levels are low (Arthun et al. 1992; Rafique et al. 1992). White-tailed deer intake averaged 15% higher when they were fed browse diets averaging 58% in vivo digestibility than when fed brome hay (*Bromus* sp.) averaging 72% in vivo digestibility (Robbins et al. 1975). Milchunas et al. (1978) found that deer fed a browse-dominated diet with 30% digestibility had an 8% higher intake than a grass-dominated diet with 43% digestibility. The browse-dominated diet had a shorter retention time than the grass-dominated diet. On mountain range in northeastern Oregon, Holechek and Vavra (1983) found that average daily gains and forage intake showed reductions, but that diet in vitro digestibility improved when cattle were forced to shift from browse- to grass-dominated diets. In southcentral Alaska, muskoxen had increased intake on grass-browse pastures compared to pastures with grass alone (Boyd et al. 1996). Higher intake was associated with increased weight gains.

In northeastern Oregon, Vavra et al. (1980) reported an 8% reduction in forage intake when cattle shifted from a grass- to a browse-dominated diet. However, forage intake per unit of dry-matter digestibility was 12% higher for the browse-dominated diet than for the grass-dominated diet. The low digestibility and high lignin values associated with the browse-dominated diet indicate cattle were consuming a large amount of woody material.

Low availability of leafy material on the shrubs may explain the reduction in diet quality and forage intake as browse consumption increased.

On the southern Nevada range, Connor et al. (1963) related forage intake by cattle to season, digestibility, and diet botanical composition. Shrub consumptions during the early, middle, and latter parts of the grazing season were 15%, 60%, and 70%, respectively. The rest of these diets was comprised of grass. Forage intakes and diet organic matter digestibilities for these periods were 3.32 kg and 62%, 4.09 kg and 70%, and 2.32 kg and 32%, respectively. The highest intake occurred in the middle period when the diet contained 60% shrubs. The reduced intake during the latter period in this study appears to have resulted from the low availability of leafy material on common winterfat (*Ceratoides lanata*) and a complete lack of other grazable forages.

More recently, Gade and Provenza (1986) studied winter diet selection and nutrition of sheep grazing crested wheatgrass (*Agropyron cristatum*) and crested wheatgrass-shrub mixtures in Utah. Sheep on the crested wheatgrass-shrub pastures consumed diets that were about one-half shrub and one-half crested wheatgrass. Big sagebrush (*Artemisia tridentata*) was the primary shrub in sheep diets. Forage intake was higher for sheep using crested wheatgrass-shrub than crested wheatgrass pasture (Table 11.6). This was attributed to higher forage availability on the crested wheatgrass-shrub pastures. When snow accumulated, shrubs were more available above the snow than crested wheatgrass. Crude protein levels were higher for sheep on the crested wheatgrass-shrub pastures, but digestibility was highest for sheep on crested wheatgrass.

Lower levels of cell wall can result in higher intakes of leafy material from palatable forbs and shrubs compared with grasses. Diet cell wall concentration has a high negative association with intake by ruminants (Mertens 1973). Forages with a high cell wall content occupy a large volume in the rumen, limiting gut capacity. The rate of breakdown of this material determines the potential intake. Forb and shrub leaves generally have thinner cell walls than those of woody material or grass leaves and stems (Spalinger et al. 1986). Forages with thin cell walls tend to break down and pass more quickly out of the rumen than do those with thick cell walls (Spalinger et al. 1986).

TABLE 11.6   Nutritional Characteristics of Sheep Grazing Crested Wheatgrass and Crested Wheatgrass-shrub Range in Central Utah

NUTRITIONAL CHARACTERISTIC	CRESTED WHEATGRASS	CRESTED WHEATGRASS-SHRUB
Forage organic matter intake (g kg $BW^{-0.75}$)[a]	27	32
Crude protein (%)	5.6	8.1
Digestibility (%)	41	33
Total fiber (%)	71	61
Lignin (%)	11.5	15.7

Source: Gade and Provenza 1986.

[a]BW = body weight.

Grazing Animal Performances

Few studies are available that quantify the comparative responses of cattle and sheep to grass-, forb-, and shrub-dominated diets. However, the potential of shrubs to serve as protein, phosphorus, and carotene supplements to grasses during forage dormancy is well recognized.

Otsyina et al. (1982) reported a preliminary grazing trial evaluating performance of sheep on dormant crested wheatgrass compared to a mixture of fourwing saltbush (*Atriplex canescens*) and crested wheatgrass. Sheep lost about 5% of their body weight on the pure grass pasture but only 1% on the grass-shrub pasture during a 20-day grazing period.

On arid grassland range in Australia, Leigh et al. (1968) found little production response (body weight and wool growth) to the presence of cottonbush (*Kochia aphylla*) and oldman saltbush (*Atriplex nummalaria*). This lack of response was attributed to adequate levels of crude protein in the grass and to the small contribution of the shrub's forage to the diet. In another Australian study, leaves from the shrub mulga (*Acacia aneura*) were successfully used to maintain sheep during drought (Gartner and Anson 1966).

In northeastern Oregon, Erickson (1974) found that mature cows preconditioned to a browse diet gained 0.37 kg per day on forested range when their diet was predominantly browse. Cattle on an adjacent meadow during the same period, consuming a grass diet, lost 0.29 kg per day. In a previous study Erickson (1974) found that cattle forced to switch abruptly from a grass to a browse diet showed large weight losses. His data suggest that preconditioning of cattle to browse diets and forested habitats can greatly improve cattle performance compared to abrupt changes.

Another northeastern Oregon study showed that weight gains of yearling heifers were highest when forbs and shrubs comprised a large part of the diet (Holechek and Vavra 1983). In this study, weight gains of cattle declined during a drought year, when they were forced to shift to a predominantly grass diet due to lack of available browse. In a separate study on adjacent ranges, Holechek et al. (1981) reported that cattle on forested pastures gained 22% more weight than cattle on grassland pastures over a 3-year period. Forbs and shrubs averaged 20% of the diet on the grassland pastures compared to 39% on the forest pastures. Shrubs consumed in these studies were deciduous, with low levels of essential oils.

On central Utah winter range, Gade and Provenza (1986) reported that sheep from crested wheatgrass-shrub pastures did not differ from those on pure crested wheatgrass pastures in weight change. However, forage intake and diet crude protein concentrations were higher on the crested wheatgrass-shrub pastures (see Table 11.6). They believed the lack of weight differences may have been due to an inadequate duration of study. Another explanation is that the volatile oils associated with big sagebrush, the primary shrub consumed, may have interfered with nutrient assimilative efficiency (Nunez-Hernandez et al. 1989).

Productivity

Many range forbs and shrubs are unpalatable and compete with desirable grasses, reducing their productivity. Studies quantifying these undesirable impacts are discussed in

Chapter 15. Some palatable forbs and shrubs may actually improve the productivity and nutritive quality of grasses. Productivity of grasses is enhanced when they are grown in association with legumes because of nitrogen fixation by bacteria associated with the legumes (Rumbaugh et al. 1982; McGinnies and Townsend 1983). Yet associations between palatable range grasses, forbs, and shrubs have received only limited study. Lorenz and Rogler (1962) found that the addition of alfalfa to crested wheatgrass in North Dakota greatly improved the productivity of crested wheatgrass. Alfalfa also increased the crude protein content of the wheatgrass (Rogler and Lorenz 1969). McGinnies and Townsend (1983) found that grass-legume plantings in northcentral Colorado were more productive and higher in nutritive quality than were grasses grown alone. Yield and protein concentration of crested wheatgrass were increased when it was grown in Utah with fourwing saltbush or various legumes (Rumbaugh et al. 1982). Both the shrub and legume species contributed to more rapid regrowth of the wheatgrass after harvesting. Foliage of the shrub and legume species contained approximately twice the protein concentration of the crested wheatgrass. In southern Idaho, Monsen (1980) found that interseeded fourwing saltbush did not reduce grass density or herbage yields, and was readily accepted as a forage species by cattle.

COMPARATIVE NUTRITION OF GRAZING ANIMALS

Body size, size of digestive system relative to body weight, type of digestive system, and mouth size and shape are primary factors determining forage selection by different range animals (Hanley 1982). Range ungulates can be divided into three groups based on their foraging habits. These groups include the grazers, which consume grass-dominated diets; the browsers, which consume primarily forbs and shrubs; and the intermediate feeders, which use near equal amounts of grasses, forbs, and shrubs (Holechek 1984). Tables 11.7 and 11.8 summarize the studies describing food habits of range ungulates. Our discussion of comparative nutrition of grazing animals follows (Holechek 1984).

The Grazers

Cattle, elk, bighorn sheep, mountain goats, musk oxen, and bison are North American ungulates considered to be grazers. However, on some ranges these ungulates, with the exception of bison and musk oxen, do consume large amounts of forbs and shrubs. This occurs primarily when green grass is unavailable. These ungulates show a strong avoidance of shrubs high in volatile oils (junipers, rabbitbrush, various sagebrushes, etc.) because they lack mechanisms to reduce the toxic effects of these substances.

The Browsers

Moose, pronghorn, mule deer, domestic goats, and white-tailed deer feed primarily on forbs and shrubs throughout the year regardless of location. With the exception of domestic goats,

TABLE 11.7 Large Herbivore Forage Selection in North America

ANIMAL	LOCATION	GRASS (%)	FORBS (%)	SHRUB (%)	REFERENCE
Cattle	Annual grassland, California	63	29	8	Van Dyne and Heady 1965
	Sagebrush, Utah	76	10	14	Cook and Harris 1968a
	Sagebrush, Wyoming	60	3	33	Krysl et al. 1984a
	Shortgrass, New Mexico	62	32	6	Thetford et al. 1971
	Salt desert, Utah	57	3	40	Cook and Harris 1968a
	Shortgrass, Nebraska	52	44	4	Streeter et al. 1968
	Desert grassland, Arizona	86	1	13	Galt 1972
	Salt desert, Nevada	26	0	74	Connor et al. 1963
	Bunchgrass, Oregon	80	14	6	Holechek et al. 1982d
	Forest, Oregon	61	16	23	Holechek et al. 1982e
	Chihuahuan desert, New Mexico	43	32	25	Rosiere et al. 1975a
	Southern pine, Florida	60	20	20	Kalmbacher et al. 1984
	Shortgrass, Colorado	57	26	17	Kautz and Van Dyne 1978
	Chaparral, Durango, Mexico	61	21	18	Gallina 1993
Sheep	Annual grassland	69	33	8	Van Dyne and Heady 1965
	Sagebrush, Utah	42	30	28	Cook and Harris 1968a
	Piñon-juniper, south/central New Mexico	38	60	2	Thetford et al. 1971
	Salt desert, Utah	29	1	70	Cook and Harris 1968a
	Shortgrass, Wyoming	82	15	3	Rice et al. 1971
	Tall forb, Montana	24	76	0	Buchanan et al. 1972
	Salt desert, north/central New Mexico	41	30	29	Holechek et al. 1986
	Shortgrass, Colorado	47	22	31	Kautz and Van Dyne 1978
	Oak grassland, central Texas	60	18	22	Bryant et al. 1979
Goat	Oak grassland, central Texas	12	20	68	McMahan 1964
	Oak grassland, central Texas	47	12	41	Malechek and Leinweber 1972
	Chaparral, southern California	–	20	–	80	Sidahmed et al. 1981
Angora goats	Oak grassland, central Texas	48	12	40	Bryant et al. 1979

Animal	Location				Reference
Spanish goats	Oak grassland, central Texas	45	13	42	Bryant et al. 1979
	Chaparral, Texas	36	5	59	Warren et al. 1984
Horse	Sagebrush, Nevada	90	5	5	Hanley and Hanley 1982
	Shrubland, Colorado	93	—	5	Hubbard and Hansen 1976
	Sagebrush, Oregon	99	—	1	Vavra and Sneva 1978
	Salt desert, north/central New Mexico	31	27	43	Stephenson et al. 1985b
	Shrubland, Wyoming	92	2	6	Olsen and Hansen 1977
	Shrubland, Wyoming	69	2	29	Krysl et al. 1984a
Feral burro	Creosotebush, Nevada	10	36	54	Browning 1960
	Mojave desert, California	4	30	61	Woodward and Ohmart 1976
Bison	Shortgrass, Colorado	94	5	1	Peden et al. 1973
	Shortgrass, Colorado	89	6	4	Kautz and Van Dyne 1978
Moose	South/central Alaska	0	0	100	Cushaw and Coady 1976
	Yellowstone Park, Wyoming	3	9	88	McMillan 1953
	Willow, aspen, Montana	0	1	99	Stevens 1970
Caribou	Tundra, Canada	29	32	39	Banfield 1954
	Tundra, Canada	55	17	28	Wilkinson et al. 1976
	Coniferous forest, Canada	13	0	87	Scotter 1967
Bighorn sheep	Mountain, Colorado	47	10	33	Todd 1975
	Alpine, Wyoming	73	27	0	Woolf 1968
	Desert, Nevada	94	3	3	Barret 1964
Mountain goat	Mountains, Montana	76	5	19	Saunders 1955
	Alpine, Alaska	36	64	0	Hjeljord 1971
	Alpine, Colorado	82	14	4	Hibbs 1966
Musk ox	Tundra, Canada	90	2	8	Wilkinson et al. 1976
Rocky Mt. elk	Coniferous forest, Alberta, Canada	97	0	3	Cowan 1947
	Sagebrush, Wyoming	79	3	18	Olsen and Hansen 1977
	Mountain shrub, north/central New Mexico	51	26	33	Stephenson et al. 1985b
Roosevelt elk	Forest grassland, California	76	2	22	Harper et al. 1967

(Continued)

TABLE 11.7   Large Herbivore Forage Selection in North America (*Continued*)

ANIMAL	LOCATION	GRASS (%)	FORBS (%)	SHRUB (%)	REFERENCE
Mule deer	Sagebrush, Montana	17	24	59	Dusek 1975
	Guadalupe Mountains, New Mexico	1	31	68	Anderson et al. 1965
	Desert mountains, Texas	13	87	0	Anderson 1949
	Sagebrush, Utah	0	100	0	Smith and Julander 1953
	Mountain shrub, north/central New Mexico	9	40	51	Stephenson et al. 1985b
	Shrubland, Wyoming	10	5	85	Goodwin 1975
	Oak woodland, New Mexico[a]	2	16	74	Boeker et al. 1972
White-tailed deer	Oak grassland, central Texas	8	31	61	Bryant et al. 1979
	Bottomlands, Montana	38	19	43	Allen 1968
	Oak grassland, central Texas	10	45	45	McMahan 1964
	Grass-shrubland, southeast Texas	23	72	5	Chamrad and Box 1968
	Chaparral, Durango, Mexico	2	13	85	Gallina 1993
Pronghorn	Mixed prairie, Alberta, Canada[b]	25	47	18	Mitchell and Smoliak 1971
	Pecos region, Texas	6	67	27	Buechner 1950
	Sagebrush, Wyoming	2	0	98	Severson and May 1967
	Semidesert, Utah	5	25	70	Beale and Smith 1970
	Salt desert, north/central New Mexico	6	49	45	Stephenson et al. 1985a
	Shortgrass, Colorado	48	49	3	Schwartz and Nagy 1976

[a]Numbers do not add to 100 due to unidentifiable plants.
[b]Numbers do not add to 100 due to rounding off.

TABLE 11.8   Seasonal Changes in Dietary Selection by Various Large Herbivores in North America

ANIMAL	SEASON	GRASSES (%)	FORBS (%)	BROWSE (%)	LOCATION	REFERENCE
Cattle	Late spring	66	27	7	Oregon (bunchgrass)	Holechek et al. 1982d
	Early summer	79	15	6		
	Late summer	90	5	5		
	Fall	89	5	6		
	Late spring	46	29	25	Oregon (coniferous forest)	Holechek et al. 1982e
	Early summer	66	13	21		
	Late summer	65	11	24		
	Fall	69	9	22		
Hereford cattle	Spring	35	40	25	New Mexico (desert grassland)	Herbel and Nelson 1966
	Summer	71	23	6		
	Fall	50	41	9		
	Winter	50	27	23		
Santa Gertrudis cattle	Spring	58	30	12	New Mexico (desert grassland)	Herbel and Nelson 1966
	Summer	81	17	2		
	Fall	49	43	8		
	Winter	65	20	15		
Sheep	Spring	37	47	16	Texas (chaparral)	McMahan 1964
	Summer[a]	51	8	31		
	Fall[b]	68	3	28		
	Winter	82	1	17		
	Spring	25	62	13	New Mexico (desert shrub)	Holechek et al. 1986
	Summer	63	19	18		
	Fall	29	14	57		
	Winter	46	24	30		
Mule deer	Spring[a]	2	30	48	New Mexico (chaparral)	Boeker et al. 1972
	Summer[a]	2	42	50		
	Fall	6	8	86		
	Winter	2	4	94		

(Continued)

TABLE 11.8   Seasonal Changes in Dietary Selection by Various Large Herbivores in North America (*Continued*)

ANIMAL	SEASON	GRASSES (%)	FORBS (%)	BROWSE (%)	LOCATION	REFERENCE
White-tailed deer	Spring	34	65	1	Texas (mixed brush)	Drawe and Box 1968
	Summer	5	71	24		
	Fall	27	66	7		
	Winter[a]	37	49	4		
	Spring[b]	38	18	43	Montana	Allen 1968
	Summer	1	54	45		
	Fall	2	17	81		
	Winter	6	29	65		
Pronghorn antelope	Spring	25	57	18	Alberta (prairie)	Mitchell and Smoliak 1971
	Summer	13	62	25		
	Fall	13	37	50		
	Winter[b]	9	47	43		
Roosevelt elk	Spring	62	4	34	California (forest)	Harper et al. 1967
	Summer	58	20	22		
	Fall	56	23	21		
	Winter	76	2	22		
Angora goats	Spring[a]	40	25	25	Texas (chaparral)	Malechek and Leinweber 1972
	Summer	65	8	27		
	Fall	47	12	41		
	Winter	47	4	49		
Spanish goats	Spring[a]	54	3	32	Texas (chaparral)	Warren et al. 1984
	Summer[b]	35	8	56		
	Fall	17	7	75		
	Winter	40	1	59		

[a]Numbers do not add to 100 due to unidentifiable plants.
[b]Numbers do not add to 100 due to rounding off.

these ungulates experience digestive upsets if forced to consume diets dominated by mature grass. This group of ungulates consumes a limited amount of grass in the spring when it is green and forbs and shrubs are unavailable. However, dry-matter grass is almost completely avoided. The smaller ruminants in this group can consume large amounts of forages high in volatile oils because their small, pointed mouthparts enable them to select the portions of these plants with the lowest levels of volatile oils (Hanley 1982) (Figure 11.3). In addition, the small ruminants chew their food to a much greater extent than do large ruminants or monogastric animals. Apparently, fine chewing of plants high in volatile oils results in release of these substances as gases, and greatly reduces their assimilation by the animal's digestive system (White et al. 1982). Fine chewing also results in more mixing of plant material with saliva. The proteins in saliva may conjugate with volatile oils and tannins, rendering them nontoxic (Robbins et al. 1991). If assimilated at high levels, the volatile oils found in many sagebrushes and junipers can be toxic to the animal (Johnson et al. 1976).

The Intermediate Feeders

Domestic sheep, burros, and caribou are considered to be intermediate feeders. These animals have the greatest capability to adjust their feeding habits to whatever forage is available. Domestic sheep are probably better adapted than any other domestic ungulate to the forage resource in the intermountain West because they will readily use grasses, forbs, or shrubs,

Figure **11.3** Mule deer, unlike cattle, can utilize juniper as an emergency food in winter because their smaller mouths and lower total forge demand permit them to select those plant parts with the lowest levels of volatile oils.

depending on availability. The primary problem with domestic sheep is that their short legs and relatively large body make them very susceptible to predation.

Digestive Systems

To understand ungulate forage selection, some knowledge of their digestive physiology is necessary. Ungulates have two basic types of digestive systems, the rumen and cecum systems (Figure 11.4). Both systems evolved to enable ungulates to digest plant fiber (plant cell walls) by microbial (bacteria and protozoa) fermentation. The fermentation processes are quite similar in both the rumen and cecum. The systems differ in that the rumen is an enlarged portion of the digestive tract that food must pass through before entering the true stomach. The cecum occurs as an enlarged portion of the large intestine that food enters after passing through the true stomach.

The rumen system has two advantages over the cecum system (Janis 1976). The process of rumination (regurgitation and retching of forage) results in considerable reduction of particle size, which provides more surface area for microbial digestion. Because food must be

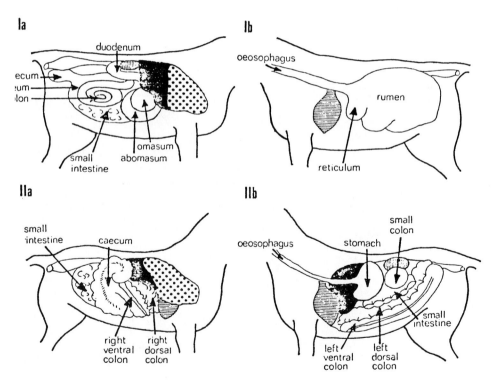

Figure **11.4** Longitudinal sections of ungulates with rumen and cecal sites of fermentation: Ia, Ruminant (cow), right view; Ib, left view; IIa, Cecal fermenter (horse), right view; IIb, left view. Key: Dotted area, lung; solid area, liver; vertical lines, kidney; horizontal lines, heart. (From Janis 1976.)

broken down to fine particle size to leave the rumen, retention of fiber is longer than in the cecum. This results in more complete digestion of fiber in the rumen than in the cecum since fiber digestion is a time-dependent process (Table 11.9). A second advantage is that in the rumen system, microbes are passed from the rumen into the abomasum, where they are digested and then absorbed, providing the animal with an important source of protein. Little microbial protein is absorbed by cecum digestors because microbial fermentation occurs after the food has passed through the stomach. However, research shows that horses will ingest their feces when their diet is low in protein, which partially compensates for the inefficient use of microbial protein-associated cecum digestion (Schurg et al. 1977).

The primary advantage of cecal digestion of fiber is that forage material can pass easily out of the cecum without any great reduction in particle size (Janis 1976). Although fiber digestion is less complete by cecum digestors than by ruminants, compensation occurs because they can consume a much greater amount of forage, since they do not have to break fiber down to a small particle size to pass it out of their system.

On the basis of the previous discussion, it is apparent that cecum digestors can subsist on lower-quality diets than ruminants (Figure 11.5). However, they must have a greater forage supply since they use the forage less efficiently (see Table 11.9). This explains why horses can survive on coarse, mature grasses better than cattle.

Smaller animals typically have higher metabolic rates (faster heartbeat) than those of larger animals. Therefore, they must eat a higher-quality diet than larger animals but require less total food due to their small size.

Large ruminants can subsist on higher-fiber diets than can small ruminants because they have lower nutrient requirements per unit body weight (Hanley 1982). Therefore, a large portion of the diet is typically comprised of highly available forage such as grasses for bison and cattle or woody material from shrubs and trees for the moose. The small ruminants, such as white-tailed deer and pronghorn, must consume diets dominated by leafy material and fruits from forbs and shrubs that have high levels of crude protein, phosphorus, and digestibility, and low levels of fiber. These animals can afford to be selective for these materials because they have a low total forage demand. However, the small size of their rumen relative to body weight necessitates selection of a high-quality diet.

TABLE 11.9   Average Daily Dry-Matter Intake and Digestibility of Three Grass Hays Fed to Horses (Mares) and Cattle (Cows)

	INTAKE (% BW)	DIGESTIBILITY	DAILY INTAKE OF DIGESTIBLE DRY MATTER (% BW)	RETENTION TIME IN INDIGESTION SYSTEM (HOURS)
Mare	2.33	54.8	1.28	8.47
Cow	1.39	61.5	0.85	20.40

Source: Johnson et al. 1982.

BW = Body weight.

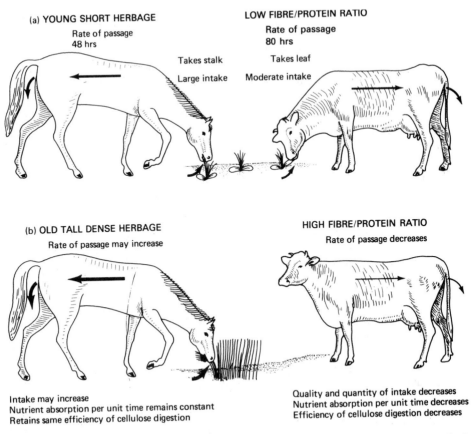

(a) YOUNG SHORT HERBAGE

Rate of passage
48 hrs

LOW FIBRE/PROTEIN RATIO

Rate of passage
80 hrs

Takes stalk   Takes leaf

Large intake   Moderate intake

(b) OLD TALL DENSE HERBAGE

Rate of passage may increase

HIGH FIBRE/PROTEIN RATIO

Rate of passage decreases

Intake may increase
Nutrient absorption per unit time remains constant
Retains same efficiency of cellulose digestion

Quality and quantity of intake decreases
Nutrient absorption per unit time decreases
Efficiency of cellulose digestion decreases

Figure **11.5** Comparison of ruminant and equid feeding strategies. (a) On this type of herbage both ruminant (cow) and equid (horse) do equally well, though both select different levels of herbage. (b) The horse continues to do well, but the ruminant cannot maintain itself on this type of herbage. (From Janis 1976.)

Management Implications

In the past, range management practices have often been geared toward replacing forbs and shrubs with pure stands of grasses. The vast areas of crested wheatgrass in the Great Basin and lovegrasses (*Eragrostis* sp.) in the Southwest support the preceding statement. Rangelands with a pure stand of grass provide good forage for cattle or in some cases elk and mule deer during active growth, but meet poorly the nutritional requirements of large or small ungulates during most of the year. Limited research in the Great Basin and the Southwest shows that inclusion of palatable forbs and shrubs in seeding mixtures with grasses can improve livestock performance during forage dormancy and provide better habitat for small wild ungulates and other wildlife species than that provided by pure stands of grasses.

Range condition has been based on the density and production of native, palatable, perennial grasses. A better criterion might be the diversity of palatable forage species (Holechek 1984). Under this criterion it might be desirable if up to 20% of the yearly forage production were comprised of palatable annuals. It is important to recognize that many annual grasses and forbs grow in periods when perennials are dormant. Annual forbs provide an important nutritional contribution to cattle, sheep, and pronghorn diets and reduce pressure on palatable perennial grasses during the growing season. In the Northwest, the introduced annual, cheatgrass (*Bromus tectorum*), provides green forage for cattle, mule deer, and sheep in the fall, winter, and early spring, when the native perennial grasses are dormant.

A large number of studies involving both domestic and wild ungulates in North America are consistent in showing that forage selection changes tremendously within and between years. The nutritive quality of various forage species also shows great fluctuations within and between years. The greater the degree of forage selection that a range provides domestic of wild ungulates, the more likely they will be to meet their nutrient needs.

HOW GRAZING ANIMALS COPE WITH PERIODS OF LOW FORAGE QUALITY

On most ranges, forage quality is adequate for weight gain, milk production, and wool production for less than 6 months of the year. On some ranges, grazing animals must exist on forage inadequate for maintenance for over 3 months of the year. Grazing animals have developed several mechanisms to cope with these long periods of low-quality forage.

During periods when range forage quality and availability are high, grazing animals have high intakes of crude protein and energy. During these periods, high rates of weight gain occur. Energy in the form of fat is stored for later periods, when forage quality and quantity are low. If grazing animals have good reserves of body fat, they can survive for 30 to 60 days with little to no food consumption. Mule deer does with high levels of fat reserves have survived periods of complete starvation for up to 64 days (de Calesta et al. 1975). However, in this same study, fawns with low fat reserves died after 33 days.

During the winter period of forage dormancy, there is evidence that native grazing animals have lower metabolic rates and consequently lower energy requirements than during the summer (Silver et al. 1969). Ruminants on low-protein diets excrete less nitrogen in the urine and feces and recycle a higher percentage of the total nitrogen intake than do those on diets with protein levels above maintenance needs (Mould and Robbins 1981). Several studies show that the potential for conserving nitrogen by urea recycling is tremendous with both wild and domestic ruminants. Soil ingestion appears to play a critical role in permitting ruminants to meet their needs for many crucial minerals, such as phosphorus, zinc, copper, and cobalt, when forage levels are inadequate (Healy et al. 1970; Mayland et al. 1975; Arthur and Allredge 1979).

ENERGY EXPENDITURE BY GRAZING ANIMALS

In contrast to confined animals, grazing animals require additional energy to meet the demands of travel. Much of the increased travel of grazing compared to confined animals is associated with searching for food and grazing. Graham (1964) estimated that grazing sheep expended 40% of their energy in standing, walking, ruminating, and eating compared to 10% for confined sheep fed forage diets. Free-ranging cattle expended 46% more energy than did stall-fed cattle in a Utah study involving crested wheatgrass range (Havstad and Malechek 1982). Grazing strategies that reduce energy expenditures associated with traveling and grazing should increase animal productivity.

SUPPLEMENTING RANGE LIVESTOCK

The major operational expense confronting the range livestock industry in most parts of the United States is that for supplemental feed. Rising costs for supplemental feed coupled with declining prices for range livestock products during the past 10 years have increased interest in ways to minimize supplementation costs without sacrificing livestock production. Our discussion of range livestock supplementation follows Holechek and Herbel (1986).

Differences in forage quality and livestock management cause supplementation to vary considerably in the United States. In the western United States, major nutritional deficiencies occur only during drought and winter forage dormancy. In contrast, soils in the southeastern United States are heavily leached due to the high rainfall, making yearlong supplementation necessary for sustained livestock productivity.

Energy, protein, phosphorus, and vitamin A are the nutrients most limiting to range livestock production. Except for emergency conditions such as after heavy snow or severe drought, energy supplements are seldom used. The energy supplements will substantially improve the performance of range animals, but they have proven uneconomical in most situations because of high cost and labor requirements. Protein and mineral supplements are most cost-effective because they generally improve forage intake and digestibility.

RANGE LIVESTOCK NUTRITIONAL GUIDELINES

Range livestock nutritional requirements are poorly understood compared to those for confined livestock. Palatable forbs and shrubs often have chemical and physical properties much different from pasture and harvested forage. Low to moderate (10% to 50% of the diet) amounts of these plants in the diet, such as globemallow (*Sphaeralcea* sp.) and verbena (*Verbena* sp.) in New Mexico, can be nutritionally advantageous, while amounts exceeding 50% of the diet are sometimes toxic. For the foregoing reasons, recommendations by the National Research Council on livestock nutritional requirements are not always applicable to range livestock. Based on several range livestock nutritional studies in the United States, Holechek and Herbel (1986) developed some guidelines (Table 11.10).

TABLE 11.10   Range Livestock Nutritional Requirements for Maintenance and
Production Based on Range Research Studies

	CRUDE PROTEIN (%)	PHOSPHORUS (%)	DIGESTIBILITY (%)
Cows			
Maintenance	6–8	0.10–0.15	40–45
Lactation	9–12	0.20–0.25	50–55
Yearling cattle			
[1 lb (.05 kg) gain/day]	8–9	0.20–0.25	45–50
Ewes			
Maintenance	7–9	0.15–0.20	45–50
Lactation	10–12	0.25–0.30	55–60

Source: Holechek and Herbel 1986.

Some range livestock operations are geared toward calf and/or lamb production because mature female animals can subsist on low-quality forages better than growing animals that have higher nutritional requirements. The alternative is to graze with yearling animals during the period of active forage growth. In the southeastern United States, where range forage quantity is high but quality is low throughout the year, cow-calf operations are used most exclusively.

Short periods of nutrient deficiency in livestock diets do not have adverse effects if followed by a high-quality diet. Research shows that mature pregnant cows in good condition can lose 10% of body weight during the winter and still produce calf crops approaching 90% if they can gain weight after parturition (Wagnon et al. 1959; Reed and Peterson 1961). This also applies to ewes. Supplementation of these animals at levels that does not permit some weight loss is a poor economic practice. Young animals that are subjected to severe undernutrition during the first 6 months of life tend to have a reduced skeleton size and are often permanently stunted (Allden 1970). However, after 6 months of age, they can show good recovery from a low nutritional plane. It is well documented that livestock show greater feed efficiency and have higher gain after periods of moderate undernutrition. This extra gain of thin compared to fat animals is commonly referred to as *compensatory gain.* When losses exceed 15% of the animal's weight in good condition, poor recovery often occurs when the animal is placed on a high nutritional plane. Animals that lose 30% or more of normal body weight will nearly always die.

From mid-gestation until parturition, cows and ewes experience their lowest protein and energy requirements. Demand for these nutrients escalates after parturition because of lactation. If forage quality is low, postpartum supplementation will generally be more effective than that in the prepartum period (Wallace 1987).

Minimizing Supplement Needs by Range Management

Judicious grazing is one of the most effective tools to minimize needs for supplemental feed. Animals on lightly to moderately (25% to 50% use) grazed ranges require less supplement because they can selectively graze and expend less energy in travel to obtain a full rumen. Grazing levels that permit high selectivity are important during forage dormancy when there is nutritive variation between forages. During active growth, all forages are generally high in nutritive quality.

On moderately grazed shortgrass range in southcentral New Mexico the response of mature cows to protein and phosphorus supplements was minimal in nondrought years (Wallace 1987). However, in drought years the response was economically advantageous (Table 11.11). Protein and phosphorus supplementation during forage dormancy was considered advantageous to replacement heifers in all years.

Seasonal suitability grazing systems have considerable potential to minimize needs for supplemental feed. These systems are discussed in detail by Holechek and Herbel (1982) and in Chapter 9.

Identification of Periods When Supplementation Is Required

Identification of what nutrients are deficient, when they are deficient, and the severity of the deficiency are major concerns of range livestock producers. Clipped samples of the available forage provide inaccurate measures of diet quality because grazing animals show high selectivity for particular plant parts and species. Esophageal fistulated animals have been useful for studying nutrient levels and trends in livestock diets. Studies have shown that the nutritional quality of livestock diets varies substantially between years on most range types and the periods when supplementation can be advantageous vary substantially both within and between years. Although fistulated animals are a good research tool, they are not practical for ranchers to monitor nutritional status because of the high cost and labor requirements for sample collection and analysis. Further, the lag time between collection of the sample and actual analysis may be too long (5 to 14 days) for any practical decision making.

In recent years, fecal analysis has shown potential as a quick, practical means for ranchers to detect periods when protein supplementation will be most advantageous economically (Lyons and Stuth 1992). Concentrations of nitrogen in the feces are related to the nutritional status of the ruminant animal (Moir 1960; Holechek et al. 1982a; Squires and Siebert 1983; McCollum 1990) (Figure 11.6).

The concentration of nitrogen in the fecal organic matter of ruminants shows a strong linear association with diet crude protein content. Research indicates that when total nitrogen concentration of the fecal organic matter drops below 1.40%, cattle undergo weight losses due to inadequate crude protein in the diet (Holechek et al. 1982a; Squires and Siebert 1983). Severe weight losses (over 0.25 kg/day) can be expected if fecal nitrogen concentrations drop below 1.30%. A fecal nitrogen concentration below 1.40% for cattle indicates that diet crude protein levels are probably lower than 6%. Once this level is reached, the forage

TABLE 11.11 Response of Mature Range Cows to Protein and Phosphorus Supplementation During Drought and Nondrought Years on Moderately Grazed Shortgrass Range in Southcentral New Mexico

| | PROTEIN SUPPLEMENTATION | | | | PHOSPHORUS SUPPLEMENTATION | | | |
| | NONDROUGHT YEARS | | DROUGHT YEARS | | NONDROUGHT YEARS | | DROUGHT YEARS | |
	SALT ONLY	PROTEIN[a] SUPPL.	SALT ONLY	PROTEIN[a] SUPPL.	SALT ONLY	PHOSPHORUS SUPPL.	SALT ONLY	PHOSPHORUS SUPPL.
Calf weaning weight (kg)	217	226	176	200	236	238	226	253
Calf crop (%)	95	95	67	83	93	95	72	67
Supplement cost ($/year)	0.74	11.58	0.88	15.77	0.74	3.10	0.88	3.87

Source: Wallace 1987.

[a]Supplemented cows were fed 3 times/week at a level of 1.05 kg/head/feeding for 130 days in non/drought and 177 days in drought years.

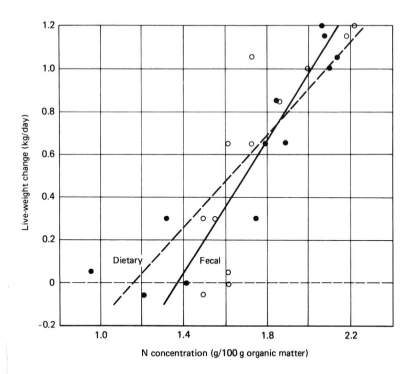

***Figure* 11.6** The relationship between live weight change and the nitrogen concentration in diet (open circles) and in the feces (solid circles) of grazing cattle in Australia. (From Squires and Siebert 1983.)

intake declines precipitously because rumen microbial needs for nitrogen are not satisfied. Increases in range forage intake (20% to 60%) and digestibility (5% to 15%) can be expected from supplemental protein when diet crude protein levels are below 6% (Cook and Harris 1968b; Rittenhouse et al. 1970; Kartchner 1981). When diet crude protein is above 6%, supplemental protein has had little to no effect on range forage digestibility and intake.

Studies conducted and reviewed by Nunez-Hernandez et al. (1992) show that animals consuming browse diets high in essential (volatile) oils (sagebrushes, rabbitbrushes, junipers) or tannins (oaks) can have elevated fecal nitrogen levels relative to diet nitrogen values. However, cattle and sheep on most ranges generally will starve before consuming forages high in essential oils and/or tannins. Browse species palatable to cattle and sheep such as fourwing saltbush, mountain mahogany (*Cercocarpus montanus*), common snowberry (*Symphoricarpos albus*), and ninebark (*Physocarpus malvaceus*) do not appear to cause elevated fecal nitrogen values (Holechek et al. 1982a; Nunez-Hernandez et al. 1989; Rafique et al. 1992; Arthun et al. 1992).

Diet and fecal phosphorus concentrations show a strong linear association for cattle (Moir 1960; Holechek et al. 1985). When fecal phosphorus concentrations drop below 0.60% in the organic matter, diet phosphorus levels below maintenance are likely. Fecal phosphorus levels above 0.85% indicate dietary phosphorus levels adequate for growing and lactating cattle.

Protein Supplementation

The two basic types of protein supplements provided to range livestock are nonprotein nitrogen (urea and biuret) and high-protein natural feeds (alfalfa hay, cottonseed meal, soybean meal). Nonprotein nitrogen sources can substitute for costlier feed proteins because rumen microorganisms can convert nitrogenous compounds into proteins. Under the right conditions about one-third of the total protein requirement can be met by nonprotein nitrogen. For nonprotein nitrogen supplements to be utilized effectively, a good energy source must be available (Clanton 1978). This approach is used with feedlot animals consuming grains but seldom with animals consuming dormant range forages. Beef cows consuming range forages have often shown a negative response to high-urea protein supplements in terms of increased weight losses, lower weaning weights, reduced reproductive performance, and even death losses. This is presumably due to buildup of nitrogen in the toxic nitrite form in the rumen. Urea and other nonprotein nitrogen sources are well utilized if accompanied by a high energy supplement. Lick wheels involving urea-molasses mixtures have been one of the most common means of providing this combination. The added energy associated with this type of supplement may depress range forage intake.

Cottonseed meal is probably the most heavily used protein plant supplement. It typically has about 40% to 45% crude protein, so fairly small amounts satisfy daily requirements. Levels of cottonseed meal from 1 lb to 2 lb (0.5 kg to 0.9 kg) per day for cows and from ¼ lb to ⅓ lb (0.1 kg to 0.2 kg) per day for ewes have given the best economic returns. These levels allow animals to meet their protein requirement and often improve intake of range forage. Higher levels may reduce intake of range forage because the excess protein will be converted into energy.

Research on blue grama range in New Mexico shows yearling heifers supplemented with high-quality alfalfa hay (18% crude protein) had weight gains comparable to those supplemented with cottonseed meal (Judkins et al. 1985a). Neither supplement affected the intake of range forage. Three to five pounds (1.4 kg to 2.3 kg) of high-quality alfalfa hay per day for cows and ½ lb to 1 lb (0.2 kg to 0.5 kg) per day for ewes should be profitable under most conditions. Which type of protein supplement to use is primarily an economic decision that depends on the comparative unit protein cost per feed. Labor and equipment requirements are other considerations.

Because of the labor cost, frequency of protein supplementation is of considerable concern to ranchers. The various studies addressing this problem have consistently shown no real differences in livestock performance between daily, alternate day, every third day, and even weekly feeding of cottonseed meal or alfalfa hay (Coleman and Wyatt 1982; Wallace 1987). A major advantage of alternate day or weekly feeding compared to daily feeding is greater opportunity for animals in poor condition to get part of the supplement.

Salt has been used to control intake of cottonseed meal. High salt consumption can have an adverse impact on ruminant protein digestion (Moseley and Jones 1974). The increased water intake associated with high salt consumption causes increased nitrogen losses in the urine. Tallow is being used to control cottonseed meal consumption by calves and yearlings; meat and fishmeal show the potential to limit cottonseed meal consumption by mature cows.

Mineral Supplementation

Phosphorus is the most limiting mineral to range livestock production in nearly all parts of the world. On western U.S. ranges phosphorus levels in livestock diets are usually adequate when forage is actively growing. During dormancy, forbs and shrubs have much higher phosphorus levels than grasses, and if included in the diet, can greatly reduce the need for supplemental phosphorus. Bone phosphorus can be mobilized during short periods (1 to 3 months) when diet phosphorus concentrations are below maintenance without adverse effects on the animal (Cohen 1975; Judkins et al. 1985b). Continuous supplementation of phosphorus throughout the year appears warranted only in the southeastern pine region. In most parts of the West, phosphorus supplement is needed only in the fall and winter. However, because the costs of supplying supplemental phosphorus are (below $1.50/cow/year), many ranchers routinely include phosphorus as part of their salt mixtures.

Salt (sodium and chlorine) is routinely provided to livestock on rangelands throughout the United States. On some western ranges it serves more as a tool to improve livestock distribution rather than as a needed nutritional supplement. California researchers found that range cattle not provided with salt performed as well as those receiving salt in terms of calf production (Morris et al. 1980). From a nutritional standpoint, the researchers questioned the practice of routinely providing range livestock with salt. In contrast, Oregon research showed that providing salt to steers on new-growth crested wheatgrass improved their performance (Wallace et al. 1963). Unlike other nutrients, sodium concentrations are lowest in new growth and highest in mature forage. Because of the low cost of salt, it seems advisable that it be provided at least during the period of active forage growth. Young growing cattle should receive about 1 lb (0.4 kg) of salt per month; 2 lb (0.9 kg) per month is recommended for lactating cows on actively growing forage; and $1 \over 2$ lb (0.2 kg) per month is usually allotted to ewes. It costs about $1.20 to $1.40 per cow to supply salt throughout the year.

Iodine deficiencies occur at several localities, particularly in the northwest. Iodized salt is cheap insurance that iodine requirements will be met. Iron, copper, cobalt, and potassium are recommended in mineral mixtures in the southern pine region, particularly in Florida. Magnesium should be included to prevent grass tetany when animals graze lush, early growth such as on wheatfields or crested wheatgrass pastures. Potassium is generally deficient in dormant warm-season grasses in Texas and should be supplemented when forage is dormant (Hinnant and Kothmann 1982).

Forage analyses generally show that minerals such as iron, copper, cobalt, zinc, and manganese are not deficient in forages from ranges in the western United States. However, studies at New Mexico State University have shown improvements in range livestock performance when these minerals were supplemented (Smith et al. 1980). Many range forages, particularly grasses, have high silica levels. Silica in forages forms complexes with other minerals, making them unavailable to the animal (Smith and Nelson 1975). Therefore, routine provision of mineral supplements throughout the year appears to be cheap insurance against deficiency.

Several feed companies provide salt-mineral mix designed for particular parts of the United States. These mixes are usually provided free of choice throughout the year and range in cost from $3 to $4 per cow per year.

In contrast to the western United States mineral deficiencies and imbalances have been one of the most limiting factors in range livestock production in tropical Latin American countries. Mineral deficiencies are also a problem in many areas of the southeastern United States because of high rainfall that leaches both the soils and forage plants. In Columbia providing salt with a complete mineral mixture doubled the weaning weight of calf per cow compared to salt alone (Miles and McDowell 1983). At the time of the Miles and McDowell (1983) study the majority of the cattle on the Ilanos pasturelands of Columbia were not receiving mineral supplements. Miles and McDowell (1983) reviewed other studies showing large increases in livestock production from mineral supplementation in Bolivia, Brazil, Panama, Peru, and Uruguay. Their review indicated calving percentages in these countries could be on average increased about 45% (from 51% to 74%) by including minerals with common salt. They provide detailed guidelines on mineral supplementation in Latin American countries.

Energy Supplementation

Energy supplements have been considered most practical under conditions of drought or heavy snow. There is evidence that the extremely high protein/energy ratios of the lush, early growth of some pasture forages has an adverse effect on livestock performance. This may be due to high levels of nonprotein nitrogen which causes a buildup of the poisonous nitrite form in the rumen. Passage rates are very high with these forages, causing a high washout of rumen microflora. Under these conditions energy supplementation can be advantageous economically.

Available energy sources vary widely by area. Barley and cracked corn are two of the more common energy supplements used under range conditions. These feeds usually depress the intake and digestibility of range forage and serve primarily as substitutes for range forage. Alfalfa hay can be a good source of energy as well as protein. At moderate levels [4 lb to 5 lb (2 kg to 2.5 kg) per cow per day] it will either improve or not affect range forage intake and digestibility. If protein as well as energy is deficient in the range animal diet, it is particularly advantageous. When lush forage is consumed with high protein levels (over 15%), one of the grain supplements would be advisable.

In contrast to protein, energy supplements must be fed daily to obtain satisfactory animal performance. Research at the USDA Livestock and Range Research Station near Miles City, Montana, showed that cattle fed cracked corn daily gained twice as much as cattle fed double the daily amount every other day (Kartchner and Adams 1982). Alternate day feeding appeared to result in rumen conditions less suitable for fiber digestion compared to feeding every day.

In another study at Miles City, the time of day when energy supplements were provided to cattle influenced livestock performance (Adams and Kartchner 1983). Two groups of steers were fed the same diet of cracked corn daily. One group was fed in the early morning, while the other group was fed in the middle of the afternoon. Steers fed in the afternoon outgained steers fed in the morning by about $\frac{1}{2}$ lb (0.2 kg). This difference is explained by reduced grazing of the group fed in the early morning compared to that of those fed in the late afternoon. The early-fed group tended to remain near the feed grounds waiting for the feed truck instead of grazing in the morning. The other group grazed in the morning and waited for feed in the afternoon. Therefore, normal grazing activities of the afternoon-fed group were unaffected. Steers fed in the late afternoon had a higher total intake of feed, which probably accounts for their extra weight gain.

Vitamin Supplementation

Vitamin A is the only vitamin that is deficient in most range livestock diets. Carotene, the precursor of vitamin A, is deficient in dormant plant material, although levels far above minimum requirements occur in green plant parts. Vitamin A deficiency is not generally a problem on ranges with a high component of palatable evergreen shrubs, or on ranges where green forage is available for over 8 months during the year (Speth et al. 1962). Vitamin A can be stored by the body, and even low levels of material from evergreen shrubs will more than meet livestock needs (Cook et al. 1954). Vitamin A can be supplied by injections or commercial range supplements.

RANGE MANAGEMENT PRINCIPLES

- Rangelands managed to maximize plant diversity will generally provide both livestock and wildlife with an optimum balance between forage quantity and quality.

- On most arid and semiarid rangelands, about 55% to 70% remaining climax vegetation (good range condition) will maximize plant diversity and livestock nutritional status.

- On most properly stocked western rangelands, protein and phosphorus are the nutrients most likely to be deficient in range forages. Generally protein supplementation is most financially effective when range grasses and forbs are dormant and palatable browse is unavailable.

- In the humid tropical areas of Latin America and Africa, provision of mineral supplements to rangeland livestock can result in large increases in their productivity.

Literature Cited

Adams, D. C., and R. J. Kartchner. 1983. Effects of time of supplementation on daily gain, forage intake and behavior of yearling steers grazing fall range. *Proc. West. Sec. Am. Soc. Anim. Sci.* 34:158–167.

Allden, W. G. 1970. The effects of nutritional deprivation on the subsequent productivity of sheep and cattle. *Nutr. Abstr. Rev.* 40:1167–1184.

Alldredge, A. W, J. F. Lipscomb, and F. W. Whicker. 1974. Forage intake rates of mule deer estimated with fallout cesium-137. *J. Wildl. Manage.* 38:508–516.

Allen, B. O. 1968. Range use, foods, condition, and productivity of white-tailed deer in Montana. *J. Wildl. Manage.* 38:508–516.

Anderson, A. W. 1949. Early summer foods and movements of mule deer in the Sierra Vieja Range of southwestern Texas. *Tex. J. Sci.* 1:45–50.

Anderson, A. E., W. A. Snyder, and G. W. Brown. 1965. Stomach content analyses related to condition in mule deer, Guadalupe Mountains, New Mexico. *J. Wildl. Manage.* 29:351–366.

Arthun, D., J. L. Holechek, J. D. Wallace, M. L. Galyean, M. Cardenas, and S. Rafique. 1992. Forb and shrub influences on steer nitrogen retention. *J. Range Manage.* 45:133–136.

Arthur, W. J., and A. W. Allredge. 1979. Soil ingestion by mule deer in northcentral Colorado. *J. Range Manage.* 32:67–71.

Banfield, A. W. F. 1954. Preliminary investigation of the barren ground caribou. II. Life history, ecology, and utilization. *Wildlife Monograph Bulletin Series 1*, No. 1013. Department of Northern Affairs and National Resources, National Parks Branch, Canadian Wildlife Service, Ottawa, Canada.

Barret, R. H. 1964. Seasonal food habits of the bighorn at the Desert Game, Nevada. In *Desert Bighorn Council Trans.,* pp. 85–93.

Beale, D. M., and A. D. Smith. 1970. Forage use, water consumption, and productivity of pronghorn antelope in western Utah. *J. Wildl. Manage.* 34:570–582.

Bines, J. A., S. Suzuki, and C. C. Balch. 1969. The quantitative significance of long-term regulation of food intake in the cow. *Br. J. Nutr.* 23:695–701.

Boeker, E. L., V. E. Scott, H. G. Reynolds, and B. A. Donaldson. 1972. Seasonal food habits of mule deer in southwestern New Mexico. *J. Wildl. Manage.* 36:56–63.

Boutouba, A., J. L. Holechek, M. L. Galyean, G. Nunez-Hernandez, J. D. Wallace, and M. Cardenas. 1990. Influence of two native shrubs on goat nitrogen status. *J. Range Manage.* 43:530–534.

Boyd, C. S., W. B. Collins, and P. J. Urness. 1996. Relationship of dietary browse to intake in captive muskoxen. *J. Range Manage.* 49:2–8.

Browning, B. 1960. Preliminary report of the food habits of the wild burro in the Death Valley National Monument. In *Desert Bighorn Council Trans.,* pp. 88–90.

Bryant, F. C., M. M. Kothman, and L. B. Merrill. 1979. Diets of sheep, angora goats, Spanish goats, and white-tailed deer under excellent range conditions. *J. Range Manage.* 32:412–417.

Buchanan, H., W. A. Laycock, and D. A. Price. 1972. Botanical and nutritive content of the summer diet of sheep on a tall forb range in southwestern Montana. *J. Anim. Sci.* 35:423–430.

Buechner, H. K. 1950. Life history, ecology, and range use of the pronghorn antelope in trans-Pecos, Texas. *Am. Midl. Nat.* 43:257–354.

Chalupa, W. 1975. Rumen bypass and protection of proteins and amino acids. *J. Dairy Sci.* 58:1198–1218.

Chamrad, A. D., and T. W. Box. 1968. Food habits of white-tailed deer in south Texas. *J. Range Manage.* 21:158–164.

Clanton, D. C. 1978. Non-protein nitrogen in range supplements. *J. Anim. Sci.* 47:765–799.

Cohen, R. D. H. 1975. Phosphorus and the grazing ruminant. *World Rev. Anim. Prod.* 11:27–43.

Coleman, S. W., and R. D. Wyatt. 1982. Cottonseed meal or small grain forages as protein supplements fed at different intervals. *J. Anim. Sci.* 55:1–17.

Connor, J. W., V. R. Bothman, A. L. Lesperance, and R. E. Kinsinger. 1963. Nutritive evaluation of summer range forage with cattle. *J. Anim. Sci.* 22:961–969.

Cook, C. W., and L. E. Harris. 1950. The nutritive content of the grazing sheep's diet on the summer and winter ranges of Utah. *Utah Agric. Exp. Stn. Bull.* 342.

Cook, C. W., and L. E. Harris. 1951. A comparison of the lignin ratio technique and the chromogen method of determining digestibility and forage consumption. *J. Anim. Sci.* 10:365–373.

Cook, C. W., and L. E. Harris. 1968a. Nutritive value of seasonal ranges. *Utah Agric. Exp. Stn. Bull.* 472.

Cook, C. W., and L. E. Harris. 1968b. Effect of supplementation on intake and digestibility of range forage. *Utah Agric. Exp. Stn. Bull.* 475.

Cook, C. W., L. A. Stoddart, and L. E. Harris. 1954. The nutritive value of winter range plants in the Great Basin. *Utah Agric. Exp. Stn. Bull.* 372.

Cook, C. W., L. A. Stoddart, and L. E. Harris. 1959. The chemical content in various portions of the current growth of salt-desert shrubs and grasses during winter. *Ecology* 40:644–651.

Cook, C. W., J. E. Mattox, and L. E. Harris. 1961. Comparative daily consumption and digestibility of summer range forage by wet and dry ewes. *J. Anim. Sci.* 20:866–870.

Cook, C. W., L. E. Harris, and M. L. Young. 1967. Botanical and nutritive content of diets of cattle and sheep under single and common use on mountain range. *J. Anim. Sci.* 26:1169–1174.

Cordova, F. J., and J. D. Wallace. 1975. Nutritive value of some browse and forb species. *Proc. West. Sec. Am. Soc. Anim. Sci.* 26:160–162.

Cordova, F. J., J. D. Wallace, and R. D. Pieper. 1978. Forage intake by grazing livestock: A review. *J. Range Manage.* 31:430–438.

Cowan, F. M. 1947. Range competition between mule deer, bighorn sheep, and elk in Jasper Park, Alberta. *Trans. N. Am. Wildl. Resour. Conf.* 12:233–237.

Cushaw, C. T. and J. Coady. 1976. Food habits of moose in Alaska: A preliminary study using rumen content analysis. *Can. Field-Nat.* 90:11–16.

de Calesta, D. S., J. G. Nagy, and J. A. Bailey. 1975. Starving and refeeding mule deer. *J. Wildl. Manage.* 39:663–669.

Donefer, E. 1969. Forage solubility measurements in relation to nutritive value. In *National Conference on Forage Quality Evaluation and Utilization,* Lincoln, NE.

Drawe, D. L., and T. W. Box. 1968. Forage ratings for mule deer and cattle on the Welder Wildlife Refuge. *J. Range Manage.* 21:225–228.

Dusek, G. L. 1975. Range relations of mule deer and cattle in prairie habitat. *J. Wildl. Manage.* 39:605–616.

Elliott, R. L., and K. Fokema. 1961. Herbage consumption studies on beef cattle. I. Intake studies on Afrikander and Mashona cows on veld grazing. *Rhod. Agric. J.* 58:49–57.

Erickson, L. R. 1974. *Livestock utilization of a clear-cut burn in northeastern Oregon.* M.S. Thesis. Oregon State University, Corvalis, OR.

Gade, A. E., and F. D. Provenza. 1986. Nutrition of sheep grazing crested wheatgrass versus crested wheatgrass-shrub pastures during winter. *J. Range Manage.* 39:527–530.

Gallina, S. 1993. White-tailed deer and cattle diets at La Michilia, Durango, Mexico. *J. Range Manage.* 46:487–493.

Galt, A. D. 1972. *Relationship of botanical composition of steer diet to digestibility and forage intake on a desert grassland.* Ph.D. Thesis. Univ. of Arizona, Tucson, AZ.

Gartner, R. J. W., and R. J. Anson. 1966. Vitamin A reserves of sheep maintained on mulga (*Acacia aneura*). *Aust. J. Exp. Agric. Anim. Husb.* 6:321–325.

Gihad, E. A. 1976. Intake, digestibility and nitrogen utilization of tropical natural grass hay by goats and sheep. *J. Anim. Sci.* 43:870–883.

Goodwin, G. A. 1975. Seasonal food habits of mule deer in southeastern Wyoming. *U.S. Dep. Agric. For. Serv. Res. Note* RM-287. Rocky Mountain Forest and Range Exp. Stn., Fort Collins, CO.

Graham, N. M. 1964. Energy costs of feeding activities and energy expenditure grazing sheep. *Aust. J. Agric. Res.* 15:969–973.

Hakkila, M. D., J. L. Holechek, J. D. Wallace, D. M. Anderson, and M. Cardenas. 1987. Diet and forage intake of cattle on desert grassland. *J. Range Manage.* 40:339–341.

Hale, O. M., R. H. Hughes, and F. E. Knox. 1962. Forage intake by cattle grazing wiregrass range. *J. Range Manage.* 15:6–9.

Hanley, T. A. 1982. The nutritional basis for food selection by ungulates. *J. Range Manage.* 35:146–152.

Hanley, T. A., and K. A. Hanley. 1982. Food resource partitioning by sympatric ungulates on Great Basin rangeland. *J. Range Manage.* 35:152–159.

Harper, J. A., J. H. Harn, W. W. Bentley, and C. F. Yocum. 1967. The status and ecology of the Roosevelt elk in California. *Wildl. Monogr.* 16:1–49.

Hart, R. H., O. M. Abdalla, D. H. Clark, M. B. Marshall, M. M. Haamid, J. A. Hager, and J. W. Waggoner Jr. 1983. Quality of forage and cattle diets on the Wyoming high plains. *J. Range Manage.* 36:46–51.

Havstad, K. M., and J. C. Malechek. 1982. Energy expenditure by heifers grazing crested wheatgrass of diminishing availability. *J. Range Manage.* 35:447–451.

Hawley, A. W. L., D. G. Peden, and W. R. Stricklin. 1981. Bison and Hereford steer digestion of sedge hay. *Can. J. Anim. Sci.* 61:165–174.

Healy, B., W. J. McCabe, and G. F. Wilson. 1970. Ingested soil as a source of microelements for grazing animals. *New Zealand. J. Agric. Res.* 13:503–521.

Hendrickson, R. E., D. P. Poppi, and D. J. Minson. 1981. The voluntary intake, digestibility, and retention time by cattle and sheep of stem and leaf fractions of a tropical legume (*Lablab purpureus*). *Aust. J. Agric. Res.* 32:389–398.

Herbel, C. H., and A. B. Nelson. 1966. Species preferences of Hereford and Santa Gertrudis cattle on southern New Mexico range. *J. Range Manage.* 19:177–181.

Hibbs, D. L. 1966. Food habits of the mountain goat in Colorado. *J. Mammal.* 48:242–248.

Hinnant, R. T., and M. M. Kothmann. 1982. Potassium content of three grass species during winter. *J. Range Manage.* 35:211–213.

Hjeljord, O. 1971. *Feeding ecology and habitat preference of the mountain goat in Alaska.* M.S. Thesis. University of Alaska, Fairbanks, AL.

Hjeljord, O., F. Sundstol, and H. Haagenrud. 1982. The nutritional value of browse to moose. *J. Wildl. Manage.* 46:333–343.

Holechek, J. L. 1984. Comparative contribution of grasses, forbs, and shrubs to the nutrition of range ungulates. *Rangelands* 6:245–248.

Holechek, J. L., and C. H. Herbel. 1982. Seasonal suitability grazing in the western United States. *Rangelands* 4:252–255.

Holechek, J. L., and M. Vavra. 1982. Forage intake by cattle on forest and grassland ranges. *J. Range Manage.* 35:737–741.

Holechek, J. L., and M. Vavra. 1983. Drought effects on diet and weight gains of yearling heifers in northeastern Oregon. *J. Range Manage.* 36:227–231.

Holechek, J. L., and C. H. Herbel. 1986. Supplementing range livestock. *Rangelands* 8:29–33.

Holechek, J. L., M. Vavra, and J. Skovlin. 1981. Diet quality and performance of cattle on forest and grassland range. *J. Anim. Sci.* 53:291–298.

Holechek, J. L., M. Vavra, and D. Arthun. 1982a. Relationships between performance, intake, diet nutritive quality and fecal nutritive quality of cattle on mountain range. *J. Range Manage.* 35:741–744.

Holechek, J. L., M. Vavra, and R. D. Pieper. 1982b. Botanical composition determination of range herbivore diets: A review. *J. Range Manage.* 35:309–315.

Holechek, J. L., M. Vavra, and R. D. Pieper. 1982c. Methods for determining the nutritive quality of range ruminant diets: A review. *J. Anim. Sci.* 54:363–376.

Holechek, J. L., M. Vavra, J. Skovlin, and W. C. Krueger. 1982d. Cattle diets in the Blue Mountains of Oregon. I. Grasslands. *J. Range Manage.* 35:109–113.

Holechek, J. L., M. Vavra, J. Skovlin, and W. C. Krueger. 1982e. Cattle diets in the Blue Mountains of Oregon. II. Forests. *J. Range Manage.* 35:239–243.

Holechek, J. L., M. L. Galyean, J. D. Wallace, and H. Wofford. 1985. Evaluation of fecal indices for predicting phosphorus status of cattle. *Grass. Forage Sci.* 40:489–492.

Holechek, J. L., J. Jeffers, T. Stephenson, C. B. Kuykendall, and S. A. Butler-Nance. 1986. Cattle and sheep diets on low elevation winter range in northcentral New Mexico. *Proc. West. Sec. Am. Soc. Anim. Sci.* 37:243–248.

Holechek, J. L., A. V. Munshikpu, L. Saiwanna, G. Nunez-Hernandez, R. Valdez, J. D. Wallace, and M. Cardenas. 1990. Influences of six shrub diets varying in phenol content on intake and nitrogen retention by goats. *Tropical Grassl.* 24:93–98.

Hubbard, R. E., and R. M. Hansen. 1976. Diets of wild horses, cattle, and mule deer in the Piceance Basin, Colorado. *J. Range Manage.* 29:389–392.

Huston, J. L., B. S. Rector, L. B. Merrill, and B. S. Ingdall. 1981. Nutritive value of range plants in the Edwards Plateau region of Texas. *Tex. Agric. Exp. Stn. Bull.* 1357.

Janis, C. 1976. The evolutionary strategy of the Equidae and the origins of ruminant and cecal digestion. *Evolution* 30:757–774.

Jefferies, N. W., and R. W. Rice. 1969. Forage intake by yearling steers shortgrass rangelands. *Proc. West. Sec. Am. Soc. Anim. Sci.* 20:343–348.

Johnson, A. E., L. F. James, and J. Spillet. 1976. The abortifacient and toxic effect of big sagebrush (*Artemisia tridentata*) and juniper (*Juniperus osteosperma*) on domestic sheep. *J. Range Manage.* 29:278–280.

Johnson, D. E., M. M. Borman, and L. R. Rittenhouse. 1982. Intake, apparent utilization and rate of digestion in mares and cows. *Proc. West. Sec. Am. Soc. Anim. Sci.* 33:294–298.

Judkins, M. B., J. J. Krysl, J. D. Wallace, M. L. Galyean, K. D. Jones, and E. E. Parker. 1985a. Intake and diet selection by protein supplemented steers. *J. Range Manage.* 38:210–214.

Judkins, M. B., J. D. Wallace, E. E. Parker, and J. D. Wright. 1985b. Performance and phosphorus status of range cows with and without phosphorus supplementation. *J. Range Manage.* 38:139–144.

Kalmbacher, R. S., K. R. Long, M. K. Johnson, and F. G. Martin. 1984. Botanical composition of diets of cattle grazing south Florida rangelands. *J. Range Manage.* 37:334–340.

Kartchner, R. J. 1981. Effects of protein and energy supplementation of cows grazing native winter range forage on intake and digestibility. *J. Anim. Sci.* 51:432–438.

Kartchner, R. J., and D. C. Adams. 1982. Effects of daily and alternate day feeding of grain supplements to cows grazing fall-winter range. *Proc. West. Sec. Am. Soc. Anim. Sci.* 33:308–311.

Kartchner, R. J., and C. M. Campbell. 1979. Intake and digestibility of range forages consumed by livestock. *Mont. Agric. Exp. Stn. U.S. Dep. Agric. Sci. Educ. Adm. Agric. Res. Bull.* 718.

Kautz, J. E., and G. M. Van Dyne. 1978. Comparative analyses of diets of bison, cattle, sheep and pronghorn antelope on shortgrass prairie in northeastern Colorado, USA. *Proc. Int. Rangel. Congr.* 1:438–442.

Krysl, L. J., M. E. Hubbert, F. B. Sowell, G. E. Plumb, J. K. Jewett, M. A. Smith, and J. W. Waggoner. 1984a. Horses and cattle grazing in the Wyoming Red Desert. I. Food habits and dietary overlaps. *J. Range Manage.* 37:72–77.

Krysl, L. J., B. F. Sowell, M. E. Hubbert, C. E. Plumb, T. K. Jewett, M. A. Smith, and J. W. Waggoner. 1984b. Horses and cattle grazing in the Wyoming Red Desert II. Dietary quality. *J. Range Manage.* 37:252–256.

Laredo, M. A., and D. J. Minson. 1973. The voluntary intake, digestibility, and retention time by sheep of leaf and stem fractions of five grasses. *Aus. J. Agric. Res.* 24:875–888.

Leigh, J. H., A. D. Wilson, and W. E. Mulham. 1968. A study of merino sheep grazing a cottonbush (*Kochia aphylla*)-grassland (*Stipa viriabilis-Danthonia caespitosa*) community on the riverine plain. *Aust. J. Agric. Res.* 19:210–218.

Long, K. R., R. S. Kalmbacher, and R. G. Martin. 1986. Diet quality of steers grazing three range sites in south Florida. *J. Range Manage.* 39:389–392.

Lorenz, R. J., and G. A. Rogler. 1962. A comparison of methods of renovating old stands of crested wheatgrass. *J. Range Manage.* 15:215–219.

Lyons, R. K., and J. W. Stuth. 1992. Fecal NIRS equations for predicting diet quality of free-ranging cattle. *J. Range Manage.* 45:238–245.

Malechek, J. C., and C. L. Leinweber. 1972. Forage selectivity by goats on lightly and heavily grazed ranges. *J. Range Manage.* 25:105–111.

Marten, G. C. 1969. Measurement and significance of forage palatability. In *National Conference on Forage Quality Evaluation and Utilization,* Lincoln, NE.

Mayland, H. F., A. R. Florence, R. C. Rosenau, V. A. Lazar, and H. A. Turner. 1975. Soil ingestion by cattle on semiarid range as reflected by titanium analysis of feces. *J. Range Manage.* 23:448–462.

McCollum, F. T. 1990. Relationships among fecal nitrogen, diet nitrogen and daily gain of steers grazing tallgrass prairie. *Oklahoma Agr. Exp.* MP-232-235.

McCollum, F. T., and M. L. Galyean. 1985. Cattle grazing blue grama rangeland II. Seasonal forage intake and digesta kinetics. *J. Range Manage.* 38:543–546.

McCollum, F. T., M. L. Galyean, L. J. Krysl, and J. D. Wallace. 1985. Cattle grazing blue grama rangeland. I. Seasonal diets and rumen fermentation. *J. Range Manage.* 38:539–543.

McGinnies, W. J., and C. E. Townsend. 1983. Yield of three range grasses grown along and in mixtures of legumes. *J. Range Manage.* 3:399–402.

McMahan, C. A. 1964. Comparative food habits of deer and three classes of livestock. *J. Wildl. Manage.* 28:798–808.

McMillan, J. F. 1953. Some feeding habits of moose in Yellowstone National Park. *Ecology* 34:102–110.

Mertens, D. R. 1973. *Application of theoretical and mathematical models to cell wall digestion and forage intake of ruminants.* Ph.D. Thesis. Cornell University, Ithaca, NY.

Milchunas, D. G., M. I. Dyer, O. C. Wallmo, and D. E. Johnson. 1978. In vivo-in vitro relationships of Colorado mule deer forage. *Colo. Div. Wildl. Spec. Rep.* 43. Fort Collins, CO.

Miles, W. H., and L. R. McDowell. 1983. Mineral deficiencies of the llanos rangelands of Columbia. *World Animal Review* 46:2–10.

Mitchell, G. J., and S. Smoliak. 1971. Pronghorn antelope, range characteristics and good habits in Alberta. *J. Wildl. Manage.* 35:238–250.

Moir, K. W. 1960. Nutrition of grazing cattle III. Estimation of protein, phosphorus and calcium in mixed diets. *Queensl. J. Agric. Sci.* 17:373–383.

Monsen, S. B. 1980. Interseeding fourwing saltbush with crested wheatgrass on southern Idaho rangelands. *Abstr. Annu. Meet. Soc. Range Manage.* 33:51.

Morris, J. G., R. E. Delmas, and J. C. Hull. 1980. Salt (sodium) supplementation of range beef cows in California. *J. Anim. Sci.* 51:722–731.

Moseley, G. and D. I. H. Jones. 1974. The effect of sodium chloride supplementation of a sodium adequate hay on digestion, production, and mineral nutrition in sheep. *J. Agric. Sci.* 83:37–42.

Mould, E. D. and C. T. Robbins. 1981. Nitrogen metabolism in elk. *J. Wildl. Manage.* 45:323–335.

Mubanga, G., J. L. Holechek, R. Valdez, and S. D. Schemnitz. 1985. Relationships between diet and fecal nutritive quality in mule deer. *Southwest Nat.* 30:573–578.

Nagy, J. G., H. W. Steinhoff, and G. M. Ward. 1964. Effects of essential oils of sagebrush on deer rumen microbial function. *J. Wildl. Manage.* 28:785–790.

Nastis, A. S. and J. C. Malechek. 1981. Digestion and utilization of nutrients on oak browse by goats. *J. Anim. Sci.* 52:283–291.

Nelson, A. B., C. H. Herbel, and H. M. Jackson. 1970. Chemical composition of forage species grazed by cattle on an arid New Mexico range. *N. Mex. Agric. Exp. Stn. Bull.* 561.

Nunez-Hernandez, G., J. L. Holechek, J. D. Wallace, M. L. Galyean, A. Tembo, R. Valdez, and M. Cardenas. 1989. Influence of native shrubs on nutritional states of goats: Nitrogen retention. *J. Range Manage.* 42:228–232.

Nunez-Hernandez, G., J. L. Holechek, D. Arthun, A. Tembo, J. D. Wallace, M. Galyean, M. Cardenas, and R. Valdez. 1992. Evaluation of fecal indicators for assessing energy and nitrogen status of cattle and goats. *J. Range Manage.* 45:143–147.

Oh, K. H., T. Sakai, M. B. Jones, and W. M. Longhurst. 1967. Effects of various essential oils isolated from Douglas-fir needles upon sheep and deer rumen microbial activity. *Appl. Microbiol.* 15:777–784.

Oh, K. H., M. B. Jones, and W. M. Longhurst. 1968. Comparison of rumen microbial inhibition resulting from relatively unpalatable plant species. *Appl. Microbiol.* 16:39–44.

Olsen, F. W., and R. M. Hansen. 1977. Food relations of wild free-roaming horses to livestock and big game in Red Desert, Wyoming. *J. Range Manage.* 30:17–20.

Otsyina, R., C. M. McKell, and G. Van Epps. 1982. Use of shrubs to meet nutrient requirement of sheep grazing on crested wheatgrass during fall and early winter. *J. Range Manage.* 35:751–754.

Parker, K. G. 1969. The nature and use of Utah range. *Utah State Univ. Ext. Circ.* 359, Logan, UT.

Peden, D. G., G. M. Van Dyne, R. W. Rice, and R. M. Hansen. 1973. The trophic ecology of *Bison bison* L. on shortgrass plains. *J. Appl. Ecol.* 11:489–498.

Pederson, J. C., and B. L. Welch. 1982. Effects of monoterpenoid exposure on ability of rumen inocula to digest a set of forages. *J. Range Manage.* 35:500–503.

Pfister, J. A., and J. C. Malechek. 1986. The voluntary forage intake and nutrition of goats and sheep in the semiarid tropics of northeastern Brazil. *J. Anim. Sci.* 63:1078–1086.

Pieper, R. D., C. W. Cook, and L. E. Harris. 1959. Effect of intensity of grazing upon nutritive content of the diet. *J. Anim. Sci.* 18:1031–1037.

Pieper, R. D., A. B. Nelson, G. S. Smith, E. E. Parker, E. J. Boggino, and S. L. Hatch. 1978. Chemical composition and digestibility of important range grass species in south-central New Mexico. *N. Mex. Agric. Exp. Stn. Bul.* 662.

Pinchak, W. E., S. C. Canon, R. K. Heitschmidt, and S. Dowheir. 1990. Effects of long-term year-long grazing at moderate and heavy rates of stocking on diet selection and forage intake dynamics. *J. Range Manage.* 43:304–309.

Poppi, D. P., D. J. Minson, and J. H. Ternouth. 1981. Studies of cattle and sheep eating leaf and stem fractions of grasses. I. The voluntary intake, digestibility, and retention time in the reticulo-rumen. *Aust. J. Agric. Res.* 32:99–108.

Powell, D. J., D. C. Clanton, and J. T. Nichols. 1982. Effect of range condition on the diet and performance of steers grazing native sandhills change in Nebraska. *J. Range Manage.* 35:96–100.

Rafique, S., J. D. Wallace, J. L. Holechek, M. L. Galyean, and D. Arthun. 1992. Influence of forbs and shrubs on nutrient digestion and balance in sheep fed grass hay. *Small Ruminant Res.* 7:113–122.

Reed, M. J., and R. A. Peterson. 1961. Vegetation, soil, and cattle responses to grazing on northern Great Plains Range. *U.S. Dep. Agric. Tech. Bull.* 1252.

Reiner, R. J., F. C. Bryant, R. D. Farfan, and B. F. Craddock. 1987. Forage intake of alpacas grazing Andean rangeland in Peru. *J. Anim. Sci.* 64:868–871.

Rice, R. W., D. R. Cundy, and P. R. Weyerts. 1971. Botanical and chemical composition of esophageal and rumen fistula samples of sheep. *J. Range Manage.* 21:121–124.

Rice, R. W., F. E. Dean, and J. E. Ellis. 1974. Bison, cattle and sheep dietary quality and food intake. *Proc. West. Sec. Am. Soc. Anim. Sci.* 25:194–197.

Rittenhouse, L. R., D. C. Clanton, and C. L. Streeter. 1970. Intake and digestibility of winter-range forage by cattle with and without supplements. *J. Anim. Sci.* 31:1215–1221.

Robbins, C. T., F. J. Van Soest, W. W. Mautz, and A. N. Moen. 1975. Feed analysis and digestion with reference to white-tailed deer. *J. Wildl. Manage.* 39:67–69.

Robbins, C. T., A. E. Hagerman, P. J. Austin, C. McArthur, and T. A. Hanley. 1991. Variation in mammalian physiological responses to a condensed tanning and its ecological implications. *J. Mammol.* 72:480–486.

Rogler, G. A., and R. L. Lorenz. 1969. Pasture productivity of crested wheatgrass as influenced by nitrogen fertilization and alfalfa. *U.S. Dep. Agric. Res. Stn. Tech. Bull.* 1402.

Rosiere, R. E., and D. T. Torell. 1985. Nutritive value of sheep diets on coastal California annual range. *Hilgardia* 53:1–17.

Rosiere, R. E., J. D. Wallace, and R. F. Beck. 1975a. Cattle diet on semidesert grassland: Nutritive content. *J. Range Manage.* 28:94–96.

Rosiere, R. E., R. F. Beck, and J. D. Wallace. 1975b. Cattle diet on semidesert grassland: Botanical composition. *J. Range Manage.* 28:94–96.

Rosiere, R. E., M. L. Galyean, and J. D. Wallace. 1980. Accuracy of roughage intake estimates as determined by a chromic oxide-in vitro digestibility technique. *J. Range Manage.* 33:237–240.

Rumbaugh, M. D., D. A. Johnson, and G. A. Van Epps. 1982. Forage yield and quality in a Great Basin shrub, grass, and legume pasture experiment. *J. Range Manage.* 36:604–609.

Saunders, J. K., Jr. 1955. Food habits and range use of the Rocky Mountain goat in the Crazy Mountains, Montana. *J. Wildl. Manage.* 19:429–437.

Schurg, W. A., D. L. Frei, P. R. Cheeke, and D. W. Holtan. 1977. Utilization of whole corn plant pellets by horses and rabbits. *J. Anim. Sci.* 45:1317–1321.

Schwartz, C. C., and J. G. Naggy. 1976. Pronghorn diets relative to forage availability in northeastern Colorado. *J. Wildl. Manage.* 40:469–478.

Scotter, G. 1967. The winter diet of barren ground caribou in northern Canada. *Can. Field-Nat.* 81:33–39.

Severson, K. E. 1982. Production and nutritive value of aspen understory, Black Hills. *J. Range Manage.* 35:786–790.

Severson, K. E., and M. May. 1967. Food preferences of antelope and domestic sheep in Wyoming Red Desert. *J. Range Manage.* 20:21–25.

Short, H. L. 1971. Forage digestibility and diet of deer on southern upland range. *J. Wildl. Manage.* 35:698–706.

Short, H. L., R. M. Blair, and C. A. Segelquist. 1974. Fiber composition and forage digestibility by small ruminants. *J. Wildl. Manage.* 38:197–202.

Sidahmed, A. E., J. G. Morris, L. J. Koong, and S. R. Radosevich. 1981. Contribution of mixtures of three chaparral shrubs to the protein and energy requirements of Spanish goats. *J. Anim. Sci.* 53:1391–1401.

Silver, H., N. F. Colovos, J. B. Holter, and J. B. Hayes. 1969. Fasting metabolism of white-tailed deer. *J. Wildl. Manage.* 33:490–498.

Sinnott, E. W. 1960. *Plant morphogenesis.* McGraw-Hill Book Company, New York.

Smith, J. G., and O. Julander. 1953. Deer and sheep competition in Utah. *J. Wildl. Manage.* 17:101–112.

Smith, G. S., and A. B. Nelson. 1975. Effects of sodium silicate added to rumen cultures on forage digestion, with interaction of glucose, urea and minerals. *J. Anim. Sci.* 41:891–899.

Smith, G. S., A. B. Nelson, and E. J. Boggino. 1971a. Digestibility of forages in vitro as affected by silica content. *J. Anim. Sci.* 33:466–472.

Smith, G. S., H. E. Kiesling, C. H. Herbel, and P. Trujillo. 1980. Evidence of copper, manganese, and zinc deficiency in beef cattle grazing semiarid range in southern New Mexico. *New Mexico State University Feeder's Day Report,* pp. 17–21.

Smith, L. W., H. K. Goering, D. R. Waldo, and C. H. Gordon. 1971b. In vitro digestion rate of forage cell wall components. *J. Dairy Sci.* 54:71–76.

Smith, L. W., H. K. Goering, and C. H. Gordon. 1972. Relations of forage composition with rates of cell wall digestion and indigestibility of cell walls. *J. Dairy Sci.* 55:1140–1147.

Spalinger, D. E., C. T. Robbins, and T. A. Hanley. 1986. The assessment of handling time in ruminants: The effect of plant chemical and physical structure and the rate of breakdown of plant particles in the rumen of mule deer and elk. *Can. J. Zool.* 64:312–321.

Speth, C. F., V. R. Bohman, H. Melendy, and M. A. Wade. 1962. Effect of dietary supplements on cows on a semi-desert range. *J. Anim. Sci.* 21:444–448.

Squires, V. R., and B. D. Siebert. 1983. Botanical and chemical components of the diet and liveweight change in cattle on semi-desert rangeland in central Australia. *Aust. Range J.* 6:28–34.

Stephenson, T. E., J. L. Holechek, and C. B. Kuykendall. 1985a. Drought effect on pronghorn and other ungulate diets. *J. Wildl. Manage.* 49:146–152.

Stephenson, T. E., J. L. Holechek, and C. B. Kuykendall. 1985b. Diet of four ungulates on winter range in northcentral New Mexico. *Southwest. Nat.* 30:55–58.

Stevens, D. R. 1970. Winter ecology of moose in the Gallatin Mountains. *Montan. J. Wildl. Manage.* 34:37–46.

Streeter, C. L., D. C. Clanton, and O. E. Hoehne. 1968. Influence of advance in season on nutritive value of forage consumed by cattle grazing western Nebraska native range. *Univ. Nebr. Agric. Exp. Stn. Res. Bull.* 227.

Thetford, F. O., R. D. Pieper, and A. B. Nelson. 1971. Botanical and chemical composition of cattle and sheep diets on pinyon-juniper grassland range. *J. Range Manage.* 24:425–431.

Todd, J. W. 1975. Food of Rocky Mountain bighorn sheep in southern Colorado. *J. Wildl. Manage.* 39:108–111.

Ulyatt, M. J. 1981. The feeding value of temperate pastures. In *Grazing animals.* Elsevier Science Publishing Co., Inc., New York.

Underwood, E. J. 1966. *The mineral nutrition of livestock.* Central Press, Aberdeen, Scotland.

Van Dyne, G. M. and G. P. Lofgren. 1964. Comparative digestion of dry annual range forage by cattle and sheep. *J. Anim. Sci.* 23:823–832.

Van Dyne, G. M. and J. H. Meyer. 1964. A method for measurement of forage intake of grazing livestock using microdigestion techniques. *J. Range Manage.* 177:204–208.

Van Dyne, G. M. and H. F. Heady. 1965. Botanical composition of sheep and cattle diets on a mature annual range. *Hilgardia* 36:465–492.

Van Dyne, G. M., O. O. Thomas, and J. L. Van Horn. 1964. Diet of cattle and sheep grazing on winter range. *Proc. West. Sec. Am. Soc. Anim. Sci.* 14:1–6.

Van Dyne, G. M., N. R. Brockington, Z. Szocs, J. Duek, and C. A. Ribio. 1980. Large herbivore subsystem. In A. I. Breymeyer and G. M. Van Dyne (Eds.). *Grasslands, systems analysis and man.* Cambridge University Press, Cambridge.

Van Soest, P. J. 1982. *Nutritional ecology of the ruminant.* O&B Books Inc., Corvallis, OR.

Van Soest, P. J., and F. E. Lovelace. 1969. Solubility of silica in forages. *J. Anim. Sci.* 29:182 (Abstr.).

Varner, L. W., L. H. Blankenship, and G. W. Lynch. 1979. Seasonal changes in nutritive value of deer food plants in south Texas. *Proc. Annu. Conf. Southeast. Assoc. Fish Wildl. Agencies* 31:99–105.

Vavra, M., and F. Sneva. 1978. Seasonal diets of five ungulates grazing the cold desert biome. *Proc. Intnl. Rangl. Cong.* 1:435–437.

Vavra, M., and J. L. Holechek. 1980. Factors influencing microhistological analysis of herbivore diets. *J. Range Manage.* 33:371–374.

Vavra, M., R. W. Rice, and R. E. Bement. 1973. Chemical composition of the diet, intake, and gain of yearling cattle under different grazing intensities. *J. Anim. Sci.* 36:411–414.

Vavra, M., W. C. Krueger, and W. P. Wheeler. 1980. Cattle grazing potential on clearcuts. *Oreg. State. Univ. Agr. Exp. Sta. Res. Rep.* 586.

Wagnon, K. A., H. R. Guilbert, and G. H. Hart. 1959. Beef cattle investigations on the San Joaquin Experimental Range. *Calif. Agric. Exp. Stn. Bull.* 765.

Wallace, J. D. 1987. Supplemental feeding options to improve livestock efficiency on rangelands. In *Achieving efficient use of rangeland resources.* Fort Keogh Res. Symp. Montana Agr. Exp. Stn., Bozeman, MT.

Wallace, J. D., F. Hubert Jr., and R. J. Raleigh. 1963. The response of yearling cattle on crested wheatgrass pasture to energy, protein, and sodium supplementation. *J. Range Manage.* 16:1–5.

Wallace, J. D., J. C. Free, and A. H. Denham. 1972. Seasonal changes in herbage and cattle diets on sandhill grassland. *J. Range Manage.* 25:100–104.

Warren, L. E., D. N. Ueckert, J. M. Shelton, and A. D. Chamad. 1984. Spanish goat diets on mixed-brush rangeland in the south Texas plains. *J. Range Manage.* 37:340–343.

Welch, B. L., and J. C. Pederson. 1981. In vitro digestibility among accessions of big sagebrush by wild mule deer and its relationship to monoterpenoid content. *J. Range Manage.* 34:497–501.

White, S. M., B. L. Welch, and J. T. Flinders. 1982. Monoterpenoid content of pygmy rabbit stomach ingesta. *J. Range Manage.* 35:107–109.

Wilkinson, D. F., C. C. Shank, and D. F. Penner. 1976. Musk-ox-caribou summer range relations on Banks Island, NWT. *J. Wildl. Manage.* 40:151–162.

Wofford, H., and J. L. Holechek. 1982. Influence of grind size on four- and forty-eight hour in vitro digestibility. *Proc. West. Sec. Am. Soc. Anim. Sci.* 33:261–263.

Woodward, S. L., and R. D. Ohmart. 1976. Habitat use and fecal analysis of feral burros (*Equis asinus*), Chemehuevi Mountains, California. *J. Range Manage.* 29:482–485.

Woolf, A. 1968. *Summer ecology of bighorn sheep in Yellowstone National Park.* M.S. Thesis. Colorado State University, Fort Collins, CO.

Yates, D. A., D. C. Clanton, and J. T. Nichols. 1982. Effect of continuous grazing on the diet of steers. *J. Range Manage.* 35:339–342.

CHAPTER 12

RANGE LIVESTOCK PRODUCTION

ECONOMICS OF RANGE LIVESTOCK PRODUCTION

In the past, "range" has been strongly associated with livestock grazing in the western United States. The traditional use of North American rangelands has been to produce meat for human consumption. However, this use, particularly in the 11 western states, is now being challenged not only by the nonranching public, but also by ranchers, as they find it difficult to make a profit. Historically, annual returns on capital investment for range livestock enterprises have been low (2% to 6%) compared to returns from investments in industry (10% to 30%). In New Mexico, annual net returns on capital investment in the best of years have averaged less than 4% (Fowler and Torell 1985; Holechek and Hawkes 1993). In most years the overall return is a little over 1% (Figure 12.1). Compared to the peak in the late 1970s, land prices in the early 1990s for ranches in the western United States dropped about 28%, operating costs increased about 60%, and prices paid for livestock increased about 11% (Workman and Evans 1993; Holechek and Hawkes 1993). Based on this scenario the economic situation for range livestock production in the western United States has been highly unfavorable. Presently, many ranchers face an economic climate that severely threatens their survival. However, there is some optimism due to recent opportunities for income from enterprises such as guest ranching, raising exotic animals (ostriches, emus, llamas, etc.), producing plants for landscaping, and fee hunting and fishing operations.

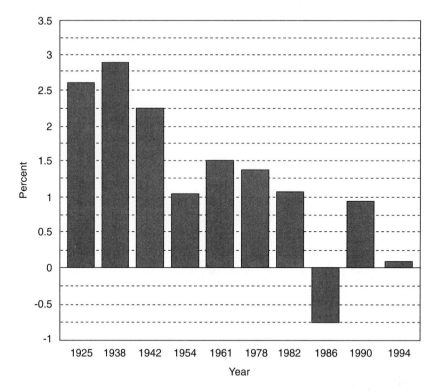

Figure **12.1** Rate of return on investment for medium-size (250-animal unit) cattle ranches in New Mexico for the 1925–1994 period. (From New Mexico Agric. Exp. Sta. Repts.)

Demand for Livestock Products

There are several reasons for the present bleak economic situation confronting ranchers in the western United States. These all basically relate to lack of demand for meat and other livestock products, coupled with rising production costs and declining value of owned capital (land and livestock).

Real cattle prices in 1996 were at their lowest levels since World War II (Table 12.1). The question confronting ranchers is why the present low real prices and what the future will hold for cattle and ranch prices. To examine this issue it is necessary to consider indirect factors such as the world economy and grain production as well as future beef demand in the United States. Our discussion follows Holechek et al. (1994).

The present low real prices for beef are partially explained by the low corn and wheat prices. Low corn and wheat prices result in low chicken and pork prices because these are the main feeds used to produce these meats. Chickens and pigs convert grains into meat more efficiently than cattle, and therefore beef becomes relatively much more costly than poultry or pork when grain prices are depressed (Godfrey and Pope 1990). Annual per capita consumption of beef has dropped from 86 lb in 1978 to 63 lb presently based on U.S.

TABLE 12.1 Cattle Prices in Relation to American Economy for the Period Between 1970 and 1996[1]

YEAR	PERCENT CHANGE IN REAL GDP[2]	PERCENT UNEMPLOYMENT	PERCENT CHANGE IN CONSUMER PRICE INDEX	DISCOUNT INTEREST RATE (%)	PRIME INTEREST RATE (%)	REAL INTEREST RATE (%)	PERCENT GAIN S&P 500 STOCK INDEX	NOMINAL CATTLE PRICES ($)[4]	REAL[3,4] CATTLE PRICES ($)
1970	0	4.9	5.6	5.95	7.91	2.31	0.1	28.40	70.12
1971	2.9	5.9	3.3	4.88	5.72	2.42	11	30.90	73.92
1972	5.1	5.6	3.4	4.50	5.25	1.85	16	35.80	80.63
1973	6.2	4.9	8.7	6.44	8.03	(0.67)	(17)	45.30	91.89
1974	(0.6)	5.6	12.3	7.83	10.81	(1.49)	(30)	36.70	68.22
1975	(0.8)	8.5	6.9	6.25	7.86	0.96	32	39.30	69.07
1976	4.9	7.7	4.9	5.50	6.84	1.94	19	36.30	59.90
1977	4.5	7.1	6.7	5.46	6.83	0.13	(12)	38.50	59.05
1978	4.6	6.1	9.0	7.46	9.06	0.06	1	52.90	72.87
1979	2.5	5.8	13.3	10.28	12.67	0.63	12	69.20	83.98
x̄	2.9	6.2	7.4	6.5	8.10	0.69	3.2	41.33	72.96
1980	(0.5)	7.1	12.5	11.77	15.27	2.77	26	64.30	70.74
1981	1.0	7.6	8.9	13.42	18.87	9.97	(10)	51.00	52.85
1982	(2.2)	9.7	3.8	11.02	14.86	11.06	15	46.99	46.99
1983	3.9	9.6	3.8	8.50	10.79	6.99	17	46.50	44.75
1984	6.2	7.5	3.9	8.80	12.04	8.14	1	46.00	42.75
1985	3.2	7.2	3.8	7.69	9.93	6.13	26	49.40	45.07
1986	2.9	7.0	1.1	6.33	8.33	7.23	15	48.50	42.69
1987	3.1	6.2	4.4	5.66	8.21	3.81	2	57.20	48.35
1988	3.9	5.5	4.6	6.20	9.32	4.72	12	62.30	50.77
1989	2.5	5.3	4.6	6.93	10.87	6.07	27	61.40	49.52
x̄	2.4	7.3	5.1	8.63	11.85	6.71	13.1	53.34	49.45

(Continued)

TABLE 12.1 Cattle Prices in Relation to American Economy for the Period Between 1970 and 1996[1] (*Continued*)

YEAR	PERCENT CHANGE IN REAL GDP[2]	PERCENT UNEMPLOYMENT	PERCENT CHANGE IN CONSUMER PRICE INDEX	DISCOUNT INTEREST RATE (%)	PRIME INTEREST RATE (%)	REAL INTEREST RATE (%)	PERCENT GAIN S&P 500 STOCK INDEX	NOMINAL CATTLE PRICES ($)[4]	REAL[3,4] CATTLE PRICES ($)
1990	0.3	5.5	6.1	6.98	10.01	3.91	(4.5)	68.00	52.02
1991	(1.2)	6.7	3.1	5.45	8.46	3.01	28	68.20	52.01
1992	2.1	7.4	2.9	3.25	6.25	3.00	4.5	65.91	48.67
1993	3.0	6.3	2.7	3.00	6.00	3.30	7.1	69.46	49.75
1994	4.0	5.8	2.6	3.88	7.20	4.60	(1.5)	63.41	44.74
1995	2.6	5.6	2.7	5.00	8.50	5.80	34.1	51.10	35.74
1996	2.9	5.3	3.3	5.13	8.25	4.95	20.3	45.78	32.02
x̄	2.0	6.1	3.3	4.67	7.81	4.08	12.57	61.69	44.99

Source: Updated from Holechek et al. 1994.

[1]*Sources:* National Agriculture Statistical Services 1945–1991; United States Department of Labor, Bureau of Labor Statistics; United States Department of Commerce, Consumer Price Index.

[2]Gross Domestic Product.

[3]Averaged across classes of cattle and adjusted for inflation using 1982 as the base year.

[4]$/100 wt (lb).

Department of Agriculture data (Figure 12.2). Although the cholesterol scare has been blamed for this drop and has caused some of it, the reduction in per capita beef consumption is primarily because of the relatively low cost of chicken and pork relative to beef (Godfrey and Pope 1990).

The other big factor is the expansion of world grain production due to improved technology. China has gone from a net importer to net exporter of wheat over the past 15 years. Russia is expected to become a grain exporter within the next 6 years assuming its free market reforms work out. As a result, this means cheaper feed for chickens and pigs.

In the United States, grain yields and total production are continuing to be boosted even though around 35 million acres of farmland have been retired since 1985 under the Conservation Reserve Program (CRP). As part of this land goes back into production it will probably adversely impact beef prices either indirectly by expanding grain supplies or directly by being used as a forage source.

An important factor is the expanding world supplies of low-grade beef from production increases in other countries, particularly Argentina and Australia. These countries are gaining world market share because their production costs are well below those in the United States. Production costs are about 62% lower in Argentina and about 34% lower in Australia compared to the United States. For this reason the United States now imports more beef than it exports.

The positives for western cattle producers are increased human population and the possibility of improved affluency in some developing countries that would allow them to afford more meat in the diet.

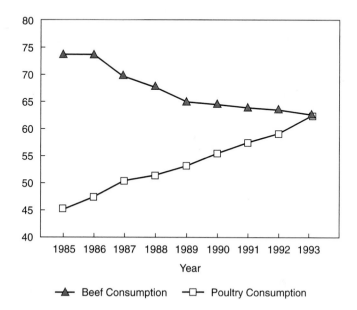

Figure **12.2** Per capita beef and poultry consumption (lb/person) in the United States (USDA 1994). (From Holechek and Hess 1996.)

The greatest improvement in living standards is occurring in the Pacific Basin (Asiatic) countries. These countries are a bright spot for U.S. cattle producers because they prefer high-quality beef and per capita consumption is increasing. Australia is interested in capturing this market, but so far the United States has had the quality advantage in producing the higher grades of beef although Australia has the cost advantage with lower grades.

In the United States the human population is growing at a low rate (1% per year). About half of this growth comes from immigrants who consume high amounts of chicken and pork because of their incomes and cultural traditions.

Based on this scenario a big increase in cattle prices over the next 5 to 10 years seems unlikely. However, there is a wild card. The United States has been experiencing disinflation since the early 1980s (see Table 12.1). This is due to increased productivity and the government switching from printing money to fund its debt in the 1970s to borrowing money to fund its debt in the 1980s (Davidson and Rees-Mogg 1993). Borrowing favors financial assets (bonds and stocks) over real estate and commodities. Debt in all sectors (consumer business, local government, and federal government) of the U.S. economy during the 1980s has led to low level economic growth in the 1990s (Davidson and Rees-Mogg 1993). If the economy slips into recession or depression the government could decide to monetize the debt (print instead of borrow the money) and stimulate the economy with massive spending. Such a program could cause money to flow into commodities (beef) and real estate (ranches) as hedges against inflation. A severe devaluation of the dollar against foreign currencies would be the outcome of this approach. A lower dollar should increase beef exports in the United States, but it could destabilize both the economy and the government (Calleo 1992; Davidson and Rees-Mogg 1993). Another problem for producers is that costs for fuel and supplemental feed could rise more than beef prices. Ranchers running extensive, low cost operations with high levels of long-term debt at low interest rates would be most likely to benefit from this type of inflationary spiral.

Importance of the West

The contribution of different regions of the United States toward the nation's production of beef is reflected in Figure 12.3. The 11 western states support about 20% of the nation's cattle and 50% of the nation's sheep. Cattle numbers are highest in the Great Plains and Southeast, with the 11 western states having the highest sheep numbers. It is often unrecognized that costs associated with cattle ranching are lower in the Great Plains than in the West (Holechek and Hawkes 1993). This is because the much lower amounts of land required per animal unit reduce property taxes, fuel costs, and labor costs, and permit better herd management in the Great Plains than in the West. Supplemental feed costs are less in the Great Plains because droughts are less frequent and severe. Also, active forage growth occurs for a longer period in the Great Plains. Although forage is generally low in quality in the Southeast and the natural vegetation is woodland, the high quantity of forage produced, flat terrain, long growing season, and ready availability of water make it a fairly efficient area for livestock production.

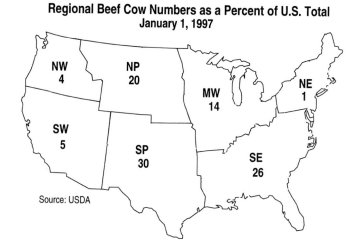

Regional Beef Cow Numbers as a Percent of U.S. Total
January 1, 1997

Source: USDA

Figure **12.3** Cattle numbers as a percentage of U.S. total. (From Cattle Fax 1997, Issue 3.)

Economic Strategies

Depending on economic conditions, either of two basic strategies can be emphasized in the management of a range livestock enterprise. These strategies involve either maximization of gross production per land unit or maximization of efficiency of production per animal unit. In the past, conventional thinking within the range profession has been geared toward maximization of gross production per land unit. This strategy has merit when the following conditions prevail (in order of importance):

1. Land values are increasing.
2. Demand for meat and prices for livestock are increasing.
3. Costs (i.e., supplemental feed, labor, veterinary supplies, and travel) associated with each animal unit are decreasing, stable, or increasing at a lower rate than property values.
4. Real interest rates (nominal interest rate minus inflation rate) on borrowed capital are low (under 3%).
5. Inflation rate of the nation's economy is high (over 5% per year).
6. Labor costs are high.

Under these conditions those range management practices that require capital invest-ment for increased grazing capacity such as brush control, seeding, or implementation of a specialized grazing system can be advantageous. The rapid increase in value of land and livestock maintains a favorable debt/owned capital ratio (below 1:2.5, i.e., debt is less than 50% of owned capital). These conditions make it financially more effective to increase grazing capacity from existing owned land than to buy more land. If increases in grazing capacity from the range improvement practice are below expectation, the consequences (increased debt) are minimized by the high rate of appreciation in owned capital (land and livestock). The preceding conditions occurred in the early to late-1970s (Holechek and Hawkes 1993; Holechek et al. 1994) (see Table 12.1).

Maximization of production efficiency per animal unit is an effective strategy when conditions occur that are opposite to those described previously. The most important of these are declining land prices, stable or declining prices for livestock, and high real interest rates on capital (over 3%). In this situation the operator who survives is the one who maintains a low debt/owned capital ratio (below 1:3) and minimizes variable costs per animal unit. This strategy involves conservative stocking rates and general avoidance of range improvement practices that involve high capital investment and high risk. Under this strategy, practices that improve efficiency of production involving low capital investment, such as better animal breeding programs, common-use grazing (livestock plus wildlife), reduction in supplemental feed costs, better detection of nonproducing animals, and reduced death loss to disease and predation would be emphasized. If expansion of herd size is desired, purchase of more land and/or implementation of practices to improve distribution will probably be financially more effective than brush control or implementation of a specialized grazing system. Conditions favoring emphasis of this strategy have characterized the 1980s and 1990s (Holechek and Hawkes 1993; Holechek et al. 1994) (see Table 12.1).

Cost of Range Livestock Production

Since 1979 costs have increased at a more rapid rate than livestock prices in the United States (Figure 12.4). This has caused ranchers to focus more on their costs and ways to reduce them.

Costs associated with range livestock production can be grouped into two categories (fixed and variable costs). *Fixed costs* are expenses that do not vary with the number of livestock raised, such as insurance, land taxes, utilities, and depreciation. *Variable costs* are those expenses that vary with number of animals, such as supplemental feed, labor, grazing fees, and transportation. A major cost associated with range livestock production is supplemental feed. For most ranching operations, better efficiency in the use of supplemental feeds has the most potential to improve total net returns. In many cases supplemental feed costs could be reduced 50% or more if ranchers had a better understanding of when, what type, and what amount of supplement was needed by their livestock. Supplemental feeding practices are discussed in detail in Chapter 11.

Two major indirect costs with considerable potential for reduction include better detection of unproductive animals and reduction in death losses. Death losses of livestock for western ranches average about 2% to 5% of the herd per year. Poison plants account for much of this loss. Management practices to reduce these losses are discussed later in the chapter. Selection of livestock, common-use grazing, breeding programs, and management in drought are other aspects of range livestock production that can be manipulated for increased net returns with minimum capital outlay that we also discuss in this chapter.

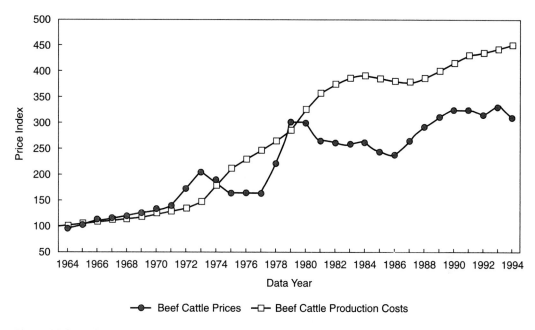

Figure 12.4 Beef cattle price index and prices paid index (production costs) for the 1964–1994 period using 1964–1968 as the base year unadjusted for inflation. (Index of prices paid by ranchers for beef production inputs as reported by USDA, National Agricultural Statistical Service reports.)

Size of Operation

Large ranches have historically been more profitable than smaller ones (Figure 12.5). Studies from New Mexico suggest that ranches smaller than 200 animal units are marginally profitable and may suffer losses as great as $60 per animal unit in some years (Torell et al. 1990; Torell and Word 1993). Low profit margins and economy of scale explain the low profitability of small compared to large ranches.

Since the 1950s, agricultural programs by the federal government have been heavily oriented toward keeping food costs low by subsidizing expansion of supply. On rangelands these programs have included government cost sharing of brush control, range seeding, grazing systems, and water development on both private and public lands. In addition, the emergency feed program (discontinued in 1996) administered by the USDA Farm Service Agency reimbursed ranchers for half of the total cost of feed purchased during droughts to replace shortfalls in native forage. These programs have all encouraged ranchers to keep rangelands stocked at maximum levels (Holechek and Hess 1995). They have lowered livestock prices and profit margins disproportionately to the increase they have caused in meat supplies (Workman et al. 1972; Knutson et al. 1990). Every 1% increase in beef supply may lower cattle prices by as much as 1.5% to 3% if other factors, such as availability of pork and chicken, remain constant.

Large ranches generally have lower average costs per animal unit of grazing capacity than small ones (economy of scale). An analysis of budgets from New Mexico ranches shows that fixed costs per animal unit are about 30% to 40% lower for medium size (250 AU) compared

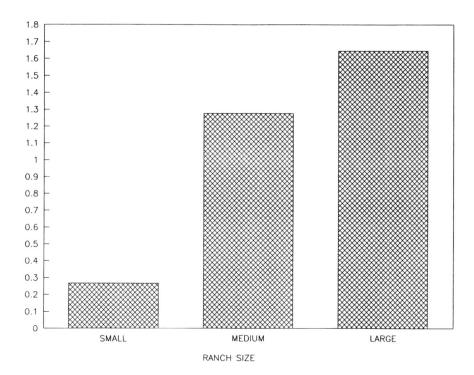

Figure **12.5** Average rate of return on investment for small (100 AU), medium (250 AU), and large (350 AU) cattle ranches in New Mexico for the 1925–1993 period. (From New Mexico Agr. Exp. Sta. Repts.)

to small (125 AU) ranches (Torell and Word 1993). Variable costs per animal unit were about 3% to 6% lower on medium compared to small ranches. When ranch size exceeds 500 animal units there is some evidence of diseconomy of scale in variable costs. In eastern New Mexico, extra-large ranches (600 AU) have 5% to 10% higher variable costs per animal unit than medium (250 AU) or large size ranches (450 AU) (Torell and Word 1993). However fixed costs and total costs were generally lower on the extra-large ranches. There did not appear to be any definite advantage in profitability of extra-large over large ranches in New Mexico.

Asset Allocation and Risk Management

Ranchers have tended to restrict their investment horizon to the ranch itself. This has been due to a lack of liquidity, lack of knowledge on alternative investments, more favorable returns from ranching in the past than present, and because ranching as a lifestyle is more important to many of them than maximization of income. However, today's ranchers are increasingly recognizing the importance of sound financial management to their survival.

Historically, stocks and bonds have given more superior returns compared to cattle ranching. Since 1900, western cattle ranches have returned about 1% to 3% on capital investment compared to 10% for stocks and 4% to 6% for bonds. In recent years the returns from stocks compared to ranches have been even more favorable than in the past (Holechek 1996a) (Table 12.2). Most western ranchers are not in the business strictly for monetary gains, but unsound financial management is one of the quickest ways for them to become ex-ranchers. Holechek et al. (1994) recommended ranchers diversify their assets, maintaining a high degree of liquidity, and keeping a major part of financial resources where they would receive the highest return. They pointed out that an important advantage of stocks and bonds is liquidity. In contrast, lack of liquidity is a disadvantage of real estate or

TABLE 12.2   Average Annual Percentage Returns for Various Types of Stock and Bond Investments in the 15-year period from 1978–1993

AVERAGE ANNUAL 15-YEAR RETURN	
CATEGORY	RETURN (%)
Stocks[1]	
S&P 500 index funds	16
Aggressive growth funds	13
Growth funds	14
Growth income funds	13
International equity funds	15
Balanced funds	13
Metals funds	13
Utilities funds	11
Bonds[1]	
Government bond funds	8
Corporate bond funds	11
High-yield (junk) funds	11
Money market funds	8
Municipal bond funds	8
Cattle Ranches[2]	
250 AU—Chihuahuan desert cattle ranch[1]	1.3% return on cattle + (−3.0% to −4.0% return on ranch purchase value)
250 AU—Shortgrass prairie cattle ranch	3.6% return on cattle + (−0.5% return on ranch purchase value)

[1]*Source:* Williamson 1995.

[2]*Sources:* Holechek 1992; Holechek and Hawkes 1993; New Mexico Agr. Exp. Sta. Reports.

investments in range improvements such as brush control, seeding, or fence for grazing systems. Historically, stocks have done best when interest rates and inflation were falling and during the pre-election and election years of the presidential cycle (Holechek 1997).

Holechek et al. (1994) suggested that ranchers invest about 25% of their discretionary assets into the ranch based on need and risk/reward ratios. Some of the options would include brush control, range seeding, specialized grazing systems, water development, herd improvement, and infrastructure repair and construction. Brush control and seeding to increase grazing capacity would make little sense if a large portion of the ranch is poorly used due to lack of water. However, it might be the best selection if forage supplies were lacking in certain seasons due to government grazing permit restrictions. It might also be appropriate if a strategic calving pasture was wanted where animals could be concentrated for better care and nutrition. Specialized grazing systems would be advantageous where distribution problems occur due to terrain and/or heterogeneity in plant communities. Ranches with limited capital resources in a desert area might choose to improve efficiency of range use and livestock productivity through better selection of livestock.

It appears that ranchers in the twenty-first century will face a more difficult and competitive environment than those of the past. Management of climatic, biological, financial, and political risks will probably be much more important to their success than their capability to increase output of livestock and other products. Some examples of these risks are given in Table 12.3. The federal government in the United States is taking a less active role in farming and ranching by discontinuing subsidies such as those for wool/mohair and emergency feed. Tomorrow's ranchers will need improved skills that permit them to accurately assess risk/reward ratios of various ranch management options as well as investment alternatives outside the ranch. Tax and estate planning will become more crucial to rancher survival. Ranchers with sophisticated financial skills will have a tremendous advantage over those who try to operate without this kind of knowledge.

TABLE 12.3   Major Risks Associated with Rangeland Livestock Production in the United States

RISK CATEGORY	EXAMPLE
Climatic risk	Drought, severe winter
Biological risk	Uncertainty in outcomes of range management practices, disease infects livestock, predation, grasshopper infestation
Financial risk	Rising interest rates, falling cattle prices, rising livestock production costs, falling land values
Political risk	Rising taxes, increased regulation, increased grazing fees on public lands, discontinuation of subsidies, increased protection for endangered species, land use restrictions
Other	Fire, theft, vandalism

Cattle Prices and Business Cycles

Cattle ranching in the western United States and many other parts of the world has been characterized by cycles of boom and bust (Holechek et al. 1994; Holechek 1996a). In the United States cattle price levels have often but not always followed the general business cycle in the country.

Since the formation of the western range livestock industry in the 1860s, there have been four basic periods of high cattle prices (Holechek et al. 1994). Each of these periods was followed by a crash in cattle and ranch values. Each period has been linked with a general economic inflation caused by a major sociopolitical event (war) that reduced supply followed by a depression or recession that occurred 7 to 10 years later due to restoration and overexpansion of supply.

Historically cattle price cycles have involved 6 to 7 up years followed by 3 to 5 down years (Holechek 1996a). The last upturn in cattle prices began in 1987 and peaked in 1993 (see Table 12.1). Nominal cattle prices increased about 50% from the bottom to the top. Between 1993 and 1996 cattle prices dropped about 35%. When adjusted for inflation cattle prices in this last cycle were the lowest since the 1930s depression.

Knowledge of business and cattle price cycles plays a critical part in successful ranching (Holechek et al. 1994). Informed ranchers will try to time their range improvements to take advantage of periods in the business cycle when interest rates are low. Conversely, after 5 or 6 years of improving cattle prices they will reduce their herds and pay down their debt levels. On the more humid ranges a strategy of making increased use of grazing capacity during the up phase of the cattle cycle followed by a period of partial destocking and range recovery during the down phase has merit for some ranchers. In the arid zones where climatic risk (drought) is much greater, keeping herd size well within grazing capacity at all times will be the best approach for most ranchers. For more detailed information on business and cattle price cycles the reader is referred to Stoken (1984), Pring (1992), Holechek et al. (1994) and Holechek (1996a).

Records and Accounts

Modern ranching is more than a job; it is a business. Therefore, it should be conducted in a businesslike manner. This means that there should be adequate records and accounts. The chief functions of ranch records and accounts are:

1. To provide information in order to analyze the ranch business. From the facts obtained, the operator may adjust current operations and develop a more effective plan of organization.

2. To provide a net worth statement, showing financial progress during the year and from year to year.

3. To furnish an accurate net income statement for use in filing tax returns.

4. To aid in making a credit statement when a loan is needed.

5. To keep production records on livestock and range units. Each individual animal must be marked so that records can be kept of each animal for each stage of its life while it is

on the ranch. Similarly, each range unit must have its own identity so that records can be kept on each pasture. These records might include weather conditions, stocking rate, and livestock production. Both livestock and plant conditions must be considered.

6. To keep a complete historical record of financial transactions and management decisions for future reference.

The reader is referred to Workman (1986) for a more detailed discussion of range economics and ranch decision making. Schools held in Albuquerque, New Mexico, and other locations by Allan Savory, Stan Parsons, and others have helped many ranchers improve the business aspects of their operations.

MANAGEMENT OF REPRODUCTIVE EFFICIENCY

Low calf or lamb crops (below 70%) characterize many ranching operations in the western United States. Calf or lamb crop refers to the percentage of breeding females in the herd that produce a salable offspring. Reasons for low calf or lamb crops are varied but commonly include lack of exposure to fertile males, failure to ovulate after parturition due to disease or low nutritional plane, loss of fetus during pregnancy due to low nutritional plane or poisonous plants, and loss of young after parturition due to predators, inadequate nutrition, poison plants, or disease.

After parturition, a cow must breed within 80 days in order to produce a calf every year (gestation period is 285 days). Sheep and goats have shorter gestation periods (145 to 155 days), so time after parturition is not as critical. If animals do not breed at the right time of the year, offspring will be dropped during unfavorable periods when forage is low in quantity and quality. An extended calving or lambing season makes it difficult for the operator to check and assist animals having difficulty with parturition. Young of varying sizes and ages bring lower prices than those that are uniform. Identification of low-producing females is difficult when young are dropped throughout the year since genetic differences between females are confounded by nutritional differences (i.e., a calf dropped in winter in the northern mixed prairie will have less opportunity for survival and will have a lower growth rate than one dropped in spring). These factors all necessitate a controlled and carefully timed breeding season.

Most range operations in the western United States are geared toward spring calving or lambing so that the mother will be on a high nutritional plane during lactation and the young calf or lamb will not be exposed to adverse climatic conditions. Some common practices to control breeding season include use of special breeding pastures, controlled estrus, and artificial insemination. On the cold desert and northern mixed prairie range types, crested wheatgrass (*Agropyron cristatum*) seedings are often used as breeding pastures in the spring because of their high productivity and nutritional value and the fact they are usually on flat terrain. This allows concentration of livestock for observation during parturition and breeding. It also allows males to detect and service females more readily. At the end of the breeding season, animals can easily be gathered and checked for pregnancy. Usually, nonpregnant animals are culled from the herd.

Confinement by use of a breeding pasture or in drylot permits use of hormones that bring all females into estrus at the same time. Artificial insemination is possible under these conditions, which reduces the expenses for bulls and permits sperm from superior bulls to be spread to more females.

The use of estrus synchronization and artificial insemination are not well suited to open range conditions during the breeding season. However, more ranch operators are finding it feasible and economically advantageous to confine their animals during a shortened (40- to 50-day) breeding season because of the improvement in calving percentages (15% to 30%) and savings in bull costs. Presently, medium-sized operations (150 to 250 AUs) in the Great Plains are making the most use of this technology.

Removal of Nonbreeding Animals

Nonbreeding animals should be removed from the herd and sold immediately after the breeding season, if possible. These animals consume forage that could otherwise be used by productive animals and place added pressure on the range. The earlier these animals are identified, the more opportunity the rancher has to market these animals when prices are most favorable and the less effect they have on variable costs (i.e., supplemental feed and veterinary supplies). New developments in pregnancy testing procedures involving rectal palpation make detection of these animals much easier and cheaper than 10 years ago.

Problems with Breeding Males

Until recent years, breeding males were selected more on the basis of their appearance than on their breeding performance under range conditions. It is generally recommended that one bull be kept for every 20 cows. For sheep (or goats) about one ram for every 25 ewes is recommended. In recent years, it has become recognized that problems relating to breeding males may affect calving or lambing percentages as much as those relating to females (Ruttle et al. 1983). Semen viability of range bulls or bucks should be checked before each breeding season and their libido (eagerness to breed) should be observed. In rugged terrain the bull's willingness to cover the pasture may have a great influence on calf crops; therefore, bulls should be checked for physical soundness before the breeding season.

Grazing Management versus Calf Crop

Calf and lamb crops on western ranges have been surprisingly low until recent years. Under good management calf crops of over 90% and lamb crops of 150% or more are possible. However, actual values for cattle have commonly been 50% to 60% and for sheep 80% to 90%. Presently calf crops on New Mexico ranges average about 75%. Stocking rate can substantially affect calf and lamb crops as well as weaning weights (Table 12.4).

TABLE 12.4 Influence of Grazing Intensity (Stocking Rate) on Cow-calf and Ewe-lamb Production

RANGE TYPE	STOCKING RATE	PERCENTAGE USE	CATTLE			REFERENCE
			CALF CROP	CALF WT. (LB)		
Chihuahuan desert (New Mexico)	Heavy	—	45	350		USDA 1936[a]
	Moderate	—	80	400		
Chihuahuan desert (New Mexico)	Heavy	50–60	75	420		Holechek 1991
	Moderate	30–35	80	450		
Sonoran desert (Arizona)	Heavy	—	55	—		USDA 1936[a]; Culley 1938
	Moderate	—	83	—		
Annual grassland (California)	Heavy	60	60	378		Bentley and Talbot 1951
	Moderate	50	78	432		
	Light	40	72	450		
Mixed prairie (Oklahoma)	Heavy	50–60	81	314		Shoop and McIlvain 1971
	Moderate	40–50	92	424		
	Light	30–40	89	437		
Mixed prairie (South Dakota)	Heavy	63	55	186		Johnson et al. 1951
	Moderate	46	60	202		
	Light	37	80	323		
Mixed prairie (Texas Rolling Plains)	Heavy	50–55	80	466		Heitschmidt et al. 1990[b]
	Moderate	40–45	83	475		
Mixed prairie (Montana)	Heavy	—	63	348		Houston & Woodward 1966
	Moderate	—	86	408		
	Light	—	92	420		
Piñon-juniper/shortgrass (New Mexico)	Heavy	60–65	91	425		Pieper et al. 1991[c]
	Moderate	40–45	93	435		
Southern pine forest (Louisiana)	Heavy	57	70	404		Pearson and Whitaker 1974
	Moderate	49	73	403		
	Light	35	82	427		

SHEEP

RANGE TYPE	STOCKING RATE	PERCENTAGE USE	LAMB CROP	LAMB WT (LB)	FLEECE WT (LB)	REFERENCE
Salt desert (Utah)	Heavy	60	88	67	9	Hutchings and Stewart 1953
	Moderate	35	79	77	10	
Chihuahuan desert (New Mexico)	Heavy	—	57	—	—	Anonymous 1938[a]
	Moderate	—	100	—	—	
Mixed prairie (Canada)	Heavy	68	53	79	—	Smoliak 1974
	Moderate	53	68	82	—	
	Light	45	73	82	—	
Annual grassland (California)	Heavy	75	75	86	6.6	Rosiere and Torell 1996
	Moderate	55	64	77	7.0	
	Light	41	78	79	6.8	
Grassland/Woodland (California)	Heavy	63	77	57	5.7	Rosiere and Torell 1996
	Moderate	49	75	68	6.2	
	Light	44	87	75	7.0	

[a]From Stoddart and Smith 1943.

[b]Cows under heavy grazing received $22.23 of supplement/year; cows under moderate grazing received $4.77 of supplement/year. The supplement was a 20% protein cube costing 9¢/lb. in 1986.

[c]Cows under both stocking rates received protein supplement in winter.

Short-duration grazing (cell) systems show much potential for improving calving percentages since they concentrate livestock. The central portion of the grazing cell associated with short-duration grazing permits better handling and observation of breeding livestock. This grazing strategy is discussed in more detail in Chapter 9.

Breeding Life of Females

Range cows usually are bred to calve at 3 years of age and are culled at around 8 to 10 years of age. Sheep lamb at 2 years of age and are usually culled at 6 years of age. Cattle can be bred to calve at 2 years of age with the right management. Early breeding of heifers extends their breeding life and reduces costs associated with keeping nonproducing animals. The major problem with early breeding of heifers is dystocia (calving difficulty). This can be minimized by breeding large yearling heifers to genetically small-bodied bulls. Yearling heifers, if bred, should receive a high nutritional plane and their calves should be weaned early. During calving they should be confined or placed in a pasture where they can be closely observed and assisted if dystocia occurs. Breeding of yearling heifers is generally not recommended for large, extensive operations where labor availability is low.

Crossbreeding

Crossbreeding involves breeding two or more types of straightbred animals with desirable characteristics to obtain hybrid vigor (heterosis). Heterosis can increase livestock productivity 10% to 15% over that of straightbred animals. The Angus-Hereford cross (black-baldy) is common in the northwestern United States, and Brahman crosses with the European breeds are common in the Southwest. The Hereford-Angus cross is considered to be hardier and more adept at using rugged terrain than are straightbreds of either breed. The Brahman-Hereford or Angus cross of the Southwest can better withstand the heat than the European breeds but has a faster growth rate than the straight Brahman. Many factors influence the advantages of crossbred animals over straightbred. Manner of crossing can have a major effect on results. Hereford sires on Angus heifers has not produced the same results as Angus sires on Hereford heifers. Some studies have shown no real advantage of crossbred over straightbred animals. Nutritional plane and environmental conditions have major influences on crossbreeding results. Local environmental conditions and the level of herd management determine the suitability of crossbreeding programs for a particular ranch.

Reproductive Disease

A major problem with flexible stocking rates is that they often involve purchase of livestock from outside the ranch, that may have disease. Some reproductive organ diseases, such as vibriosis, are not readily detectable and can quickly spread through the herd. All new animals brought into the breeding herd should be checked by a veterinarian and vaccinated for diseases such as vibriosis, Bang's disease, and anaplasmosis. Because vibriosis is difficult to detect and widely dispersed through the beef population in the United States, annual in-oculation of cows and heifers 30 to 60 days before breeding is often recommended when calving percentages are low or there is a likelihood of exposure.

ANIMAL SELECTION

Different animal species are adapted to different types of rangelands. These differences are reflected in the influence of rangeland on the animal and the effects of the animal on range-land. Some of the important factors affecting animal selection are (1) type of vegetation, (2) topography, (3) water requirements, (4) predators, (5) pests and diseases, and (6) eco-nomic and social conditions.

Topography

Different types of animals have differing abilities to utilize forages on rangelands with var-ious terrains. Domestic animals developed under differing environments. Animals can ad-just somewhat to different environments, but they can use range forage better when the topography is similar to that where they originated.

Sloping terrain is often rockier than level areas. Large animals such as cattle are poorly adapted for rocky areas. Sheep and goats are better adapted to grazing steep topography. Their smaller size and surefootedness enable them to graze on steeper terrain with less dif-ficulty than that experienced by larger animals.

Cattle and horses graze level land more easily, although rolling rangeland can be uti-lized by them (see Chapter 10). Where the range is mountainous, cattle congregate on more level areas, particularly valley bottoms, leaving the steeper portions lightly utilized. Cattle can traverse rough terrain but they are often reluctant to do so; therefore, bottoms are often heavily used. The range as a whole may have adequate forage, but the uneven terrain re-sults in uneven utilization of the forage. This can result in adverse effects on the livestock and the range. The range manager must evaluate the effects of terrain on utilization of the range and make needed adjustments. Topographic influences on livestock distribution are discussed in more detail in Chapter 10.

Climate

It is well recognized that different livestock types are adapted to different climates. European cattle (*Bos taurus*) are best adapted to cooler climates, while African or zebu cat-tle (*Bos indicus*) are suited to hot, humid climates. Sheep and goats are better adapted to desert climates than are cattle. Sheep have lower water requirements than cattle and less water turnover (Macfarlane 1972). The wool on sheep helps to insulate them from heat and panting further dissipates absorbed heat by their bodies. In humid areas cattle are generally more resistant than sheep to the associated disease and parasite problems.

Type of Vegetation

Forage preferences of grazing animals are a major consideration in selecting the type or types of animals for a particular ranch. Foraging habits are discussed in detail in Chapter 11. Cattle and horses are well suited to ranges dominated by tall, coarse grasses. Sheep are adapted to ranges with a diversity of grasses, forbs, and shrubs while goats are best suited to shrubland ranges. In certain cases, exotic animals have the potential to use plants unpalatable to native animals or livestock (see Chapter 13).

Water Requirements

Livestock vary greatly in their water requirements, watering frequency, and distance they will travel from water. Table 12.5 shows the daily water intake and frequency of watering for various types of livestock in arid areas. For best results, cattle should be watered daily but sheep and goats can be watered every 2 to 3 days. Horses may require water daily if vegetation is mature and dry but can go without water up to 3 days when vegetation is succulent. Sheep have gone without water for up to 11 days under hot conditions without adverse effects when the moisture content of the feed was high (Lynch 1974).

Salinity content of the water and forage can affect watering frequency. Sheep grazing saltbush (*Atriplex* sp.) have been observed to drink twice daily (Squires and Wilson 1970). Cattle have been observed to drink twice daily in New Mexico when the water was high in total dissolved solids (TDS) (over 4,000 mg/L TDS). Most animals can tolerate TDS levels up to 3,000 ppm TDS. Beyond this level the tolerance varies greatly between animals. Sheep and goats have a higher tolerance than cattle to salt. Young, pregnant, and lactating animals have less salt tolerance than do nonproducing animals.

When water is low in either quantity or quality, animal production can be reduced. Forage intake is well associated with water intake (Hyder et al. 1968). When water intake is reduced, animal gain drops due to reduced forage consumption. On salt desert range Hutchings (1946)

TABLE 12.5   Daily Intake and Frequency of Watering of Range Animals in the Dry Season

ANIMAL	DAILY INTAKE (GALLONS)	DAILY INTAKE (LITERS)	FREQUENCY OF DRINKING
Sheep	1.0–1.5	4–5	Once every 2 days
Goats	1.0–1.5	4–5	Once a day
Asses and donkeys	2.5–4.0	10–15	Once a day
Horses	4.0–8.0	20–30	Once or twice a day
Cattle	8.0–11.0	30–40	Once a day or once every 2 days
Camels	15.0–22.0	60–80	Once in 4 or 5 days

Source: Baudelaire 1972, from Stoddart et al. 1975.

found that sheep watered daily, every second day, and every third day gained 1.5 kg, 0.36 kg, and lost 2.7 kg, respectively. In Kenya, Musimba et al. (1987) found that zebu cattle watered every other day and every third day had forage intake reductions of 48% and 50%, respectively, compared to cattle watered every day during the dry season. These differences, however, were greatly reduced in the wet season. In his study, watering frequency had small influence on weight changes over the entire study period. However, cattle during the dry season lost weight under the intermittent water regime, while those watered daily maintained weight. On northeastern Oregon mountain range, daily weight gains in late spring were 0.5 kg per day for cattle with a good supply of clean water, compared to -0.1 kg for cattle watering from a small pond contaminated by urine and feces (Holechek 1980). Forage intake by the animals with contaminated water was severely reduced compared to those with clean water. These studies all point out the critical importance of a good supply of clean water to range livestock production.

A final point concerns temperature. As temperature increases, livestock water requirements increase. Above 30°C (85°F), livestock water requirements increase sharply. This is important to recognize when planning water provision for livestock in desert areas.

Predators

Keeping predators within tolerable limits permits herbivores (both domestic and wildlife) to utilize fully areas that might otherwise be denied to them. Predators, particularly coyotes (discussed in Chapter 13), make the raising of sheep and/or goats infeasible in some locations. Without the disrupting influence of predators, livestock can be manipulated more effectively. Sheep can be open-herded or grazed under fence without herding and cattle can be fully distributed over the range without fear of intolerable losses.

Control of predator numbers can also be important to grazing wild animals. When wildlife populations are low, predators may be a depressive factor and thus contribute to less-than-proper forage use. On the other hand, overkill of predators may encourage population explosions among big game rabbit and rodent species to the detriment of the forage.

Extreme positions on predator control can be detrimental to proper forage use by herbivores. Total protection and eradication or near-eradication of predators may create excessive impacts on the rangeland ecosystems. One of the most challenging rangeland problems requiring research involves properly applied predator control. Chapter 13 provides a more complete discussion of predator control and impacts on range livestock production.

Pests and Diseases

In some circumstances, the presence of pests can have a major influence on livestock selection. Biting insects make it difficult to raise cattle in certain parts of Africa. The tsetse fly also renders large areas of Africa unsuitable for cattle (see Chapter 13). In these areas, game ranching may be more practical than ranching for domestic livestock. On high mountain meadow ranges in the United States, blood sucking insects heavily stress cattle. These areas can be better used by sheep because of the protection provided by their wool. Some diseases associated with particular areas affect one type of livestock but not another.

Economic and Social Considerations

Many parts of the western United States are better suited to sheep than cattle. However, cattle are raised if the operator prefers cattle over sheep. Cattle are associated with the cowboy persona and the western way of life. Many ranchers are willing to forgo larger profits possible with sheep to maintain the cowboy lifestyle associated with cattle.

Cattle do have some advantages over sheep assuming that the ranch is equally suited to both animals. Primarily, these involve lower labor requirements and reduced predator problems. Sheep have the advantage over cattle of producing two crops (wool and lamb). The price for lambs has generally exceeded that for calves. Goats have yielded higher economic returns than either cattle or sheep singly in central Texas, where oaks dominate the vegetation. However, in this range type, a combination of cattle and wildlife will presently yield the highest economic returns (see Chapter 13). Common-use grazing is discussed as follows.

COMMON-USE GRAZING

Rangelands produce forage for more than one type of animal—usually cattle, sheep, goats, and wildlife. Common-use grazing—the use of rangelands by more than one type of animal—may substantially increase production from rangelands. Such use often promotes better distribution of animals, harvesting of more of the available plant species, and more uniform use of rangelands (Smith 1965).

Suitability of Vegetation

The mix of plants and animals available is important in the utilization of a range ecosystem. The similarities and differences in diets of the various species of grazing animals are important in the use of the plant materials in the system. Forage selection by grazing animals is influenced by the interaction of (1) forage quality and availability, (2) animal-prehensile-grazing ability (teeth, lips, and mouth structure), (3) topography, (4) animal agility, (5) chemical and physical plant properties, and (6) animal competition (Baker and Jones 1985).

Management practices of controlling grazing pressure and mixing animal species influence diet selection, vegetation changes, and animal foraging. When range forage availability is low, as during drought, there are, with some exceptions, more similarities in the diets of various animal species. Permitting animals to select the most nutritious vegetation is vital to successful production systems. Knowledge of each range unit, characteristics of the animal and plant species, and weather are essential to determine the most desirable mixture of grazing animals in any given ranch situation (Baker and Jones 1985).

On high-elevation summer rangeland in southwestern Utah, sheep removed less grass and more forbs and shrubs than were removed by cattle (Ruyle and Bowns 1985). Cattle were reluctant to browse mountain snowberry (*Symphoricarpos oreophilus*) even when herbaceous plants were greatly reduced (Figure 12.6). Cattle and sheep in the same range unit used a greater variety of forage than did either species in separate units.

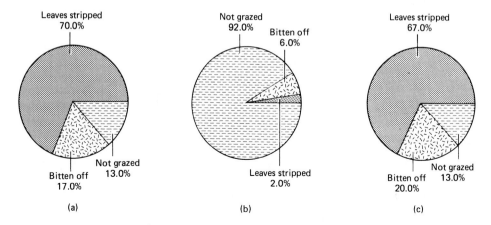

Figure 12.6 Mountain snowberry (*Symphoricarpos oreophilus*) use by (a) sheep, (b) cattle, and (c) mixed species, depicting the browsing methods. Sheep stripped the leaves from the stems, whereas cattle took entire stems. Common-use grazing resulted in most of the browsed stems stripped of leaves, similar to those in sheep treatments. (From Ruyle and Bowns 1985.)

Economics

Common-use grazing is a desirable range management and land conservation practice (Ospina 1985). Economically, several factors to consider are (1) producers can improve cash flow by combining animal species; (2) risk is reduced when production is diversified; and (3) some undesirable plants are reduced by sheep and goats. Conversely, when more than one animal species is grazed, additional costs are realized and management practices become more complex. The reader is referred to Baker and Jones (1985) and Heady and Child (1994) for more detailed information on common-use grazing.

LIVESTOCK MANAGEMENT DURING DROUGHT

Drought is a fact of life for stockmen using most rangelands. As aridity increases, severity and frequency of drought tend to increase. Climatic records show ranchers can expect drought conditions to prevail about 3 years out of every 10 years on most western rangelands. Learning to live with drought is the major challenge confronting ranchers in most of the western United States, Australia, and many parts of Africa. This challenge is best met by advanced planning. This planning process is shown in Figure 12.7.

Reducing livestock in accordance with forage availability rather than holding livestock and providing them harvested feeds has generally been the best drought strategy financially and biologically because harvested feed costs have increased and cattle prices have declined as drought increased in severity (Boykin et al. 1962; Holechek 1996b). The more ranchers invest in purchased feed and the lower livestock prices become, the more reluctance there is to sell livestock. This has often resulted in severe rangeland degradation and

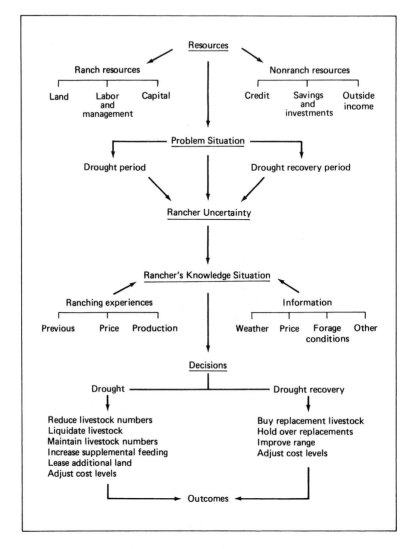

Figure **12.7** Drought decision theory. (Reprinted by permission from *Ranch Economics* by James R. Gray, © 1968 by Iowa State University Press, 2121 South State Avenue, Ames, Iowa 50010.)

in some cases death of livestock when the rancher ran out of money to purchase additional feed but did not want to sell livestock at giveaway prices. Ranchers who maintain livestock on harvested feed should confine them to avoid damage to the range.

Some ranchers have held livestock on pastures where forage is depleted, without supplementation, in hopes of rainfall. They need to keep in mind that once animals lose 15% to 25% of normal body weight their recovery will be slow and costly (Young and Scrimshaw 1971). Animals losing 30% or more of normal body weight will nearly always die (de Calesta et al. 1975). Ranchers who allow their livestock to get into poor condition may find it difficult to sell them at any price. Excessive weight loss by livestock should be avoided by either selling them or providing them with maintenance feed.

Forage Production and Grazing Management

Good grazing management is the key to rancher survival during and after drought. The fundamentals of grazing management are discussed in Chapters 8 and 9. A point worth reiterating is that the basic cow herd should be based on forage production in years of average rainfall. Forage not used in above-average rainfall years provides carryover feed for use in years when rainfall is below average. Plants that are healthy and in high vigor produce more feed in drought years than do those in low vigor. Ranches conservatively grazed during drought recover much more quickly after drought than do those receiving heavy use.

Many inexperienced ranchers have underestimated the degree to which drought can reduce forage production. Drought can reduce forage production by more than 50% compared to the annual average (Table 12.6). Pastures heavily grazed show greater reductions than those lightly or moderately grazed (Table 12.7).

One common thread that binds the various drought management papers together is the advocacy of conservative stocking before, during, and after drought (Lantow and Flory 1940; Reynolds 1954; Klipple and Costello 1960; Boykin et al. 1962; Paulsen and Ares 1962). From both vegetation and financial standpoints, this appears to be the key to drought survival. Boykin et al. (1962) evaluated survivors of the 1950s drought in the southern Great Plains from the standpoint of their ranch management practices. The four ranchers studied firmly believed that conservative stocking was the critical element in their survival.

Conservative stocking involves 30% to 40% use of the current year standing crop of the primary forage species (Klipple and Costello 1960; Paulsen and Ares 1962). Forage plants on conservatively or lightly stocked ranges actually seem to do better during and

TABLE 12.6 Herbage Production (kg/ha) on Heavily, Moderately, and Lightly Grazed Shortgrass Prairie in Colorado Compared to the 5-year Average

GRAZING INTENSITY	DROUGHT YEAR	5-YEAR AVERAGE	DROUGHT YEAR AS PERCENT OF AVERAGE
Heavy			
54% use	281	536	52
Moderate			
37% use	520	690	75
Light			
21% use	549	736	75

Source: Klipple and Costello 1960.

TABLE 12.7 Fluctuation in Annual Forage Production in Response to Weather Changes for Various Ranges under Moderate Grazing

RANGE TYPE	LOCATION	VARIATION (KG/HA)	YEARS	REFERENCE
Northern mixed prairie	Alberta, Canada	140–773	1951–1969	Smoliak 1974
	South Dakota	1,390–2,700	1952–1955	Lewis et al. 1956
	Montana	250–1,780	1927–1934	Campbell 1936
	Alberta, Canada	96–925	1930–1983	Smoliak 1986
Shortgrass	Nebraska	1,033–1,866	1958–1967	Burzlaff and Harris 1969
	Kansas	150–2,815	1940–1942	Weaver and Albertson 1944
	Colorado	520–879	1949–1953	Klipple and Costello 1960
Salt desert	Utah	90–505	1935–1947	Hutchings and Stewart 1953
Desert grassland	New Mexico	133–357	1939–1953	Paulsen and Ares 1962
	New Mexico	5–1,509	1959–1977	Herbel and Gibbens 1996
	Arizona	222–858	1954–1964	Martin 1975
Annual grassland	California	1,492–2,580	1912–1948	Bentley and Talbot 1951
	California	2,561–7,081	1979–1983	Rosiere 1987
Southern pine forest	Louisiana	912–2,675	1961–1972	Pearson and Whitaker 1974
Coniferous forest-bunchgrass	Colorado	370–500	1942–1957	Smith 1967

after drought than those on areas with no grazing (Paulsen and Ares 1962; Ganskopp and Bedell 1981). In drought, residue or stubble height may be a more appropriate criterion than utilization standards if new growth is minimal. Grazing should be discontinued if average plant heights fall below 1 1/2 to 2 inches on shortgrasses, such as blue grama, or 4 to 6 inches on midgrasses such as sand dropseed. These same height guidelines apply to initiation of grazing on new growth after rainfall has occurred.

Livestock Confinement

Confinement of cows during drought can relieve stress on plants and improve livestock production during drought. However if confinement lasts more than a few months, harvested feed costs can become excessive relative to value of the livestock (Boykin et al. 1962; Holechek 1996b). Herbel et al. (1984) found that calving percentages were higher for part-year confinement (spring) than with yearlong placement on rangeland (86% versus 72%) during a 5-year study on southern New Mexico range. Production per cow was higher under part-year confinement than with yearlong placement on rangeland (134 kg versus 110 kg). Their study showed that during some drought years confinement of cows in spring and early weaning of calves in late May could be economically effective. The higher the price of calves and the lower the cost of feed, the more feasible partial confinement becomes. Confinement during drought does increase labor costs but reduces vehicle costs.

Cactus as Feed in Drought

In hot, arid climates in many parts of the world, cactus (*Opuntia* sp.) comprises a large part of the vegetation and can be an important animal fodder. Cactus has four-to fivefold greater efficiency in converting water to dry matter than that of most grasses (Kluge and Ting 1978). Nutritionally, cactus is low in protein but rich in digestible carbohydrates, water, and vitamins (Shoop et al. 1977). In many parts of the world, including the southwestern United States, Engelmann prickly pear (*Opuntia engelmannii*) and cholla (*Opuntia imbricata*) cactus are used as emergency drought food for cattle (Figure 12.8). Various propane torches known as "pearburners" have been used to singe the spines off cactus so that cattle can eat the pads. Based on recent figures, both from south Texas, it costs about $0.35 per animal unit per day to maintain animals on prickly pear compared to $0.78 if relief corn (government aid) is used or $1.09 for alfalfa (Russell 1985). If cattle must be carried for an extended period, an additional $0.24 worth of protein supplement (cottonseed) is required when cactus is fed.

Ranchers in Texas are now planting prickly pear in rows to increase the efficiency of burning spines (Russell 1985). Some ranchers are so pleased with the performance of their cattle on cactus that they are using it in normal years. Although spineless varieties of prickly pears are available, they are subject to considerable herbivory by wildlife.

One study is available from Colorado that evaluated cattle performance when they were fed plains prickly pear (*Opuntia polyacantha*). Singed prickly pear increased dry matter consumption 43% and weight gain 72% when added to a basal ration formulated to

Figure **12.8** Cholla cactus makes an excellent cattle feed during winter and in drought in New Mexico when the spines are burned off.

TABLE 12.8 Comparative Protein and Energy Concentrations of Singed Prickly Pear, Grass Hay Pellets, and Alfalfa Hay Fed to Cattle in Winter in Colorado

CONSTITUENT	PRICKLY PEAR	HAY PELLETS	ALFALFA HAY
Crude protein, (%)	5.3	5.7	16.8
Digestible protein, (%)	3.4	3.0	11.4
Gross energy, (Mcal/kg)	4.75	4.01	4.87
Digestible energy, (Mcal/kg)	2.61	2.08	2.64

Source: Shoop et al. 1977.

represent the best possible winter feed on shortgrass range (Tables 12.8 and 12.9). During 84 days of high prickly pear consumption, cattle suffered no digestive problems. The cattle actually seemed to prefer the prickly pear over the hay pellets, as it was usually consumed first. Total daily dry-matter intake was 2.53 kg of prickly pear plus 5.86 kg of basal rotation, for a total daily dry-matter intake of 3.0% body weight. This level of intake is greater than that for a high-quality alfalfa hay.

Drought Management Principles

Advance planning is the key to drought survival. Following is a summary of important principles and practices in dealing with drought (Hanselka and White 1986):

1. Adequate forage must be left to provide a reserve for the inevitable periods of drought.
2. Healthy vigorous perennial grasses with a good root system can maintain production longer into a drought and recover more quickly once rainfall occurs.
3. Light rains are more effective if some litter and plant residues remain.

TABLE 12.9   Average Weight Gain and Feed Intake of Heifers Fed a Singed Prickly Pear Ration and Heifers Fed the Basal Diet

ITEM	BASAL RATION PLUS PRICKLY PEAR	BASAL RATION
Heifers (numbers)	6	6
Initial weight 9/2/74 (kg)	254	242
Final weight 11/25/75 (kg)	310	275
Daily feed intake (kg)		
Basal ration		
Hay pellets	5.15	4.98
Cottonseed meal (41% crude protein)	0.30	0.30
Crested wheatgrass hay	0.41	0.41
Subtotal	5.86	5.69
Prickly pear	2.53	—
Total	8.39	5.69
Daily gain (kg)	0.67	0.39
Feed intake efficiency (kg/kg gain)	12.5	14.6

Source: Shoop et al. 1977.

4. There will be a few preferred plants by each class of livestock that can serve as a barometer of use and health of a range for determining when management adjustments are needed.

5. Realizing that drought is inevitable, a drought plan should be developed. Flexibility in forage use, livestock numbers, livestock classes, marketing strategies, and so on, result in better management decisions.

6. Maintain a percentage of the livestock herd as a readily marketable class of stock.

7. Distribute livestock to use existing forage uniformly.

8. Store silage, hay, and other feeds while plentiful and inexpensive.

9. Encourage prickly pear growth in certain pasture areas and configurations to allow ease of "burning" at an economical price.

10. Seed abandoned fields and barren range areas to adapt forage plants.

11. During a recognized drought

    a. Top priority should be on range recovery once the drought breaks.

    b. When reducing the herd gradually, cull low producers and older animals with the least reproductive potential.

c. Curtail replacement animal development.

d. Sell light offspring earlier than normal.

e. Determine the amount of money that can be spent on animal feed purchases.

f. Utilize a drylot for all—concentrate feeding to maintain livestock. This reduces needless energy expenditure in the search for food and allows forage species to utilize light rains fully. Plants can then be grazed after sufficient growth or dormancy occurs.

POISONOUS PLANT PROBLEMS

The average annual range livestock death loss from poisonous plants is estimated at 2% to 5% (Nielsen 1978; James et al. 1992). Besides death losses, other direct costs include weight loss, lengthened calving intervals, abortions, and other reproductive problems. Indirect costs to the livestock industry are additional fencing, herding, supplemental feeding, medical needs, and loss of forage.

Symptoms of poisoning in livestock can be confused with those caused by disease or other toxic agents. Some plants may be grazed without any trouble most of the time but may be extremely toxic at certain times. Poisoning may occur in lush pastures at one time and under conditions of drought at another. Several species of plants may have the same poisonous ingredient and cause similar symptoms. One single species may be poisonous in several ways, so that the symptoms vary, depending on which ingredient is most active at a particular time.

Livestock losses from poisonous plants are often related to poor management, poor range condition, or type of animals. Animals most commonly graze poisonous plants on poor-condition range that lacks palatable forage (Figure 12.9). Most poisonous species are unpalatable but are eaten because an animal is hungry and the plants are readily available. Some, however, are relished by certain animals and may be taken in preference to other forage. Some poisonous plants are succulent when desirable species are dormant. Animals new to an area may be poisoned in a pasture already being grazed safely by the same type of livestock that are adjusted to the area. Small amounts of one plant may be lethal shortly after consumption, while the toxic substance from other plants is cumulative, and those plants must be grazed for some time before symptoms of poisoning appear. Even a sudden change in the weather may cause a species that is being grazed safely to become toxic.

Toxic Agents in Plants

Alkaloids. Of the many toxic substances in plants, alkaloids are the most powerful. They are generally distributed throughout the plant and commonly remain after a plant matures or freezes. Poisoned animals show symptoms of nervous disorders, such as trembling muscles, intoxication, salivation, bloating, and difficulty in breathing. Poisoning is often acute,

***Figure* 12.9** Heavy grazing forces animals to use plants otherwise avoided and increases death losses. Annual death losses on ranches in the western United States from poisonous plants average 2% to 5%.

but if a lethal amount is not consumed, the animal usually recovers completely in a short time. There are no antidotes for most alkaloids, so prevention of poisoning and sick care are the only remedial measures.

Plants that contain alkaloids include western false hellebore (*Veratrum californicum*), lupines (*Lupinus* spp.), locoweeds (*Astragalus* spp.), larkspurs (*Delphinium* spp.), water hemlock (*Cicuta* spp.), poison hemlock (*Conium* spp.), tansy ragwort (*Senecio jacobaea*), threadleaf and Riddell's groundsels (*S. longilobus* and *S. riddellii*), sneezeweed (*Helenium hoopesii*), bitterweed (*Hymenoxys odorata*), pingue (*Hymenoxys richardsonii*), broomweed (*Gutierrezia microcephala*), nightshades (*Solanum* spp.), and ergot (*Claviceps* spp.). *Gutierrezia microcephala* also contains saponins, another category of toxic compounds (Ralphs et al. 1991).

Locoweeds and larkspurs occur over large areas in the western United States. Livestock death losses from ingesting these plants have been well over 5% on some ranges. Ralphs (1995) found that herbicides (picloram) could be effective in long-term control of tall larkspur (*Delphinium barbeyi*) in the subalpine zone of Utah. Herbicidal management of larkspur was economically effective when it occurred in thick stands (Nielsen et al. 1994). Manipulation of the ruminal environment using carbachol (carbamylcholine chloride), a saliva stimulant, and salt-mineral supplementation did not reduce cattle susceptibility to larkspur toxicosis in Colorado (Pfister and Manners 1995). However animals can be trained to avoid plants (larkspur and loco) that contain alkaloids (Provenza 1995, 1996).

Locoweed poisoning has been a serious problem on shortgrass ranges in Colorado and New Mexico. Locoweed poisoning occurs mostly in the spring before the growth of warm-season grasses (Ralphs et al. 1993). Pastures free of locoweed or with light infestations should be reserved for spring grazing. Locoweeds can be effectively controlled with herbicides (Ralphs and Ueckert 1988).

Glycosides. A second group of substances found in plants that cause poisoning are known as glycosides. These compounds are normally not toxic, but under special conditions, they are broken down and toxic substances are formed. A common glycoside found in plants under range conditions is one that yields hydrocyanic (prussic) acid on hydrolysis. Natural events such as freezing, wilting, or crushing may cause the release of prussic acid (HCN) within plants, but enzyme activity by the rumen microflora can also release the HCN in the paunches of cattle, sheep, and goats. Death results within a few minutes after a toxic amount of HCN is absorbed. Water causes an increased release of HCN, so animals may be found dead near water. Because HCN blocks the release of oxygen from the blood, the venous blood of poisoned animals is bright red for some time after death. Signs of toxication include nervousness, abnormal respiration, trembling, blue coloration of the lining of the mouth, and spasm or convulsion ending in death (Majak 1992).

Nitrates. High levels of nitrates in plants frequently cause livestock poisoning. Common weedy plants such as pigweeds (*Amaranthus* spp.), Russianthistle (*Salsola kali*), and lambsquarter (*Chenopodium album*) are most frequently involved, but many other plants also cause nitrate poisoning. Plants growing on nitrogen-rich soils or under drought conditions and lowered light tend to have a high nitrate content. Also, plants treated with herbicides such as 2,4-D may accumulate excess nitrate and at the same time become more palatable, thus increasing the possibility of poisoning. Plants containing more than 1.5% nitrate can kill livestock and death occurs rapidly after symptoms appear. In nitrate poisoning the blood is unable to carry oxygen, so the ultimate cause of death is asphyxiation, as in HCN poisoning. In nitrate poisoning, however, the blood is a dark, chocolate color. Also, the mucous membranes have a bluish tinge, and the whites of the eyes are brownish.

Some plant species, such as johnsongrass, may cause HCN or nitrate poisoning, depending on conditions. Diagnosis of the poisonous principle is critical for treatment, because the treatments used for HCN poisoning are lethal to animals suffering from nitrates.

Oxalates. Oxalates are found in plants in soluble and insoluble forms. Soluble oxalates are toxic to livestock; insoluble oxalates are not. There are two principal oxalate-producing plants in North America. They are halogeton (*Halogeton glomeratus*) and greasewood (*Sarcobatus vermiculatus*). These plants are readily grazed by cattle and sheep. They have caused a few deaths in cattle but many deaths of sheep. Halogeton poisoning in sheep is characterized by depression, weakness, and salivation. There is increased heart rate, and finally, coma and death.

Photosensitization. Some plants contain photosensitizing substances that can cause a swelling of the head and ears and sloughing of the skin in light-colored animals exposed to sunlight. These substances poison the liver and prevent the breakdown of certain pigmented materials during digestion. The pigmented substances are absorbed and circulated in the peripheral circulatory system and, when exposed to sunlight, irritate the skin, especially in light-colored animals. Range plants causing photosensitization include buckwheat (*Polygonum fagopyrum*), St. Johnswort (*Hypericum perforatum*), sacachuista (*Nolina* *microcarpa*), lechuguilla (*Agave lecheguilla*), goathead (*Tribulus terrestris*), and horsebrushes (*Tetradymia glabrata* and *T. canescens*). The horsebrushes may cause photosensitization only in sheep when they have browsed black sagebrush (*Artemisia nova*) (James and Johnson 1976).

Miscellaneous Poisonous Plants

Ponderosa pine (*Pinus ponderosa*), junipers (*Juniperus* spp.), and broom snakeweed (*Gutierrezia sarothrae*) cause abortions in cattle when they are grazed during pregnancy. Tannic acid is the toxic substance in the developing buds and leaves of Gambel and shinner oaks (*Quercus gambelii* and *Q. havardii*). An alcohol in burroweed (*Haplopappus tenuisectus*) and rayless goldenrod (*Haplopappus heterophyllus*) produces "trembles" when these species are grazed for 2 or 3 weeks. Resins contained in milkweeds (*Asclepias* spp.) and spurges (*Euphorbia* spp.) affect muscular and nervous tissues in livestock.

Plants growing on soils with more than 2 ppm of selenium may accumulate toxic levels of this element. Some selenium-accumulating plants are locoweeds, asters (*Aster* spp.), and saltbushes (*Atriplex* spp.).

Tansy mustard (*Descurainia pinnata*) and mesquite (*Prosopis juliflora*) cause paralyzed tongue or "wooden tongue" when large amounts are eaten. Inkweed (*Drymaria pachyphylla*), desert marigold (*Baileya multiradiata*), and annual goldeneye (*Viguiera annua*) cause heavy losses in some years. The reader is referred to Kingsbury (1964), Keeler et al. (1978), and James et al. (1988) for more detailed information on poisonous plants effects on range livestock.

Preventing Livestock Poisoning

Sound animal and grazing management is the best prevention of livestock poisoning from range plants. Few antidotes to plant toxins are available, but the land manager can prevent most losses with the following practices (Stoddart et al. 1975):

1. Animals must have an adequate supply of good-quality range forage. Most poisonous plants are relatively unpalatable, so animals seldom consume enough to cause toxicity, except when they are hungry. Range livestock may need supplemental feed in the early spring or other times when the quality of range forage is poor to prevent consumption of evergreen or early growing poisonous plants (Taylor and Ralphs 1992).

2. Livestock must have an adequate supply of salt and other minerals. Animals on deprived diets may eat poisonous plants with high salt or high phosphorus content.

3. Animals should be fed before they are placed on a range unit with many poisonous plants. This applies to animals being trailed, after hauling, or being worked. A land manager should avoid trailing hungry animals through, or holding animals on, dense stands of poisonous plants.

4. Livestock should not be moved to range units with an abundance of poisonous plants. Animals may be poisoned when grazing unfamiliar range, even though animals accustomed to the area are not poisoned.

5. A land manager should know when poisonous plants are growing on a range unit. It is important for the manager to know which parts of the plant are poisonous, when they are most toxic, conditions for livestock toxicity, the type of livestock affected, symptoms of poisoning, treatment for poisoned animals, and methods of plant control.

6. Sometimes animals must be removed from range units that are treated with herbicides to control the poisonous plants. Some plants become more palatable or poisonous after treatment with herbicides.

7. It is important to place livestock on a given range unit during the proper season, when the desirable plants are at the proper stage of growth, and to avoid poisonous plants when they are most toxic. Fencing may be necessary to keep livestock off areas infested with poisonous plants during critical periods, or the plants may need to be eradicated from some range units.

8. If possible, prevent the introduction of noxious plants. Before the introduction of new animals into a ranch or allotment, hold them in a pen for about 48 hours to permit foreign seeds to pass through the animal.

9. Absorption of toxic substances may be modified by feeding binding agents such as clay, resins, and indigestible fibers, or by pharmaceuticals (Smith 1992).

10. Some animals can be trained to avoid certain plants (Provenza et al. 1992).

RANGE MANAGEMENT PRINCIPLES

■ Livestock production is generally more profitable in the humid compared to the arid range types. This is because of lower infrastructure requirements and longer periods when forage is actively growing in the more humid types. This means ranchers must focus more on passive (low cost) and less on active (high cost) management practices in the arid areas.

■ Livestock problems relating to drought, poisonous plants, and poor reproductive performance are greatly minimized by proper stocking of the range.

■ Periodic drought is one of the biggest challenges confronting ranchers on arid and semiarid rangelands. Advance planning is the key to drought survival. In semiarid and arid areas, conservative stocking is one of the best approaches to minimizing biological and financial risks associated with drought.

■ Success in modern ranching depends heavily on managing various risks. These risks can be divided into climatic, biological, financial, and political categories. Ranch management in the twenty-first century will probably be more oriented toward managing risk than expansion of supply.

Literature Cited

Anonymous. 1938. Ganado area lambs again set record. *Southwest News* 1:6.

Baker, F. H., and K. R. Jones (Eds.). 1985. *Proc. Conference on Multispecies Grazing.* Winrock International Institute, Morrilton, AR.

Baudelaire, J. P. 1972. Water for livestock in arid zones. *World Anim. Rev.* 3:1–9.

Bentley, J. R., and M. W. Talbot. 1951. Efficient use of annual plants on cattle ranges in the California foothills. *U.S. Dep. Agric. Circ.* 870.

Boykin, C. C., J. R. Gray, and D. P. Caton. 1962. Ranch production adjustments to drought in eastern New Mexico. *New Mexico Agr. Expt. Sta. Bull.* 470.

Burzlaff, D. F., and L. Harris. 1969. Yearling steer gains and vegetation changes of western Nebraska rangeland under three rates of stocking. *Nebr. Agric. Exp. Stn. Bull.* 505.

Calleo, D. P. 1992. The bankrupting of America: How the federal budget is impoverishing the nation. William Morrow and Company, Inc., New York.

Campbell, R. S. 1936. Climatic fluctuations. In U.S. Department of Agriculture Forest Service (Ed.), *The western range.* U.S. Congress 74th 2nd Session, Senate Doc. 199.

Cattle Fax. 1997. *Beef cattle numbers in the United States.* Cattle Fax 24:(3)1.

Culley, M. J. 1938. An economic study of cattle business on a southwestern semidesert range. *U.S. Dept. Agr. Circ.* 448.

Davidson, J. D., and W. Rees-Mogg. 1993. *The great reckoning.* Simon & Schuster, New York.

de Calesta, D. S., J. G. Nagy, and J. A. Bailey. 1975. Starving and refeeding mule deer. *J. Wildl. Manage.* 39:663–669.

Fowler, J. M., and L. A. Torell. 1985. *The financial position of the New Mexico range livestock industry, 1940–1984.* New Mexico State University Range Improvement Task Force, Dept. 20, Las Cruces, NM.

Ganskopp, D. C., and T. Bedell. 1981. An assessment of vigor and production of range grasses following drought. *J. Range. Manage.* 34:137–141.

Godfrey, E. B., and C. A. Pope, III. 1990. The case for removing livestock from public lands. In *Current issues in rangeland resource economics.* Oregon State Univ. Ext. Serv. Spec. Rep. 852.

Gray, J. R. 1968. *Ranch economics.* Iowa State University Press, Ames, IA.

Hanselka, C. W., and L. White. 1986. Rangeland in dry years: Drought effects on range, cattle, and management. In R. D. Brown (Ed.), *Livestock and wildlife management during drought.* Ceasar Kleberg Wildlife Research Institute, Texas A&M University, Kingsville, TX.

Heady, H. F., and R. D. Child. 1994. *Rangeland ecology and management.* Westview Press, San Francisco, CA.

Herbel, C. H., and R. P. Gibbens. 1996. Post-drought vegetation dynamics on arid rangelands of southern New Mexico. *New Mexico Agr. Exp. Sta. Bul.* 776.

Herbel, C. H., J. D. Wallace, M. D. Finkner, and C. C. Yarbrough. 1984. Early weaning and part-year confinement of cattle on arid rangelands of the Southwest. *J. Range Manage.* 37:127–130.

Heitschmidt, R. K., J. R. Conner, S. K. Canon, W. E. Pinchak, J. W. Walker, and S. L. Dowhower. 1990. Cow/calf production and economic returns from yearlong continuous, deferred rotation and rotational grazing treatments. *J. Prod. Agric.* 3:92–99.

Holechek, J. L. 1980. *The effects of vegetation type and grazing system on the performance, diet, and intake of yearling cattle.* Ph.D. Thesis. Oregon State University, Corvallis, OR.

Holechek, J. L. 1991. Chihuahuan desert rangeland, livestock grazing, and sustainability. *Rangelands* 13:115–120.

Holechek, J. L. 1992. Financial aspects of cattle production in the Chihuahuan desert. *Rangelands* 14:145–149.

Holechek, J. L. 1996a. Drought and low cattle prices: Hardship for New Mexico ranchers. *Rangelands* 18:11–13.

Holechek, J. L. 1996b. Drought in New Mexico: Prospects and management. *Rangelands* 18:225–227.

Holechek, J. L. 1997. The stock market: What ranchers should know. *Rangelands* 19(3):14–17.

Holechek, J. L., and J. Hawkes. 1993. Desert and prairie ranching profitability. *Rangelands* 15:104–109.

Holechek, J. L., J. Hawkes, and T. Darden. 1994. Macroeconomics and cattle ranching. *Rangelands* 16:118–123.

Holechek, J. L., and K. Hess, Jr. 1995. The emergency feed program. *Rangelands* 17:133–136.

Holechek, J. L., and K. Hess Jr. 1996. Grazing lands: Prices, value, and the future. *Rangelands* 18:102–105.

Houston, W. R., and R. R. Woodward. 1966. Effects of stocking rates on range vegetation and beef cattle production in the Northern Great Plains. *U.S. Dept. Agr. Tech. Bull.* 1357.

Hutchings, S. S. 1946. Drive the water to the sheep. *Natl. Wool Grow.* 36:10–11.

Hutchings, S. S., and G. Stewart. 1953. Increasing forage yields and sheep production on intermountain ranges. *U.S. Dep. Agric. Circ.* 925.

Hyder, D. N., R. E. Bement, and J. J. Norris. 1968. Sampling requirements of the water-intake method of estimating forage intake by grazing cattle. *J. Range Manage.* 21:392–397.

James, L. F., and A. E. Johnson. 1976. Some major plant toxicities of the western United States. *J. Range Manage.* 29:356–363.

James, L. F., D. B. Nielsen, and K. E. Panter. 1992. Impact of poisonous plants on the livestock industry. *J. Range Manage.* 45:3–8.

James, L. F., M. H. Ralphs, and D. B. Nielsen (Eds.). 1988. *The ecology and economic impact of poisonous plants on livestock production.* Westview Press, Inc., Boulder, CO.

Johnson, L., L. R. Albeer, R. O. Smith, and A. L. Moxon. 1951. Cows, calves, and grass: Effects of grazing intensities on beef, cow, and calf production on mixed prairie vegetation on western South Dakota ranges. *S. Dak. Agric. Exp. Stn. Bull.* 412.

Keeler, R. F., K. R. Van Kampen, and L. F. James (Eds.). 1978. *Effects of poisonous plants on livestock.* Academic Press, Inc., New York.

Kingsbury, J. M. 1964. *Poisonous plants of the United States and Canada.* Prentice-Hall, Inc., Englewood Cliffs, NJ.

Klipple, G. E., and D. F. Costello. 1960. Vegetation and cattle responses to different intensities of grazing on shortgrass ranges on the central Great Plains. *U.S. Dep. Agric. Tech. Bull.* 1216.

Kluge, M., and I. P. Ting. 1978. *Crassulacean acid metabolism: Analysis of an ecological adaptation.* Springer-Verlag Inc., New York.

Knutson, R. D., J. B. Penn, and W. T. Boehm. 1990. *Agricultural and food policy.* 2d ed. Prentice-Hall, Englewood Cliffs, NJ.

Lantow, J. L., and E. L. Flory. 1940. Fluctuating forage production: Its significance in proper range and livestock management on southwestern ranges. *Soil Conservation* 6:1–8.

Lewis, J. K., G. M. Van Dyne, L. R. Albee, and F. W. Whetzall. 1956. Intensity of grazing: Its effect on livestock and forage production. *S. Dak. Agric. Exp. Stn. Bull.* 459.

Lynch, J. J. 1974. Behavior of sheep and cattle grazing in the more arid zones of Australia, pp. 37–49. In A. D. Wilson (Ed.), *Studies of the Australian arid zone. 2. Animal production.* CSIRO, Melbourne, Australia.

Macfarlane, W. V. 1972. *Prospects for new animal industries: Functions of mammals in the arid zone.* Proc. South Australia Water Resources Foundation, Adelaide, S. Australia.

Majak, W. 1992. Metabolism and absorption of toxic glycosides by ruminants. *J. Range Manage.* 45:67–71.

Martin, S. C. 1975. Biology and management of southwestern semidesert grass-shrub ranges. *U.S. Dep. Agric. For. Serv. Res. Pap.* RM-156. Fort Collins, CO.

Musimba, N. K. R., R. D. Pieper, J. D. Wallace, and M. L. Galyean. 1987. Influence of watering frequency on forage consumption and steer performance in southeastern Kenya. *J. Range Manage.* 40:412–415.

Nielsen, D. B. 1978. The economic impact of poisonous plants on the range livestock industry in the 17 western states. *J. Range Manage.* 31:325–328.

Nielsen, D. B., M. H. Ralphs, J. O. Evans, and C. A. Cali. 1994. Economic feasibility of controlling tall larkspur on rangelands. *J. Range Manage.* 47:369–372.

Ospina, E. 1985. A proposal for a research agenda on the economics of multispecies grazing. pp. 216–218. In F. H. Baker and R. K. Jones (Eds.), *Proc. Conference on Multispecies Grazing.* Winrock International Institute, Morrilton, AR.

Paulsen, H. A., and F. N. Ares. 1962. Grazing values and management of black grama and tobosa grasslands and associated shrub ranges of the Southwest. *U.S. Dep. Agric. Tech. Bull.* 1270.

Pearson, H. A., and L. B. Whitaker. 1974. Forage and cattle responses to different grazing intensities on southern pine range. *J. Range Manage.* 27:444–446.

Pfister, J. A., and G. D. Manners. 1995. Effects of carbachol administration in cattle grazing tall larkspur-infested range. *J. Range Manage.* 48:343–349.

Pieper, R. D., E. E. Parker, G. B. Donart, and J. D. Wright. 1991. Cattle and vegetational response to four-pasture and continuous grazing systems. *New Mexico Agr. Exp. Sta. Bull.* 56.

Pring, M. J. 1992. *The all-season investor.* John Wiley & Sons, Inc., New York.

Provenza, F. D. 1995. Postingestive feedback as an elementary determinant of food preference and intake in ruminants. *J. Range Manage.* 48:2–17.

Provenza, F. D. 1996. Acquired aversions as the basis for varied diets of ruminants foraging on rangelands. *J. Anim. Sci.* 74:2010–2020.

Provenza, F. D., J. A. Pfister, and C. D. Cheney. 1992. Mechanisms of learning in diet selection with reference to phytotoxicosis in herbivores. *J. Range Manage.* 45:36–45.

Ralphs, M. H. 1995. Long term change in vegetation following herbicide control of larkspur. *J. Range Manage.* 48:459–464.

Ralphs, M. H., D. Graham, R. Molyneux, and L. F. James. 1993. Seasonal grazing of locoweeds by cattle in northeastern New Mexico. *J. Range Manage.* 46:416–420.

Ralphs, M. H., J. E. Bowns, and G. D. Manners. 1991. Utilization of larkspur by sheep. *J. Range Manage.* 44:619–622.

Ralphs, M. H., and D. W. Ueckert. 1988. Herbicide control of locoweeds: A review. *J. Range Manage.* 2:460–465.

Reynolds, H. G. 1954. Meeting drought on southern Arizona rangelands. *J. Range Manage.* 7:33–40.

Rosiere, R. E. 1987. An evaluation of grazing intensity influences on California annual range. *J. Range Manage.* 40:160–166.

Rosiere, R. E. and D. T. Torell. 1996. Performance of sheep grazing California annual range. *J. Sheep and Goat Res.* 12:49–57.

Russell, C. 1985. Cactus, ecology, and range management during drought. In R. D. Brown (Ed.), *Livestock and wildlife management during drought.* Ceasar Kleberg Wildlife Research Institute, Kingsville, TX.

Ruttle, J. L., D. Bartlett, and D. Hallford, 1983. Fertility characteristics of New Mexico range bulls. *New Mexico Agr. Exp. Sta. Bul.* 705.

Ruyle, G. S., and J. E. Bowns. 1985. Forage use by cattle and sheep grazing separately and together on summer range in southwestern Utah. *J. Range Manage.* 38:299–302.

Shoop, M. C. and E. H. McIlvain. 1971. Why some cattlemen overgraze and some don't. *J. Range Manage.* 24:252–257.

Shoop, M. C., A. J. Alford, and H. F. Mayland. 1977. Plains prickly pear is a good forage for cattle. *J. Range Manage.* 30:12–17.

Smith, A. D. 1965. Determining common use grazing capacities by application of the key species concept. *J. Range Manage.* 18:196–201.

Smith, D. R. 1967. Determining common-use grazing capacities by application of the key species concept. *J. Range Manage.* 18:196–201.

Smith, D. R. 1967. Effects of cattle grazing on a ponderosa pine-bunchgrass range in Colorado. *U.S. Dep. Agric. For. Serv. Tech. Bull.* 1371.

Smith, G. S. 1992. Toxification and detoxification of plant compounds by ruminants: An overview. *J. Range Manage.* 45:25–30.

Smoliak, S. 1974. Range vegetation and sheep production at three stocking rates on *Stipa-Bouteloua* prairie. *J. Range Manage.* 27:23–26.

Smoliak, S. 1986. Influences of climatic conditions on production of *Stipa-Bouteloua* prairie over a 50-year period. *J. Range Manage.* 39:100–103.

Squires, V. R., and A. D. Wilson. 1970. Distance between food and water supply and its effect on drinking frequency and good water intake of Merino and Border Leicester X Merino sheep. *Aust. J. Agric. Res.* 22:283–290.

Stoddart, L. A., and A. D. Smith. 1943. *Range Management* (1st ed.). McGraw-Hill, New York.

Stoddart, L. A., A. D. Smith, and T. W. Box. 1975. *Range Management.* 3d ed. McGraw-Hill Book Company, New York.

Stoken, D. A. 1984. *Strategic investment timing.* MacMillan Publishing Company, New York.

Taylor, C. A. Jr., and M. H. Ralphs. 1992. Reducing livestock losses from poisonous plants through grazing management. *J. Range Manage.* 45:9–12.

Torell, L. A., A. Williams, and B. A. Brockman. 1990. Range livestock cost and return estimates for New Mexico, 1986. *New Mexico Agr. Exp. Sta. Res. Rep.* 639.

Torell, L. A., and W. R. Word. 1993. Range livestock and cost estimates for New Mexico, 1991. *New Mexico Agr. Exp. Sta. Res. Rep.* 670.

United States Department of Agriculture–Forest Service. 1936. *The western range.* U.S. Congress 74th, in session, Senate Doc. 1991.

United States Department of Agriculture (USDA). 1994. *Agricultural statistics* 1994. U.S. Government Printing Office, Washington, DC.

Weaver, J. E., and F. W. Albertson. 1944. Nature and degree of recovery of grassland from the great drought of 1933 to 1940. *Ecol. Monogr.* 14:393–479.

Williamson, G. W. 1995. *The 100 best mutual funds you can buy.* Adams Media Corporation. Halbrook, MA.

Workman, J. P. 1986. *Range economics.* Macmillan Publishing Company, New York.

Workman, J. P., and S. G. Evans. 1993. Utah ranches—An economic snapshot. *Rangelands* 15:253–255.

Workman, J. P., S. L. King, and H. R. Hooper. 1972. Price elasticity of demand for beef and range improvement decisions. *J. Range Manage.* 25:337–340.

Young, V. R., and W. S. Scrimshaw. 1971. The physiology of starvation. *Sci. Am.* 225:14–21.

CHAPTER 13

RANGE WILDLIFE MANAGEMENT

D uring the past 20 years, influences of range management practices and livestock grazing on wildlife have become increasing concerns to those in the range profession in the United States. This concern has resulted from a rapidly changing society. Reduced demand for meat, coupled with greater affluency and a higher human population, has greatly increased the demand for recreation on public rangelands. Low economic returns from livestock, coupled with public willingness to pay for hunting privileges, have caused ranchers in many localities (Texas, California, and South Dakota) to incorporate fee hunting into their operations. Some ranches generate more income from the sale of hunting privileges than from livestock. The trend toward game ranching is expected to continue, particularly if the affluency of our society increases and agriculture productivity stays ahead of population growth. In developing countries where increases in food production lag far behind human population growth rates, the outlook for rangeland wildlife is less optimistic.

In this chapter we will provide an overview of range wildlife management. We refer the reader to Krausmann (1996) for highly detailed coverage of the various aspects of this subject.

BASIC CONCEPTS CONCERNING WILDLIFE HABITAT

Wildlife populations are regulated by the availability of food, water, and cover. The closer these basic components of wildlife habitat (habitat is the natural abode of a plant or animal, including all biotic, climatic, and edaphic factors affecting life) occur together, typically the greater diversity in wildlife species and total numbers. Range management practices and livestock grazing can promote habitat diversity or destroy habitat diversity, depending on their application.

The term *edge effect* is often used in discussing wildlife habitat. Edge effect refers to wherever two habitat types come together (Figure 13.1). The more edge effect a particular area has, the higher the wildlife population, because most wildlife species require several types of vegetation to meet their needs. Extensive seedings of crested wheatgrass (*Agropyron cristatum*) (monocultures) and pure stands of big sagebrush (*Artemisia tridentata*) induced by heavy livestock grazing in the intermountain United States are nearly devoid of wildlife. In some cases range improvement practices have created several different plant communities in close proximity and have interspersed with water developments. This arrangement of vegetation has increased wildlife numbers and diversity. However, in other cases range improvement practices have reduced wildlife populations by replacing diverse native vegetation with a monoculture of one species over huge areas.

Livestock can diversify wildlife habitat by opening up dense stands of vegetation and changing plant composition by selective grazing. Positive benefits from controlled livestock grazing compared to no grazing are most likely to occur in the more humid, wet rangeland types [over 500 mm (20 in.) annual precipitation], where tall grasses without periodic defoliation shade out nutritious forbs and impede mobility of small mammals and birds. Positive benefits to wildlife from grazing influences become questionable in the more arid and desert ranges, where shading of desirable forbs and excessive accumulation of vegetation are seldom a problem. However, controlled light to moderate grazing will generally not damage wildlife habitat even in arid areas. This will be discussed later in the chapter.

Diversity in both plant species and plant communities over short distances is the key to healthy wildlife populations. Livestock grazing, if carefully controlled, is often compatible and sometimes beneficial to the habitat needs of desirable wildlife. Many range management practices can be used to improve habitat for wildlife as well as increase forage for livestock. The application of these practices is discussed next.

GRAZING EFFECTS ON RANGELAND WILDLIFE

Livestock Grazing Affects Wildlife Habitat

Direct Impacts. Livestock affect wildlife habitat directly by removal and/or trampling of vegetation that could otherwise be used for food and cover. Such effects may benefit wildlife by opening dense stands of herbaceous or shrubby vegetation that retard wildlife movement. Different types of livestock may affect the vegetation of an area much differently because of differences in food habits. When defoliation is carried to an extreme, the number of wildlife species will decline because of loss of diversity in food and cover.

Size, pattern, and location of defoliated areas determine both how and to what extent wildlife is affected. Moderate grazing (30% to 50% use of current year's herbage production) in broken terrain generally results in heavy use of lowland areas close to water and light use of

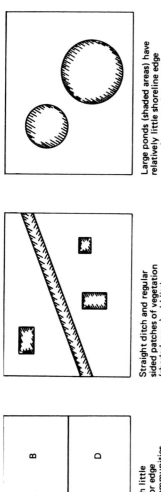

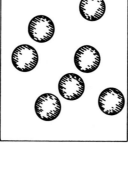

Large ponds (shaded areas) have relatively little shoreline edge per surface area of water.

Numerous small ponds create a large proportion of edge with no loss of total surface area.

Straight ditch and regular sided patches of vegetation (shaded areas) add little edge effect.

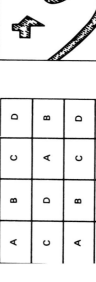

Meandering ditch and irregularly shaped patches greatly increase the amount of edge for wildlife use.

Habitat diversity with little interspersion and poor edge effect. Four plant communities (letters) meet just once in this case.

A	B	C	D
C	D	A	B
A	B	C	D
C	D	A	B

Habitat diversity with good interspersion and edge effect. Four plant communities now have many more contact points, without reducing the total area occupied by each.

Figure **13.1** Schematic diagrams comparing poor interspersion and edge effect (top row) with greatly improved habitat conditions (bottom row). (From Robinson and Bolen 1984.)

upland areas removed from water sources (Cook 1966; Roath and Krueger 1982). Such grazing favors a mosaic of grazed and ungrazed vegetation, although the best wildlife habitat near and around the watering areas will receive heavy use (Roath and Krueger 1982). In flat terrain, moderate grazing results in heavy use or trampling near water sources, with a gradient of decreasing vegetation use as distance from water decreases (Valentine 1947; Holscher and Woolfolk 1953). Kirsch and Kruse (1972) and Owens and Myres (1973) suggest that bison (*Bison bison*) on the North American plains produced the same effect as described previously. This results in diversity of vegetation structure but to a lesser extent than in broken terrain. If water is well distributed in flat terrain, utilization of the pasture by livestock will generally be quite uniform (Jensen and Schumacher 1969). Although uniform use of vegetation is desirable from the standpoint of maximizing livestock production, it can be undesirable to wildlife because of reduced habitat diversity, lack of heavy escape cover, and greater social interaction between domestic and wild species (Brown 1978; Mackie 1978). However, in many situations water development for livestock has been beneficial to wildlife in the western United States because it has permitted the use of areas from which wildlife were previously excluded due to lack of free water. Wood et al. (1970) reported development of water for livestock increased mule deer (*Odocoileus hemionus*) numbers in southcentral New Mexico. Development of water for livestock in southeastern Oregon benefited several game and nongame species (Heady and Bartolome 1977). Crawford and Bolen (1976) reported that stock ponds may increase survival of lesser prairie chickens (*Tympanuchus cupido*) during periods of drought.

Indirect Impacts. Indirect impacts of livestock grazing on wildlife habitat are much less understood than are direct effects. The primary indirect impact on wildlife habitat that results from livestock grazing is change in vegetation composition. Our summary of large herbivore food habits (see Table 11.7) shows in general that cattle prefer grasses, domestic sheep prefer forbs, and goats prefer shrubs. However, several of the studies were contradictory to the preceding statement. A careful examination of the studies shows that many perennial forbs, when available, were preferred by all these types of livestock, and many of those forbs preferred by livestock were also preferred by wild ungulates. Klebenow (1980) emphasized the critical role of forbs in the nutrition of wild ungulates.

With some exceptions (Wood 1969), heavy grazing reduces plant species diversity (Ellison 1960). As range condition declines and plant diversity decreases, the nutritive value of wild ungulate diets declines because they are forced to be less selective (Figure 13.2) (Buechner 1950; Pederson and Harper 1978; Bryant et al. 1981). Evidence is available that ungrazed plant communities have less plant species diversity than those that are lightly or moderately grazed (Johnston 1961; Wood 1969; Campbell et al. 1973; Smith 1996). However, this depends on the type of ecosystem under consideration. Moderate to light cattle grazing favors more forb production in most plant communities (Riegel et al. 1963; Penfound 1964; Hazell 1967; Potter and Krenetsky 1967; Krueger and Winward 1974), although exceptions exist (Brown and Schuster 1970). In contrast, moderate to light sheep grazing severely reduces forbs if grazing occurs during the growing season (Mueggler 1950; Laycock 1967; Jensen et al. 1972; Bowns and Bagley 1986) (Figure 13.3). During the dormant season, sheep grazing generally favors grasses and forbs but reduces shrubs (Laycock 1967; Jensen et al. 1972).

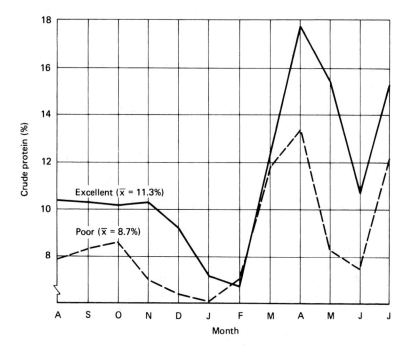

***Figure* 13.2** Percentage of crude protein in diets selected by deer from pastures in excellent and poor range condition at the Sonora Research Station. (From Bryant et al. 1981, Fig. 3.)

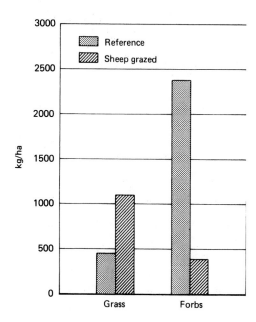

***Figure* 13.3** Aboveground biomass (kg/ha) of forbs and grasses within reference areas on mountain range in Utah grazed by sheep and lightly grazed by cattle and horses. (From J. E. Bowns and C. F. Bagley 1986, Fig. 4.)

Grazing a range by only one species of animal tends to cause a trend away from one vegetation type to another. A vegetation trend that results from one species is usually undesirable for the animal causing the trend. Therefore, common-use (grazing the range by more than one type of animal) generally results in higher animal and vegetation productivity than single use.

Heavy mule deer use on winter range in Utah greatly increased grasses and decreased shrubs over an 11-year period (Smith 1949). Heavy cattle grazing on a nearby area during the same period produced a heavy stand of shrubs. Common-use grazing of these ranges would have been beneficial to both animals. Under pristine conditions, pronghorn antelope (*Antilocapra americana*) and bison appear to have been beneficial grazing companions because the bison utilized the grasses while pronghorn used the forbs and shrubs (England and De Vos 1969; Kautz and Van Dyne 1978). Under light to moderate grazing intensities cattle and pronghorn share a similar relationship today (England and De Vos 1969; Yoakum 1975; Kautz and Van Dyne 1978).

Overgrazing tends to open perennial grassland vegetation types and allow invasion by annual forbs and grasses (Ellison 1960). Annuals are important foods of many species of rodents and lagomorphs (Wood 1969). Populations of several small mammal species appear to fluctuate in response to the availability of these food items (Wood 1969).

Virtually all the gallinaceous game birds depend heavily on annuals and/or early successional forbs for food (Martin et al. 1951; Grenfell et al. 1980). Annual grasses and forbs are particularly important to this category of wildlife during the winter because they have large seeds that are high in energy, unlike most perennial grasses. Grazing plays an important role in maintaining these species in the plant community composition (Ellison 1960).

The importance of insects, particularly grasshoppers, in the diets of birds is well recognized. Insects provide critical high-protein food for many game bird species during the summer brood-rearing period (Martin et al. 1951; Grenfell et al. 1980). The impact of grazing on total insect numbers and diversity is not well established; however, grasshoppers are much more abundant on moderately or heavily grazed ranges than on ungrazed or lightly grazed ranges (Nerney 1958; Holmes et al. 1979).

Food and cover requirements of one wildlife species or group are often directly opposite those of another. Vegetation requirements for cover of many wildlife species are often much different than those for feeding. These requirements may also vary drastically between seasons for some wildlife species. Therefore, diversity in vegetation structure, vegetation composition, and terrain favors the highest diversity and density of wildlife. If carefully controlled, livestock grazing can be a useful tool to obtain and maintain habitat diversity. Readers are referred to Severson and Urness (1994) for a recent detailed discussion on the use of livestock to improve wildlife habitat.

Social Interactions between Domestic Livestock and Wildlife

Social interactions between domestic livestock and wildlife are difficult to study because factors such as terrain, forage availability, water distribution, and vegetation structure confound results. Most of the work concerning social interactions with livestock has involved wild ungulates. White-tailed deer (*Odocoileus virginianus*), mule deer, bighorn sheep (*Ovis canadensis*), moose (*Alces alces*), and elk (*Cervus elaphus*) appear to have some social aversion to livestock, but with the exception of bighorn sheep, this appears to be of minor importance if stocking rates are light or moderate. Darr and Klebenow (1975) showed that white-tailed deer preferred pastures grazed by cattle over those grazed by sheep. Movement of cattle into unused pastures caused movement of elk to ungrazed or lightly grazed pastures in Oregon and Montana (Skovlin et al. 1968; Komberec 1976). Yoakum (1975) concluded that pronghorn had no aversion to feeding with cattle or sheep. Given a choice, North American deer and elk appear to prefer pastures unoccupied by livestock (Figure 13.4) (Skovlin et al. 1976; Reardon et al. 1978; Ragotzkie and Bailey 1991). Bighorn sheep appear to be relatively intolerant of livestock and have, in some instances, abandoned areas occupied by livestock (Morgan 1971; Gallizioli 1977). Moose also have reportedly abandoned areas when use by livestock was initiated (Denniston 1956; Schladweiler 1974).

There are no data available showing a definite social aversion to livestock by nongame mammals, nongame birds, or game birds. In Montana sharp-tailed grouse (*Pediocetes phasianellus*) did not avoid grazed pastures when ungrazed pastures were available (Nielson 1978). Similar findings have been reported for lesser prairie chickens in New Mexico (Davis et al. 1979), wild turkeys (*Meleagris gallopavo*) in New Mexico (Jones 1981), and sage grouse (*Centrocerus urophasianus*) in Nevada (Klebenow 1980).

Transmission of Disease to Wild Ungulates by Livestock

Several diseases prevalent in domestic livestock also occur in wild ungulates. However, the importance of disease in the regulation of wild ungulate populations and the interrelationships

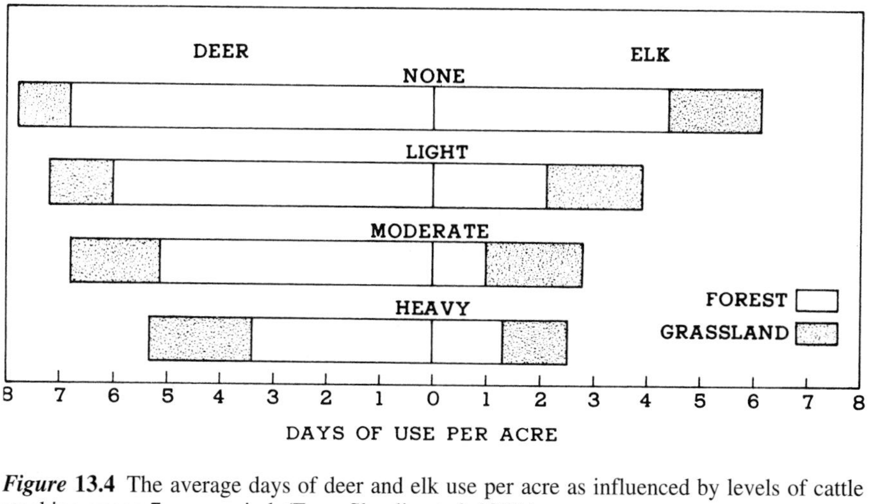

Figure **13.4** The average days of deer and elk use per acre as influenced by levels of cattle stocking over a 7-year period. (From Skovlin et al. 1976.)

with domestic livestock are presently not well understood (Mackie 1978). Generally, it appears that livestock indirectly contribute to disease problems in wild ungulates primarily through overgrazing (Allen 1974). When overgrazing occurs, lack of forage causes nutritional stress that results in greater susceptibility to disease. Because animals often concentrate where forage remains and they are in a weakened physiological condition, disease spreads more quickly than would otherwise occur. Bighorn sheep have been more adversely impacted by domestic livestock (particularly sheep) and livestock diseases than any other North American wild ungulate (Krausman et al. 1996). It has been recommended that livestock not be grazed on bighorn sheep habitat.

Operational Impacts of Livestock Grazing on Wildlife

Fencing, alteration of vegetation through brush control, water development, predator control, and fertilization often have much greater impact on wildlife than do the presence and activities of livestock per se.

Fencing directly influences wildlife movement (primarily wild ungulates), and indirectly affects cover and food available to wildlife by controlling livestock distribution. Most of the direct fencing problems have concerned pronghorn antelope and domestic sheep. Because antelope seldom jump sheep-proof fences, considerable mortality has occurred where special crossings were not used (Buechner 1950; Newman 1966; Spillett et al. 1967; Mapston 1972). Recent research has shown that antelope guards can be effective in restrict-

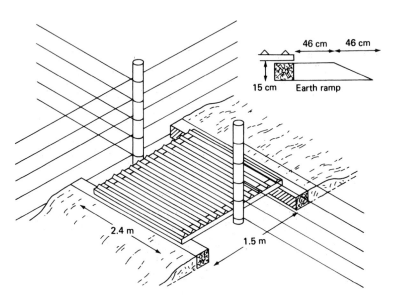

Figure 13.5 Characteristics of the horizontal antelope grill. (Adapted from Mapston et al. 1970.)

ing livestock and permitting pronghorn passage (Figure 13.5). Detailed information on construction and placement of pronghorn pass structures is given in Gross et al. (1983).

An important, often unrecognized aspect of fencing is that it can create edge effect or habitat variety, and consequently, increase the number of wildlife species in an area. Wiens and Dyer (1975) reported that for this reason alone, regardless of other impacts, fencing is beneficial to many bird species. Fences change the microclimate of the immediate area they occupy, and this often results in a different vegetation than that in the pastures they separate. Therefore, three or more plant communities often result from fencing, where only one would exist otherwise. Fences also provide roost sites for many bird species (Wiens and Dyer 1975).

In recent years, the application of cell-type, short-duration grazing has in some cases reduced the availability of water to wildlife. Under this strategy, peripheral water facilities are fenced or deactivated to force livestock to visit the cell center and facilitate herd movement. Prasad and Guthery (1986) compared wildlife use of water facilities under cell-type short-duration and continuous grazing at two locations in southcentral Texas. Their results indicated that livestock water at the center of a short-duration grazing cell was largely unavailable to white-tailed deer, collared peccaries (*Tayassu tajacu*), and wild turkeys. These species may have avoided cell centers due to increased human and livestock activity and the concentration of fencing. Based on these results, it appears that peripheral water facilities should be preserved to maintain wildlife populations. These can be fenced so that livestock rotation is not disrupted but wildlife are permitted access.

GRAZING METHODS FOR WILDLIFE ENHANCEMENT

When carefully controlled, grazing can be a useful tool for the enhancement of wildlife habitat. However, the frequency, intensity, and timing of livestock grazing for maximum wildlife benefits may be much different from that which would be used for maximum livestock production. In many cases, burning and/or mowing give better results than livestock grazing when the primary goal is wildlife habitat improvement. Other situations may require a combination of grazing, burning, and/or mowing for best results. The wildlife manager must keep in mind that each situation is somewhat different and therefore requires a separate analysis and a different prescription. Generalizations are difficult to make because habitat needs vary tremendously among wildlife species.

Livestock grazing at any intensity may be harmful to a few species, such as bighorn sheep. However, most wildlife species are tolerant and some may be benefited by grazing at light to moderate intensities. Limited grazing appears to increase diversity in small mammal and upland nongame bird populations. Heavy grazing is highly detrimental to most ground-nesting birds, but some of these species can be benefited by carefully controlled grazing. With some exceptions, grazing at even light intensities appears harmful to nesting waterfowl. However, rest-rotation grazing can be used to ameliorate this problem. Mule deer and white-tailed deer in some localities have benefited from plant composition changes that have resulted from overgrazing. However, in many cases overgrazing has adversely affected these wild ungulates by reducing food availability. Because social antagonism exists among mule deer, bighorn sheep, moose, and elk, with livestock, specialized grazing systems that leave part of the area unoccupied by livestock can be beneficial.

The remainder of our discussion on this topic is devoted to livestock grazing management practices that can be used for maintenance and improvement of wildlife habitat.

Wild Ungulates

Specialized livestock grazing systems can be a useful tool to minimize livestock grazing impacts on wild ungulates and can be used to improve wild ungulate habitat. Our discussion of this subject follows Holechek et al. (1982).

In Oregon, Skovlin et al. (1968, 1976) found that mule deer and elk preferred deferred-rotation cattle grazing over season-long grazing because deferred-rotation grazing provided elk and mule deer with pastures free from cattle disturbance. Yeo et al. (1993) reported similar findings with mule deer and elk on cattle range in Idaho under a rest-rotation system. In southcentral Texas, Reardon et al. (1978) found that a seven-pasture, rapid-rotation grazing system supported higher white-tailed deer densities than did continuous or Merrill three-herd/four-pasture grazing systems (Table 13.1). They concluded that deer prefer to stay in periodically deferred pastures, and the more frequent the deferment, the higher the preference for the system. Merrill et al. (1957) reported that white-tailed deer made greater use of deferred-rotation pastures than they did of those grazed continuously at higher rates.

In Oregon, cattle grazing appeared to improve elk winter range (Anderson and Scherzinger 1975). The grazing strategy involved using cattle as a tool to top-off grasses in the late spring and early summer so that the forage remaining cured in a highly nutritious state. Care was taken to ensure that cattle were removed before the end of the growing season so that plenty of high-quality forage remained for the elk. On the Bridge Creek Wildlife

TABLE 13.1   Livestock Stocking Rates, Deer Densities, and Hunting Revenues from Seven Systems of Grazing at Sonora Research Station, 1971–1975

GRAZING SYSTEM	LIVESTOCK STOCKING RATE (ACRES/AU)	DEER DENSITY (ACRES/DEER)	HUNTING REVENUES (INCOME/ACRE)
Moderate, continuous	16.0	18	$0.82
Heavy, continuous	11.9	20	0.59
Merrill: Control	16.0	15	2.25
Merrill: Aerial-sprayed	12.8	16	0.88
Merrill: Front-end grubbed	12.8	16	0.62
Merrill: Root-plowed	12.8	15	0.79
Seven-pasture short-duration	10.7	10	1.79

Source: From Reardon et al. 1978.

Management Area, there were approximately 320 elk counted annually between 1961 and 1964, with no cattle grazing. Cattle grazing was initiated to improve forage quality in 1964, and by 1974 the elk had increased to approximately 1,190.

In contrast, a southeastern Washington study showed that spring cattle grazing reduced winter elk use compared to areas ungrazed by cattle (Skovlin et al. 1983). Both studies were conducted on similar range, where Idaho fescue (*Festuca idahoensis*) was the key species for both cattle and elk. The inconsistency between Anderson and Scherzinger (1975) and Skovlin et al. (1983) could be explained by factors other than forage quality that may have influenced elk movement into the Bridge Creek Wildlife Management Area. However, more recent studies by Pitt (1986), Rhodes and Sharrow (1990), and Jourdonnais and Bedunah (1990) do indicate that controlled grazing can increase range use by elk through improved forage quality.

Spring grazing of cattle or sheep on mule deer range has been an effective tool to increase the browse available to mule deer (Smith et al. 1979). This grazing strategy results in livestock using primarily understory grasses and shrubs which compete with bitterbrush (*Purshia tridentata*). The productivity of the bitterbrush is increased because of reduced competition. Care is taken to remove sheep from the area before the understory forage species mature or there is excessive utilization of the bitterbrush. In British Columbia (Willms et al. 1981) and Oregon (Leckenby 1968), light to moderate summer grazing of bunchgrass range by cattle improved the attractiveness of these ranges to mule deer in the spring and fall because nutritious new growth was more available than on ungrazed range.

Upland Game Birds

Results from studies examining specialized grazing system impacts on upland game birds have been inconsistent. Because different upland game bird species have different habitat requirements, the most effective grazing methods depend on the game bird species and the type of vegetation and terrain involved. Rest-rotation and deferred-rotation grazing systems are beneficial to most game bird species because they provide pastures free from disturbance during the nesting and other critical seasons. However, this benefit may be offset if heavy use occurs in the grazed pastures (Buttery and Shields 1975). We will provide some examples of livestock grazing interactions with upland game birds but refer the reader to Guthery (1996) for more detailed coverage of this subject.

In northern Nevada, rest-rotation grazing has been effective in improving sage grouse habitat (Klebenow 1980; Neel 1980). Forbs are important foods of sage grouse in the summer (Klebenow and Gray 1968; Klebenow 1969). Neel (1980) found that rest-rotation management improved both forb abundance and the entire grazing allotment under study. Klebenow (1980) reported in Nevada that moderately grazed meadows were actually more attractive to sage grouse than were those receiving protection. However, because of lack of cover, overgrazed meadows were not preferred.

Sharp-tailed grouse may not adapt to conventional grazing systems (Sisson 1976). In Montana, rest-rotation grazing appeared detrimental to sharp-tailed grouse because the birds did not adjust to changing grazing patterns (Nielson 1978). This grazing system did not improve range condition or wildlife habitat. It is likely that the stocking rate was too heavy for maintenance of an adequate vegetation residue in Nielson's (1978) study. In a more recent North Dakota study the density of successful sharp-tailed grouse nests was similar on rotation-grazed and nongrazed areas (Kirby and Gross 1995).

Bobwhite quail (*Colinus virginianus*) numbers were higher under a high intensity-low frequency (HILF) grazing system than under continuous or a Merrill three-herd/four-pasture system during a 2-year study in south Texas (Hammerquist-Wilson and Crawford 1981). In this study continuous grazing appeared superior to the Merrill system. They attributed the higher numbers of quail in the HILF and continuous pastures to the greater amounts of bare ground and tall forbs and the lesser amounts of grass that occurred in these pastures compared to Merrill system pastures. In another south Texas study during drought, short-duration grazing provided better nesting and protective cover for quail than did continuous grazing even though the stocking rate was higher for the short-duration system (Campbell-Kissock et al. 1984). Stoddard (1931) reported that few bobwhite quail occurred in dense stands of grass in the southeastern United States. Dense stands of grass support few preferred quail food items in the form of forbs (Jackson 1969; Kiel 1976). Bare ground in conjunction with tall forbs appears to provide the most desirable surface for movement and feeding by bobwhite quail (Hammerquist-Wilson and Crawford 1981).

Livestock grazing in conjunction with fire are useful tools for providing bobwhite quail with optimal habitat in the southeastern United States. Reid (1954) reported that under light or moderate cattle grazing, there is little overlap of quail and cattle diets on longleaf pine (*Pinus palustris*) range in the southeastern United States. He found that light or moderate cattle grazing maintained important quail foods and permitted free movement of quail.

In southcentral New Mexico, wild turkeys showed no preference for either grazed or ungrazed rest-rotation pastures (Jones 1981). However, in south Texas a Merrill three-herd/four-pasture grazing system appears beneficial because it provides wild turkeys with better nesting habitat than does continuous grazing (Glazener 1967; Merrill 1975). The New Mexico study by Jones (1981) was conducted in mountainous terrain where overall livestock use was moderate, and practically no livestock use occurred on steep slopes that were used by turkeys for nesting. The Texas studies were conducted in relatively flat topography where pasture use was comparatively uniform.

Limited livestock grazing has potential as a tool for maintaining Attwater's and greater prairie chicken (*Tympanuchus cupido*) habitat. However, overgrazing has been a bigger factor limiting prairie chicken populations than has undergrazing (Hamerstrom and Hamerstrom 1961). Attwater's prairie chickens avoid matted, thick cover of ungrazed coastal prairie (Lehman 1941, Kessler and Dodd 1978). It appears that carefully controlled rotational grazing by livestock, burning, and mowing can be used to maintain optimal cover conditions. This strategy may also benefit greater prairie chickens in the eastern part of their range (Grange 1948; Westemeier 1972).

Mearns quail (*Cyrtonyx montezumae*), scaled quail (*Callipepla squamata*), greater prairie chickens in the western part of their range, lesser prairie chickens, bobwhite quail in the southwestern United States, and sharp-tailed grouse in the western part of their range in the United States are very sensitive to livestock grazing because they live in relatively flat, grassland areas that undergo periodic drought (Brown 1978). In near-average or above-average years of precipitation, light or moderate grazing does not severely affect these birds in most localities (Brown 1978). However, in drought years serious population declines occur because of reduced cover and food (Hamerstrom and Hamerstrom 1961; Brown 1978). These declines can be greatly magnified by grazing that is not well controlled (Brown 1978). Brown suggested that interspersion of ungrazed exclosures through grazed pastures could be effective in maintaining cover for these game birds during critical times of the year and during drought. However, in Texas, Webb and Guthery (1982) found implementation of grazing exclosures expensive and of questionable value to bobwhite quail. A better approach appears to be balancing grazing intensity with game bird cover needs.

Populations of many upland game bird species associated with riparian zones, such as California quail (*Lophortyx californicus*), ringneck pheasants (*Phasianus cochicus*), ruffed grouse (*Bonasa umbellus*), Gambel's quail (*Lophortyx gambelii*), and bobwhite quail could probably be enhanced by temporary or permanent fencing of sections along waterways. This practice has shown promise for improvement of habitat for pheasants, California quail, ruffed grouse, and many nongame wildlife species (Winegar 1977; Duff 1979).

Upland game birds with high mobility in rugged terrain, such as chukar partridge (*Alectoris chukar*), mountain quail (*Oreortyx pictus*), white-tailed ptarmigan (*Lagopus leucurus*), and blue grouse (*Dendragapus obscurus*) may benefit from specialized grazing systems involving deferment. These birds can usually find adequate cover (i.e., chukar partridge use rocky hillsides) if grazing is light or moderate, but in the spring they need a good vegetative cover and freedom from disturbance for nesting and brood-rearing. A grazing system that allows part of the range to be deferred during the nesting and brood-rearing season should be effective.

The influence of short-duration versus continuous grazing on upland game bird nesting success has been a recent concern. In south Texas, continuously grazed and short-duration pastures stocked at the same rate had nearly equal cover and dispersion of suitable nest sites for both wild turkey and bobwhite quail (Table 13.2) (Bareiss et al. 1986). Grazing treatment

TABLE 13.2 Availability of Nesting Converts for Bobwhites and Wild Turkeys in Rangeland under Continuous Grazing (CG) and Short-duration Grazing (SDG) on Two Study Areas in South Texas, March 1984

SPECIES	AREA	PERCENTAGE COVERAGE		COVERTS/30 M		FREQUENCY[a]	
		CG (x)	SDG (x)	CG (x)	SDG (x)	CG	SDG
Bobwhite	Welder	7.2	8.0	3.9	5.2	92	96
	Encino	3.3	2.6	2.0	1.5	74	66
Turkey	Welder	0.9	0.9	0.2	0.3	24	26
	Encino	0.1	0.3	0.1	0.3	4	12

Source: From Bareiss et al. 1986 (Table 1).

[a]Percentage of lines 30 meters in length with one or more nesting coverts. A total of 50 lines were sampled.

did not affect the loss rate of artificial nests. Koerth et al. (1983) found that trampling of clay pigeon targets, placed to simulate ground nests, was similar between the two grazing systems. If stocking rates are not excessive, short-duration grazing apparently does not adversely affect upland game bird nesting activities compared to continuous grazing.

Waterfowl

Livestock grazing has been quite detrimental to many species of waterfowl when not carefully controlled. Grazing systems show potential for amelioration of these adverse effects. In some situations livestock grazing can be used for enhancement of waterfowl habitat.

Wetlands supporting an interspersion of cover and open water are preferred by waterfowl (Kantrud 1990). Without fire, grazing, or some other type of disturbance, tall robust plants such as bullrushes (*Scirpus* spp.) and cattails (*Typha* spp.) in extensive, unbroken stands often dominate wetlands. Some grazing systems may hold promise to improve duck production over ungrazed areas particularly if grazing is delayed until May (Sedivec et al. 1990). By this time at least one-half of duck nests have been initiated.

In Montana, rest-rotation grazing effectively increases waterfowl production compared with season-long grazing (Gjersing 1975; Mundinger 1976). The rested and spring grazed pastures under this system accumulate enough vegetation to provide conditions suitable for good waterfowl production the following year. Under the rest-rotation grazing system 43 broods were produced per year compared to 10 broods per year under season-long grazing (Gjersing 1975). In south Texas, carefully planned grazing that provides for deferment, particularly during the growing season, can be used successfully to control the effect of cattle grazing on shoreline vegetation and maintain good stands of waterfowl plant foods (Whyte and Cain 1981; Whyte et al. 1981). Evans and Krebs (1977) suggested that spring deferment of grazing around stock watering ponds in the northern Great Plains would improve nesting use by waterfowl and shorebirds. They also reported that implementation of rest-rotation grazing would be less costly than pond fencing. Their recommendations were based on a 7-year study of waterfowl and shorebirds on man-made stock watering ponds in South Dakota.

Excessive accumulations of vegetation are sometimes detrimental to waterfowl (Kirsch and Kruse 1972). Limited grazing or burning every 1 to 3 years can be effectively used to increase blue-winged teal (*Anas discors*) production in Iowa and South Dakota (Kaiser et al. 1979). Native plant communities in high condition with matted mulch had the highest nest success and density in the study by Kaiser et al. (1979). Either excessive rest or overgrazing favors degradation of habitat by causing Kentucky bluegrass (*Poa pratensis*) invasion into the plant community. Burning, resting, and haying, as well as grazing, were mentioned as tools that could be used to maintain optimal conditions for blue-winged teal.

Nongame Wildlife

There is no universal grazing approach that will benefit all wildlife species. Some species are most abundant under heavily grazed conditions while others thrive under lightly to ungrazed conditions (Buttery and Shields 1975; Bock et al. 1984; Baker and Guthery

1990) (Table 13.3). Research from the Chihuahuan desert of New Mexico showed that wildlife diversity was higher on a moderately grazed range in good ecological condition compared to a more lightly grazed range in near climax condition (Smith et al. 1996). This was attributed to more diversity in plant species and vegetation structure on the good condition range. On the Welder Wildlife Refuge in eastern Texas, Baker and Guthery (1990) found densities of different ground-foraging bird species responded inconsistently to grazing intensity and soil type (Figure 13.6). Mourning doves were favored by heavy grazing but eastern meadowlark densities were highest under moderate grazing. On a semidesert grassland site in southeastern Arizona grazing compared to grazing exclusion

TABLE 13.3   Total Numbers of Birds Flushed in Grazed versus Ungrazed Semidesert Grassland in Southeastern Arizona

	NUMBERS PER TREATMENT			
	SUMMER		WINTER	
SPECIES	GRAZED	UNGRAZED	GRAZED	UNGRAZED
Scaled quail	48	5	35	0
Mourning dove	20	7	—	—
Northern flicker	5	0	—	—
Cassin's kingbird	8	3	—	—
Western kingbird	2	1	—	—
Horned lark	101	24	6	3
Northern mockingbird	22	9	—	—
Loggerhead shrike	2	4	0	2
Blue grosbeak	2	2	—	—
Cassin's sparrow	0	66	0	42
Chipping sparrow	—	—	7	21
Brewer's sparrow	—	—	82	4
Vesper sparrow	—	—	103	90
Black-throated sparrow	2	2	19	0
Lark sparrow	40	4	—	—
Grasshopper sparrow	0	53	13	65
Eastern meadowlark	18	13	17	15
Total birds counted	270	193	247	242

Source: From Bock et al. 1984.

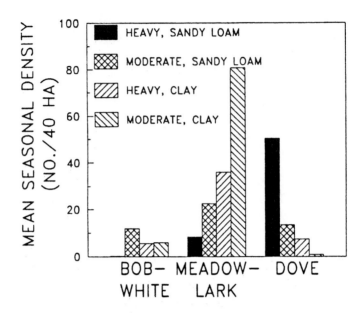

Figure 13.6 Mean seasonal densities of three bird species on experimental pastures differing in soil type and grazing pressure, San Patricio County, Texas, 1984–1985. (From Baker and Guthery 1990, Fig. 2.)

appeared to favor birds as a class over rodents (Bock et al. 1984). However, different species within each group differed substantially in their occurrences on grazed and ungrazed areas.

Most of the adverse effects of grazing on nongame wildlife occur in riparian zones. Riparian habitats are much more important to birds than are surrounding uplands (Tubbs 1980). Research by Fitzgerald (1978) demonstrates the importance of riparian zones to small mammals, reptiles, and amphibians. The key to improving and maintaining riparian habitats for wildlife is to prevent excessive use of the vegetation by livestock. In many cases this is difficult to do without complete livestock exclusion. However, some grazing practices can be used to maintain and/or improve habitats. These are discussed by Skovlin (1984), Elmore and Kauffman (1994), Ohmart (1996), and in Chapter 9.

PROVISION OF FORAGE TO BIG GAME

In the past and currently, forage allocation to domestic livestock and big-game animals has been a controversial problem. Many early beliefs and solutions to this problem were based on intuitive reasoning rather than experimental evidence. Recently, investigations have discredited many early beliefs concerning competition and forage preferences. Some generalities regarding big-game and livestock competitive relationships will be reviewed following the discussion of Holechek (1980).

Three possible interactions between domestic and big-game animals exist: food, space, and social interactions. On most rangelands, the most critical interaction area is food. Competition for food can exist only under the following three conditions (Smith and Julander 1953):

1. Domestic and big-game animals are using the same area.

2. Forage plants are in short supply.

3. Both domestic and big-game animals are using the same forage plants.

To interpret and relieve possible forage competition problems between domestic and big-game animals, the following information must be determined:

1. The key forage plants for both species.

2. The degree of use on key species.

3. The ability of wild herbivores to switch to other foods.

4. Key areas where dual use occurs.

5. Repeatability of dual use on key areas from year to year.

Once this information has been obtained, definite management programs for different areas can be formulated. It is important to recognize that it is the amount of overlap in different animal diets on given areas that determines the potential for competition. Degree of use is also important since a plant species that is not consumed under moderate grazing may be fully utilized under heavy grazing. Maintenance and improvement of the forage resource must always be a primary consideration when forage is allocated to different animals. There are five important concepts that can be of much value in forage allocation to different animals. These concepts are the following:

1. Animals (cattle and sheep) with the broadest food habit adaptability tend to be the most successful under restricted forage availability. This is because they can regulate their diet to what is available.

2. Large ruminants have the ability to alter their food habits substantially. These animals have a large rumen volume/body weight ratio. Therefore, quantity is more important than quality. This is why elk can out compete deer when the two animals share the same winter range.

3. Severe disturbances can force animals to use forage or habitats not normally used. These disturbances are generally climatic, such as drought, cold, snow, or flood.

4. Forage availability to animals prior to critical periods may be more important than availability of forage during the critical period. Research is available indicating that fall forage for deer is more important than winter forage on northern ranges (Julander et al. 1961; Ullrey et al. 1967; Verme 1967). This is because the animals can go for long periods (40 days) with little forage intake if they have high body fat reserves (Ozoga and Verme 1970; de Calesta et al. 1975).

5. Population size can alter animal habitat use. When populations expand because of low mortality or high natality to the limit of their range, some animals are forced into marginal habitats that would not normally be occupied. This increases the potential for competition.

It must be emphasized that under proper livestock grazing minimal competition occurs between big-game animals and livestock in most situations. This is because during most of the year game animals use the more rugged and inaccessible areas where livestock use is light (Miller and Krueger 1976; Stuth and Winward 1977; Yeo et al. 1993; Sheehy and Vavra 1996). In the critical period during the winter, game animals such as deer and elk concentrate in lower areas that receive heavy livestock use. These are the areas where forage allocation is critical. On important big-game winter ranges that belong to the public, in most situations management should be primarily for game rather than livestock. This does not mean, however, that livestock grazing must be eliminated from these ranges. In some cases livestock can be used as a valuable tool to enhance these areas for big game, as has been discussed previously.

In British Columbia, Willms et al. (1979) reported that the greatest potential for competition between deer and cattle was in the spring, when both cattle and deer made heavy use of grasses. However, this was negated by the fact that deer and cattle selected different grass species at light to moderate grazing intensities. Vavra and Sneva (1978) reported similar findings on common-use cattle and deer range in southeastern Oregon. Willms et al. (1979) found that moderate or heavy fall grazing by cattle made the spring forage more attractive to deer by removing mature growth. Deer spring use on pastures not grazed by cattle in the fall was 35%, while heavily grazed pastures received 56% use.

Competition between cattle and elk is more of a problem than between deer and cattle or between cattle and pronghorn. This is because cattle and elk often have similar diets when found on the same ranges (Wisdom and Thomas 1996). However, competition is minimized because cattle and elk usually use the same ranges at different times (Skovlin et al. 1968; Miller and Krueger 1976; Sheehy and Vavra 1996). On common-use cattle and elk ranges, forage allocation depends on public demand for harvestable elk and on whether the range is public or private. The purchase of elk wintering areas by wildlife departments in conjunction with controlled elk harvest appears to be a reasonable solution to elk damage problems on private lands. This type of program has been effective in Oregon, Washington, and Colorado; also, Wyoming has a winter feeding program for some of its elk herds. Winter feeding programs are very expensive and may lead to deterioration of both summer and winter range if elk numbers are not controlled. However, winter feeding does have some utility when elk or deer are limited by forage availability only during the 3- or 4-month winter pinch period. Range improvement practices such as fertilization, reseeding, brush control, and burning can be applied to big-game wintering units to increase forage production. In a few cases where only winter range is limited and little potential exists to provide natural forage, winter feeding may have value. Big-game and livestock numbers should always be regulated so that the forage resource is maintained or enhanced. Current reviews on forage allocation to big game and livestock are provided by Wisdom and Thomas (1996) and Peek and Krausman (1996).

IMPACTS OF BRUSH CONTROL ON WILDLIFE

Many studies have evaluated brush control impacts on wildlife habitat. These were reviewed comprehensively by Holechek (1981) and in Krausman (1996). Range and wildlife managers now generally recognize that variety and quality of cover are as important to wildlife as forage, variety, quantity, and quality. For this reason, brush control projects that increase habitat diversity and edge effect are generally beneficial. Those that reduce the plant species and community diversity over large, continuous blocks are detrimental. Just as there is an upper limit on how much brush can be removed from an area without detrimental effects on wildlife, there is an upper limit on the amount of brush that can occupy an area without reducing wildlife populations. The distribution or pattern of brush in an area is far more important than the quantity of brush.

   Use of the following guidelines can ensure wildlife population maintenance and enhancement in developing brush control projects.

1. Identify resident wildlife and the areas they presently inhabit.

2. Determine the ecological requirements of resident wildlife.

3. Determine those factors most limiting to resident wildlife.

4. Determine what critical habitats, if any, may be destroyed by the proposed brush control project.

5. Determine the longevity of the proposed project.

6. Evaluate the impacts of similar brush control projects in areas where they have been applied.

7. Coordinate the project with the needs of resident wildlife as much as possible.

8. Monitor the response of resident wildlife to the brush control treatment after its application.

   Considerable research has been done on the response of white-tailed and mule deer to brush control. Both species depend on brush for food and cover and when brush is completely removed from large areas, deer use declines (Scotter 1980). However, if sufficient brush is available to meet deer cover requirements, quantity and quality of available forbs become the second-most-limiting factor to deer.

   Opening up dense stands of brush with fire, mechanical treatment, or herbicides generally increases the availability of forbs and palatable browse to deer. But if brush control is to benefit deer, it should be done so that strips or blocks of brush remain. Diversity of food and cover types over relatively short distances is the key to enhancing mule deer populations in sagebrush areas. Generally, deer habitat can be improved most effectively by controlling brush over small areas of 2 ha to 16 ha (5 to 40 acres). Removal of more than half the brush over a large area appears to be detrimental to deer (Scotter 1980). Beasom and Scifres (1977) found that spraying mesquite in alternating strips did not adversely affect population of white-tailed deer.

Brush control treatments applied to areas of 200 ha (500 acres) or more may be particularly detrimental to pronghorn. Herbicides that eliminate many forbs and shrubs should be used cautiously if pronghorn habitat is to be maintained or enhanced.

Yoakum (1979) recommended that brush control projects for pronghorns be less than 400 ha (1,000 acres) in size and maintain at least 5% to 10% shrub cover on the control area. He suggested spraying, chaining, or prescribed burning, rather than plowing to minimize the loss of native plants. When reseeding, mixtures including alfalfa (*Medicago sativa*) or other palatable forbs should be used instead of a single grass species. More antelope were seen on areas seeded to grasses and forbs than on adjacent shrub-dominated areas in Oregon when these practices were applied (Yoakum 1979).

Elk depend less on forbs and browse for forage than do either pronghorn or deer. Elk also depend less on brush for cover. However, cover becomes more important as disturbances by man increase (Marcum 1975).

Research shows that herbicides can be a valuable tool for improving elk habitat, particularly in areas with dense stands of brush and little understory. Spraying Gambel oak (*Quercus gambelii*) with 2,4,5-TP in western Colorado, for example, resulted in a 73% increase in use of the area by elk (Kufeld 1977). Similar results were reported in Wyoming when sagebrush range was sprayed with 2,4-D (Wilbert 1963). In both studies, areas of 28 ha (70 acres) or less were sprayed, resulting in greater habitat diversity as well as increased forage availability. Results from these studies may have been different if large areas had been treated.

Impacts on Birds

As with wild ungulates, brush control can either benefit or harm bird populations, depending on how it is accomplished. Generally, the same considerations that apply to wild ungulates also apply to birds.

The impact of big sagebrush control on sage grouse has been a subject of considerable controversy. Big sagebrush provides essential food and cover for sage grouse. In Idaho, spraying big sagebrush with 2,4-D greatly reduced the number of nesting grouse (Klebenow 1970). Sage grouse did not return to the location for nesting until 5 years after spraying. Brooding use was less affected, although the number of broods declined during the 5 years after spraying. Ten years after spraying, the area was back to full nesting use. The same study showed that sage grouse avoided dense stands of big sagebrush.

The loss of forbs, important sage grouse foods during the summer, as well as the loss of nesting habitat, should be a primary concern when herbicides are used to control big sagebrush. However, forbs will generally return to prespray abundance within 5 years (Klebenow 1969). Large block removal of sagebrush generally reduces sage grouse numbers (Klebenow 1969). However, spraying areas with a sagebrush cover in excess of 30% in small blocks or strips can benefit sage grouse (Rogers 1964).

Fire may be a better tool than herbicides to control sagebrush from the standpoint of sage grouse (Klebenow 1972). Patches of sagebrush can be left unburned. Also, fire is less destructive to forbs than are herbicides such as 2,4-D.

Maintenance of sage grouse habitat requires cooperation between state wildlife agencies and whatever land management agency plans sagebrush control (Braun et al. 1977).

Maps should be made of areas used by sage grouse for brooding grounds, nesting, and wintering. Control should not be applied where sagebrush cover is less than 20% and where slopes are greater than 20%. Nor should control measures be applied within a 2-mile (1.61 km) radius of brooding grounds, on known wintering and nesting areas, or within 100 yards (90 m) of streams and meadows.

An alternative to spraying constant amounts of herbicide over large blocks to eradicate all brush is to spray with variable amounts of herbicide over small blocks. This type of scheme was used by Scifres and Koerth (1986) in southcentral Texas. This strategy greatly enhanced diversity in vegetation structure and species composition compared to that existing prior to treatment. Edge effect was also greatly amplified in their study. This type of spray program appears to have much potential to enhance wildlife populations on big sagebrush ranges in western states.

GAME RANCHING

There are two basic types of game ranching operations: (1) fee-for-hunting enterprises, where game animals are harvested for sport, and (2) the raising of native animals for the production of meat and other products. Each type of operation will be discussed as follows.

Fee-Hunting Enterprises

Economic returns from fee hunting can exceed those from livestock in some localities. Fee-hunting operations have been most successful where the following criteria are met:

1. Opportunity to hunt on public lands is restricted.

2. Hunting as a recreational pursuit is culturally acceptable and prestigious.

3. A large, affluent human population occurs in close proximity to rangelands capable of producing high populations of game animals.

4. Game animals and domestic livestock are complementary in their use of the range forage resource.

All of the foregoing criteria are met in central Texas, where fee hunting has been most successful. Ramsey (1965) reported that returns over a 6-year period were $38.60 per animal unit of deer (6 deer = 1 animal unit) and $28.22 per animal unit of livestock (1 cow = 1 animal unit) at the Kerr Wildlife Management Area in central Texas. Net income per deer harvested was over $50 per deer at the Sonora Research Station in Texas between 1971 and 1975 (Reardon et al. 1978). Presently, ranchers in central Texas can net about $10 to $15 per acre on white-tailed deer compared to $6 to $8 per acre on livestock. Because food habits of cattle complement those of white-tailed deer, the returns from livestock and white-tailed deer are additive if animal numbers are kept in balance with the forage supply. Light to moderate cattle grazing in the central Texas area is considered necessary to maintain ideal conditions for white-tailed deer. In addition to white-tailed deer, mule deer, wild turkey, bobwhite quail, scaled quail, mourning doves (*Zenaida macroura*), whitewing doves (*Zenaida asiatica*), pheasants, javelina, and several exotic big-game animals are hunted under a fee system in Texas.

In the Chihuahuan desert of southwest Texas, Butler and Workman (1993) found the typical fee-hunting enterprise provided a total net revenue of about $7,900. Average annual net grazing returns per livestock animal unit were smaller on fee-hunting ranches but fee-hunting revenue offset the difference. The fee-hunting enterprises also reduced risk by providing a second source of cash returns.

Teer (1975) describes the various types of hunting leases used in the United States. These include season leases, day leases, broker leases, and direct charge for the animal.

Season leases provide a hunter or group of hunters with the exclusive hunting privileges for specified game species for one season. This has been the common type of hunting lease in the western United States. Presently, rates vary from about $300 to over $3,000 per hunter per season, depending on the quality of the hunting and the facilities provided. Some advantages of this system are that it requires low labor inputs, income is usually known in advance, and the landowner usually can get to know the hunters personally.

The day-hunting system involves a fee for each day the hunter spends afield. The fees are often staggered so that they are highest in the early part of the hunting season, when game is more available and less wary. Day-hunting leases have commonly been used for waterfowl, doves, quail, and other species of upland birds. This system is becoming more popular for big game because it usually maximizes net returns per unit area of land. The major disadvantages of this system are that it requires high labor inputs by the landowner, and there is usually poor familiarity with the hunter. Presently, day leases for game birds range from $50 to $70 per day for doves to over $300 per day in some of the better quail and waterfowl hunting areas. Day leases for deer in Texas range from $100 to $300 per day depending on the area and what part of the hunting season is involved.

The hunting "broker" or hunting "outfitter" system has been common for big-game animals (particularly elk) in the Rocky Mountain region of the United States and for waterfowl along the Texas coast. Hunting rights on tracts of land owned by several landowners are acquired by the outfitter, who runs the area as a shooting preserve. This system involves the least effort on the part of the landowner but usually gives the lowest returns per unit area.

Direct charge for the animal harvested is becoming increasingly popular for big game. Fees for a trophy elk in New Mexico are as high as $7,000, while a trophy white-tailed deer in Texas will bring $2,000 to $4,000 (White 1987).

Native game animals are owned by the state rather than the landowner in the United States. However, landowners can get around this problem since they can control who hunts, and how they hunt, on their property. Fences that restrict big-game animals to the landowner's property are commonly used on Texas ranches where direct charge is used.

Game Animals as a Source of Meat

The husbandry of game animals as a source of meat has not been widely practiced in the United States but is developing in parts of Africa. The fundamental concept has been that native animals are better adapted to local conditions than are domestic livestock. However, in recent years, demand for game meat has increased because of its low fat content compared to that of livestock (White 1987). In areas with temperate, moderate climates such as the United States and western Europe, domestic cattle, sheep, and goats are well adapted to environmental conditions. In hot, humid and cold, subarctic climates, native animals have definite advantages over cattle, sheep and goats. The classic example is the husbandry of reindeer in northern Scandinavia by Lapps. In Africa, native ungulates are much better adapted to the stresses of heat, disease, low forage quality, and poor water availability than are domestic livestock.

Although the introduction of cattle into Africa dates back several hundred years, they have not fully adapted to the African environment. Zebu cattle (*Bos indicus*) are best adapted to African conditions. With the most advanced management in Africa, annual calf crops can exceed 80%, with death losses of about 5% (Skovlin 1971). However, under existing conditions, calf crops are below 50% and annual mortality is 15% to 20%. During the drought in the early and mid-1980s, over 70% of the cattle in parts of northern Africa died due to water deprivation and lack of forage.

The native African antelope uses water more efficiently than do domestic cattle. These antelopes have lower losses of water through sweating and excretion of body wastes than do cattle (Kyle 1972; Price 1985). By feeding at night when relative humidity is highest, the African antelopes are able to consume forage with more moisture than if feeding occurred during the daytime, as in the case of cattle. Some of these antelopes can obtain all or nearly all their moisture from the forage they consume (Taylor 1968).

Disease resistance is another major advantage of native ungulates over cattle in Africa. In most parts of Africa, cattle require periodic vaccination, dipping, or spraying to keep them free of diseases such as anthrax, rinderpest, anaplasmosis, and contagious bovine pleuropneumonia. Tsetse flies and the associated trypanosome parasite excluded cattle from nearly 40% of Africa until recent years, when control was possible with insecticides. The aforementioned diseases and parasites have only a small influence on the health of native ungulates.

Probably the best argument for native ungulates over domestic livestock rests on the fact that native species utilize the range forage resources most efficiently. The highest production of meat per unit area occurs when several ungulates with different foraging habits share a common rangeland. Based on studies of several ungulates in northcentral Tanzania, Lamprey (1963) suggested that differences in use of terrain and food types resulted in complementary rather than a competitive relationship among the various animals. Some of the animals were grazers, whereas others were browsers or intermediate feeders. Animals ate different parts of particular plant species. For example, zebras (*Equus burchelli*) ate the fibrous, upper stems of grasses, while the small antelope used the more digestible leaves. Thomson's gazelles (*Gazella thomsonii*) selected new shoots from the base of the grasses (leaves). Animals used different levels of vegetation (Figure 13.7). Giraffes (*Giraffa camelopardalis*) used browse at higher levels on particular tress than did the shorter-browsing antelopes. In contrast to the above situation, cattle typically use only a few forage species and the more convenient parts of the range.

The African antelopes appear superior to domestic livestock with regard to growth, live-weight gains, and age at which they first breed. African cattle breed at 2 to 3.5 years,

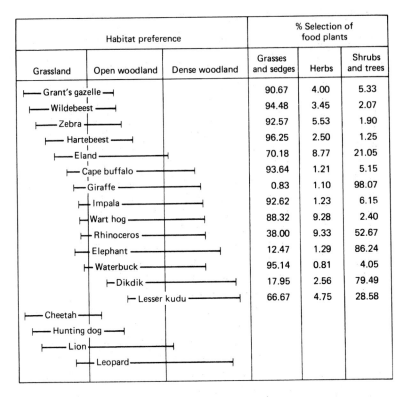

Habitat preference			% Selection of food plants		
Grassland	Open woodland	Dense woodland	Grasses and sedges	Herbs	Shrubs and trees
├── Grant's gazelle ──┤			90.67	4.00	5.33
├── Wildebeest ──┤			94.48	3.45	2.07
├── Zebra ─┼─┤			92.57	5.53	1.90
├── Hartebeest ──┤			96.25	2.50	1.25
├── Eland ──────────┤			70.18	8.77	21.05
├── Cape buffalo ─────┤			93.64	1.21	5.15
├── Giraffe ─────┤			0.83	1.10	98.07
├─ Impala ──────┤			92.62	1.23	6.15
├─ Wart hog ──────┤			88.32	9.28	2.40
├─ Rhinoceros ─────┤			38.00	9.33	52.67
├─ Elephant ────────┤			12.47	1.29	86.24
├─ Waterbuck ──────┤			95.14	0.81	4.05
├─ Dikdik ─┼────────┤			17.95	2.56	79.49
├─ Lesser kudu ──────┤			66.67	4.75	28.58
├── Cheetah ─┼─┤					
├── Hunting dog ──┤					
├── Lion ─┼─┤					
├─ Leopard ──────┤					

Figure **13.7** Habitat preference and food preference based on frequency observations along transects in two areas of Tanzania. (Adapted from Lamprey 1963 and Heady 1975.)

while sheep and goats breed at just under 1 year (Pratt and Gwynne 1977). Gazelle and impala (*Aepyceros melampus*) breed when under 1 year old, and topi (*Damaliscus korrigum*), kongoni (*Alcelaphus buselaphus*), and wildebeest (*Connochaetes taurinus*) breed when just over 1 year old. Growth rate and live-weight gain data indicate that various wild animals reach an economically marketable size more quickly than do domestic stock (Talbot et al. 1961; Kyle 1972).

Casebeer (1978) reported on the success of game-cropping activities with the Masai herdsmen in Kenya. He developed a program in which a portion (40%) of the revenues from hunting, game viewing, sale of animals to zoos, and the harvest of animals for meat, skins, and hides was transferred back to the herders rather than going entirely to the government. This program was effective in increasing the monetary income of the herders, and other tribes asked that the program be introduced to their lands. Because of the costs of harvesting, butchering, and refrigerating animals, it was found impossible to sell the meat at less than the price of beef. Therefore, the idea of harvesting wild animals to supply protein to the low-income native population was economically infeasible without government subsidization. The game ranching program was terminated by the Kenyan government in 1977 when all game hunting was banned in an effort to stop poaching. Based on results from this study, it does appear that game ranching can be economically successful in parts of Africa. Research in Kenya has indicated that the raising of game animals for meat can be more profitable than raising livestock under some conditions (Hopcraft 1988).

Most recently studies in Zimbabwe have challenged claims that more efficient resource use and greater profits can result from game ranching (safari-style hunting) than from cattle production in semiarid African savannas (Kreuter and Workman 1994; Kreuter and Workman 1996). It was found that cattle ranches in areas with sparse wildlife populations produced the greatest profits per hectare but only cattle plus wildlife ranches were profitable in areas with abundant wildlife (Kreuter and Workman 1996). The authors did mention that their results did not account for government policy effects on ranch profits. They also stated the long-term biological effects of differential stocking pressures on livestock and wildlife ranches were not taken into account.

In the United States, there has been interest in raising bison on a commercial basis. Compared to cattle, bison have a more efficient digestive system and can exist on lower-quality feed (Peden et al. 1974; Hawley et al. 1981). During grazing, bison disperse more and are less selective than cattle (Peden et al. 1974). Therefore, they use rangeland more efficiently and have less influence on plant species diversity. Feedlot trials have shown that in the summer bison make better weight gains than cattle on low- to medium-quality forages (Hawley et al. 1981). However, the reverse occurs during the winter (Table 13.4). Unlike cattle, bison have decreased metabolic rates and rates of gain when provided with high-quality feed during the winter (Hawley et al. 1981). This is indicative of differences in adaptive strategies in wild versus domestic ruminant species. Other studies with deer (Silver et al. 1969; Ozoga and Verme 1970; Kirkpatrick et al. 1975) and elk (Westra and Hudson 1981) in winter show that intake and metabolic rate are depressed even when high-quality feed is

TABLE 13.4   Feed Intake, Digestibility, and Average Daily Gain (ADG) of Bison and Hereford Steers Fed Native Sedge Hay

SEASON	HAY CRUDE PROTEIN (%)	BODY WEIGHT (KG)	DAILY FEED INTAKE DM/W[a] (%)	DIGESTIBILITY (%)	ADG (KG/DAY)
			BISON		
Summer	7.8	244	1.6	52	0.42
Winter	8.5	314	1.6	51	0.04
			CATTLE		
Summer	7.8	310	1.4	45	0.05
Winter	8.5	375	2.0	47	0.31

Source: Adapted from Hawley et al. 1981.

[a]DM/W = kg feed dry-matter intake per kilogram of body weight.

provided. Beef cattle have been selected for rapid rate of gain when level of nutrition is not a major limitation. In wild animals, survival under conditions of low forage quantity and quality is of primary importance. The failure of bison to respond to a high nutritional plane throughout the year is a disadvantage compared to cattle if high-quality harvested forages are to be fed during the winter. Other problems with the commercial production of bison are that they have delayed sexual maturity (they breed at 2 to 3 years compared to 1.5 years for cattle), and they are more difficult than cattle to restrain and handle.

Probably the major limitation to raising game animals as a source of meat in Africa rests in the difficulty of slaughter and meat processing. The animals must be shot, eviscerated, and cooled before the meat can be processed. Slaughter facilities under field conditions are generally crude and often result in considerable loss of meat to heat spoilage and insects. Refrigerated trucks and trailers, which are used in handling kangaroo meat in Australia, can ameliorate some of this problem. After slaughter, transport problems further hamper the marketing of meat since Africa generally has a poor road system. Because of these problems, game meat was as costly as that from domestic livestock in Kenya (Casebeer 1978).

Most of the problems associated with game animal harvesting could be solved if the animals could be driven to the abattoir and provided with feed until slaughter. This has been possible with gazelle, zebra, and wildebeest in some locations. Dr. David Hopcraft, on his ranch in Kenya, has developed procedures that solve many of the problems associated with game animal harvests.

Another problem is that the sale of game meat, skins, hair, and horns is highly regulated to prevent the destruction and wanton slaughter of these animals by poachers. On the whole, it appears that game animals in both Africa and the United States will become more important as objects of sport hunting rather than as sources of meat. The reader is referred to White (1987) for a more detailed discussion of big-game ranching in the United States.

PROBLEMS WITH WILD HORSES AND BURROS

Wild horses (*Equus cabalus*) and burros (*Equus asinus*) were first introduced onto western ranges in the United States by the Spanish explorers in the late-sixteenth and early-seventeenth centuries. By the early-eighteenth century these animals had become abundant and were exerting a significant impact on rangeland vegetation. By the mid-nineteenth century their numbers in the West were estimated at 6 million (Cook 1975). Both horses and burros have proven well adapted to western ranges.

After the Taylor Grazing Act was passed in 1934 to control grazing on public land in the West, the need to control wild horses was recognized by the newly formed Grazing Service (now the Bureau of Land Management). Numbers of free-roaming horses were reduced by the enforcement of astray livestock laws (Cook 1975).

During the 1950s the capture of wild horses for use in dog and cat food became profitable. The capturing of wild horses on public lands on a large-scale basis was possible because the federal government had no legal jurisdiction over them. In addition, they were generally considered undesirable by ranchers and federal land managers. Motorized vehicles such as helicopters, airplanes, snowmobiles, pickups, and motorcycles were often used to run horses into box canyons or corrals. In some cases the horses were run until exhaustion. The often inhumane and indiscriminate means of capturing wild horses and, in some cases, wild burros for pet food led to the passage of the Wild Free Roaming Horse and Burro Act of 1971. Under this law, the public cannot capture or harass wild horses or burros on public lands. The law does provide for reduction measures when wild horse or burro population exceed range-carrying capacity. However, all methods that involve killing of the animals are forbidden. After capture, the horses can be given away but the federal agency must retain title to the animal. The recipient of the animal must make sure that both animal and its offspring receive good care (sale of the animal is forbidden). Under such constraints, population reductions are both difficult (in some cases impossible) and expensive.

Since passage of the 1971 Wild Horse and Burro Act, populations of both animals have shown drastic increases on many ranges (Wolfe 1980; Allen et al. 1981). These increases have had disastrous consequences on rangeland grazing capacity and native ungulate populations at some locations (Carothers et al. 1976; Miller 1983). Livestock numbers have had to be reduced to accommodate wild equine population buildups.

Because they have a cecal digestive system (see Chapter 11) and can cover longer distances than can domestic or native ruminants, wild horses and burros can remain in good health under forage conditions fatal to ruminants (Seegmiller and Ohmart 1981). Both wild horses (Miller 1981) and burros (Buechner 1960) will drive away livestock and native ungulates from watering and feeding areas. Wild horses and burros are capable of cropping forage much more closely than can wild or domestic ruminants since they have both upper and lower incisor teeth (ruminants have only lower incisors).

The Wild Horse and Burro Act of 1971 shifted the status of these animals from one extreme to another. Cook (1975) suggested that the act should be modified to liberalize methodology available to federal agencies for control (permit use of aircraft), permit sale and title transfers of excess animals, and permit complete removal of animals from areas where they jeopardize native ungulates (particularly bighorn sheep) and cannot be managed. These modifications seem essential for the sustained welfare of the rangeland, native wildlife, and the wild horses and burros. The reader is referred to Douglas and Leslie (1996) for a more complete discussion of wild horses, burros, and other feral ungulates on rangelands of the United States.

PROBLEMS WITH SMALL MAMMALS

Rodent and rabbit populations show considerable response to domestic livestock grazing. Some species are favored by heavy livestock grazing, whereas others show population declines (Fagerstone and Ramey 1996). Rodents and rabbits have been considered both symptoms and causes of rangeland degradation. Based on recent research, both can be true, depending on the situation. Jackrabbits and prairie dogs (*Cynomys ludovicianus*) have received the most attention by range researchers because of their high populations and their heavy use of range forages that might otherwise be used by livestock.

Jackrabbits

Jackrabbit populations, particularly those of black-tailed jackrabbits (*Lepus californicus*), increase with heavy livestock grazing (Taylor et al. 1935; Phillips 1936; Brown 1947). Under moderate grazing intensities, jackrabbits favor succession toward more grasses (Daniel et al. 1993a), while on heavily degraded ranges more forbs and shrubs are favored (Bond 1945).

Food habits of jackrabbits are quite similar to those of domestic sheep. Data from Sparks (1968) in Colorado show that black-tailed jackrabbits make considerable use of forbs throughout the year but grasses are used primarily in the spring and summer, when they are green and succulent (Table 13.5). Other studies involving jackrabbit foods have shown similar findings. Jackrabbits will use shrubs when grasses and forbs are unavailable (Bear and Hansen 1966; Currie and Goodwin 1966; Daniel et al. 1993a). Shrubs considered unpalatable to livestock, such as big sagebrush, mesquite (*Prosopis* sp.), rabbitbrush (*Crysothamnus* sp.), and creosotebush (*Larrea tridentata*) are heavily consumed by jackrabbits at some locations. As grass availability declines, the impact of jackrabbit defoliation on the grasses becomes increasingly severe. On a highly deteriorated range in New Mexico, jackrabbits removed nearly all of the standing crop of perennial grass.

Although there has been considerable speculation about the impact of jackrabbits on range vegetation, few data are available documenting their influences. In a study of several exclosures protected and not protected against jackrabbits for various periods in northern Utah, Rice and Westoby (1978) found that protection against jackrabbits had no definite effect on vegetation succession. In southcentral New Mexico plots with long-term exclusion of jackrabbits had lower cover of both shrubs and grasses than plots where jackrabbits were not excluded (Gibbens et al. 1993).

Feeding trials conducted by Arnold (1942) show that jackrabbits consume about 6.5% of their body weight per day on a dry-matter basis. This is over three times the daily rate of most ruminants on range forages. Vorhies and Taylor (1933) derived estimates of the number of jackrabbits required to consume as much forage as cattle and sheep. They assumed a jackrabbit daily dry-matter intake of 6% of body weight; thus their calculations indicate that 30 and 148 black-tailed jackrabbits eat as much forage as one sheep and one cow, respectively. Antelope jackrabbits (*Lepus alleni*) had equivalency figures of 15 and 74, re-

TABLE 13.5   Percentage Composition of Black-tailed Jackrabbit Diets throughout the Year in Colorado

PLANT TYPE	SEPTEMBER 28	DECEMBER 1	FEBRUARY 8	APRIL 29	JUNE 29	AUGUST 1
Grasses	18	57	7	87	57	63
Forbs	43	26	52	23	43	36
Shrubs	38	17	40	0	0	0

Source: Adapted from Sparks 1968.

spectively, for sheep and cattle. On salt desert range in Utah, Currie and Goodwin (1966) found that black-tailed jackrabbits consumed or wasted 260 grams of forage compared to 1,500 grams for sheep. Only 5.8 black-tails were required to equal one sheep. Regardless of the figures used, it is obvious that jackrabbits remove considerable forage that could otherwise be used by livestock and big-game animals.

Jackrabbit populations show great fluctuations between years due to weather and forage conditions and migration from other areas. During favorable periods, their numbers can build up quite rapidly due to their high reproductive potential. Goodwin (1960) reported populations as high as nine jackrabbits per hectare when they were concentrated in winter. Control is generally effective when densities exceed one jackrabbit per hectare.

Although shooting has been widely used for jackrabbit control, this method has been only partially effective. Rabbit drives which involve rounding up jackrabbits and killing them by various means has been somewhat successful during periods when jackrabbits are extremely abundant.

A good control measure is the use of strychnine-treated grain or alfalfa hay during the winter when the ground is covered with snow. In the spring and summer, strychnine-treated salt placed in holes in the ground has given good results. Prevention of excessive jackrabbit populations by maintaining the range in good condition is the best means of avoiding a jackrabbit problem (Daniel et al. 1993b) (Table 13.6).

Prairie Dogs

The prairie dog has been the most highly disliked of all range rodents by ranchers because of its extensive use of land area and forage. Prairie dogs completely denude areas adjacent to their homes, with vegetation showing less use with greater distance from the safety of the burrow (Figure 13.8). Prairie dogs promote a high component of forbs and tend to eliminate grasses and shrubs by their foraging activities (Koford 1958). Food habit studies show that prairie dogs show a strong preference for grass (Fagerstone et al. 1977; Uresk 1984).

In the tallgrass prairie region, bison and/or cattle grazing may be necessary to maintain suitable conditions for prairie dogs. Osborn and Allen (1949) documented the demise of a prairie dog town in Wichita Mountains National Wildlife Refuge, Oklahoma, after heavy cattle grazing pressure was eliminated. The prairie dogs were evidently unable to retard the successional advance of the tallgrass climax. The loss of visibility associated with the encroachment of the tallgrasses on the town evidently caused the prairie dog to

TABLE 13.6   Density of Black-tailed Jackrabbits (Rabbits/ha) on Good and Fair Condition Chihuahuan Desert Ranges Averaged Across Two Years

RANGE CONDITION	SUMMER	FALL	WINTER	SPRING	AVERAGE
Good	0.42	0.46	0.53	0.67	0.69
Fair	0.83	0.62	0.50	0.98	0.98

Source: Adapted from Daniel et al. 1993b.

***Figure* 13.8** Ranchers generally dislike prairie dogs because they heavily defoliate areas surrounding their homes.

be highly vulnerable to predators. In the more arid shortgrass country of eastern Colorado, Montana, Wyoming, and New Mexico, it appears that prairie dogs can maintain a low vegetation without livestock or bison grazing influences.

Although prairie dog towns covered several million hectares in the early twentieth century, they are now rarely seen in most parts of the Great Plains because of intensive control programs implemented over several years. These programs have undoubtedly improved range condition and increased forage available to livestock. However, prairie dogs are very interesting animals from the naturalist's viewpoint, consequently, an effort has been made in recent years to maintain colonies on public lands.

Positive Impacts

The burrowing herbivorous small mammals, such as prairie dog, ground squirrels (*Citellus* sp.), and pocket gophers (*Thomomys* sp.), have some positive effects on the range ecosystem that often go unrecognized. These animals often speed soil development and mineral cycling. It has been estimated that mountain pocket gophers bring to the surface as much as 11 to 15 metric tons of soil per hectare in some locations (Ellison 1946; Ingles 1952). In Colorado, Koford (1958) estimated that 9 metric tons of soil per hectare were raised to the surface by prairie dogs. Thorp (1949) believed that as much as one-third of his study area near Akron, Colorado, may have been converted from silt loam to loam through rodent and badger activities. Prairie dogs move mineral-rich soil from deep layers and mix it with organic-rich soil near the surface (Koford 1958). Burrowing rodents relieve soil compaction by their digging activities and contribute organic matter in the form of feces, urine, and decay of their bodies. The reader is referred to Fagerstone and Ramey (1996) for a more detailed discussion of rodents and rabbits in rangeland ecosystems.

PROBLEMS WITH INSECTS

Insects can more severely overgraze ranges than can domestic livestock or small mammals. These negative effects have occurred through destruction of plant roots, direct consumption of forage, reduction in forage nutritional value and palatability, and destruction of seed.

Mormon crickets, range caterpillars, black grass bugs, and harvester ants have caused the greatest damage.

Of all the insects, grasshoppers (primarily *Melanoplus* sp.) have had the greatest influence on range vegetation. The composition and density of cover on native grasslands largely determines the distribution and abundance of grasshoppers (Anderson 1964). Several studies show that grasshoppers are more numerous on heavily grazed pastures than on moderately to lightly grazed pastures (Table 13.7) (Smith 1940; Nerney 1958; Holmes et al. 1979). This is because most grasshopper species prefer ranges with a sparse stand of grass and a high forb component. Arizona studies have shown that grasshoppers consume about equal amounts (30% of diet) of perennial grasses, annual grasses, and forbs (Nerney 1958, 1960). During drought years grasshoppers can consume up to 99% of all vegetation.

Grasshoppers cause considerable damage beyond the consumption of forage. They eat only part of the grass stems and blades they cut. They can graze closer than livestock and often destroy the growing points of grasses by eating into the crown of the plant.

Grasshopper populations fluctuate considerably from year to year. The interaction of several biotic and abiotic factors causes the periodic grasshopper outbreaks. Temperature (Edwards 1960) and precipitation (Symmons 1959) are the abiotic factors that most influence grasshopper outbreaks.

Presently, spraying with the insecticide malathion is recommend for control of grasshopper infestations (USDA 1967). Control appears economically advantageous when grasshopper numbers exceed 30 per square meter (Shewchuk and Kerr 1993).

TABLE 13.7   Mean Numbers of Grasshoppers per Fifty Sweeps on Fields at Stavely Substation, Alberta, Canada. Subjected to Four Rates of Grazing, 1970–1976

	INTENSITY OF GRAZING OF FIELD				
SPECIES OF GRASSHOPPER	LIGHT	MODERATE	HEAVY	VERY HEAVY	MEAN
Melanoplus dawsoni	46.6	52.4	90.0	132.0	80.3
Chorthippus longicornis	61.9	47.0	40.6	23.0	43.1
Camnula pellucida	18.9	14.6	18.7	27.7	20.0
Melanoplus sanguinipes	14.0	9.0	1.7	6.1	7.7
Melanoplus gladstoni	2.9	3.4	4.6	8.4	4.8
Melanoplus bivittatus	3.0	4.7	2.1	2.3	3.0
Neopodismopsis abdominalis	4.9	2.0	3.0	1.0	2.7
Melanoplus infantalis	1.3	1.6	.7	5.1	2.2
Aeropedellus clavatus	1.0	3.3	0.4	1.0	1.4
	154.5	138.0	161.8	206.6	

Source: From N. D. Holmes et al. 1979 (Table 1).

In the western United States, the Mormon cricket (*Anabrus simplex*) has been the most important insect pest. The most severe and widespread outbreaks occurred during the 1930s with a few more localized outbreaks being reported after 1940. Forbs with fleshy, succulent leaves are preferred foods for this insect, but they will consume almost any plant if preferred foods are unavailable. When conditions become favorable populations explode and they migrate in huge bands (they are flightless), destroying most of the vegetation in their path. They are capable of traveling approximately 0.8 km per day and will travel about 40 km in a season. Fence barriers of poison bait placed in front of migrating bands are the main method of control.

Unlike the grasshopper and the Mormon cricket, range caterpillars (*Hemileuca oliviae*) feed almost exclusively on grasses. It has been most damaging in northeastern New Mexico, although infestations have occurred in southeastern Colorado and the western edge of the Texas Panhandle. Outbreaks have occurred at 10- to 20-year intervals and have lasted 6 to 12 years. Infestations in New Mexico have reduced the livestock carrying capacity by 30% to 50% (Hewitt et al. 1974). Range caterpillars will eat down into the crown of the grass, wasting large portions of leaves and stems. Since meristematic tissue is destroyed, the impact is far more severe than heavy livestock grazing. The grazing of remaining plant parts is discouraged by the molted skin and spines of the active larvae, therefore, infested pastures are rendered completely unsuitable for grazing.

Large-scale seedings of various wheatgrasses have greatly increased forage production for livestock on North American ranges since 1940. However, black grass bugs (*Labops hesperius*) have infested these wheatgrass monocultures, causing considerable reduction in forage productivity and nutritional value (Campbell et al. 1984). Native rangelands are not greatly affected by this insect (Todd and Kamm 1974; Higgins et al. 1977) (Figure 13.9). Black grass bugs lay eggs in the stems of wheatgrass, where they hatch and the larvae feed (Todd and Kamm 1974). Chemically controlling this insect is not usually economically feasible because of the relatively high application costs on large areas of low land value (Campbell et al. 1984). Because the eggs overwinter in the wheatgrass straw, burning of the straw in early spring before growth can greatly reduce infestation problems (Todd and Kamm 1974). Grazing practices that do not permit accumulation of straw can also be effective. The development of grass cultivars through genetic engineering that are insect resistant due to chemical properties shows potential to reduce damage on seeded grass stands from black grass bugs as well as other insects (Campbell et al. 1984). Selected arthropods are also effective in controlling black grass bugs (Araya and Haws 1988).

Only about one-tenth of 1% of the hundreds of thousands of insects identified in the world are classified as harmful to man and his activities (Haws 1982). Several insects are beneficial to rangelands. Beneficial insect activities include decomposition of organic matter, digging holes in the soil that facilitate aeration and water infiltration, reduction in weeds, pollination of desirable plants, destruction of injurious insects, and production of useful products such as dyes, waxes, honey, and so on (Haws 1982). As with small mammals, most problems from insects on rangeland have been caused by human activities such as overgrazing or extensive land clearing and revegetation with monocultures that reduce habitat diversity. These practices generally destroy habitat for beneficial insects but create favorable conditions for those that are undesirable. The reader is referred to Watts et al. (1989) for detailed coverage of rangeland entomology.

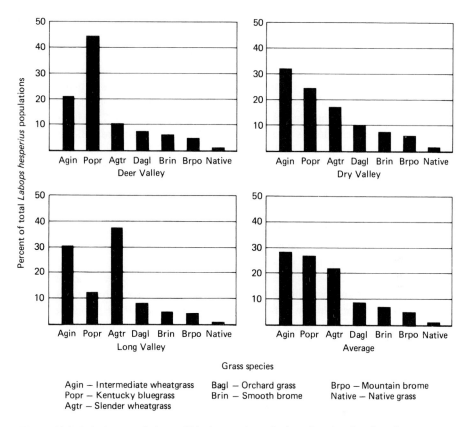

Figure **13.9** Relative populations of black grass bugs for introduced and native plant communities at three locations in Utah. (From K. M. Higgins et al. 1977, Fig. 3.)

PROBLEMS WITH PREDATORS

Predators, particularly coyotes (*Canis latrans*), have considerable influence on the range livestock industry in the United States. This subject is covered in detail by Andelt (1996). Black bears (*Ursus americanus*), golden eagles (*Aquila chrysaetos*), bobcats (*Lynx rufus*), foxes (*Vulpes* sp.), and mountain lions (*Felis concolor*) are other important predators on livestock. Grizzly bears (*Ursus horribilis*) and wolves (*Canus lupus*) presently occur in such low numbers that they cause little loss of livestock. However, they were major predators on livestock prior to the early twentieth century. Selection of the proper class and type of animal is one of the four basic principles in range management. Predators are a major consideration when the type of animal is selected. Predators have caused many ranchers to switch from sheep and goats to cattle over the last 50 years.

Sheep losses to coyotes in the 15 western states were estimated at 8.1% of lambs and 2.5% of adult sheep in 1974 (USDI 1978). Mortality from causes other than predation was 15.1% of the lambs and 7.9% of the adult sheep. The distribution of these losses among ranchers was highly uneven, with 45% of the sheep ranchers reporting no lamb losses but another 10% reporting losses over 20%.

Herding of sheep can greatly reduce losses to predators. However, there has been an increasing shortage of sheepherders over the last 30 years. DeLorenzo and Howard (1976) reported 15.6% and 12.1% losses of lamb crops to predators in two consecutive years in New Mexico for unherded sheep. More than 16% of the total flock was killed by predators annually in Montana when sheep were not herded (Henne 1975; Munoz 1976). In contrast, total annual death losses for a band of herded sheep in the Great Basin of the United States was 3.8% (McAdoo and Klebenow 1978). In the McAdoo and Klebenow (1978) study, coyotes accounted for 90% of the losses, bobcats for 2%, and 8% of the loss to predators was undetermined. Of the sheep attacked by predators, 93% were healthy and 7% were in poor condition.

Based on these data there is no reason to believe that predators select unfit and unhealthy individuals or that these animals are more susceptible as prey. In fact, the more active, healthier animals can be expected to be more frequently attacked since coyotes are attracted by movement of prey (Fox 1971).

Predation losses to livestock vary considerably between years and seasons on most ranges due to the availability of natural prey (Table 13.8). Rabbits and rodents are the main foods of coyotes (Sperry 1941; Ferrel et al. 1953; Grier 1957). Populations of these animals fluctuate considerably between years, depending on range conditions. Drought years result

TABLE 13.8   Average Predator Losses and Verified Predator Losses as a Percentage of Lambs Docked in Southwestern Utah, 1972-1975

YEAR	SEASON	TOTAL PREDATOR LOSS	VERIFIED PREDATOR LOSS	TOTAL PREDATOR LOSS	VERIFIED PREDATOR LOSS
1972	Spring	0.75	0.48	0.87	0.68
	Summer	6.21	1.25	7.54	0.32
1973	Spring	0.56	0.51	0.77	0.78
	Summer	4.14	0.97	3.77	1.96
1974	Spring	0.86	0.69	0.91	0.69
	Summer	4.96	1.87	4.34	1.35
1975	Spring	NA	NA	0.95	0.65
	Summer	NA	NA	1.94	1.35
Mean	Spring	0.72	0.56	0.88	0.70
	Summer	5.19	1.36	4.40	1.25

Source: From R. G. Taylor et al. 1979 (Table 3).

NA = Not Available.

in low prey populations, while the abundance of food following wet years favors high prey populations (Turkowski and Vahle 1977). In Texas, higher lamb losses to predators occurred during periods of low prey availability compared to those when availability was high (Pearson and Caroline 1981). A Nevada study showed that coyote predation on domestic sheep was highest in areas with the lowest natural prey/predator ratio (Kauffield 1977). Knowledge of the predator-prey relationship can be invaluable in reducing predation losses. Intensive herding practices and predator control are most advantageous during periods when the natural prey base is low.

Time of the year can influence the effectiveness of predator control programs. On central Texas ranges heaviest losses occurred from October to May when small lambs were present and natural prey was probably least available (Pearson and Caroline 1981). Predator control efforts were most successful during winters.

Terrain and vegetation composition can greatly influence predator numbers and ease of control. Irregular, brushy terrain provides high-quality habitat for coyotes, bobcats, and cougars. In Texas, predator losses in these areas were much higher than in the flatter, more grassy areas (Pearson and Caroline 1981). Woven-wire fencing to impede coyote travel, coupled with intensive aerial gunning periods, have made it economically possible to raise sheep in the flat eastern shortgrass country of New Mexico and the Edwards Plateau area of Texas.

Predators, mainly coyotes, killed 33% and 16% of the known goat kid crop on untreated and treated pastures, respectively, in south Texas (Guthery and Beasom 1978). Because predators apparently were responsible for most unknown losses, the true predation losses were estimated as high as 95% and 59%, respectively, of the known kid crop (Figure 13.10). The net kid crop under intense predator control was 27 times greater than that under no control. However, the kid crop under treatment was only 13.5% because predation losses were still high. Coyotes killed 25% of the nannies in the untreated pasture but none in the treated pasture. Nonpredator losses of untreated nannies was 10%. This study showed that intense localized predator control in regions with high coyote densities using traps, snares, and poison bait could curtail predation of adult goats, but would be insufficient to prevent heavy kid losses.

There are four basic methods of predator control: trapping, shooting, poisoning, and den hunting (Wagner 1972). Since coyotes will not enter a box or cage, trapping nearly always involves steel leg-hold traps with a bait or scent as an attractant. Shooting methods usually involve calling coyotes to the concealed hunter or gunning them from aircraft. Den hunting is practical in the spring when the pups are in the den and can be destroyed by asphyxiation using carbon monoxide cartridges or digging them out. Poisoning is practiced by poisoned meats or by use of a "coyote getter." The coyote getter is a small pipe embedded in the ground with a scent attractant which releases a spring loaded cartridge containing cyanide into the mouth of the coyote.

Poisoning has been the most controversial form of coyote control. Strychnine was one of the earliest poisons but has been largely replaced because of its nonselectivity. The compound 1080 (sodium monofluoracetate) was heavily used as a bait poison in the 1950s and 1960s. Because of public concern over loss of nontarget animals, President Nixon by executive order banned all poisons for predator control on federal land in 1972, followed by a restriction on the availability of chemical toxicants for state and private use in animal control by the U.S. Environmental Protection Agency.

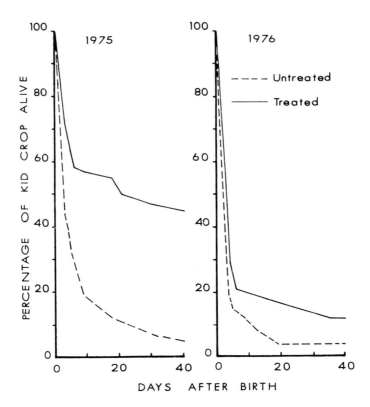

Figure 13.10 Forty-day survivorship of Angora kids in untreated (no predator control) and treated pastures, Zavala County, Texas. (From F. S. Guthery and S. L. Beasom 1978, Fig. 2.)

The long-term effectiveness of coyote control programs has been a subject of controversy. Data presented by Wagner (1972) do indicate that 1080 lowered coyote densities when it was applied in the northwestern states. However, there was no evidence that 1080 lowered sheep losses to predators when the period before application (1940–1949) was compared with the period of widespread application (1950–1970). The present policy is for the U.S. Department of Agriculture to restrict predator control activities to those ranchers suffering high known predator losses.

Since 1972, several nonlethal techniques have been applied to reduce coyote predation problems. The aversive conditioning agent, lithium chloride, reduced sheep predation losses in a Saskatchewan, Canada study (Jelinski et al. 1983) but other cases of less favorable results also exist. Aversive conditioning involves treating bait with an illness-inducing drug (lithium chloride) that causes gastrointestinal discomfort. The coyote then generalizes this association to live animals (sheep), thus suppressing predatory attacks. A number of chemosterilants and repellents also show potential. One of the more interesting approaches is the use of Komondor guard dogs trained to attack captive coyotes and stay within fenced sheep pastures (Linhart et al. 1979). Bonding of sheep to cattle and other animals shows promise as a tool to reduce sheep losses from predation (Hulet et al. 1987).

THREATENED AND ENDANGERED WILDLIFE

Since 1966, when President Johnson signed the Endangered Species Preservation Act, there has been increased concern for birds and mammals threatened with extinction. This support resulted in successive strenthenings of the original legislation culminating in the Endangered Species Act of 1973 signed by President Nixon. The original 1966 legislation only involved provisions for protection of about 70 species inhabiting federal lands but has now evolved to the point where thousands of species are covered across private and public lands. Although the Endangered Species Act of 1973 has had considerable public support it has many critics who complain it has cost the public millions of dollars from environmental assessments, mitigation fees, education, reduction in property values, and compromises in land development and management. On western rangelands the Endangered Species Act has generated great controversies such as the reintroduction of wolves in New Mexico, the reintroduction of masked bobwhite quail in southern Arizona, and the protection of the desert tortoise in the Mojave Desert of California (Carrier and Czech 1996).

The biggest problem confronting ranchers in regard to the Endangered Species Act is that if a threatened or endangered species occurs on their property or grazing allotment it is a liability rather than an asset. This has lead to considerable resentment against the Endangered Species Act by property owners. Reformists have advocated alternative approaches that would reward landowners for protection of threatened and endangered species habitat through tax breaks, income from a biodiversity trust funded from public land user fees, and other incentives (Anderson and Leal 1991; Holechek and Hess 1995). Without question the maintaining of endangered species and biological diversity will provide great challenges and opportunities for rangeland managers in the twenty-first century. While a detailed discussion of this subject is beyond the scope of our book it is well covered by Anderson and Leal (1991) and Carrier and Czech (1996).

RANGE MANAGEMENT PRINCIPLES

- Maintaining rangelands in late seral ecological condition (about 55% to 75% of climax vegetation) will maximize both plant and wildlife diversity in most grassland and shrubland range types.

- Moderately grazed rangelands generally support more diverse wildlife populations than those that are heavily grazed or ungrazed.

- Problems with insects, jackrabbits, prairie dogs, and other undesirable wildlife are generally caused by heavy grazing.

- Specialized grazing systems can be effective in enhancement of populations of some wildlife species.

Literature Cited

Allen, B. H. 1988. Wildlife recreation and other resources. In *The Vale rangeland rehabilitation program: An evaluation.* U.S. Dept. Agr. Forest Serv. PNW-RB-157.

Allen D. L. 1974. *Our wildlife legacy.* 3d ed. Funk & Wagnalls, Inc., New York.

Allen, D. L., L. Erickson, E. R. Hall, and W. M. Schiraa. 1981. *A review and recommendations on animal problems and related management needs in units of the National Park System.* Report to Secretary of the Interior. Mimeograph.

Andelt, W. F. 1996. Carnivores. In P. R. Krausman (Ed.), *Rangeland wildlife,* pp. 135–151. The Society for Range Management, Denver, CO.

Anderson, E. W., and R. J. Scherzinger. 1975 Improving quality of winter forage for elk by cattle grazing. *J. Range Manage.* 28:2–7.

Anderson, N. L. 1964. Some relationships between grasshopper and vegetation. *Ann. Entomol. Soc. Am.* 57:736–742.

Anderson, T. L., and D. L. Leal. 1991. *Free market environmentalism.* Westview Press, Boulder, CO.

Araya, J. E., and B. A. Haws. 1988. Arthropod predation of black grass bugs (*Hemiptera:Miridae*) in Utah ranges. *J. Range Manage.* 41:100–103.

Arnold, J. R. 1942. Forage consumption and preferences of experimentally fed Arizona and antelope jackrabbits. *Univer. Ariz. Tech. Bull.* 98:50–86.

Baker, D. L., and F. S. Guthery. 1990. Effects of continuous grazing on habitat and density of ground foraging birds in south Texas. *J. Range Manage.* 43:2–6.

Bareiss, L. J., P. Schulz, and F. S. Guthery. 1986. Effects of short-duration and continuous grazing on bobwhite and wild turkey nesting. *J. Range Manage.* 39:259–260.

Bear, G. D., and R. M. Hansen. 1966. Food habits, growth, and reproduction of white-tailed jackrabbits in southern Colorado. *Colo. Agric. Exp. Stn. Tech. Bull.* 90. Fort Collins, CO.

Beasom, S. L., and C. J. Scifres. 1977. Population reactions of selected game species to aerial herbicide applications in south Texas. *J. Range Manage.* 30:138–143.

Bock, C. E., J. H. Bock, W. R. Kenney, and V. M. Hawthorne. 1984. Responses of birds, rodents, and vegetation to livestock exclosure in a semidesert grassland site. *J. Range Manage.* 37:239–243.

Bond, R. M. 1945. Range rodents and plant succession. *Trans. N. Am. Wildl. Nat. Resour. Conf.* 10:229–233.

Bowns, J. E., and C. F. Bagley. 1986. Vegetation responses to long-term sheep grazing on mountain ranges. *J. Range Manage.* 39:431–437.

Braun, C. E., T. Britt, and R. O. Wallestad. 1977. Guidelines for maintenance of sage grouse habitats. *Wildl. Soc. Bull.* 5:99–106.

Brown, D. E. 1978. Grazing, grassland, cover, and gamebirds. *Trans. N. Am. Wildl. Nat. Resour. Conf.* 43:477–485.

Brown, H. L. 1947. Coaction of jackrabbit, cottontails, and vegetation in a mixed prairie. *Kans. Acad. Sci. Trans.* 50:28–44.

Brown, J. W., and J. L. Schuster. 1970. Effects of grazing on a hardland site in the southern high plains. *J. Range Manage.* 23:418–424.

Bryant, F. C., C. A. Taylor, and L. B. Merrill. 1981. White-tailed deer diets from pastures in excellent and poor range condition. *J. Range Manage.* 34:193–201.

Buechner, H. K. 1950. Life history, ecology, and range use of the pronghorn antelope in Trans-Pecos Texas. *Am. Midl. Nat.* 43:257–354.

Buechner, H. K. 1960. The bighorn sheep in the United States, its past, present, and future. *Wildl. Monogr.* 4.

Butler, L. D., and J. P. Workman. 1993. Fee hunting in the Texas Trans-Pecos area: A descriptive economic analysis. *J. Range Manage.* 46:38–43.

Buttery, R. F., and P. W. Shields. 1975. Range management practices and bird habitat values. *Proc. Symposium on Management of Forest and Range Habitats of Nongame Birds. U.S. Dep. Agric. For. Serv. Gen. Tech. Rep.* WO-1.

Campbell, H., D. K. Martin, P. E. Ferkovich, and B. K. Harris. 1973. Effects of hunting and some other environmental factors on scaled quail in New Mexico. *Wildl. Monogr.* 34.

Campbell, W. F., B. A. Haws, K. H. Asay, and H. D. Hansen. 1984. A review of black grass bug resistance in forage grasses. *J. Range Manage.* 37:365–370.

Campbell-Kissock, L., L. H. Blankenship, and L. D. White. 1984. Grazing management impacts on quail during drought in the north Rio Grande Plain, Texas. *J. Range Manage.* 37:442–447.

Carothers, S. W., M. E. Stitt, and R. R. Johnson. 1976. Feral asses on public lands: An analysis of biotic impact, legal considerations, and management alternatives. *Trans. N. Am. Wildl. Nat. Resour. Conf.* 41:396–406.

Carrier, W. D., and B. Czech. 1996. Threatened and endangered wildlife and livestock interactions. In P. R. Krausman (Ed.), *Rangeland wildlife,* pp. 39–51. The Society for Range Management, Denver, CO.

Casebeer, R. L. 1978. Coordinating range and wildlife management in Kenya. *J. For.* 76:374–375.

Cook, C. W. 1966. Factors affecting utilization of mountain slopes by cattle. *J. Range Manage.* 19:200–204.

Cook, C. W. 1975. Wild horses and burro: A new management problem. *Rangeman's J.* 2:19–21.

Crawford, J. A., and E. G. Bolen. 1976. Effects of land use on lesser prairie chicken populations in west Texas. *J. Wildl. Manage.* 40:96–104.

Currie, P. O., and D. L. Goodwin. 1966. Consumption of forage by black-tailed jackrabbits on salt-desert ranges in Utah. *J. Wildl. Manage.* 30:304–311.

Daniel, A., J. L. Holechek, R. Valdez, A. Tembo, L. Saiwana, M. Fusco, and M. Cardenas. 1993a. Range condition influences on Chihuahuan desert cattle and jackrabbit diets. *J. Range Manage.* 46:296–302.

Daniel, A., J. L. Holechek, R. Valdez, A. Tembo, L. Saiwana, M. Fusco, and M. Cardenas. 1993b. Jackrabbit densities on fair and good condition Chihuahuan desert range. *J. Range Manage.* 46:524–529.

Darr, G. W., and D. A. Klebenow. 1975. Deer, brush control, and livestock on the Texas Rolling Plains. *J. Range Manage.* 28:115–119.

Davis, C. A., T. Z. Riley, R. A. Smith, R. A. Suminski, and M. J. Wisdom. 1979. *Habitat evaluation of lesser prairie chickens in eastern Chaves County, New Mexico.* N. Mex. Agric. Exp. Stn.

de Calesta, D. S., J. G. Nagy, and J. A. Baily. 1975. Starving and refeeding mule deer. *J. Wildl. Manage.* 39:663–669.

DeLorenzo, D. G., and V. W. Howard, Jr. 1976. *Evaluation of sheep losses on a range lambing operation without predator control.* New Mexico Agriculture Experiment Station Final Report to the U.S. Fish and Wildlife Service, Denver Research Center, Denver, CO.

Denniston, R. H. II. 1956. Ecology, behavior and population dynamics of the Wyoming or Rocky Mountain moose, *Alces alces shirasi. Zoologica* 41:105–118.

Douglas, C. L., and D. M. Leslie, Jr. 1996. Feral animals on rangelands. In P. R. Krausman (Ed.), *Rangeland wildlife,* pp. 281–295. The Society for Range Management, Denver, CO.

Duff, D. A. 1979. Riparian habitat recovery on Big Creek, Rich County, Utah. In *Forum-grazing and riparian stream ecosystems,* pp. 91–92. Trout Unlimited, Denver, CO.

Edwards, R. L. 1960. The relationship between grasshopper abundance and weather conditions in Saskatchewan, 1930–1935. *Can. Entomol* 92:406–408.

Ellison, L. 1946. The pocket gopher in relation to soil erosion on mountain range. *Ecology* 33:177–186.

Ellison, L. 1960. Influence of grazing on plant succession of rangelands. *Bot. Rev.* 26:1–78.

Elmore, W., and B. Kauffman. 1994. Riparian and watershed systems: Degradation and restoration. In *Ecological implications of livestock herbivory in the West.* Society for Range Management, Denver, CO.

England, R. E., and A. De Voss. 1969. Influence of animals on pristine conditions on the Canadian grasslands. *J. Range Manage.* 22:87–94.

Evans, K. E., and R. R. Krebs. 1977. Avian use of livestock watering ponds in western South Dakota. *U.S. Dep. Agric. For. Serv. Gen. Tech. Rep.* RM-35, Fort Collins, CO.

Fagerstone, K. A., H. T. Tietjen, and G. K. La Voie. 1977. Effects of range treatment with 2,4-D on prairie dog diet. *J. Range Manage.* 30:57–61.

Fagerstone, K. A., and C. A. Ramey. 1996. Rodents and lagomorphs. In P. R. Krausman (Ed.), *Rangeland wildlife,* pp. 83–133. The Society for Range Management, Denver, CO.

Ferrel, C. M., H. R. Leach, and D. F. Tillotson. 1953. Food habits of the coyote in California. *Calif. Fish Game* 39:301–341.

Fitzgerald, J. P. 1978. Vertebrate associations in plant communities along the South Platte River in southeastern Colorado. In *Proc. Lowland River and Stream Habitat Symposium,* pp. 73–88. Greeley, CO.

Fox, M. W. 1971. *Behavior of wolves, dogs, and related canids.* Harper & Row, Publishers, Inc., New York.

Gallizioli, S. 1977. Overgrazing on desert bighorn ranges. *Trans. Desert Bighorn Counc.* 21:21–23.

Gibbens, R. P., K. M. Havstad, D. D. Bullheimer, and C. H. Herbel. 1993. Creosotebush vegetation after 50 years of lagomorph exclusion. *Oecologia* 94:210–217.

Gjersing, F. M. 1975. Waterfowl production in relation to rest-rotation grazing. *J. Range Manage.* 28:37–42.

Glazener, W. C. 1967. Management of the Rio Grande turkey. In *The wild turkey and its management,* pp. 453–493. The Wildlife Society, Washington, DC.

Goodwin, D. L. 1960. Seven jackrabbits equal one ewe. *Farm Home Sci.* 21:38–39.

Grange, W. B. 1948. Wisconsin grouse problems. *Wis. Conserv. Rep. Publ.* 338.

Grenfell, W. E., B. M. Browning, and W. E. Stienecker. 1980. Food habits of California upland game-birds. *Calif. Dep. Fish Gam Rep.* 80-1.

Grier, T. T. 1957. Coyotes in Kansas. *Kans. Agric. Exp. Stn. Bull.* 393.

Gross, B. D., J. L. Holechek, D. Hallford, and R. D. Pieper. 1983. Effectiveness of antelope pass structures in restriction of livestock. *J. Range Manage.* 36:22–24.

Guthery, F. S. 1996. Upland gamebirds. In P. R. Krausman (ed.), *Rangeland wildlife,* pp. 59–71. The Society for Range Management, Denver, CO.

Guthery, F. S., and S. L. Beasom. 1978. Effects of predator control on angora goat survival in south Texas. *J. Range Manage.* 31:168–173.

Hamerstrom, F. N. Jr. and F. Hamerstrom. 1961. Status and problems of North American grouse. *Wilson Bull.* 73:284–294.

Hammerquist-Wilson, M. M., and J. A. Crawford. 1981. Response of bobwhites to cover changes within three grazing systems. *J. Range Manage.* 41:213–218.

Hawley, A. W. L., D. G. Peden, and W. R. Stricklin. 1981. Bison and Hereford steer digestion of sedge hay. *Can. J. Anim. Sci.* 61:165–174.

Haws, A. B., 1982. An introduction to beneficial and injurious rangeland insects of the western United States. *Utah Agric. Exp. Stn. Spec. Rep.* 23.

Hazell, D. B. 1967. Effect of grazing intensity on plant composition, vigor, and production. *J. Range Manage.* 20:261–264.

Heady, H. F. 1975. *Rangeland management.* McGraw-Hill Book Company, New York.

Heady, H. F., and J. Bartolome. 1977. The Vale rangeland rehabilitation program. The desert repaired in southeastern Oregon. *U.S. For. Serv. Res. Bull.* PNW-70, Portland, OR.

Henne, D. R. 1975. *Domestic sheep mortality on a western Montana ranch.* M.S. Thesis. University of Montana, Missoula, MT.

Hewitt, G. B., E. W. Huddleston, R. S. Lavigne, D. N. Ueckert, and J. G. Watts. 1974. *Rangeland entomology.* Range Science Series No. 2. Society for Range Management, Denver, CO.

Higgins, K. M., J. E. Borons, and B. A. Haws. 1977. The black grass bug (*Labops hesperius uhler*): Its effect on several native and introduced grasses. *J. Range Manage.* 30:380–384.

Holechek, J. L. 1980. Concepts concerning forage allocation to livestock and big game. *Rangelands* 2:158–160.

Holechek, J. L. 1981. Brush control impacts on rangeland wildlife. *J. Soil Water Conserv.* 26:265–270.

Holechek, J. L. 1992. Financial aspects of cattle production in the Chihuahuan desert. *Rangelands* 14:145–149.

Holechek, J. L., R. Valdez, R. Pieper, S. Schmemnitz, and C. Davis. 1982. Manipulation of grazing to improve or maintain wildlife habitat. *Wildl. Soc. Bull.* 10:204–210.

Holechek, J. L., and K. Hess Jr. 1995. Government policy influences on rangeland conditions in the United States: A case example. *Environmental Monitoring and Assessment* 37:179–187.

Holmes, N. D., D. S. Smith, and A. Johnston. 1979. Effect of grazing by cattle on the abundance of grasshoppers on fescue grassland. *J. Range Manage.* 32:310–312.

Holscher, C. E. and E. J. Woolfolk. 1953. Forage utilization by cattle on northern Great Plains ranges. *U.S. Dep. Agric. Circ.* 918.

Hopcroft, D. 1988. An ecological approach to natural ranching. *Proc. Intn'l Wildlife Ranching Symp.* 1:73–82. Las Cruces, NM.

Hulet, C. V., D. M. Anderson, J. N. Smith, and W. L. Shupe. 1987. Bonding of sheep to cattle as an effective technique for predation control. *Applied Anim. Behav. Sci.* 19:19–25.

Ingles, L. G. 1952. The ecology of the mountain pocket gopher, *Thomomys monticola. Ecology* 33:87–95.

Jackson, A. S. 1969. A handbook for bobwhite quail management in the west Texas Rolling Plains. *Tex. Parks Wildl. Rep. Bull.* 48.

Jelinski, D. E., R. C. Rounds, and J. R. Jowsey. 1983. Coyote predation on sheep, and control by aversive conditioning in Saskatchewan. *J. Range Manage.* 6:16–20.

Jensen, C. H., A. D. Scotter, and G. W. Scotter. 1972. Guidelines for grazing sheep on rangelands used by game in winter. *J. Range Manage.* 25:346–352.

Jensen, P. N., and C. M. Schumacher. 1969. Changes in prairie plant composition. *J. Range Manage.* 22:57–60.

Johnston, A. 1961. Comparison of lightly grazed and ungrazed range in the fescue grassland of southwestern Alberta. *Can. J. Plant Sci.* 41:615–622.

Jones, K. 1981. *Effects of grazing and timber management on Merriam's turkey habitat in mixed conifer vegetation of southcentral New Mexico.* M.S. Thesis. New Mexico State University, Las Cruces, NM.

Jourdonnais, C. S., and D. J. Bedunah. 1990. Prescribed fire and cattle grazing on an elk winter range in Montana. *Wildl. Soc. Bull.* 18:232–240.

Julander, O., W. L. Robinette, and D. A. Jones. 1961. Relation of summer range condition to mule deer herd productivity. *J. Wildl. Manage.* 25:54–60.

Kaiser, P. H., S. S. Berlinger, and L. H. Fredrickson. 1979. Response of blue-winged teal to range management on waterfowl production areas of southeastern South Dakota. *J. Range Manage.* 32:295–299.

Kantrud, H. A. 1990 Effects of vegetation manipulation on breeding waterfowl in prairie wetlands— A literature review. In Can livestock be used as a tool to enhance wildlife habitat? *U.S. Dept. Agr. Forest Serv. Gen. Tech. Rep.* RM-194.

Kauffield, J. 1977. *Availability of natural prey and its relationship to coyote predation on domestic sheep.* M.S. Thesis. University of Nevada, Reno, NV.

Kautz, J. E., and G. M. Van Dyne. 1978. Comparative analyses of diets of bison, cattle, sheep, and pronghorn antelope on shortgrass prairie in northeastern Colorado. *Proc. Int. Rangel. Congr.* 1:438–443.

Kessler, W. B., and J. D. Dodd. 1978. Response of coastal prairie vegetation and Attwater prairie chickens to range management practices. *Proc. Int. Rangel. Congr.* 1:473–476.

Kiel, W. M., Jr. 1976. Bobwhite quail population characteristics and management implications in south Texas. *Trans. N. Am. Wildl. Nat. Resour. Conf.* 41:407–420.

Kirby, D. R., and K. L. Gross. 1995. Cattle grazing and sharptailed grouse nesting success. *Rangelands* 17:124–127.

Kirkpatrick, R. L., D. E. Buckland, W. A. Abler, and W. A. Scanlon. 1975. Energy and protein influences on blood urea nitrogen of white-tailed deer. *J. Wildl. Manage.* 39:692–698.

Kirsch, L. M. and A. D. Kruse. 1972. Prairie fires and wildlife. *Proc. Tall Timbers Fire Ecol. Conf.* 12:289–305.

Klebenow, D. A. 1969. Sage grouse nesting and brood habitat in Idaho. *J. Wildl. Manage.* 33:649–662.

Klebenow, D. A. 1970. Sage grouse versus sagebrush control in Idaho. *J. Range Manage.* 23:649–662.

Klebenow, D. A. 1972. The habitat requirements of sage grouse and the role of fire management. *Proc. Tall Timbers Fire Ecol. Conf.* 12:305–315.

Klebenow, D. A. 1980. Impacts of grazing systems on wildlife. In *Proc. Grazing Management Systems Southwestern Rangelands Symposium,* New Mexico State University, Las Cruces, NM.

Klebenow, D. A. and G. M. Gray. 1968. The food habits of juvenile sage grouse. *J. Range Manage.* 21:80–83.

Koerth, B. H., W. M. Webb, F. C. Bryant, And F. S. Guthery. 1983. Cattle trampling of simulated ground nests under short-duration and continuous grazing. *J. Range Manage.* 36:385–387.

Koford, C. B. 1958. Prairie dogs, whitefaces and blue grama. *Wildl. Monogr.* 3.

Komberec, T. J. 1976. *Mule deer population ecology, habitat relationships, and relations to livestock grazing management in "breaks" habitat of eastern Montana and range relationships of mule deer, elk, and cattle in a rest-rotation grazing system during winter and spring.* Mont. Res. Proj. W-120, R-6,7, Study N-136-14, Job 2.

Krausman, P. R., Ed. 1996. *Rangeland wildlife.* The Society for Range Management, Denver, CO.

Krausman, P. R., R. Valdez, and J. A. Bissonette. 1995. Bighorn sheep and livestock. In P. R. Krausman (Ed.), *Rangeland wildlife,* pp. 237–243. The Society for Range Management, Denver, CO.

Kreuter, U. P., and J. P. Workman. 1994. Government policy effects on cattle and wildlife ranching in Zimbabwe. *J. Range Manage.* 47:264–269.

Kreuter, U. P., and J. P. Workman. 1996. Cattle and wildlife ranching in Zimbabwe. *Rangelands* 18:44–47.

Krueger, W. C., and A. H. Winward. 1974. Influence of cattle and big game grazing of understory structure of a Douglas fire-ponderosa pine-Kentucky bluegrass community. *J. Range Manage.* 30:53–57.

Kufeld, R. C. 1977. Improving Gambel's oak ranges for elk and mule deer by spraying with 2,4,5-TP. *J. Range Manage.* 30:53–57.

Kyle, R. 1972 Will the antelope recapture Africa? *New. Sci.* 53:640–643.

Lamprey, H. F. 1963. Ecological separation of the large mammal species in the Tarangire Game Reserve, Tanganyika. *East Afr. Wildl. J.* 1:63–92.

Laycock, W. A. 1967. How heavy grazing and protection affect sagebrush-grass ranges. *J. Range Manage.* 20:206–213.

Leckenby, D. A. 1968. Influences of plant communities on wintering mule deer. *Proc. Conf. West. Assoc. State Game Fish Comm.* 48:1–8.

Lehman, V. W. 1968. Influences of plant communities on wintering mule deer. *Procl Conf. West. Assoc. State Fish Comm.* 48:1–8.

Linhart, S. B., R. T. Sterner, T. C. Carrigan, and D. R. Henne. 1979. Komondor guard dogs reduce sheep losses to coyotes: A preliminary evaluation. *J. Range Manage.* 32:238–241.

Mackie, R. J. 1978. Impact of livestock grazing on wild ungulates. *Trans. N. Am. Wildl. Nat. Resourc. Conf.* 43:462–476.

Mapston, R. D. 1972. Guidelines for fencing an antelope range. *Proc. Biennial Antelope States Workshop.* 5:167–170.

Mapston, R. D., R. S. Zobell, K. B. Winter, and W. D. Dooley. 1970. A pass for antelope in sheep-tight fences. *J. Range Manage.* 23:457–459.

Marcum, C. L. 1975. *Summer-fall habitat selection and use by a western Montana elk herd.* Ph.D. Thesis. University of Montana, Missoula, MT.

Martin, A. C., H. S. Zim, and A. L. Nelson. 1951. *American wildlife and plants.* McGraw-Hill Book Company, New York.

McAdoo, J. R., and D. A. Klebenow. 1978. Predation on range sheep with no predator control. *J. Range Manage.* 31:111–114.

Merrill, L. B. 1975. Effects of grazing management practices on wild turkey habitat. *Proc. Natl. Wild Turkey Symp.* 3:108–112.

Merrill, L. B., J. G. Teer, and O. C. Wallmo 1957. Reaction of deer populations to grazing practices. *Tex. Agric. Prog.* 3:10–12.

Miller, R. 1981. Male aggression, dominance and breeding behavior in Red Desert feral horses. *Z. Tierpsychol.* 57:340–351.

Miller, R. 1983. Habitat use of feral horses and cattle in Wyoming's Red Desert. *J. Range Manage.* 36:195–199.

Miller, R. F. and W. C. Krueger. 1976. Cattle use on summer foothill rangelands in northeastern Oregon. *J. Range Manage.* 29:367–372.

Morgan, J. L. 1971. *Ecology of the Morgan Creek and East Fork of the Salmon River bighorn sheep herds and management of bighorn sheep in Idaho.* M.S. Thesis. Utah State University, Logan, UT.

Mueggler, W. F. 1950. Effects of spring and fall grazing by sheep on vegetation of the upper Snake River plains. *J. Range Manage.* 3:308–315.

Mundinger, J. G. 1976. Waterfowl response to rest-rotation grazing. *J. Wildl. Manage.* 40:60–68.

Munoz, J. R. 1976. *Cause of sheep mortality at the Cook Ranch, Florence, Montana, 1975–1976.* Annual Report to U.S. Fish and Wildlife Commission, Serv. Contract 14-16-0008-1135. Denver Research Center, Denver, CO.

Neel, L. A. 1980. *Sage grouse response to grazing management in Nevada.* M.S. Thesis. University of Nevada, Reno, NV.

Nerney, N. J. 1958. Grasshopper infestations in relation to range condition. *J. Range Manage.* 11:247.

Nerney, N. J. 1960. Grasshopper damage on shortgrass rangeland of the San Carlos Apache Indian Reservation, Arizona. *J. Econ. Entomol.* 53:640–646.

Newman, J. L. 1966. Effects of woven wire fence with cattle-guards on a free-roaming antelope population. *Proc. Antelope States Workshop* 3:62–64.

Nielson, L. S. 1978. *The effects of rest-rotation grazing on the distribution of sharptailed grouse.* M.S. Thesis. Montana State University, Bozeman, MT.

Ohmart, R. D. 1996. Historical and present impacts of livestock grazing on fish and wildlife resources in western riparian habitats. In P. R. Krausman (Ed.), *Rangeland wildlife,* pp. 245–281. The Society for Range Management, Denver, CO.

Osborn, B., and P. F. Allen. 1949. Vegetation of an abandoned prairie dog town in tallgrass prairie. *Ecology* 30:322–332.

Owens, R. A., and M. T. Myres. 1973. Effects of agriculture upon populations of native passerine birds of an Alberta fescue grassland. *Can. J. Zool.* 51:687–713.

Ozoga, J. J., and L. J. Verme. 1970. Winter feeding habits of penned white-tailed deer. *J. Wildl. Manage.* 34:431–439.

Pearson, E. W., and M. Caroline. 1981. Predator control in relation to livestock losses in central Texas. *J. Range Manage.* 34:435–442.

Peden, D. G., G. M. Van Dyne, R. W. Rice, and R. M. Hansen. 1974. The trophic ecology of *Bison bison* L. and shortgrass plains. *J. Appl. Ecol.* 11:489–497.

Pederson, J. C., and K. T. Harper. 1978. Factors influencing productivity of two mule deer herds in Utah. *J. Range Manage.* 31:105–110.

Peek, J. M., and P. R. Krausman. 1996. Grazing and mule deer. In P. R. Krausman (Ed.), *Rangeland wildlife,* pp. 183–192. The Society for Range Management, Denver, CO.

Penfound, W. T. 1964. The relation of grazing to plant succession in the tallgrass prairie. *J. Range Manage.* 17:256–260.

Phillips, P. 1936. The distribution of rodents in overgrazed and normal grasslands in central Oklahoma. *Ecology* 17:673–679.

Pitt, M. D. 1986. Assessment of spring defoliation to improve fall forage quality of bluebunch wheatgrass. *J. Range Manage.* 39:175–181.

Potter, L. D., and J. C. Krenetsky. 1967. Plant succession with released grazing on New Mexico rangelands. *J. Range Manage.* 29:145–151.

Prasad, N. L. N. S. and F. S. Guthery. 1986. Wildlife use of livestock water under short-duration and continuous grazing. *Wildl. Soc. Bull.* 14:450–454.

Pratt, D. J., and M. D. Gwynne. 1977. *Rangeland management and ecology in East Africa.* R. E. Krieger Publishing Co., Inc., Huntington, New York.

Price, M. R. S. 1985. Game domestication for animal production in Kenya: Feeding trials with oryx, zebu cattle and sheep under controlled conditions. *J. Agric. Sci.* 104:367–374.

Ragotzkie, K. E., and J. A. Bailey. 1991. Desert mule deer use of grazed and ungrazed habitats. *J. Range Manage.* 44:487–491.

Ramsey, C. W. 1965. Potential economic returns from deer as compared with livestock in the Edward Plateau region of Texas. *J. Range Manage.* 18:247–250.

Reardon, P. O., L. B. Merrill, and C. A. Taylor, Jr. 1978. White-tailed deer preferences and hunter success under various grazing systems. *J. Range Manage.* 31:40–42.

Reid, V. H. 1954. Multiple land use: Timber, cattle, and bobwhite quail. *J. For.* 52:575–578.

Rhodes, B. D., and S. H. Sharrow. 1990. Effect of grazing by sheep on the quantity and quality of forage available to big game in Oregon's Coast Range. *J. Range Manage.* 43:235–237.

Rice, B., and M. Westoby. 1978. Vegetative responses of some Great Basin shrub communities protected against jackrabbits or domestic stock. *J. Range Manage.* 31:28–34.

Riegel, D. A., F. W. Albertson, G. W. Tomanek, and F. E. Kinsinger. 1963. Effects of grazing and protection on a twenty-year-old seeding. *J. Range Manage.* 16:60–63.

Roath, L. R., and W. C. Krueger. 1982. Cattle grazing influence on a mountain riparian zone. *J. Range Manage.* 35:100–104.

Robinson, W. L., and E. G. Bolen. 1984. *Wildlife ecology and management.* Macmillan Publishing Company, New York.

Rogers, G. E. 1964. Sage grouse investigations in Colorado. *Game Res. Div. Tech. Publ.* 16. Colorado Game, Fish and Parks Department, Denver, CO.

Schladweiler, P. 1974. *Ecology of shiras moose in Montana.* Mimeograph. Montana Department of Fish and Game, Helena, MT.

Scifres, C. J. 1980. *Brush management.* Texas A and M University Press, College Station, TX.

Scifres, C. J., and B. H. Koerth. 1986. Habitat alterations in mixed brush from variable rate herbicide patterns. *Wildl. Soc. Bull.* 14:355–356.

Scotter, G. W. 1980. Management of wild ungulate habitat in the western United States and Canada. *J. Range Manage.* 33:16–28.

Sedivec, K. K., T. A. Messmer, W. T. Barker, K. F. Higgens, and D. R. Hertel. 1990. Nesting success of upland nesting waterfowl and sharp-tailed grouse in specialized grazing systems in south-central North Dakota. In Can livestock be used as a tool to enhance wildlife habitat. *U.S. Dep. Agr. Forest Serv. Gen. Tech. Rep.* RM-194.

Seegmiller, R. F., and R. D. Ohmart. 1981. Ecological relationships of feral burros and desert bighorn sheep. *Wildl. Monogr.* 78.

Severson, K. E. and P. J. Urness. 1994. Livestock grazing: A tool to improve wildlife habitat. In *Ecological implications of livestock herbivory in the West.* Society for Range Management, Denver, CO.

Sheehy, D. P., and M. Vavra. 1996. Ungulate foraging areas on seasonal rangeland in north-eastern Oregon. *J. Range Manage.* 49:16–24.

Shewchuk, B. A., and W. A. Kerr. 1993. Returns to grasshopper control on rangelands in southern Alberta. *J. Range Manage.* 46:458–462.

Silver, H., N. F. Colovos, J. B. Holter, and H. H. Hayes. 1969. Fasting metabolism of white-tailed deer. *J. Wildl. Manage.* 33:490–498.

Sisson, L. 1976. *The sharp-tailed grouse in Nebraska.* Nebraska Game, Fish and Parks Commission, Lincoln, NE.

Skovlin, J. M. 1971. Ranching in east Africa: A case study. *J. Range Manage.* 24:263–270.

Skovlin, J. M. 1984. Impacts of grazing on wetlands and riparian habitat. In National Research Council/National Academy of Sciences (Eds.), *Developing strategies for rangeland management.* Westview Press, Inc., Boulder, CO.

Skovlin, J. M., P. J. Edgerton, and R. W. Harris. 1968. The influence of cattle management on deer and elk. *Trans. N. Am. Wildl. Nat. Resour. Conf.* 33:169–181.

Skovlin, J. M., R. W. Harris, G. S. Strickler, and G. A. Garrison. 1976. Effects of cattle grazing methods on ponderosa pine-bunchgrass range in the Pacific northwest. *U.S. Dep. Agric. For. Serv. Tech. Bull.* 1531.

Skovlin, J. M., P. J. Edgerton, and B. R. McConnell. 1983. Elk use of winter range as affected by cattle grazing, fertilizing and burning in southeastern Washington. *J. Range Manage.* 36:184–189.

Smith, A. D. 1949. Effects of mule deer and livestock upon a foothill range in northern Utah. *J. Wildl. Manage.* 13:421–423.

Smith, C. C. 1940. The effect of overgrazing and erosion upon the biota of the mixed grass prairie of Oklahoma. *Ecology* 21:381–397.

Smith, G., J. L. Holechek, and M. Cardenas. 1996. Wildlife numbers on excellent and good condition Chihuahuan Desert rangelands: An observation. *J. Range Manage.* 49:489–493.

Smith, J. G., and O. Julander. 1953. Deer and sheep competition in Utah. *J. Wildl. Manage.* 17:101–112.

Smith, M. A., J. C. Malechek, and K. O. Fulgham. 1979. Forage selection by mule deer on winter range grazed by sheep. *J. Range Manage.* 32:40–46.

Sparks, D. R. 1968. Diet of black-tailed jackrabbits on sandhill rangeland in Colorado. *J. Range Manage.* 21:203–208.

Sperry, C. C. 1941. Food habits of the coyote. *U.S. Dep. Inter. Wildl. Res. Bull.* 4.

Spillett, J. J., J. B. Low, and D. Sill. 1967. Livestock fences-how they influence pronghorn antelope movement. *Utah Agric. Exp. Stn. Bull.* 470. Logan, UT.

Stoddard, H. L. 1931. *The bobwhite quail, its habits, preservation and increase.* Charles Scribner's Sons, New York.

Stuth, J. W., and A. H. Winward. 1977. Livestock-deer relations in the lodgepole pine-pumice region in central Oregon. *J. Range Manage.* 30:110–116.

Symmons, P. 1959. The effect of climate and weather on the numbers of the red locust in the Rukwa Valley outbreak area. *Bull. Entomol. Res.* 50:507–521.

Talbot, L. M., H. P. Ledger, and W. J. A. Payne. 1961. The possibility of using wild animals for animal protection on the semiarid tropic of east Africa. *Proc. Int. Congr. Anim. Husb.* 8:205–210.

Taylor, C. R. 1968. The minimum water requirements of some east African bovines. In *Comparative nutrition of wild animals.* Academic Press Inc. (London) Ltd., London. Symp. Zool. Soc. Lond. 21:195–206.

Taylor, R. G., J. P. Workman, and J. E. Bowns. 1979. The economics of sheep predation in southwestern Utah. *J. Range Manage.* 32:317–322.

Taylor, W. D., C. T. Vorhies, and P. B. Lister. 1935. The relation of jackrabbits to grazing in southern Arizona. *J. For.* 33:490–498.

Teer, J. G. 1975. Commercial uses of game animals on rangelands of Texas. *J. Anim. Sci.* 40:1000–1008.

Thorp, J. 1949. Effects of certain animals that live in soils. *Sci. Monthly* 68:180–191.

Todd, J. G., and J. A. Kamm. 1974. Biology and impact of a grass bug (*Labops hesperius Uhler*) in Oregon rangelands. *J. Range Manage.* 23:453–457.

Tubbs, A. A. 1980. Riparian bird communities of the Great Plains. In Management of western forests and grasslands for nongame birds proceedings. *U.S. Dep. Agric. For. Serv. Gen. Tech. Rep.* INT-86. Ogden, UT.

Turkowski, F. J., and J. R. Vahle. 1977. Desert rodent abundance in southern Arizona in relation to rainfall. *U.S. Dep. Agric. Rocky Mt. For. Range Exp. Stn. For. Serv. Res. Note* RM-346.

Ullrey, D. E., W. G. Youatt, H. E. Johnson, L. D. Fay, and B. E. Brent. 1967. Digestibility of cedar and jack pine browse for the white-tailed deer. *J. Wildl. Manage.* 31:448–454.

Uresk, D. W. 1984. Black-tailed prairie dog food habits and forage relationships in western South Dakota. *J. Range Manage.* 37:325–330.

U.S. Department of Agriculture (USDA). 1967. Suggested guide for the use of insecticides to control insects affecting crops, livestock, households, stored products, forests, and forest products. *U.S. Dep. Agric. Agric. Handb.* 331.

U.S. Department of Interior (USDI). 1978. *Predator damage in the West: A study of coyote management alternatives.* U.S. Department of Interior, Fish and Wildlife Service.

Valentine, K. A. 1947. Distances from water as a factor in grazing capacity of rangeland. *J. For.* 45:749–754.

Vavra, M., and F. Sneva. 1978. Seasonal diets of five ungulates grazing in the cold desert biome. *Proc. Int. Rangel. Congr.* 1:435–437.

Verme, L. J. 1967. Influence of experimental diets on white-tailed deer reproduction. *Trans. N. Am. Wildl. Nat. Resour. Conf.* 32:405–420.

Vorhies, C. T. and W. P. Taylor. 1933. The life histories and ecology of jackrabbits in relation to grazing in Wyoming. *Ariz. Agric. Exp. Stn. Tech. Bull.* 49.

Wagner, F. H. 1972. Coyotes and sheep—Some thoughts on ecology, economics and ethics. *Utah State Univ. Faculty Honro Lect.* 44.

Watts, J. G., G. B. Hewitt, E. W. Huddleston, A. G. Kinzer, R. J. Larigne, and D. N. Ueckert. 1989. *Rangeland entomology,* Second edition, Range science Series Number 2. The Society for Range Management, Denver, CO.

Webb, W. M. and F. S. Guthery. 1982. Response of bobwhite to habitat management in northwest Texas. *Wildl. Soc. Bull.* 10:142–147.

Westemeier, R. L. 1972. Prescribed burning in grassland management for prairie chickens in Illinois. *Proc. Tall Timbers Fire Ecol. Conf.* 12:317–338.

Westra, R. and R. J. Hudson. 1981. Digestive function of Wapiti calves. *J. Wildl. Manage.* 45:148–155.

White, R. J. 1987. *Big game ranching in the United States.* Wild Sheep and Goat International Publishing Co., Mesilla, NM.

Whyte, R. J., and B. W. Cain. 1981. Wildlife habitat on grazed or ungrazed small pond shorelines in south Texas. *J. Range Manage.* 34:64–68.

Whyte, R. J., N. J. Silvy, and B. W. Cain. 1981. Effects of cattle on duck food plants in southern Texas. *J. Wildl. Manage.* 45:512–516.

Wiens, J. A., and M. I. Dyer. 1975. Rangeland avifaunas: Their composition energetics and value in the ecosystem. In Proc. management of forest and range habitats for nongame birds. *U.S. Dep. Agric. For. Serv. Gen. Tech. Rep.* WO-1.

Wilbert, D. E. 1963. Some effects of chemical brush control on elk distribution. *J. Range Manage.* 16:74–78.

Willms, W. A., A. McLean, R. Tucker, and R. Ritchey. 1979. Interactions between mule deer and cattle on big sagebrush range in British Columbia. *J. Range Manage.* 32:299–304.

Willms, W., A. W. Baily, A. McLean, and R. Tucker. 1981. The effects of fall defoliation on the utilization of bluebunch wheatgrass and its influence on the distribution of deer in spring. *J. Range Manage.* 34:16–21.

Winegar, H. H. 1977. Camp creek channel fencing-plant, wildlife, soil, and water responses. *Rangeman's J.* 4:10–13.

Wisdom, M. J., and J. W. Thomas. 1996. Elk. In P. R. Krausman (Ed.), *Rangeland wildlife,* pp. 157–183. The Society for Range Management, Denver, CO.

Wolfe, M. J., Jr., 1980. Feral horse demography: A preliminary report. *J. Range Manage.* 33:354–360.

Wood, J. E. 1969. Rodent populations and their impact on desert rangelands. *N. Mex. Agric. Exp. Stn. Bull.* 555, Las Cruces, NM.

Wood, J. E., T. S. Bickle, W. Evans, J. C. Germany, and V. W. Howard, Jr. 1970. The Fort Stanton mule deer herd. *N. Mex. Agric. Exp. Stn. Bull.* 567, Las Cruces, NM.

Yeo, J. J., J. M. Peek, W. T. Wittinger, and C. T. Kvale. 1993. Influence of rest-rotation cattle grazing on mule deer and elk habitat use in eastcentral Idaho. *J. Range Manage.* 46:245–251.

Yoakum, J. D. 1975. Antelope and livestock on rangelands. *J. Anim. Sci.* 40:985–988.

Yoakum, J. D. 1979. Managing rangelands for pronghorn. *Rangelands* 1:146–148.

CHAPTER 14

RANGE MANAGEMENT FOR MULTIPLE USE

Multiple use involves the harmonious use of the range for more than one purpose (Society for Range Management 1989). Most public ranges in the United States are managed under a multiple-use philosophy in which an attempt is made to accommodate all legitimate rangeland uses demanded by society. Increasingly, ranchers are finding that they can derive significant income from the sale of products other than livestock, such as hunting privileges, wildlife viewing, horseback riding, wood, and ornamental plants. It is seldom possible to maximize the benefits from any single use under the multiple-use concept. However, most rangelands will provide the greatest benefits to society when managed for several uses rather than for a single use.

In this chapter, we discuss the relationships between livestock grazing, water production, wood production, and recreation. Management practices directed toward optimizing these four rangeland uses are emphasized.

RANGELAND HYDROLOGY

Water is the primary limiting factor to plant production on most rangelands of the world. All water used on rangelands is derived from precipitation. A portion of the precipitation moves laterally off the site to streams, ponds, lakes, reservoirs, and oceans. This water is referred to as *surface runoff* or *overland flow* (Figure 14.1). Another portion is retained on

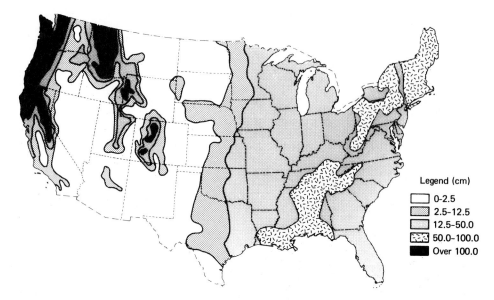

Legend (cm)
- 0-2.5
- 2.5-12.5
- 12.5-50.0
- 50.0-100.0
- Over 100.0

***Figure* 14.1** Average annual runoff in the United States. (From U.S. Water Resources Council 1978.)

the site by the process of infiltration (movement of water into the soil profile) and becomes available for plant growth. Part of the moisture infiltrating the soil moves deep below the soil surface and recharges underground water supplies that serve as sources for wells, seeps, springs, and streams. The major portion of precipitation (about 70%) received on range-lands evaporates and returns to the atmosphere as water vapor.

In arid to semiarid areas such as the western United States, rangelands are the primary sources of water for household use, irrigation of farmland, and industry. Range management practices can have tremendous influence on the quality and quantity of water available for these uses. The failure to apply sound range management practices in the past has resulted in high economic loss and great human hardship due to flooding of towns, farmland, and homes; silting of reservoirs; salinization of farmland; erosion of farmland; and loss of graz-ing capacity on rangelands. The type and amount of vegetation has a strong influence on the disposition of precipitation. Grazing management, in turn, has a considerable influence on these vegetation characteristics. Lack of water is the major limitation on human population growth and economic development in the western United States. We believe that in the near future range management practices will be geared primarily toward water production rather than forage production in portions of the southwestern United States.

The Hydrologic Cycle

The process by which energy from the sun causes water from the land and oceans to va-porize into the atmosphere, condense as the result of cooling due to uplift and/or movement over land, and return to the earth as precipitation is termed the *hydrologic cycle* (Figure 14.2). The primary source of moisture in the air is from oceans. Other important

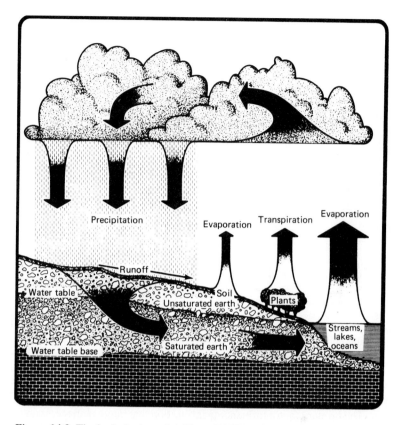

***Figure* 14.2** The hydrologic cycle. (From CAST 1982.)

sources are from transpiration (moisture released from plant parts, mainly leaves) and evaporation from streams, lakes, and land surfaces. Man can influence the hydrologic cycle by modification of vegetation and soil. These modifications, which can have a major impact on both water yield and quality, are discussed later in the chapter.

Infiltration

After a raindrop reaches the soil surface, it can infiltrate the soil, evaporate, or become a part of overland flow. Once in the soil, water movement is defined as *percolation.* The primary factors influencing infiltration rate are intensity of precipitation, amount and kind of vegetation cover, and soil surface properties (texture, structure, and organic matter) (Figure 14.3).

When infiltration is high, much of the precipitation is stored in the soil for plant use, and part of it may penetrate to groundwater, where it can be recovered from springs or wells. Conversely, high rates of surface runoff contribute to soil loss and flooding.

Fine-textured soils (clays) generally have lower infiltration rates than do coarse-textured soils (sands). However, some fine-textured soils have infiltration rates comparable to those of

VEGETATION TYPE

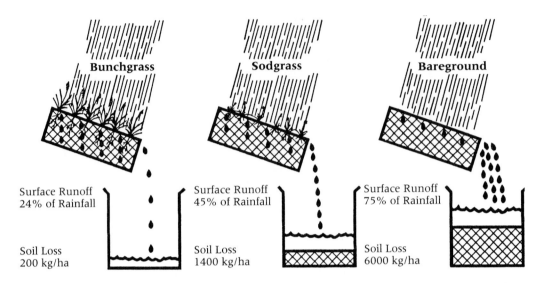

***Figure* 14.3** Influence of vegetation type on sediment loss, surface runoff, and rainfall infiltration from 10 cm of rain in 30 minutes. (Adapted from W. H. Blackburn et al. 1986 by Knight 1993.)

coarse texture. This is due to cementing of the fine particles into aggregates that act like larger particles.

The primary factor influencing infiltration that range managers can control is vegetation cover. When raindrops fall on unprotected soil, they dislodge soil particles and remove the soil surface. As plant cover declines, infiltration decreases (Table 14.1). This reduces soil moisture available for forage production, and contributes to desertification of arid areas.

Water repellency has been recognized in the western United States in many soils (DeBano 1969; DeBano and Rice 1973). It affects the hydrologic cycle by reducing infiltration rates and has been found under a variety of climates and vegetation types (DeBano 1969). Water repellent soils are present with and without burning, but they are often associated with a burn. On burned watersheds the water repellent layer occurs just below the surface of the soil. The thickness of the layer depends on the intensity of the fire and the type and amount of mulch. Plants with high levels of volatile oils are most often associated with water repellency (Heede et al. 1988; DeBano 1989; Dennis 1989).

Runoff and Erosion

Surface runoff is initiated when the amount of precipitation exceeds the infiltration and storage capacity of the soil. The primary factor influencing runoff is the amount of vegetation available to retard water movement over the soil surface (Figure 14.4). Runoff declines as soil cover increases.

TABLE 14.1 Standing Air-dry Herbage, Mulch, and Rate of Water Intake on Heavily, Moderately, and Lightly Grazed Watershed in South Dakota

GRAZING INTENSITY	TOTAL HERBAGE (LB/ACRE)	MULCH (LB/ACRE)	WATER INTAKE RATE (IN/HR)
Heavy	900	456	1.05
Moderate	1,345	399	1.69
Light	1,869	1,100	2.95

Source: Data from Rauzi and Hanson 1966.

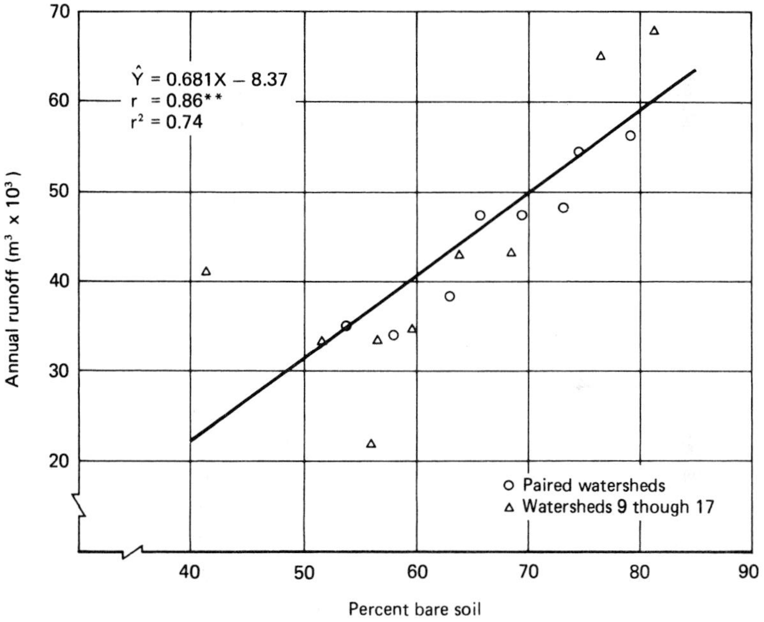

Figure **14.4** Regression line for percentage of bare soil versus average annual runoff from 17 watersheds in western Colorado. (From Branson and Owen 1970, cited in Branson et al. 1981.)

Erosion and sediment (suspended mineral materials in water) deposition are major problems caused by excessive runoff. Maximum sediment yields occur at about 250 mm (10 in.) of annual precipitation. A decrease occurs below 250 mm precipitation because of lack of runoff to transport the sediment. Above 250 mm precipitation, the decrease is due to increased cover of vegetation. Sediment is economically significant because it is detrimentally deposited on land and plants, reduces reservoir storage capacities, causes increased flood hazard, and pollutes stored water supplies (Schlesinger et al. 1989; Higgins et al. 1989; Walling and Webb 1996).

Geologic erosion is normal erosion for a natural environment undisturbed by man. Disturbances by man, such as overgrazing, logging, farming, and road construction, can cause accelerated erosion, which proceeds at a higher rate than normal geologic erosion. Differentiating the two types of erosion is a challenging problem to the range manager. Rangelands often have high rates of geologic erosion because of steep slopes, aridity, thin soils, and a sparse vegetation cover.

Accelerated erosion occurs when man's activities destroy the vegetation cover that retards soil loss from the forces of water and wind (Figure 14.5). The inverse relationship between accelerated erosion and plant cover is well established (Osborn 1956; Marston 1958; Thurow et al. 1986). Accelerated erosion is the most severe consequence of overgrazing, due to the fact that replenishment of lost soil is a slow process. Several hundred years are required to form an inch of soil, therefore, losses of soil result in nearly permanent reductions in grazing capacity. The best protection against erosion is to establish and maintain a good vegetative cover. During the initial phases, changes in grazing management can often bring accelerated erosion under control. However, in the more advanced stages, costly measures such as mechanical structures and revegetation are often necessary.

GRAZING IMPACTS ON WATERSHEDS

Livestock affect watershed properties by consumption of plant parts and through the physical action of their hooves. Reduction in the plant cover can increase the impact of raindrops, decrease soil organic matter and soil aggregates, and increase soil crusts. The primary effect of hoof action is compaction of the soil surface. Removal of cover and soil compaction reduce water infiltration rates, increase runoff, and increase erosion. A possible positive effect of hoof action is increased mineral cycling by incorporation of mulch into the soil surface, where it can be more quickly broken down by soil organisms. In temperate areas, excessive accumulations of mulch can result in delayed warming of the soil and subsequent delay in plant growth in the spring. Interception of precipitation and evaporation before infiltration into the soil is another disadvantage of excessive mulch accumulation. However, excessive mulch accumulation is seldom a problem in arid areas with less than 400 mm (16 in.) annual precipitation. The real problem in these areas is a lack of mulch and vegetation to protect the soil surface.

Infiltration and Grazing

Grazing and browsing animals reduce water infiltration by removing protective plant materials and compacting the soil surface by hoof action (Figure 14.6). The negative impact of heavy grazing on water infiltration is well documented (Johnson 1962; Rhoades et. al 1964; Rauzi and Smith 1973; Thurow et al. 1986; Weltz and Wood 1986a; Pluhar et al.

FORCES
OF EROSION   UNPROTECTED   PROTECTED

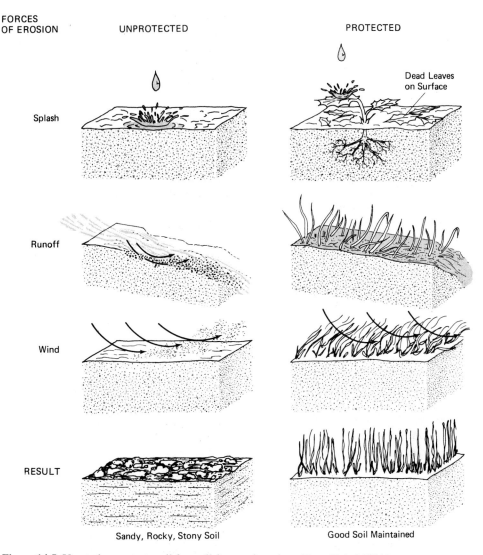

Figure **14.5** Vegetation protects soil from all forms of erosion. (From Nebel 1981.)

1987; Naeth et. al 1990). After reviewing grazing impacts on infiltration, Gifford and Hawkins (1978) concluded:

1. Ungrazed plots have higher infiltration rates than those of grazed plots.

2. Moderate and light grazing intensities have similar infiltration rates.

3. Heavy grazing causes definite reductions in infiltration rates over moderate and light grazing intensities.

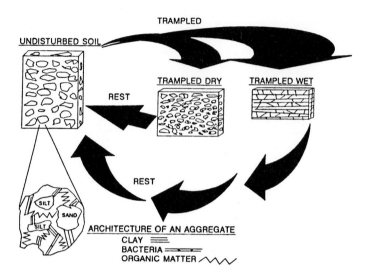

Figure 14.6 Conceptual architecture of a soil aggregate and the changes in soil aggregate structure caused by trampling under wet and dry conditions. (From Taylor et al. 1993.)

Although it has been speculated that under some conditions the hoof action of grazing animals will loosen the surface of compacted or crusted soils, actual research shows just the opposite effect (Warren et al. 1986a,b,c). Several studies in Texas and New Mexico are consistent in showing the concentration of hoof action under short-duration grazing reduced infiltration compared to continuous grazing (McCalla et al. 1984a; Thurow et al. 1986; Weltz and Wood 1986a; Pluhar et al. 1987; Weltz et al. 1989) (Table 14.2; see also Figure 9.10). In Wyoming, a short-term study on sandy loam soils showed no definite difference between short-duration grazing and moderate continuous grazing in terms of water infiltration or destruction of soil crusts (Abdel-Magib et al. 1987). Heavy stocking consistently reduced infiltration during the grazing season, but this appeared to be alleviated by winter freeze-thaw activities.

Research is available comparing infiltration rates of other specialized grazing systems with moderate continuous grazing (Wood and Blackburn 1981b; Gamougoun et al. 1984; Abdel-Magib et al. 1987; Pluhar et al. 1987). These studies are consistent in showing that grazing systems other than short-duration grazing have little influence on infiltration rate but that reductions occur when stocking rates are increased from moderate to heavy.

A few studies have evaluated the influences of grazing on soil structure. In New Mexico, lightly grazed, heavily grazed, and severely grazed ranges had pore spaces of 68%, 51%, and 46%, respectively (Flory 1936). In South Dakota, lightly grazed, moderately grazed, and heavily grazed watersheds had pore spaces of 10.6%, 8.4%, and 7.7%, respectively (Rauzi and Hanson 1966). In northcentral Texas, Wood and Blackburn (1984) found heavy grazing degraded soil structure by reducing the percentage of water-stable aggregates compared to moderate grazing. It was also found that heavy grazing increased soil

TABLE 14.2   Infiltration Rates and Sediment Production for Two Types of Plant Communities in Five Grazing Treatments[a]

TREATMENT	INFILTRATION RATE (MM/HR)		SEDIMENT PRODUCTION (KG/HA)	
	MIDGRASS	SHORTGRASS	MIDGRASS	SHORTGRASS
Short-duration				
(14 pastures)				
Before grazing	95	75	37	63
After grazing	64	55	105	105
(42 pastures)				
Before grazing	81	86	41	61
After grazing	85	79	75	53
Merrill three-herd/four-pasture				
Before grazing	86	80	28	45
After grazing	81	68	71	54
Moderate continuous	89	85	35	30
Exclosure	88		23	

Source: From Pluhar et al. 1987.

[a]The stocking rate was the same for all treatments.

compaction (bulk density) more than did moderate grazing. In the same study, different grazing systems (continuous, Merrill three-herd/four-pasture, and high intensity-low frequency) had a similar effect on water-stable aggregates and bulk density.

Runoff and Grazing

Livestock grazing increases runoff by reducing infiltration. Increased surface runoff due to heavy grazing is usually associated with water quality degradation because of increases in sediment and other pollutants (animal wastes, agricultural chemicals, decayed vegetation, etc).

It is well documented that heavy grazing increases runoff compared to moderate grazing (Dunford 1949; 1954; Liacos 1962b; Sharp et al. 1964; Hanson et al. 1970) (Figure 14.7). However, protected areas generally have the least runoff. Branson and Owen (1970) showed that runoff increases as vegetative cover and mulch decreases and the amount of bare soil increases. Other studies have documented the inverse relationship between runoff and vegetation cover (Marston 1952; Hanson et al. 1970).

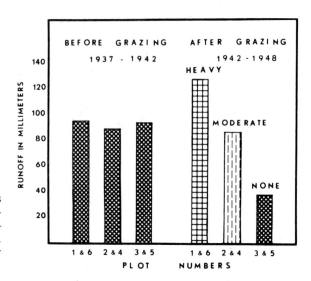

***Figure* 14.7** Runoff for bunchgrass rangeland in Colorado prior to grazing (1937–1942) and after (1942–1948) heavy and moderate grazing. (Adapted from Dunford 1949 by Branson et al. 1981.)

Limited data indicate that moderate or light grazing can increase groundwater and runoff compared to no grazing, without having a detrimental impact on the watershed or water quality (Liacos 1962a,b; Hanson et al. 1970; Lusby 1970). Under moderate or light grazing intensities, adequate vegetation is maintained to protect the site, but excessive vegetation that causes water losses by transpiration and evaporation is removed. More research is needed on the potential of controlled grazing to increase water yields. Information regarding influences on groundwater would be particularly useful for the southwestern United States, where many underground aquifers are being rapidly depleted.

Erosion and Grazing

Sediment yields in runoff and streamwaters are commonly used as a measure of erosion on rangelands. Heavy grazing accelerates erosion by reducing the mulch and plant cover that protects the soil and retards overland flow. Several studies have documented higher erosion under heavy grazing intensities than under moderate intensities (Dunford 1949; McCalla et al. 1984b; Thurow et al. 1986; Weltz and Wood 1986b; Pluhar et al. 1987) (Figure 14.8). It is also well documented that moderate grazing increases erosion over the ungrazed condition (Dunford 1949; Wood and Blackburn 1981a; Gamougoun et al. 1984; Thurow et al. 1986; Weltz and Wood 1986b; Pluhar et al. 1987). However, in nearly all cases these differences were small and statistically unimportant. Therefore, we conclude that moderate grazing will not cause watershed damage on most rangelands. In many cases watershed recovery can be accomplished by changes in grazing management. In New Mexico, Aldon (1963, 1964) reported that sediment loads were reduced more than 75% due only to better livestock control (change from season-long to winter grazing) and reduced grazing intensities.

The influence of short-duration grazing on sediment production compared to other grazing methods has been a concern. Available research is consistent in showing that short-duration grazing increases sediment production compared to moderate continuous grazing (McCalla

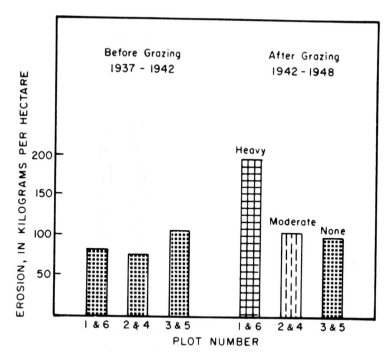

***Figure* 14.8** Average erosion from plots subject to different grazing intensities before grazing (1937–1942) and after grazing (1942–1948) on pine-bunchgrass range in Colorado. (Adapted from Dunford 1949 by Branson et al. 1981.)

et al. 1984b; Thurow et al. 1986; Weltz and Wood 1986b; Pluhar et al. 1987) (Figure 14.9). The reduced vegetation standing crop and cover associated with short-duration grazing in the studies cited previously appeared to cause the higher sediment production.

The most detailed evaluation of hydrologic responses under short-duration grazing was reported by Warren et al. (1986a,b,c). They studied infiltration and sediment production on a silty clay soil in Texas using a short-duration grazing system with moderate, double-moderate, and triple-moderate stocking rates. Short-duration grazing at all intensities reduced infiltration and increased sediment production compared to no grazing (Warren et al. 1986c) (Table 14.3). These deleterious effects were increased as stocking rate increased. The damage was augmented when the soil was moist at the time of trampling. Thirty days of rest was insufficient to allow hydrologic recovery. Another part of the study evaluated seasonal changes in infiltration and sediment production under short-duration grazing at a moderate stocking rate (Warren et al. 1986a). The infiltration rate declined and sediment production increased following the short-term intense grazing periods inherent to this system. These effects were most severe during drought and dormancy, due to reduced vegetation standing crop. It was also found that there was no definite hydrologic advantage of increased stocking density via manipulation of pasture size and numbers (Warren et al. 1986b).

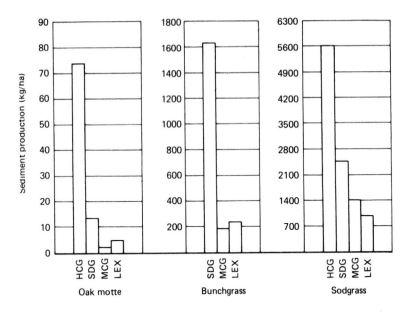

***Figure* 14.9** Sediment production for heavy continuous (HCG), short-duration (SDG), moderate continuous (MCG), and livestock enclosure (LEX) treatments, Edwards Plateau, Texas. (From Thurow et al. 1986.)

TABLE 14.3   Infiltration Rate and Sediment Production in Relation to Stocking Rate and Soil Water Content at the Time of Trampling on the Edwards Plateau, Texas

STOCKING RATE	TRAMPLED DRY	TRAMPLED MOIST
	INFILTRATION RATE (MM/HR.)	
0	166	160
1 X	140	133
2 X	121	99
3 X	117	96
	SEDIMENT PRODUCTION (KG/HA)	
0	976	2,007
1 X	2,827	2,875
2 X	3,438	4,274
3 X	4,788	5,861

Source: Adapted from Warren et al. 1986c.

1 X = moderate stocking rate, 2 X = twice moderate stocking rate, 3 X = triple moderate stocking rate.

Sediment production under various other specialized grazing systems has been compared with moderate continuous grazing (Wood and Blackburn 1981a; Gamougoun et al. 1984; Pluhar et al. 1987). As in the case of infiltration, these studies show little difference between grazing systems other than short-duration grazing.

Although limited data are available, the amount of vegetation required to protect different types of rangeland needs more study. Complete protection of the soil requires about 550 kg/ha of plant material (Osborn 1956). On mountainous areas, a ground cover of at least 65% is required to prevent excessive erosion (Packer 1951; Marston 1952). Ground cover levels of 30% to 40% appear adequate for flat, arid areas with low-intensity storms (Branson et al. 1981). Coarse soils, steep slopes, and intense storms all increase ground cover requirements for adequate site protection. Ground cover levels as low as 20% will protect most soils from wind erosion (Branson et al. 1981).

The key to maintaining healthy hydrological conditions on rangelands is through grazing practices that develop and maintain a good plant cover. Perennial grasses, because of their high basal area and excellent soil binding properties, play the critical role in watershed stability. Moderate stocking rates in conjunction with practices that promote even livestock distribution over the range are the best approaches to maintaining a good perennial grass cover. The success of any grazing program geared toward watershed maintenance and enhancement is best measured by the residue of living and dead vegetation (mulch) it maintains on the site throughout the year. A good residue of forage left ungrazed prior to initiation of new growth in the spring may appear to be a waste. However, in the long run, the operator will be rewarded by higher forage production and less variation in the forage crop between years due to increased soil moisture and mineral supplies available for plant growth. In Chapter 8 we discussed the positive impacts of moderate stocking rates on economic returns and provided guidelines for determining minimum vegetation standing crop residues for different rangeland types in the United States.

Water Quality and Grazing

Fecal wastes from livestock grazing can be a sizable pollution problem in range watershed management. Fecal coliform bacteria counts in water have been used as an indicator of infectious bacterial contamination (Wadleigh 1968). However, the coliform bacteria themselves are not pathologically harmful. Livestock operations have caused increased coliform bacterial pollution in rangeland streams (Kunkle and Meiman 1967; Skinner et al. 1974; Jawson et al. 1982; Gary et al. 1983; Tiedemann et al. 1987). The extent of the bacterial pollution depends largely on livestock numbers, timing of grazing, frequency of grazing, and access to the stream. In central Oregon, livestock grazing elevated coliform bacteria levels in streamwater compared to areas where livestock were not present (Tiedemann et al. 1987) (Figure 14.10). Fecal coliform bacteria levels tended to increase as intensity of livestock use increased. This study presented evidence that livestock removal may not provide an immediate solution to elevated fecal coliform bacteria levels in streamwater. Coliform bacteria can apparently live through the winter in soil sediment and animal feces after livestock are removed. In northern Utah, fecal coliform and streptococcus bacteria counts were increased in

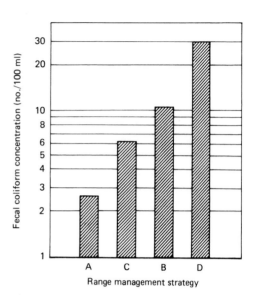

Figure **14.10** Fecal coliform bacteria loadings for four range management strategies for the period 1979–1984 in central Oregon. (A = no grazing, B = grazing without management to improve livestock distribution, C = grazing with management to improve livestock distribution, D = grazing with cultural practices to increase forage and improve livestock distribution. (From Tiedemann et al. 1987.)

streamwaters below areas grazed by cattle and sheep (Darling and Coltharp 1973). In Wyoming, livestock grazing, recreational activities, and wildlife all contributed to increased bacterial loads in streamwaters (Hussey et al. 1986).

Grazing strategies that disperse rather than concentrate livestock appear best when fecal contamination of streamwaters is a concern. Practices that improve livestock distribution and attract livestock away from streamside areas are recommended based on the Tiedemann et al. (1987) study. Additional effects of management on water quality are given by Binkley and Brown (1993).

Treated sewage sludge as a soil amendment on rangelands does not appear to deleteriously affect the quality of runoff water (Aguilar and Loftin 1991). In some situations, applications of municipal sewage sludge increases herbage yields and reduces detrimental runoff and sedimentation (Aguilar et al. 1994). Additional effects of management on water quality are given by Binkley and Brown (1993).

Fish Habitat and Grazing

Although riparian (streamside) zones comprise a small percentage of most rangelands, their management is extremely critical to rangeland health and associated fish and wildlife populations (Thomas 1986). As discussed previously in this and other chapters, these zones usually receive a disproportionate amount of the grazing pressure. Fish populations have been severely affected by uncontrolled grazing in riparian zones. A summary of 20 studies by Platts (1982) was consistent (with one exception) in showing that grazing degraded the streamside environment and the local fishery. Heavy grazing degrades fish habitat by widening the stream channel through trampling and destruction of streamside vegetation (Figure 14.11a,b). This results in fewer undercut banks, less overhanging vegetation, shal-

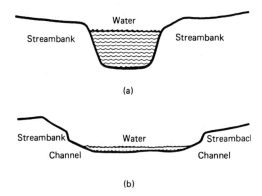

***Figure* 14.11** Typical stream channel cross section in (a) lightly grazed and (b) heavily grazed units. (From Platts 1981.)

lower water, and warmer water temperatures. These changes are all unfavorable for trout. Trout have difficulty surviving when water temperatures are above 18°C (65°F). Higher sediment, lower oxygen content, reduced cover, higher water temperatures, and lower food associated with heavily grazed and trampled banks create an unfavorable environment for trout and other desirable fish. Platts (1981) reported that a heavily grazed meadow was five times as wide as a section where no grazing occurred. Large increases in trout production (over 150%) have occurred where grazing was light or eliminated (Bowers et al. 1979). These authors presented the following management recommendations to benefit trout production:

1. Implement grazing systems that will create and/or maintain good trout habitat. Avoiding grazing riparian areas in the spring and summer months is particularly critical.

2. Fence both easily damaged and important trout streams.

3. Use salt and water developments to attract livestock away from the riparian zones.

In northeastern Oregon, Bryant (1982) found that neither judicious placement of salt nor water developments appreciably reduced summer use of mountain riparian zones. The most practical solution in many cases appears to be fencing of the riparian zones and delaying use of the areas until late summer or early fall. When the meadow is considered properly used, the rancher can be called to gather the animals and haul them home. These ranges are seldom grazed in winter due to snow. In the same situations on public lands near large urban centers where fishing is the primary land use, we believe discontinuation of livestock grazing may be the most practical approach. In these situations we advocate livestock grazing permit holders receive fair monetary compensation for loss of their privileges.

Riparian Areas and Range Management

Riparian areas are the most productive sites on rangelands. They have a greater diversity of plant and wildlife species than adjoining ecosystems. Healthy riparian systems purify water as it moves through the vegetation by removing sediment and retaining water in streambanks. Maintaining proper amounts of herbaceous vegetation is a critical part of increasing sediment deposition and enhancing channel restoration in small stream systems (Clary et al. 1996). Many wildlife species are dependent upon the diverse habitat niches in riparian areas. They also serve as a focal point for many forms of recreation (Swanson 1989; DeBano and Schmidt 1989; Abell 1989; Schmidt 1991; Tellman 1993). It was suggested by Howery et al. (1996) that selective culling of cattle can effectively decrease the use of riparian areas. Other grazing strategies for riparian zones are discussed in Chapter 9 and by Skovlin (1984), Elmore and Kauffman (1994), Ohmart (1996), and Masters et al. (1996a,b).

MANIPULATION OF WATER YIELD

The major objective of most mechanical rangeland treatments is to improve vegetative production by increasing moisture storage and reducing soil erosion. Soil characteristics (texture, structure, consistency, and moisture-holding capacity), climate, type of vegetation, and equipment used are the principal variables that determine treatment impact on any watershed. Although the various mechanical treatments are discussed briefly, we refer the reader to Branson et al. (1981) and Vallentine (1989) for comprehensive coverage of this subject.

Contour Trenches and Furrows

Contour trenches were initially applied on rangeland for controlling floods and soil erosion in northern Utah. The principle of contour trenching—to retain water on the site where precipitation occurs, thus preventing overland flow, erosion, and sedimentation—has proved effective, and many eroded flood-source areas have been trenched.

Vegetation-Type Conversion

Several studies reviewed by Branson et al. (1981) show that conversion of shrubland or woodland to herbaceous vegetation can greatly increase water yields. This is because grasses and forbs generally transpire much less water than do woody plants. Water yield for rangelands has been improved with chemical and mechanical brush control (Wood and Wood 1988; Griffin and McCarl 1989; Lacey et al. 1989; Wood et al. 1991; Sturges 1993; Martin and Morton 1993). Methodologies for vegetation-type conversion are discussed in Chapter 15.

Contour furrows differ from trenches in being narrower and shallower. They have been effective in retaining water on the site and improving forage production in the northern mixed prairie country of Wyoming and Montana.

Pitting

Pitting on rangeland consists of forming small basins or pits in the soil to catch and hold precipitation and runoff water. This practice has been used successfully in the northern mixed prairie and in mined land reclamation in the western United States.

Ripping

Ripping is used to shatter compacted soil profiles that inhibit moisture penetration and root development (Branson et al. 1981). Ripping to a depth of 30 cm to 90 cm has been effective in improving vegetation composition and productivity in the shortgrass prairie. Deep ripping (chiseling) has increased forage production in the northern Great Plains. It has been financially most effective on sites dominated by shallow-rooted plants (shortgrasses) with potential for the more productive midgrasses (Lacey et al. 1995).

Chiseling

Chiseling improves forage production in the northern Great Plains. Improvement is most economical on productive sites where forage production is limited by shallow-rooted plants (Lacey et al. 1995).

Water Harvesting

Water harvesting is the process of collecting and storing precipitation for beneficial uses from land that has been treated to increase runoff (Myers 1964). The process has been used in Israel to grow olives, apricots, and other crops. In the United States and Australia, water harvesting is used to provide drinking water for livestock and wildlife and to create wildlife habitat on arid ranges (Fink and Ehrler 1986). Water harvesting structures in the United States are called rain traps, catchment basins, paved drainage basins, trick tanks, and guzzlers.

Waterspreading

Waterspreading is a technique that involves diversion of water from drainages onto the surrounding landscape through a system of dikes, dams, or ditches (Branson et al. 1981). The diversion of runoff water with earthen banks to areas favorable for cultivation dates back to the early inhabitants of the Middle East and South America. Waterspreading on rangelands has three main functions: (1) increasing forage production by spreading of floodwater, (2) reducing erosion in drainageways, and (3) reducing downstream flooding and sedimentation. Stream channels (dry most of the time but flowing for short periods) generally provide the water supply for waterspreading schemes. Waterponding has been used in Australia to reclaim bare areas (Cunningham 1987).

Waste Disposal on Rangelands

Increasingly, western rangelands are being considered as disposal sites for various waste from large cities. There has been particular interest in using rangelands for dispersal of treated sewage sludge. In some situations applications of municipal sewage sludge can increase herbage yields and reduce detrimental runoff and sedimentation (Aguilar et al.

1994). Treated sewage sludge as a soil amendment on rangelands does not appear to adversely affect the quality of runoff water (Aguilar and Loftin 1991).

TIMBER PRODUCTION AND GRAZING

Importance of Forest Grazing

Forests support large herbivores on many continents of the world (Child et al. 1984). In most of these forests production of forest products influences grazing, and vice versa. Often, the managers of these forests must make decisions concerning trade-offs between wood products and grazing. In some cases grazing and wood production are competitive and in other cases complementary.

Most forested grazing areas in the United States are located in the West and Southeast (Chapter 4). In the West many of these areas serve as seasonal summer ranges which are grazed 2 to 4 months. In some areas of the Southwest, ponderosa pine (*Pinus ponderosa*) forest may be grazed for longer periods because of lack of restrictions from snow cover and other climatic extremes.

Forest ranges are often managed for several products, such as wood, forage, browse, recreational values, watershed values, and mineral resources. Such multiple-use management does not mean that every hectare must be devoted to all the uses. However, multiple-use can be applied at the management unit level. For example, block clear-cutting is the most appropriate silvicultural system for Douglas fir (*Pseudotsuga menziesii*) forests in the Pacific Northwest. Such a system can provide forage for livestock and game animals, extend the snow melt period, and stabilize overland flow in the spring if the cutting scheme maintains a diversity of age classes in clear-cut stands over the management unit.

Multiple-use management often requires information concerning products of uses as a function of some condition or situation regarding the range. For example, tree density influences timber production, forage production (or livestock numbers), browse production, water infiltration and water runoff, and so on. If the manager knows how tree density influences these other products or values, he or she is in a good position to decide on optimum tree density. Multiple-use management often involves trade-offs among different uses or products.

Forest Management versus Herbage Production

As forests mature, the individual trees become larger and the crowns of neighboring trees may eventually form a closed canopy. When the canopies become closed, little sunlight penetrates to the forest floor. Thus many mature forests have only limited understory and represent poor habitats for grazing and browsing animals (Adams 1975). Consequently, any practice that opens the stand will generally improve conditions for livestock and game (Table 14.4).

Considerable research has been conducted on the relationship between tree overstory and herbaceous production (Bartlett and Betters 1983). Although the general relationship

TABLE 14.4   Influence of Different Levels of Timber Removal on Deer Use (Days/Acre) in Mixed Oak-Pine Forest in Virginia

PERCENTAGE BASAL AREA REMOVED	NUMBER OF YEARS AFTER CUTTING			
	1	2	3	4
30	10	12	15	17
40	15	16	19	23
50	20	23	27	31
60	27	32	37	43
70	37	44	51	59
80	51	60	70	82

Source: Adapted from Patton and McGinnies 1964.

remains similar, the relationship between overstory cover and herbaceous production varies from site to site. Wolters (1981) found nearly a linear decrease in herbaceous production as basal area of longleaf pine (*Pinus palustris*) increased on grazed and ungrazed stands. Clary (1975) showed a curvilinear relationship between herbage production and ponderosa pine basal area in Arizona (Figure 14.12). Herbage production decreased sharply as ponderosa pine stocking increased initially and then gradually leveled off.

Since the natural tendency of forests is to increase in density and overstory canopy cover, they have been termed *transitory range* (Spreitzer 1985). Transitory range was defined as "forested lands that are suitable for grazing for a limited time following a complete or partial forest removal" (Spreitzer 1985). Thus cutting or burning of the forest may promote development of understory vegetation suitable for livestock grazing or game production. The example of secondary succession on ponderosa pine forests from Colorado shows several pathways depending on the influence of burning or grazing (Figure 14.13). Most of the grass stages would be suitable for livestock grazing. The oak (or other shrub) stage might be more suitable for wildlife production. In Arizona, the short-lived half-shrub stage would not be particularly suitable for livestock grazing since broom snakeweed (*Gutierrezia sarothrae*) is unpalatable and can be poisonous but the other stages would be favorable for livestock production.

Studies in eastern Oregon also demonstrated the influence of overstory on understory productivity in mixed coniferous forests (Hedrick 1975). Herbage production was 588 kg/ha in open stands compared to 100 kg/ha on heavily shaded stands (Table 14.5). Logging also served to increase productivity on herbaceous vegetation and browse.

Grazing Impacts on Timber Production

Damaging Effects. Foresters have long been concerned with the impact of livestock on developing tree seedlings. This damage may take several forms: girdling, trampling, and defoliation. Lewis (1980a) found that cattle girdling of slash pine in Florida resulted in little mortality. Mortality occurred only when trees were completely girdled.

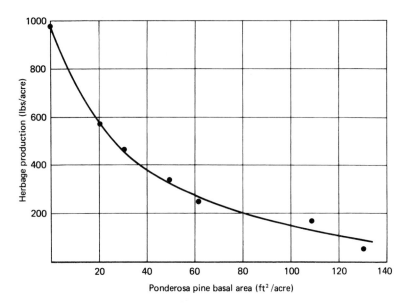

***Figure* 14.12** Relationship between understory herbage production and ponderosa pine basal area in northern Arizona. (From Clary 1975.)

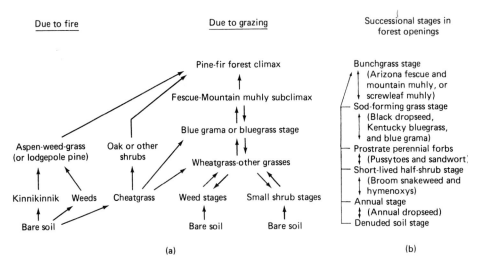

***Figure* 14.13** Secondary succession on western forests: (a) ponderosa pine forests in Colorado (from Currie 1975) and (b) ponderosa pine forests in Arizona. (From Clary 1975.)

TABLE 14.5   Annual Herbage and Browse Yields (Kg/ha) in the Mixed Coniferous Forests of Eastern Oregon as Influenced by Logging and Degrees of Shading

TYPE OF PRODUCTION AND SHADING	TREATMENT	
	LOGGED	UNLOGGED
Herbage		
Open	773	594
Moderate shade	348	303
Dense shade	134	101
Browse	213	78

Source: From Hedrick 1975.

TABLE 14.6   Density (Trees/acre) of Slash Pine Surviving after Planting in Louisiana

DATE	GRAZING INTENSITY			
	NONE	LIGHT	MODERATE	HEAVY
May (first year)	809	811	866	660
October (first year)	745	714	776	615
Second year	731	653	764	565
Fifth year	719	618	754	559

Source: From Pearson et al. 1971.

Hardwoods are apparently damaged more by sheep than cattle, but the reverse is true for conifers (Adams 1975). Young seedlings are more susceptible to damage than are older trees (Lewis 1980b,c; Pearson 1980) (Table 14.6). Protection of individual seedlings, reducing grazing pressure for the first growing season, restricting late winter and early spring grazing, and judicious placement of salt and supplemental feed can minimize damage to southern pine seedlings (Pearson 1980; Spreitzer 1985). However, trampling damage may still occur on western mixed conifer forests (Eissenstat et al. 1982).

Beneficial Effects. Grazing on understory grass or browsing on shrubby species can be beneficial to small tree seedlings by reducing competitive pressures. Often small tree seedlings do not become established readily in face of competition from established grass and shrubs. Consequently, livestock and game use of these areas can benefit establishment of tree seedlings.

MANAGEMENT SYSTEMS FOR TREES, LIVESTOCK, AND WILDLIFE

Western Coniferous Forests

Western forests are greatly influenced by elevation (Figure 14.14). In the central and southern Rocky Mountains, the ponderosa pine-bunchgrass type is very important for livestock grazing (Clary 1975; Currie 1975). Often, these stands are fairly open and support dense grass cover and some shrubby species.

Management problems relating to western coniferous forests often relate to canopy closure of the tree overstory and uneven distribution of livestock grazing. Silvicultural systems, either selective-cutting or clear-cutting, open the stand and promote development of understory species.

Several factors influence grazing distribution of forested rangeland (Glendening 1944; Skovlin 1965):

1. Stock water distribution

2. Slope steepness (Figure 14.15)

3. Trails and access routes

4. Density of trees

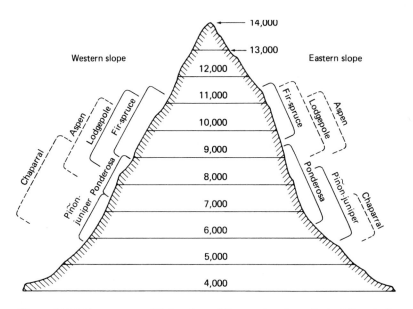

***Figure* 14.14** Forest types of Colorado at different elevations. Dashed lines indicate several types. (From Miller and Choate 1964.)

5. Botanical composition of understory

6. Range condition

Skovlin (1965) listed several practices that are useful in improving livestock distribution on western mountain rangelands:

1. Fencing

2. Water developments

3. Distribution of salt

4. Construction of trails

5. Range riding

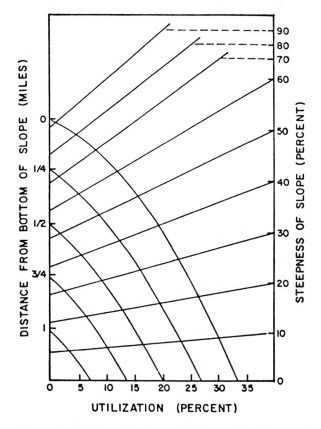

***Figure* 14.15** Effect of steepness and distance from bottom of slope on percentage grazing use of mountain muhly. (From Glendening 1944.)

Various grazing systems have been implemented on western coniferous forests, but few have been researched (Clary 1975). National forest allotments were being grazed under the following systems in Arizona (Clary 1975):

GRAZING SYSTEM	PERCENTAGE BASED ON AREA
Continuous	9
Rotation	15
Deferred	19
Rest-rotation	42
Deferred-rotation	11
No indication	4

In central Colorado, Currie (1976) reported that complete protection from grazing or seasonal grazing did not promote improvement of ponderosa pine-bunchgrass range. Judicious use of 2,4-D and fertilizer improved conditions for livestock grazing by changing species composition and increasing herbage production. In the Blue Mountains of Oregon, cattle performance and vegetation responded to stocking rate and a deferred-rotation grazing system (Skovlin et al. 1976) (see Chapter 9). In some cases introduced species such as Russian wildrye (*Elymus junceus*) and big bluegrass (*Poa ampla*) can be used in seasonal grazing schemes to improve livestock production (Currie 1969).

Piñon-Juniper Woodlands

Both distribution and tree density of piñon-juniper woodlands have increased over the last 200 years (West et al. 1975; Springfield 1976). Consequently, during the period following World War II, considerable effort was expended in piñon-juniper control projects to increase forage production for livestock. The need for this control has been a controversial subject. Lanner (1977) questioned the general assumption of encroachment of trees into grassland, suggesting that in some cases the trees were simply becoming reestablished following cutting for wood, posts, fuel, and so on. Large-scale piñon-juniper control programs were often destructive to wildlife habitat. Environmental groups brought esthetic concerns to the forefront. Now most piñon-juniper control programs are conducted only on "invasion" stands and in strips to blend in with the landscape and furnish food and cover for wildlife (Williamson and Currier 1971).

Southeastern Pine Forests

Silvicultural systems in the southeastern United States play a critical role in forage production for livestock. Clear-cutting in patches or small blocks opens up forest stands and stimulates understory production. Cutting patterns can be scheduled to provide both timber and forage. In some cases competing shrubs must be controlled to ensure tree regeneration.

Herbaceous forage grows rapidly on southern pine forest rangelands. However, there are some nutritional problems relating to livestock production (see Chapters 4 and 11). Operators who graze these ranges yearlong often provide protein supplement, either in liquid form or as cottonseed cake (if labor is relatively cheap) throughout the year (Grelen 1978). Cattle usually have some Brahman breeding in their background for resistance to parasites, high temperatures, and humidity (Grelen 1978).

Fire has been used as a tool for southern forests for many years, although historically fires were often set indiscriminately for a variety of reasons. Now prescribed burning under certain conditions is a recognized silvicultural tool on longleaf pine-bluestem and longleaf pine-wiregrass ranges. Fire can be used to control undesirable shrubs and succulents, such as gallberry (*Ilex glabra*) and saw palmetto (*Serenoa repens*) (Grelen 1978). Fire also removes coarse standing dead herbage and stimulates palatable new growth of grasses.

RECREATIONAL USE OF RANGELANDS

Importance of Rangeland Recreation

Recreational use of rangelands in the United States has accelerated tremendously during the past 20 years. This increase is demonstrated by Table 14.7, showing the change in visitor days on Forest Service land. Much of the increase has resulted from the rapid human population increase in the 11 contiguous western states combined with rising affluency and leisure time. In some areas, such as southern California, central Colorado, western Arizona, central Texas, and central Oregon, recreation has become a far more important use of rangeland than livestock grazing. If present trends of urbanization continue in the 11 western states, it appears likely the recreational value will exceed livestock grazing value on most rangelands within the next 15 to 20 years.

Recreational use of rangelands includes a variety of activities (Table 14.8). The economic values of many of these activities, such as camping, hiking, and water skiing, are difficult to quantify. Some forms of recreation, such as Jeeping (off-road vehicle travel) and trail biking, can be as destructive as overgrazing if uncontrolled.

Subdividing the West

The sale of recreational home sites has had the most severe impact of all the rangeland recreational uses. Literally thousands of recreational home sites are being sold every year in each of the western states. Although the total area lost each year is relatively small [about 1 million acres (404,694 ha)] compared to the total land base, this trend is having a major impact on the local range livestock industry.

Large-scale conversion of rangeland to housing is occurring in the mountain valleys of the intermountain West. These lands are critical in terms of providing forage for livestock during

TABLE 14.7   Total Recreational Use (1,000 Visitor Days) on National Forest System Lands in 1980, 1990, and 1992 for the 11 Contiguous Western States

	1980	1990	1992
Arizona	17,745	19,038	25,544
California	57,533	66,007	67,614
Colorado	22,449	25,204	29,053
Idaho	10,797	11,819	13,087
Montana	8,577	9,704	11,046
Nevada	2,364	3,278	3,360
New Mexico	5,843	7,704	8,603
Oregon	18,527	21,036	19,898
Utah	14,661	12,744	18,413
Washington	12,892	22,451	18,740
Wyoming	5,540	6,609	7,516
Total	176,328	207,594	222,874

Source: USDA 1980, 1992.

TABLE 14.8   Types of Recreational Uses of Rangelands

ACTIVITIES WITH MINOR IMPACTS ON RANGELAND	ACTIVITIES WITH MAJOR IMPACTS ON RANGELAND
Hiking	Hunting
Camping	Horseback riding
Fishing	Trail biking
Skiing	Motorcycling
Boating	Jeeping
Rock hounding	Dune-buggy riding
Tubing	Home building
Canoeing	
Mountain climbing	
Relic hunting	
Bird watching	
Picnicking	

the winter when snow makes grazing infeasible at the higher elevations. Without winter base property, grazing of surrounding summer range, typically in federal ownership, becomes impractical. The people who build homes in the mountain valleys generally come from the urban areas and use these second homes primarily for vacation purposes. They heavily use the surrounding summer ranges for recreational pursuits. Under these conditions, recreation replaces livestock production as the primary use on the remaining federally owned rangeland.

The subdivision of mountain valleys has severe negative effects on big-game populations, particularly mule deer and elk. These animals are directly affected by the loss of important winter range. Indirectly they are subjected to greater stress on remaining range as the result of the increased human activity, such as snowmobiling, woodcutting, and hiking, that goes with subdivisions.

To maintain esthetic values, clean air, wildlife populations, and other amenities that make mountain valleys so desirable, it appears the restrictions on private land use will be necessary. Otherwise, the very qualities that make these areas so attractive will be destroyed.

Practices that provide economic incentives to landowners to retain their land in agriculture show some promise. State legislation that has reduced property taxes for land kept in agriculture coupled with zoning laws that restrict subdivision in farming and ranching areas have kept a good balance between development and agriculture in Oregon. Similar approaches to those used in Oregon will probably be applied to other western states as they become more urbanized.

Ranching on the Urban Interface

Large zones of rapid urbanization have occurred around many western cities in the 1990s. The problems of ranching on the urban-rangeland interface are considered by Huntsinger and Hopkinson (1996). Ranching on the urban-rangeland interface becomes more difficult as a result of marauding dogs, vandalism, trespass, carelessness with gates and fences, increased liability costs, and introduction of exotic plants (Hart 1991; Forero et al. 1992). Management options such as prescribed burning, predator control, and weed control are often restricted. Many people moving to rural areas do not understand rural customs and activities. Stray livestock can cause property damage and vehicle accidents. Commuter traffic may be blocked by slow ranch vehicles. The new neighbors may complain about livestock odors and the threat of pollution from pesticide or fertilizer applications. For these reasons and several others, better planned development is critical to ranchers as well as society at large. Huntsinger and Hopkinson (1996) believe environmentalists should be more tolerant in their rangeland appearance expectations while ranchers should rethink their concept of property ownership. They mentioned that debates over the ecological impacts of grazing, wildlife management imperatives and riparian zone restoration become moot when grass is replaced by concrete.

Huntsinger and Hopkinson (1996) point out that sustaining rangeland ecosystems in the future will be as much a social process as an ecological one. They provide examples from Marin County, California, of how successful planning and alliance building between ranchers and environmentalists prevented development of a large farming-ranching area.

Near San Francisco in this case, a combination of zoning, conservation easements, tax relief for farmers and ranchers, community leadership, and recognition of the heritage value of rural lifeways all played a part in the success. They noted that a similar pattern was emerging in other parts of the West. They believe carefully planned ranching can play a vital role in conserving many now threatened biodiverse rangeland landscapes. This planning process should include ecological, social and economic factors. Although environmentalists and ranchers differ in many beliefs, Huntsinger and Hopkinson (1996) point out both groups generally want to preserve open space and share a love of nature. When environmentalists were given better exposure to ranchers and their way of life, many came to see ranching as compatible with their goals. Huntsinger and Hopkinson (1996) believe range professionals can do much to help differing groups (ranchers and environmentalists) find "cultural convergence" by providing sound information on ecological systems and management options. Without question, the issue of rangeland and ranching preservation will be one of the biggest challenges facing range managers in the twenty-first century.

Scenic Beauty and Range Management

The impact of range management practices on scenic beauty in the United States has been an important concern in recent years because of the public's increased sensitivity to the environment and its greater role in the decision-making processes on public lands. A study of 241 dispersed recreationists on the Malhuer National Forest in eastern Oregon evaluated the public's response to range management activities (Sanderson et al. 1986). Photographs of selected ecosystems, range management practices, and management intensities were used to elicit attitudes of dispersed recreationists concerning management of the range resource. Recreationists were broken into categories of fishermen, hunters, and campers.

Recreationists responded most favorably to photos showing environmental scenes where livestock were grazed but range management practices were least visible (Table 14.9). Here, livestock numbers were within grazing capacity of the pasture and a minimum of fences were used. No methodology was used to distribute livestock evenly through the pasture. Extensive management of the environment and livestock was rated second. Here, the goal was full utilization of forage by livestock by range management methods that promote even distribution of livestock over the range. Intensive management of the environment and livestock was ranked least appealing. Here, the goal was to maximize forage production for livestock consistent with environmental constraints, including multiple use. Cultural practices were used to increase forage production and practices that improve livestock distribution were fully applied.

Fishermen rated photographs lower than campers. They considered grazing near riverbanks, herbicide spraying, alteration of upstream vegetation, and improved river access to recreationists as unacceptable management practices. They reacted more favorably toward fences than did other groups because they considered fences vital to maintenance of the fishery by excluding cattle.

Hunters gave the photographs higher ratings than did campers or fishermen. This appeared to have been due to the high hunting success in the area. Hunters generally re-

TABLE 14.9   Mean Photo Rating Scores for Recreationists Using the Malhuer National Forest in Eastern Oregon (Scale: 1 = Least Scenic, 5 = Most Scenic)[a]

RESPONDENTS	MANAGEMENT INTENSITY			
	ENVIRONMENTAL	EXTENSIVE	INTENSIVE	MEAN
Ponderosa pine				
Fishermen	3.3	2.9	2.3	2.8
Hunters	4.0	3.9	3.8	3.9
Campers	4.0	3.4	2.9	3.5
Overall mean	3.8	3.4	3.0	3.4
Mountain meadow				
Fishermen	4.0	3.3	3.3	3.5
Hunters	4.8	4.0	4.0	4.3
Campers	4.6	4.2	4.3	4.4
Overall mean	4.5	3.8	3.4	4.1
Mountain grassland				
Fishermen	4.0	2.8	2.5	3.1
Hunters	4.2	4.0	3.5	3.9
Campers	4.6	4.0	3.0	3.9
Overall mean	4.3	3.6	3.0	3.6
All ecosystems				
Fishermen	3.8	3.0	2.7	3.2
Hunters	4.3	4.0	3.8	4.0
Campers	4.4	3.9	3.4	3.9
Overall mean	4.2	3.6	3.3	3.7

Source: From H. R. Sanderson, R. A. Meganck, and K. C. Gibbs, 1986. Range management and scenic beauty as perceived by dispersed recreationists. *J. Range Manage.* 39:464–470, Table 1.

[a]An additional 21 respondents indicated other activities as their primary purpose for visiting the Malhuer National Forest.

sponded adversely to practices restricting access, such as road closure and establishment of more wilderness areas.

Campers, in general, felt that cattle were more appropriate for the mountain meadow and mountain grassland ecosystems than in the forest. They tended to consider open spaces as areas for grazing and forest as areas for camping. Generally, they did not mind seeing cattle from a distance on open areas, but they did not want cattle present near camping sites in the forest.

This study indicated several relationships between range management activities, recreationists, and the scenic qualities of rangeland. These are summarized as follows:

1. Recreationists do perceive differences in the visual quality of rangelands.
2. Different types of recreationists differ in their perception of visual quality.
3. The greater the familiarity with national forests, as measured by the number of prior visits, the greater the willingness to accept intensive management.
4. The public, in general, is not aware of the requirements for efficiently managing forest-range environments for increased forage.

The last relationship merits discussion. Many recreationists asked why cows were grazing in a national forest and why certain portions of a stream were fenced off. There was a failure to understand that the fences were to control livestock. Recreationists were asked if they had heard the term "multiple-use." The majority (84%) were unfamiliar with the term. When participants were asked if they knew what the concept "more intensive management of the range resource" meant, 91% of the users had no idea of its meaning. About 1% properly defined it as the application of techniques to improve the quality and quantity of range forage. The remainder believed that it meant the practice of grazing a greater number of cattle per unit of land. Based on this study, it appears that the public is tolerant of range management activities but generally has a poor understanding of what they involve.

More recently, visitor perceptions about cattle grazing on national forest land were evaluated in Colorado (Mitchell et al. 1996). The number of visitors indicating range livestock added to their stay (34%) was no different than the number stating a negative relationship (33%). Visitors in dispersed campsites tended to be more critical of grazing than those in developed campgrounds. The authors speculated this might be due to the fact that dispersed campsites were not fenced, and visitors were objecting to consequences of camping where cattle had previously been. This study indicated that if livestock are kept out of developed campgrounds and adjacent riparian areas used for fishing there is little objection to them. It was concluded that if possible, livestock should be managed so they are away from dispersed camping areas during times of high recreational demand.

Public Opinion and Management of Federal Rangelands

Public opinions on grazing issues have become an important concern of Federal rangeland managers. A survey conducted by Brunson and Steel (1994) showed U.S. residents place little confidence in livestock, mining, and energy extraction groups. They tend to believe more protection should be given to wildlife and fish resources (Table 14.10). In another survey, Brunson and Steel (1996) found only weak opinion differences on range issues between the eastern and western United States. However there were definite attitude differences between urbanized areas and rural regions which could eventually result in political

TABLE 14.10 Public Attitudes and Beliefs About Range Management Policies and Environmental Conditions on Federal Rangelands

STATEMENT	STRONGLY DISAGREE	DISAGREE	NEUTRAL	AGREE	STRONGLY AGREE
			(%)		
ATTITUDES TOWARD FEDERAL RANGE MANAGEMENT POLICIES					
Livestock grazing should be banned on federal rangelands	11	10	45	18	16
More rangeland wilderness areas should be established	10	5	14	24	47
Livestock grazing should be permitted in rangeland wilderness areas	31	19	20	19	11
Greater protection should be given to fish such as salmon	6	8	10	28	48
More should be done to protect rare plant communities	9	4	12	24	51
Greater efforts should be given to protect wildlife	3	4	7	23	63
Endangered species laws should be set aside to preserve ranching jobs	45	20	17	10	10
Federal range policy should emphasize livestock grazing	19	24	32	11	14
Ranchers should pay more than they do now to graze livestock on federal rangelands	7	7	19	29	38
The economic vitality of local communities should receive highest priority when making rangeland decisions	16	25	22	15	23
BELIEFS ABOUT ENVIRONMENTAL CONDITIONS OF FEDERAL RANGELANDS					
Most federal range is overgrazed by cattle and/or sheep	12	14	14	30	30
Soil erosion is only a minor problem on federal rangelands	30	33	13	13	10
Populations of most wildlife species on federal rangelands have remained constant or are increasing	44	30	14	8	4
The quality of water from federal rangelands has decreased markedly in the past 50 years	3	4	7	23	63
The extent of overgrazing on federal rangelands has decreased markedly in the past 50 years	34	31	18	9	8
Loss of streamside vegetation is a serious range problem	5	3	10	32	51

Source: From Brunson and Steel 1994.

power loss for range constituencies in the western United States. This survey was consistent with Sanderson et al. (1986) in showing that the public has a poor understanding of range management practices and polices. Public attitudes were likely to derive more from people's general opinion on the environment than from reasoned thought and informed belief. Respondents tended to believe either that rangeland conditions were declining and needed protective action or that rangeland conditions were improving and uniformly good.

Surveys by Sanderson et al. (1986), Brunson and Steel (1994) and Brunson and Steel (1996) all indicate the need to better inform the public on range management issues. At the same time many new and divergent interpretations of multiple-use management on public lands are being advanced that must be considered by modern range mangers (Kessler et al. 1992; Holechek 1993; Holechek and Hess 1995; Box 1995; Vavra 1996; Kessler 1996). Many range scientists have come to believe that more sustainable environmentally sensitive range livestock production systems are needed as alternatives to traditional systems (Box 1995; Heitschmidt 1996; Vavra 1996). Meeting this need will be an important challenge for range scientists and managers in the twenty-first century.

Recreation and Ranching

Advantages to Ranchers. The urbanization of the western United States does have some advantages to the ranching industry. Returns from livestock on western ranches as a percentage of capital investment have been low (they have averaged about 1% to 3% during the past 25 years). However, the increased human population in the West provides considerable opportunity for ranchers who are willing to diversify their enterprise. The potential of fee hunting has been discussed previously (Chapter 13). Other forms of recreation, such as packing trips, horseback riding, sight-seeing trips, dude ranching, and fishing are proving to be highly lucrative in certain areas. Many ranches have built special ponds stocked with fish and charge either daily fishing fees or so much per fish. They often also rent cabins and provide meals to vacationing tourists. These enterprises in some cases bring in far more net income than does sale of livestock. It appears that recreational enterprises will increasingly displace livestock as the main source of income from ranches in many parts of the western United States.

Disadvantages to Ranchers. On public rangeland, ranchers using these ranges generally view recreation negatively, often with good reason. Increased human activity can cause reduced animal performance due to disturbance. Increased incidences of vandalism, fire, and losses of livestock from theft, traffic collision, and shooting usually accompany heavier recreational use. In some cases, camping occurs around livestock watering points and livestock not used to human activity may be reluctant to water when this occurs.

Recreation Problems on Public Lands

Many ranchers and other segments of the public feel that federal agencies should charge fees for recreational use of grazing lands. These fees could than be used to mitigate some of the negative effects of recreation on livestock production. In reality, many recreational uses, particularly those forms involving off-road vehicle travel, if unregulated, affect the range more severely than does livestock grazing. To prevent range destruction in the future, regulation of recreation on pubic rangelands will be increasingly necessary. From the authors' point of view, it seems only reasonable that recreationists using public land, as well as the rancher, be charged a fee for their activities.

Many of the basic principles used in controlling livestock grazing can also be used in controlling recreational use (Heady and Vaux 1969). Proper stocking rate can be modified to proper numbers of people or vehicles on a particular piece of rangeland for a particular recreational activity. Proper distribution and timing of use is just as important with most recreational activities as it is with grazing. This is particularly true for off-road vehicle travel (Figure 14.16).

Off-road vehicle travel is a recreational pursuit that can have considerable impact on rangeland soils and vegetation. This was evaluated on a northern Great Plains range in southeastern Montana (Payne et al. 1983). These researchers studied the influence of different numbers of trips at different times of the year on soil and vegetation characteristics using a typical off-road vehicle (four-wheel-drive Chevrolet Blazer). When soils were dry, they found that up to eight trips in the same tracks did little long-term damage to living plants or the soil (Figure 14.17). However, above eight trips there was a strong likelihood of damage carrying over into following years. Damage to both soils and vegetation was much more severe with wet than with dry soils. It was recommended that drivers be encouraged not to follow other tracks when vegetation is actively growing because one to two trips had a negligible effect on soils and vegetation compared to repeated trips (8 to 32). However, this recommendation is reversed for dry, mature vegetation (winter use), because breakage became an important factor. If the range was not being reserved for winter use, they recommended that drivers be encouraged not to follow other tracks. When soils are wet, off-road travel should not occur because of severe damage to both soils and vegetation.

In a Nevada study, the comparative influences of motorcycle, four-wheel-drive truck, and no traffic on infiltration rate and sediment production were studied (Eckert et al. 1979). In this study infiltration rates were decreased and sediment production was increased by traffic (Table 14.11). The four-wheel-drive truck had a much greater impact than did the motorcycle.

Off-road vehicle travel on federal lands in the United States has become highly regulated during the past 10 years. It seems likely that this form of range recreational use will be even more restricted in coming years if recent trends in public use of federal land continue.

Figure **14.16** Off-road vehicle travel can be as destructive as uncontrolled grazing if vehicle numbers, timing of travel, and frequency of travel are not regulated.

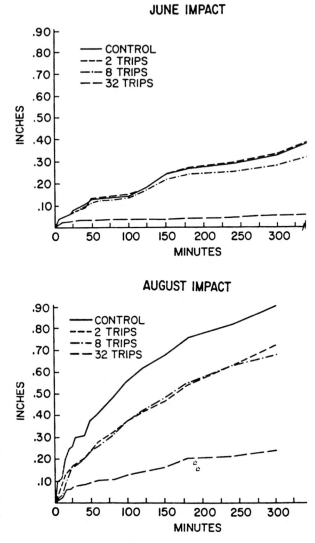

Figure 14.17 Infiltration rates resulting from four levels of off-road vehicle impact in June and August 1976. (From G. F. Payne, J. W. Foster, and W. C. Leininger, 1983. Vehicle impacts on northern Great Plains range vegetation. *J. Range Manage.* 36:327–331, Fig. 1.)

Conflict Resolution in Multiple-Use Decisions

In recent years public land management agencies have become heavily burdened with resolving controversies among various interest groups regarding natural resource use and management. Capability in conflict resolution has become of critical importance in managing public rangelands. Coordinated Resource Management Planning (CRMP) has been widely used in the western United States to deal with multiple-use conflicts. CRMP brings together public and private interests to resolve multiple-use and land management conflicts (Anderson 1991). It involves reasoned-scientific analysis, bargaining, and compromise.

TABLE 14.11 Mean Terminal Infiltration Rate and Sediment Production on Coppice and Interspace Soils in Response to Traffic Treatments[a]

TREATMENT AND SOIL	TERMINAL INFILTRATION (CM/HR)		SEDIMENT PRODUCTION (KG/HA)	
	1975	1976	1975	1976
Motorcycle				
Coppice	2.8	2.5	9	36
Interspace	0.2	0.2	376	509
Truck				
Coppice	2.0	2.2	85	45
Interspace	1.1	0.1	461	857
Control				
Coppice	3.2	3.1	4	9
Interspace	1.4	0.4	81	244

Source: From Eckert et al., 1979, Table 2.

[a]Simulated precipitation was applied at a rate of 3.4 cm/hr for 30 minutes.

The players have a common goal in correcting past land management mistakes and making informed decisions on management alternatives. It has been effective in resolving many conflicts on public rangelands (Anderson 1991; McClure 1992; Grizzle 1992). However, quantitative decisions based on economic tradeoffs are usually not possible because many natural resource products on public lands are free and their values vary among interest groups (Anderson 1981). CRMP guidelines that have developed from 40 years of experience are available from Cleary and Phillippi (1993).

Another approach called dispute resolution is discussed by Torell (1994). It involves bringing together parties in disagreement to participate in joint decision-making processes that seek win/win solutions and avoid litigation. A third party person is commonly utilized to assist in resolving conflicts. A study of environmental disputes found that 78% of the cases where alternative dispute resolution techniques were used resulted in settlement (Bingham 1986).

RANGE MANAGEMENT PRINCIPLES

■ The most important aspect of range watershed management is to maintain critical vegetation residues throughout the year so the soil is protected from dislodgement associated with rainfall. Proper stocking rate is the most important part of range watershed management.

- Livestock grazing reduces plant cover that protects the soil. This can cause soil erosion and compaction. Some specialized grazing system will overcome these effects if stocking rates are not excessive.

- Grazing strategies that keep livestock well dispersed and away from streamside areas minimize water quality problems associated with grazing.

- Recreational use of rangeland can be as damaging as overgrazing. Basic principles involved in controlling recreation are similar to those in controlling livestock. These include proper numbers, proper timing, and proper distribution of recreational users.

Literature Cited

Abdel-Magib, A. H., G. E. Schuman, and R. H. Hart. 1987. Soil bulk density and water infiltration as affected by grazing systems. *J. Range Manage.* 40:307–310.

Abell, D. L. (Coord.). 1989. Proceedings of the California Riparian Systems Conference: Protection, management, and restoration for the 1990s. *U.S. Dep. Agric., Pacific Southw. For. & Range Exp. Sta. Gen. Tech. Rep.* P8W-110.

Adams, S. N. 1975. Sheep and cattle grazing the forests: A review. *J. Appl. Ecol.* 12:143–152.

Aguilar, R. D., and S. R. Loftin. 1991. Sewage sludge application in semiarid grasslands: Effects on runoff and surface water quality. *In Proc. 36th Annual Water Conf., N. Mex. Water Resour. Res. Inst. Rep.* 265, pp. 101–111.

Aguilar, R., S. R. Loftin, T. J. Ward, K. A. Stevens, and J. R. Gosz. 1994. Sewage sludge application in semiarid grasslands: Effects on vegetation and water quality. *N. Mex. Water Resources Research Inst. Rep.* 285, pp. 1–75.

Aldon, E. F. 1964. Ground-cover changes in relation to runoff and erosion in west-central New Mexico. *U.S. Dep. Agric. For. Serv. Note* RM-34.

Aldon, E. F. 1963. Sediment production as affected by changes in ground cover under semiarid conditions on the Rio Puerco drainage in west central New Mexico. *Trans. Am. Geophys.* 44:872–873.

Anderson, E. W. 1991. Innovations in coordinated resource management planning. *J. Soil & Water Cons.* 46:411–414.

Anderson, T. L. 1981. *Multiple use management: Pie slicing vs. pie enlarging.* Workshop on political and legal aspects of range management. Natl. Res. Council, Jackson Hole, WY.

Bartlett, E. T., and D. R. Betters (Eds.). 1983. Overstory-understory relationships in western forests. *Colo. State Univ. Agric. Exp. Stn. West. Reg. Res. Publ.* 1, Ft. Collins, CO.

Bingham, G. 1986. *Resolving environmental disputes, a decade of experience,* p. 43. The Conservation Foundation, Washington, DC.

Binkley, D., and T. C. Brown. 1993. *Management impacts on water quality of forests and rangelands.* U.S. Dep. Agric. Rocky Mtn. For. & Range Exp. Sta. Gen. Tech. Rep. RM-239.

Blackburn, W. H., T. L. Thurow, and C. A. Taylor Jr. 1986. Soil erosion on rangeland. In *Use of cover, soil, and weather data in rangeland monitoring,* pp. 31–39. Symp. Proc., Soc. Range Manage., Denver, CO.

Bowers, W. B., A. Hosford, A. Oakley, and C. Bond. 1979. *Wildlife habitats in managed rangelands in the Great Basin of northeastern Oregon.* U.S. Dep. Agric. For. Serv. Gen. Tech. Rep. PNW-84.

Box, T. W. 1995. A viewpoint: range managers and the tragedy of the commons. *Rangelands* 17:83–85.

Branson, F. A., and J. R. Owen. 1970. Plant cover, runoff, and sediment yield relationships on Mancos shale in western Colorado. *Water Resour. Res.* 6:783–790.

Branson, F. A., G. F. Gifford, K. G. Renard, and R. F. Hadley. 1981. *Rangeland hydrology. Range Science Series No. 1.* Society for Range Management, Denver, CO.

Brunson, M. W., and B. S. Steel. 1994. National public attitudes towards federal rangeland managements. *Rangelands* 16:77–81.

Brunson, M. W., and B. S. Steel. 1996. Sources of variation in attitudes and beliefs about federal rangeland management. *J. Range Manage.* 49:69–75.

Bryant, L. D. 1982. Response of livestock to riparian zone exclusion. *J. Range Manage.* 35:780–785.

Child, R. D., H. F. Heady, W. C. Hickey, R. A. Peterson, and R. D. Pieper. 1984. *Arid and semiarid lands: Sustainable use and management in developing countries.* Winrock International Institute, Morrilton, AR.

Clary, W. P. 1975. Range management and its ecological basis in the ponderosa pine type of Arizona: The status of our knowledge. *U.S. Dep. Agric. For. Serv. Res. Pap.* RM-158. Ft. Collins, CO.

Clary, W. P., C. I. Thornton, and S. R. Abt. 1996. Riparian stubble height and recovery of degraded streambanks. *Rangelands* 18:137–140.

Cleary, C. R., and D. Phillippi. 1993. *Coordinated resource management guidelines.* Society for Range Management, Denver, CO (11 units paged separately).

Council for Agricultural Science and Technology (CAST). 1982. Water use in agriculture: Now and for the future. *Counc. Agric. Sci. Tech. Rep.* 95.

Cunningham, G. M. 1987. Reclamation of scaled land in western New South Wales. *J. Soil Conserv. New South Wales* 43:52–61.

Currie, P. O. 1969. Use seeded range in your management. *J. Range Manage.* 22:432–434.

Currie, P. O. 1975. Grazing management of ponderosa pine-bunchgrass ranges of the central Rocky Mountains: The status of our knowledge. *U.S. Dep. Agric. For. Serv. Res. Pap.* RM-159. Ft. Collins, CO.

Currie, P. O. 1976. Recovery of ponderosa pine-bunchgrass range through grazing and herbicide or fertilizer treatments. *J. Range Manage.* 29:444–448.

Daddy, F., M. J. Trlica, and C. D. Bonham. 1988. Vegetation and soil water differences among big sagebrush communities with different grazing histories. *Southw. Natur.* 33:413–424.

Darling, L. A., and G. B. Coltharp. 1973. *Effects of livestock grazing on the water quality of mountain streams.* Proc. Symposium on Water-Animal Relations, Southern Idaho State College, Twin Falls, ID.

DeBano, L. F. 1969. *Observations on water-repellent soils in the western United States. Symposium on Water-Repellent Soils, University of California, Riverside.* Proc. May 6–10, 1969, pp. 17–29.

DeBano, L. F. 1989. Effects of fire on chaparral soils in Arizona and California and postfire management implications, pp. 55–62. In N. H. Berg (Coord.), *Proc., Symp., Fire and Watershed Management. U.S. Dep. Agric. Pacific Southw. For. & Range Exp. Sta. Gen. Tech. Rep.* PSW-109.

DeBano, L. F., and R. M. Rice. 1973. Water-repellent soils: Their implications in forestry. *J. For.* 71:220–223.

DeBano, L. F., and L. J. Schmidt. 1989. Improving southwestern riparian areas through watershed management. *U.S. Dep. Agric., Rocky Mtn. For. & Range Exp. Sta. Gen. Tech. Rep.* RM-182.

Dennis, N. 1989. The effects of fire on watersheds: A summary, pp. 92–94. In N. H. Berg (Coord.), *Proc., Symp., Fire and Watershed Management. U.S. Dep. Agric., Pacific Southw. For. & Range Exp. Sta. Gen. Tech. Rep.* PSW-109.

Dunford, E. G. 1949. Relation of grazing to runoff and erosion on bunchgrass ranges. *U.S. Dep. Agric. For. Serv. Note* RM-7.

Dunford, E. G. 1954. Surface runoff and erosion from pine grasslands of the Colorado front range. *J. For.* 59:923–927.

Eckert, R. W. Jr., M. K. Wood, W. H. Blackburn, and F. F. Peterson. 1979. Impact of off-road vehicles on infiltration and sediment production of two desert soils. *J. Range Manage.* 32:394–398.

Eissenstat, D. M., J. E. Mitchell, and W. W. Pope. 1982. Trampling damage by cattle on a northern Idaho forest plantation. *J. Range Manage.* 35:715–716.

Elmore, W., and B. Kauffman. 1994. Riparian and watershed systems: Degradation and restoration. In *Ecological implications of livestock herbivory in the West.* Society for Range Management, Denver, CO.

Fink, D. H., and W. L. Ehrler. 1986. Runoff farming. *Rangelands* 8:53–54.

Flory, E. L. 1936. Comparison of the environment and some physiological responses of prairie vegetation. *Ecology* 17:67–103.

Forero, L., L. Hutsinger, and W. J. Clawson. 1992. Land use change in three San Francisco Bay Area counties: Implications for ranching at the urban fringe. *J. Soil & Water Conserv.* 47:475–480.

Gamougoun, N. D., R. P. Smith, M. K. Wood, and R. D. Pieper. 1984. Soil, vegetation, and hydrologic responses to grazing management at Fort Stanton, New Mexico. *J. Range Manage.* 37:538–541.

Gary, H. L., S. R. Johnson, and S. L. Ponce. 1983. Cattle grazing impact on surface water quality in Colorado front range streams. *J. Soil Water Conserv.* 38:124–128.

Gifford, G. F., and R. H. Hawkins. 1978. Hydrologic impact of grazing on infiltration: A critical review. *Water Resour. Res.* 14:305–313.

Glendening, G. E. 1944. Some factors affecting cattle use of northern Arizona pine-bunchgrass ranges. *U.S. Dep. Agric. For. Serv. Southwest For. Range Exp. Stn. Res. Rep.* 6.

Grelen, H. E. 1978. Forest grazing in the south. *J. Range Manage.* 31:244–250.

Griffin, R. C., and B. A. McCarl. 1989. Brushland management for increased water yield in Texas. *Water Resour. Bull.* 25:175–186.

Grizzle, G. 1992. Coordinated resource management planning in New Mexico. *Rangelands* 14:272–274.

Hanson, C. L., A. R. Kuhlman, C. J. Erickson, and J. K. Lewis. 1970. Grazing effects on runoff and vegetation on western South Dakota rangeland. *J. Range Manage.* 23:418–420.

Hart, J. 1991. *Farming on the edge.* University of California Press, Berkeley (174 pp).

Heady, H. F. and H. J. Vaux. 1969. Must history repeat. *J. Range Manage.* 22:209–210.

Hedrick, D. W. 1975. Grazing mixed conifer forest clearcuts in northeastern Oregon. *Rangeman's J.* 2:6–9.

Heede, B. H., M. D. Harvey, and J. R. Laird. 1988. Sediment delivery linkages in a chaparral watershed following a wildfire. *Environ. Manage.* 12:349–358.

Heitschmidt, R. K., R. E. Short, and E. E. Grings. 1996. Ecosystems, sustainability, and animal agriculture. *J. Anim. Sci.* 74:1395–1405.

Higgins, D. A., S. B. Maloney, A. R. Tiedemann, and T. M. Quigley. 1989. Storm runoff characteristics of grazed watersheds in eastern Oregon. *Water Resour. Bull.* 25:87–100.

Holechek, J. L. 1993. Policy changes on federal rangelands: A perspective. *J. Soil & Water Cons.* 48:166–174.

Holechek, J. L., and K. Hess, Jr. 1995. Government policy influences on rangeland conditions in the United States: A case study. *Environ. Monitoring and Assess.* 37:175–187.

Howery, L. D., F. D. Prooenza, R. E. Banner, and C. B. Scott. 1996. Differences in home range and habitat use among individuals in a cattle herd. *Appl. Anim. Beh. Sci.* 49:305–320.

Huntsinger, L., and R. Hopkinson. 1996. Viewpoint: Sustaining rangeland landscape: A social and ecological process. *J. Range Manage.* 46:167–173.

Hussey, M. R., Q. D. Skinner, and J. C. Adams. 1986. Changes in bacterial populations in Wyoming mountain streams after ten years. *J. Range Manage.* 39:369–370.

Jawson, M. D., L. F. Elliott, K. E. Saxton, and D. H. Fortier. 1982. The effect of cattle grazing on indicator bacteria in runoff from a Pacific northwest watershed. *J. Environ. Qual.* 11:621–627.

Johnson, A. 1962. Effects of grazing intensity and cover on the water intake rate of fescue grassland. *J. Range Manage.* 15:79–82.

Kessler, W. B. 1996. Requiem for range? Will our profession survive into the next century? *Rangelands* 18:238–242.

Kessler, W. B., H. Salwasser, C. W. Cartwright, and J. Caplan. 1992. New perspectives for sustainable natural resource management. *Ecol. Appl.* 2:221–225.

Knight, R. W. 1993. Managing stocking rates to prevent adverse environmental impacts. In *Managing livestock stocking rates on rangeland,* pp. 97–107. TX. Agric. Ext. Serv., College Station, TX.

Kunkle, S. H. and J. R. Meiman. 1967. Water quality on mountain watersheds. *Hydrology Paper No. 21.* Colorado State University, Fort Collins, CO.

Lacey, J., R. Carlstrom, and K. Williams. 1995. Chiseling rangeland in Montana. *Rangelands* 17:164–166.

Lacey, J. R., C. B. Marlow, and J. R. Lane. 1989. Influence of spotted knapweed (*Centaurea maculosa*) on surface runoff and sediment yield. *Weed Tech.* 3:627–631.

Lanner, R. M. 1977. The eradication of piñon-juniper woodland. *West. Wildl. Spring.* pp. 12–17.

Lewis, C. E. 1980a. Simulated cattle injury to planted slash pine: Girdling. *J. Range Manage.* 33:337–340.

Lewis, C. E. 1980b. Simulated cattle injury to planted slash pine: Combinations of defoliation, browsing, and trampling. *J. Range Manage.* 33:340–345.

Lewis, C. E. 1980c. Simulated cattle injury to planted slash pine: Defoliation. *J. Range Manage.* 33:345–348.

Liacos, L. G. 1962a. Soil moisture depletion in the annual grass type. *J. Range Manage.* 15:67–72.

Liacos, L. G. 1962b. Water yield as influenced by degree of grazing in the California winter grasslands. *J. Range Manage.* 15:34–42.

Lusby, G. C. 1970. Hydrologic and biotic effects of grazing vs. non-grazing near Grand Junction, Colorado. *J. Range Manage.* 23:256–260.

Marston, R. B. 1952. Ground cover requirements for summer storm runoff control on aspen sites in northern Utah. *J. For.* 50:303–307.

Marston, R. B. 1958. Parrish Canyon, Utah: A lesson in flood sources. *J. Soil Water Conserv.* 13:165–167.

Martin, S. C., and H. L. Morton. 1993. Mesquite control increases grass density and reduces soil loss in southern Arizona. *J. Range Manage.* 46:170–175.

Masters, L. S. Swanson, and W. Burkhardt. 1996a. Riparian grazing management that worked, I. Introduction and winter grazing. *Rangelands* 18:192–195.

Masters, L., S. Swanson, and W. Burkhardt. 1996b. Riparian grazing management that worked, II. Rotation with and without rest and riparian pastures. *Rangelands* 18:196–200.

McCalla, G. R. II, W. H. Blackburn, and L. B. Merrill. 1984a. Effects of livestock grazing on infiltration rates, Edwards Plateau of Texas. *J. Range Manage.* 37:265–269.

McCalla, G. R. II, W. H. Blackburn, and L. B. Merrill. 1984b. Effects of livestock grazing on sediment production, Edwards Plateau of Texas. *J. Range Manage.* 37:291–295.

McClure, N. R. 1992. Ranchers and resources roping benefits of CRM. *Rangelands* 14:249–250.

Miller, R. L., and G. A. Choate. 1964. The forest resources of Colorado. *U.S. Dep. Agric. For. Serv. Resour. Bull.* INT-3. Ogden, UT.

Mitchell, J. E., G. N. Wallace, and M. D. Wells. 1996. Visitor perceptions about cattle grazing on national forest land. *J. Range Manage.* 49:81–86.

Myers, L. E. 1964. Harvesting precipitation. *Int. Assoc. Sci. Hydrol. Berkeley Calif. Publ.* 65:343–351.

Naeth, M. A., R. L. Rothwell, D. S. Chanasyk, and A. W. Bailey. 1990. Grazing impacts on infiltration in mixed prairie and fescue grassland ecosystems of Alberta. *Can. J. Soil Sci.* 70:593–605.

Nebel, B. J. 1981. *Environmental science: The way the world works.* Prentice-Hall, Englewood Cliffs, NJ.

Ohmart, R. D. 1996. Historical and present impacts of livestock grazing on fish and wildlife resources in western riparian habitats. In P. R. Krausman (Ed.), *Rangeland wildlife,* pp. 245–287. The Society for Range Management, Denver, CO.

Osborn, B. 1956. Cover requirements for the protection of range site and biota. *J. Range Manage.* 9:75–80.

Packer, P. E. 1951. An approach to watershed protection criteria. *J. For.* 49:639–644.

Patton, D. R., and B. S. McGinnies. 1964. Deer browse relative to age and intensity of timber harvest. *J. Wildl. Manage.* 28:458–463.

Payne, G. F., J. W. Foster, and W. C. Leininger. 1983. Vehicle impacts of northern Great Plains range vegetation. *J. Range Manage.* 36:327–331.

Pearson, H. A. 1980. Livestock in multiple-use management of southern forest range. In R. D. Child and E. K. Byington (Eds.). *Proc. Symposium Southern Forest Range and Pasture Resource.* Winrock International Livestock Research and Training Center, Morrilton, AR.

Pearson, H. A., L. B. Whitaker, and V. L. Duval. 1971. Slash pine regeneration under regulated grazing. *J. For.* 69:744–746.

Platts, W. S. 1981. Effects of sheep grazing on a riparian-stream environment. *U.S. Dep. Agric. For. Serv. Res. Note* INT-307.

Platts, W. S. 1982. Livestock and riparian-fishery interactions: What are the facts? *In Trans. 47th North American Wildlife and Natural Resources Conference,* pp. 507–515. Wildlife Management Institute, Washington, DC.

Pluhar, J. J., R. W. Knight, and R. K. Heitschmidt. 1987. Infiltration rates and sediment production as influenced by grazing systems in the Texas Rolling Plains. *J. Range Manage.* 40:240–243.

Rauzi, F. C., and C. L. Hanson. 1966. Water intake and runoff as affected by intensity of grazing. *J. Range Manage.* 19:351–356.

Rauzi, F. C., and F. M. Smith. 1973. Infiltration rates: Three soils with grazing levels in northeastern Colorado. *J. Range Manage.* 26:126–129.

Rhoades, E. D., L. F. Locke, H. M. Taylor, and E. H. McIlvain. 1964. Water intake on a sandy range as affected by 20 years of differential stocking rates. *J. Range Manage.* 17:185–190.

Sanderson, H. R., R. A. Meganck, and K. C. Gibbs. 1986. Range management and scenic beauty as perceived by dispersed recreationists. *J. Range Manage.* 39:464–470.

Schlesinger, W. H., P. J. Forteyn, and W. A. Reiners. 1989. Effects of overland flow on water relations, erosion, and soil water percolation on a Mojave Desert landscape. *Soil Sci. Soc. Am. J.* 53:1567–1572.

Schmidt, R. H. 1991. Defining and refining value for riparian systems. *Rangelands* 13:80–82.

Sharp, A. L., J. J. Bond, J. W. Nueberger, A. R. Kuhlman, and J. K. Lewis. 1964. Runoff as affected by intensity of grazing on rangeland. *J. Soil Water Conserv.* 19:103–106.

Skinner, Q. D., J. C. Adams, P. A. Richard, and A. A. Beetle. 1974. Effect of summer use of mountain watershed on bacterial water quality. *J. Environ. Qual.* 3:329–335.

Skovlin, J. M. 1965. Improving cattle distribution on western mountain rangelands. *U.S. Dep. Agric. Farmers' Bull.* 2212.

Skovlin, J. M. 1984. Impacts of grazing on wetlands and riparian habitat. In National Research Council/National Academy of Sciences (Eds.). *Developing strategies for rangeland management.* Westview Press, Inc., Boulder, CO.

Skovlin, J. M., R. W. Harris, G. S. Strickler, and G. A. Garrison. 1976. Effects of cattle grazing methods on ponderosa pine-bunchgrass range in the Pacific northwest. *U.S. Dep. Agric. For. Serv. Tech. Bull.* 1531.

Society for Range Management. 1989. *A glossary of terms used in range management.* 3d ed. Society for Range Management, Denver, CO.

Spreitzer, P. N. 1985. Transitory range: A new frontier. *Rangelands* 7:33–34.

Springfield, H. W. 1976. Characteristics and management of southwestern piñon-juniper ranges: The status of our knowledge. *U.S. Dep. Agric. For. Serv. Gen. Tech. Rep.* RM-39.

Sturges, D. L. 1993. Soil-water and vegetation dynamics through 20 years after big sagebrush control. *J. Range Manage.* 46:161–169.

Swanson, S. 1989. Priorities for riparian management. *Rangelands* 11:228–230.

Taylor, C., N. Garza, and T. Brooks. 1993. Grazing systems on the Edwards Plateau of Texas: Are they worth the trouble? I. Soil and vegetation response. *Rangelands* 15:53–57.

Tellman, B., H. J. Cortner, M. G. Wallace, L. F. DeBano, and R. H. Hamre (Coord.). 1993. Riparian management: Common threads and shared interests. *U.S. Dep. Agric., Rocky Mtn. For. and Range Exp. Sta. Gen. Tech. Rep.* RM-226.

Thomas, A. E. 1986. Riparian protection/enhancement in Idaho. *Rangelands* 8:224–227.

Thurow, T. L., W. H. Blackburn, and C. A. Taylor, Jr. 1986. Hydrologic characteristics of vegetation types as affected by livestock grazing systems, Edwards Plateau, Texas. *J. Range Manage.* 39:505–508.

Tiedemann, A. R., D. A. Higgins, T. M. Quigley, H. R. Sanderson, and D. B. Marx. 1987. Responses of fecal coliform in streamwater to four grazing strategies. *J. Range Manage.* 40:322–329.

Torell, D. J. 1994. Viewpoint: Alternative dispute resolution in public land management. *J. Range Manage.* 47:70–74.

United States Department of Agriculture (USDA). 1980. *Report of the Forest Service: Fiscal Year 1985.* United States Department Agriculture, Forest Service. Washington, DC.

United States Department of Agriculture (USDA). 1992. *Report of the Forest Service: Fiscal Year 1992.* United States Department of Agriculture, Forest Service. Washington, DC.

United States Water Resources Council (USWRC). 1978. *The Nation's Water Resources, 1975–2000.* Second National Water Assessment. Vol. I. Summary. Supt. Doc. 052-045-00051-7. U.S. Government Printing Office, Washington, DC.

Vallentine, J. F. 1989. *Range development and improvements.* 3d ed. Brigham Young University Press, Provo, UT.

Vavra, M. 1996. Sustainability of animal production systems: An ecological perspective. *J. Anim. Sci.* 74:1418–1423.

Wadleigh, C. H. 1968. Wastes in relation to agriculture and forestry. *U.S. Dep. Agric. Misc. Publ.* 1065.

Walling, D. E., and B. W. Webb. 1996. Erosion and sediment yield: A global overview. *Int. Assoc. Hydrol. Sci.* 236:3–19.

Warren, S. D., W. H. Blackburn, and C. A. Taylor, Jr. 1986a. Effects of season and stage of rotation cycle on hydrologic condition of rangeland under intensive rotation grazing. *J. Range Manage.* 39:486–491.

Warren, S. D., T. L. Thurow, W. H. Blackburn, and N. E. Garza. 1986b. Soil hydrologic response to number of pastures and stocking density under intensive rotation grazing. *J. Range Manage.* 39:500–505.

Warren, S. D., W. H. Blackburn, and C. A. Taylor, Jr. 1986c. The influence of livestock trampling under intensive rotation grazing on soil hydrologic conditions. *J. Range Manage.* 39:491–496.

Weltz, M., and M. K. Wood. 1986a. Short duration grazing in central New Mexico: Effects on infiltration rates. *J. Range Manage.* 39:365–368.

Weltz, M., and M. K. Wood. 1986b. Short-duration grazing in central New Mexico: Effects on sediment production. *J. Soil Water Conserv.* 41:262–266.

Weltz, M., M. K. Wood, and E. E. Parker. 1989. Flash grazing and trampling: Effects on infiltration rates and sediment yield on a selected New Mexico range site. *J. Arid Environ.* 16:95–100.

West, N. E., K. H. Rea, and R. J. Tausch. 1975. Basic synecological relationships in piñon-juniper woodlands. In G. M. Gifford and F. E. Busby (Eds.). *The piñon-juniper ecosystem.* Utah Agric. Exp. Stn., Utah State University, Logan, UT.

Wilcox, B. P., M. S. Seyfried, K. R. Cooley, and C. L. Hanson. 1991. Runoff characteristics of sagebrush rangelands: Modeling implications. *J. Soil and Water Conserv.* 46:153–158.

Williamson, R. M., and W. F. Currier. 1971. Applied landscape management in plant control. *J. Range Manage.* 24:2–6.

Wolters, G. L. 1981. Timber thinning and prescribed burning as methods to increase herbage on grazed and protected longleaf pine ranges. *J. Range Manage.* 34:494–497.

Wood, J. C., and M. K. Wood. 1988. Influence of piñon-juniper and sagebrush control on infiltration rates and runoff water quality in northern New Mexico. *N. Mex. J. Sci.* 28:7–20.

Wood, M. K., and W. H. Blackburn. 1981a. Sediment production as influenced by livestock grazing in the Texas Rolling Plains. *J. Range Manage.* 34:228–231.

Wood, M. K., and W. H. Blackburn. 1981b. Grazing systems: Their influence on infiltration rates in the Rolling Plains of Texas. *J. Range Manage.* 34:331–335.

Wood, M. K., and W. H. Blackburn. 1984. Vegetation and soil responses to cattle grazing systems in the Texas Rolling Plains. *J. Range Manage.* 37:298–303.

Wood, M. K., E. L. Garcia, and J. M. Tromble. 1991. Runoff and erosion following mechanical and chemical control of creosotebush. *Weed Tech.* 5:48–53.

CHAPTER 15

MANIPULATION OF RANGE VEGETATION

The manipulation of range vegetation by practices other than control of grazing, such as with herbicides, mechanical control, and fertilization, received great emphasis in the United States in the period from the mid-1950s to the late 1970s. However, in recent years, grazing management has received increased attention, while management practices geared toward causing major changes in range vegetation have been deemphasized. This has partially resulted from the high supply of meat and low prices for livestock in the 1980s and 1990s that has made cost reduction management more attractive to ranchers than practices that may substantially increase rangeland forage productivity but at great cost and risk. Reduced land prices in the 1980s made the purchase of more land an attractive alternative to vegetation manipulation when increased grazing capacity was desired. Pressure from environmental groups coupled with reduced government spending on conservation has greatly curbed brush control projects. Based on present trends, we do not foresee a shift back toward heavy emphasis on manipulation of range vegetation in the western United States until well into the next century, if then. Therefore, we have emphasized grazing management rather than vegetation manipulation in this book. However, we do recognize that vegetation manipulation is the only practical way to increase forage for livestock and to improve wildlife habitat on some ranges. Further, for some ranchers this may still be the best way to increase economic returns. In this chapter we cover the fundamentals of vegetation manipulation on rangelands, but refer the reader to Scifres (1980), Wright and Bailey (1982), Vallentine (1989), and Heady and Child (1994) for both excellent and detailed reviews on the various aspects of this subject. The influences of vegetation manipulation on rangeland wildlife are covered in a new book edited by Krausman (1996).

CONTROL OF UNWANTED PLANTS

Across the vast rangelands of the world, many ecosystems are changing because of rapid invasion by alien plants. These immigrants have arrived mostly in the last 150 years and compete well with native plants, increase soil erosion, and can transform wetlands, grasslands, and hillsides into stands of unwanted plants. Some examples in North America are: halogeton (*Halogeton glomeratus*), russianthistle (*Salsola* spp.), leafy spurge (*Euphorbia esula*), spotted knapweed (*Centawrea maculosa*), and rush skeletonweed (*Chondrilla juncea*). In addition, shrubs that have been in the flora for several thousand years are increasing on rangelands (e.g., see Buffington and Herbel 1965; Gibbens et al. 1996). A perspective on present weed problems, biological and chemical control technologies, and regulatory policies was presented by Antognini et al. (1995).

It is useful to conceptualize in a diagram the practices used in range management (Figure 15.1). Each land manager determines the desired level of productivity based on economic, political, cultural, and social factors and the availability of technology. Examples of extensive practices for rangelands are manipulation of grazing and the use of fire. These practices may require fencing and/or water developments, but they generally do not cause a major change in the present vegetation as do some of the more intensive manipulative practices. Although extensive practices are not costly, the opportunities to substantially increase production are usually low.

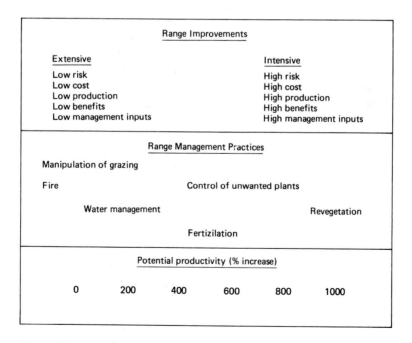

***Figure* 15.1** Types of range management practices and potential productivity increases. (From C. H. Herbel 1983.)

Drastic manipulations of range ecosystems are sometimes required or desired. The invasion of unwanted plants, severe droughts, past abuses by grazing animals, or the desire by the operator to change botanical composition or productivity, on all or part of the range unit, can make practices to revegetate with useful plants desirable. The risk associated with revegetation is high because the practice may not give the desired effects even when properly done. Control of unwanted plants, revegetation, and/or fertilization can result in production increases of two- to tenfold within 1 to 3 years. High management inputs are required once these risky, costly practices are used if the land manager wishes to realize a reasonable return on his or her investment.

Few, if any, land managers use intensive practices exclusively on a unit of rangeland. Rather, some combination of beneficial practices is applied whereby both intensive and extensive principles are utilized. In the northern Great Plains of Canada and the United States, this may mean seeding part of the range unit with Russian wildrye (*Elymus junceus*) and crested wheatgrass (*Agropyron cristatum*), and using nitrogen fertilizer on both native and introduced species. On the semiarid southern Great Plains of the United States and associated grasslands of Mexico, a useful strategy may include seeding weeping lovegrass (*Eragrostis curvula*), side oats grama (*Bouteloua curtipendula*), wheat (*Triticum aestivum*), and/or sudangrass (*Sorghum sudanense*) on the more productive sites. Rangeland with thinner soils and rougher topography is usually left in native vegetation. In the arid portions of the southwestern United States and northern Mexico, range productivity on some sites can be greatly increased by control of mesquite (*Prosopis glandulosa*) and tarbush (*Flourensia cernua*), and seeding with Lehmann lovegrass (*Eragrostis lehmanniana*) and fourwing saltbush (*Atriplex canescens*). On rangelands infested with big sagebrush (*Artemisia tridentata*), control of the sagebrush and seeding with crested wheatgrass often results in much greater productivity and soil stability than can be obtained from native range. These seeded ranges are particularly useful as calving and breeding pastures in the spring. In some instances, the composition of plant species may be manipulated to improve wildlife habitat, while at the same time maintaining or improving livestock production. The use of various practices changes with time as dictated by economic, political, social conditions, or as improved technology becomes available. Land managers must be flexible and innovative in planning operations on a range unit. Practices that work on one range unit may be entirely unsuited for the range unit next to it, or on the range unit 250 km distant. Differences in opinion over management objectives can and do lead to serious conflicts.

The most economical methods for reclaiming deteriorated grazing lands is through use of methods not requiring planting of desirable species. This may be accomplished by control of unwanted plants, concentrating moisture or harvesting precipitation, and/or by grazing management. Remote sensing techniques offer a cost-effective means to distinguish and measure infestations of noxious plants (Everitt et al. 1995). Proper grazing use of desirable plants is very important. If natural revegetation is not feasible, planting of desirable vegetation may be needed. The machinery is sometimes expensive or difficult to obtain, therefore, the use of inexpensive and innovative techniques must be maximized. Whether natural revegetation or artificial planting is used depends on the residual vegetation and the manager's desires. Plants propagate by seed and natural means (rhizomes, stolens, tillers, etc.). For natural revegetation to be effective, there must be a residue of desirable plants to take over and dominate the site. It must be recognized that reduction of unwanted plants may not result in a stand of wanted plants if they do not already exist on the site. In instances where undesirable vegetation is competing severely with the establishment of desirable vegetation, it is necessary to reduce or eliminate such undesirable vegetation. Natural revegetation often is quite rapid on sandy soils following control of unwanted plants. Conversely, natural revegetation often is very slow on medium- to heavy-textured soils.

The control of unwanted plants is necessary to make more water available for the reproduction and production of desirable vegetation. This may be accomplished by chemical, biological, or mechanical means; by judicious use of fire; or by use of different animal species. Plant control in range management is simply the reduction of unwanted or undesirable plants that have invaded or increased in a plant community. Plants "out of place," or the movement of certain species out of their normal range or habitat, is one of the major problems on rangelands of the arid and semiarid regions. Leaving scattered oak mottes on clearings facilitated deer use, promoted bird use, and afforded a more aesthetically pleasing landscape (Rollins et al. 1988).

Fire Control

Rangeland burning may be designed to fit one or more of the following objectives: (1) increased or improved livestock forage by reducing or eliminating some competing plants; (2) reduced litter and increased growth of desirable forage plants; (3) improved wildlife habitat; (4) fuel reduction with a long-range goal of lowering the chance of catastrophic wildlife in forest and chaparral types; (5) improved visibility of livestock after planned burning of certain ranges, especially in piñon-juniper, chaparral, and desert shrub types; (6) reduced labor costs of handling cattle and horses; and (7) reduced predation on sheep due to better visibility of sheep and predators.

Management of a prescribed burn differs in several ways from emergency treatment usually applied following wildfires. Because prescribed fires are planned with specific goals, under constraints that will assure minimum damage to plant cover and soils, much less injury to the site occurs. With preplanned funding for range improvements (including seeding where needed) and a grazing strategy to accommodate the prescribed burn, there should be little disruption of livestock operations. In contrast, after wildfires, management alternatives often are limited to minimizing expected damage from flooding and erosion and repair or replacement of range improvements, and so on (Jordan 1981; Baumgartner et al. 1989; White and Hanselka 1989; Krammes 1990).

Effects on Soil. Depending on its intensity, fire may have far-reaching effects on soil characteristics, erosion, water yield, and plant succession. Removal of the litter cover often exposes mineral soil and subjects the surface to rain impact. Sealing the soil surface increases overland flow and attendant soil loss. In grassland or mixed grass-shrub areas, however, there is seldom enough litter to fuel catastrophic fires. The light, flashy burns characteristic of grasslands seldom modify surface soil structure or seriously inhibit infiltration as do intense burns in forest, woodland, or chaparral. Grassland burns do usually inhibit the growth of woody plants.

Burning may significantly affect nutrients in range soils. Burning sometimes increases the supply of nitrogen, phosphorus, and sulfur available for plant growth. Nitrogen frequently may be limiting, especially on brush-supporting soils. Addition of even small amounts of available nitrogen may have a profound effect on revegetation (Hobbs et al. 1991).

Certain soil physical properties may also be adversely affected by burning. On forest ranges where slash and litter make heavier fuels, or in dense chaparral, intense fires may decrease soil aggregates and porosity and increase bulk density. These problems are most severe in the first or second postfire years, then disappear within about 4 years. A temporary increase in overland flow and erosion may be expected where fire changes soil structure, but this effect is usually brief.

Some moderately permeable soils may develop resistance to wetting as a result of burning. This water-repellent characteristic has been observed mostly in chaparral, woodland, and forest types where much fuel has accumulated. Apparently, the hydrophobic material is moved downward in the soil, where it condenses on soil particles, greatly retarding percolation. Infiltration may also be reduced if this water-resistant layer is at or near the soil surface. In any case, the nonwettable character may last a year or more, and may be a prime factor in the typically high runoff and erosion rates following fires on steep slopes.

Pattern in Vegetation. Most ranges subjected to random or intermittent burning do not have vegetation uniformity. Vegetation under such conditions is a variety of species, ages, and density classes. Burn intensity varies greatly from spot to spot, and "skips" may occur where fuels were lighter or where perverse winds altered the fire path. The resulting habitat diversity may be highly beneficial to wildlife, livestock, and landscape esthetics. These benefits may be reduced or nonexistent, however, following catastrophic wildfires.

Burning favors fire-tolerant species. Certain species, such as desert ceanothus (*Ceanothus greggii*) and manzanita (*Arctostaphylos* sp.), may even be wholly dependent on fire and tend to disappear from the plant community in the absence of burning. Seeds from these plants require heat scarification to germinate. Other species of forbs seldom found on unburned ranges may become abundant for a few postfire years, then abruptly decline. Certain species, such as shrub live oak (*Quercus turbinella*), mesquite, and alligator juniper (*Juniperus deppeana*), may adapt to occasional burning by vigorous production of crown and/or root sprouts. However, young seedlings or spouts are more susceptible to burning damage, and frequent fires can keep them in check.

Large bunchgrasses are damaged more than smaller bunchgrasses because more fuel is present, fire duration is longer, and heat penetration is deeper into plant tissue (Engle et al. 1993). The presence of soil moisture retards heat buildup in the base of grass plants, however, the most effective control of shrubs usually occurs when the soil is dry. Grasses tend to be more fire resistant than shrubs because of protected growing buds, while dormant grasses are more resistant than those having active buds. Fires are more effective on nonsprouting shrubs such as sagebrush than on resprouting shrubs such as creosotebush (*Larrea tridentata*).

Burning may affect palatability and availability of forage for both livestock and wildlife. Cattle tend to congregate on recent burns, largely because of accessibility of the tender, succulent new growth and the temporary communities of forbs that commonly develop in the early postfire periods. Utilization of weeping lovegrass, for example, increased more than 50% after a winter burn (Klett et al. 1971).

Fires in productive grasslands may occur rather frequently (5- to 10-year intervals); in less productive areas, burning may be possible only in unusual years. Because grass crowns regenerate quickly after burning, light fuels soon accumulate and many areas can be reburned in as little as 1 to 3 years. On chaparral ranges, however, this is not so. Usually, fuel accumulates more slowly, and, once burned, such areas may be relatively "fireproof" for 15 to 20 years. Ponderosa pine (*Pinus ponderosa*) ranges may be reburned on a 6- to 10-year schedule. In the Great Basin, cheatgrass (*Bromus tectorum*) provides fuel for range fires (Billings 1994).

Topography. As the steepness of the slope increases, the rate of fire spread increases. The upslope rate is favored because of convection (the rising of hot air currents) and the more intense radiation that occurs on the upslope side of the fire. Winds tend to move upslope during the day and downslope during the night. Vegetation in saddles is more readily burned because saddles are subject to wind tunnel effects.

Fuel. Sufficient fuel must be present to carry a ground fire, and under ideal conditions this should be at least 700 kg/ha. The surface/volume ratio affects the rate of combustion; consequently, a high surface area of small-sized fuel releases heat quickly (i.e., has a high intensity). Larger fuels burn with less intensity but for longer. Intensity and duration are the controlling influences on plants and animals. Fire duration is related directly to total heat yield and fuel quantity, and indirectly to intensity. Heat penetration into soil, bark, and other plant tissue increases as fire duration increases. In addition to size and amount of fuel, environmental temperatures and moisture levels need to be prescribed within rather narrow limits.

Weather. Weather before, during, and after a burn must be considered. Of the three elements necessary for combustion (oxygen, fuel, and heat), heat can be manipulated by selection of the weather conditions for the burn. Before combustion can occur, the fuel must be raised to the appropriate temperature. All moisture must be removed before this temperature is reached or else temperatures remain near the boiling point of water. Thus moisture, relative humidity, and temperatures are prescribed elements of weather preceding and during a burn. While the cooling effect of water or moisture on a fire is elementary knowledge, its overall effect on a large fire is much more difficult to predict.

Air Pollution. Air quality near large urban centers is apt to be marginal at best. Any land management activity that tends to add significantly to the pollution is undesirable. Prescribed burning of extensive areas can be planned when fire weather forecasters predict airflow away from urban centers to avoid stagnation and/or mixing with polluted air layers. Most range burning involves relatively light fuels, and unless areas are very large, the air pollution (mostly from particulates) tends to be rather transitory.

Smoke from man-made fires is not all bad. Particulate matter from fires has been a principal source of condensation nuclei, which are necessary for the production of clouds and precipitation. Also, they have served as natural cleansing agents, purifying the air, soil, and water of undesirable chemicals and toxicants for millions of years. Recently, particulates given off from forest and grassland fires consist largely of charcoal and ash with great diversity in form, structure, and porosity, and absorptive capacity. These particulates are very different from those apparently nonabsorbent ones produced by the combustion engine and by burning oil, rubber, and plastics.

Water Quality. Range burning has not seriously affected water quality. Water quality studies have shown that initial conversion of chaparral to a grass-shrub range, for example, may increase nitrate in runoff to high levels temporarily, but these peaks do not last long and apparently are confined to the first or second year. Nitrate concentration then drops to a relatively low level (in the range of 10 to 16 ppm), and occasional reburning of the light fuels has little or no effect on nitrates. No significant increases have been documented in phosphorus, calcium, or total dissolved salts.

Erosion. Wildfire often result in massive sheet and gully erosion, but this is much less of a problem with prescribed burns. Prescribed burns tend to be less intense and some residue is left for site protection.

Even the most carefully planned and executed range burn may produce sediment. Sediment yield declines rapidly, however, and, even after wildfire, yield drops to near preburn levels within 3 to 5 years (Pase and Lindenmuth 1971). Early establishment of a good grass cover, and subsequent conservative grazing, assures soil stability and low sediment yields on moderate slopes.

Effects on Wildlife. Well-planned prescribed burns may benefit wildlife (Wade and Lewis 1987; Klinger et al. 1989; Thompson et al. 1991; Riggs et al. 1996). By maintaining a variety of vegetation, including brush islands of various sizes and ages, a highly diverse habitat can be provided. Habitat diversity encourages wildlife species diversity. Leaving shrubby areas of adequate size as escape cover and providing a number of seral stages of postburn vegetation can benefit both game and nongame wildlife.

Aesthetics. People who have only a limited understanding of the role of fire in the development of natural communities see a newly burned forest or range area as land management at its worst. They find the bare, blackened soil with occasional burned stubs of mesquite or borroweed (*Haplopappus tenuisectus*) or shrub live oak profoundly disturbing. Like a plowed field or a homesite under construction, the disturbance does not last long. It should be noted that many seral plant communities, including some of the most colorful, exist only during the brief postfire recovery period.

Planning and Burning. The steps taken to develop a prescribed burn are planning, preparation, burning, and postburn management. Technically, qualified help must be available. Planning provides for adequate preparation, which in turn promotes a successful burn. The area must have the potential of being improved by this practice. The costs of burning, such as labor, rental of equipment, supplies, and short-term loss of forage, should be considered. Areas up to 400 ha approach maximum cost-effectiveness and manageability for daily burns.

Preparation includes the construction of firebreaks, fuel-break systems, and the placement of fire lines. For more positive control, fire lines are commonly started on ridgetops. Fires should not be started in valley bottoms but rather, 10 m to 100 m upslope. Various topographic features, such as streams, roads, and old burns, may be used to advantage to control burns. Up-to-the-minute spot weather forecasts are needed to determine wind velocity, relative humidity, and temperature. Wind speeds of 5 to 15 km/hr are desirable. Finally, public information is important. It is better to have an informed public rather than having to justify the fire to a concerned public after the fact.

The actual burning can be conducted by several ignition designs; for example, a helitorch is used to burn redberry juniper (*Juniperus pinchotii*) in the Texas Rolling Plains (Masters et al. 1986). The general principle is to have cool fires on the perimeter and hot fires in the center. Most designs start the fire 15 m to 30 m inside the fire line. The sequence of ignition varies according to soil, topography, and weather conditions. It should be remembered that one fire tends to draw another fire to it through convection and radiation processes, therefore, there is no substitute for experienced help in conducting a prescribed burn.

Postburn management objectives must be part of the prescribed burn project. A single fire seldom attains the management objective. Herbicide treatments might be necessary to control sprouting of unwanted plants. When feasible, grazing by deer or goats after the fire can suppress sprouting. Seeding may be necessary to obtain the desired cover and erosion control. Deferment from grazing will be necessary until stand establishment, which usually involves a 1- to 2-year period.

Increased Forage Production. Increased in forage production is probably the main reason for range burning. Burning may be insurance against some future decrease in carrying capacity due to woody plant encroachment, or planned as conversion from a brush to a grass-shrub range to increase forage production (Florence 1987). Benefits may last 5 to 10 years, but seldom longer.

Improved Forage Quality. Forage quality may be improved, because of higher nutrient content and digestibility or improved species composition (Figure 15.2). The removal of mature forage makes new growth more available to grazing or browsing animals.

Better Utilization and Livestock Distribution. Utilization and livestock distribution may be improved. Reduction or retardation of woody plant growth permits better distribution of stock and more uniform forage utilization. Another benefit is the reduction in personnel needed to work livestock (Scifres 1987; Ueckert et al. 1988).

Increased Water Yield. Increased water yields have resulted from burning of some chaparral ranges. Most range burning, however, improves on-site water (soil moisture) savings and use.

Improved Wildlife Habitat. Wildlife can benefit from the increased diversity of habitats following well-planned range burns. Wildlife food may be increased. Nongame birds and mammals also may benefit from the increase in habitat diversity caused by careful range burning (Wood 1988; Thompson et al. 1991; Riggs et al. 1996).

Figure 15.2 Range in Florida is burned periodically to reduce old growth and to increase the quality and availability of forage.

Mechanical Control

Considerations in selection of mechanical methods, such as grubbing or root plowing to remove unwanted plants; are cost, availability of equipment, the size and stand of the plants to be eliminated, whether the target plants have sprouting or nonsprouting characteristics, soil conditions, and the type of terrain.

1. *Size and stand of the target plants.* The best time to employ hand-grubbing is during early invasion of unwanted plants, before the stand of desirable species becomes greatly reduced. Hand-grubbing of small shrubs (up to 90 cm in canopy diameter) is an economical control method when the stand is relatively thin, usually less than 80 plants/ha. With sprouting species, the root must be severed below the budding zone. Cabling or chaining is most effective in controlling even-aged, mature shrubs or small trees with stem diameters of 8 cm or more. Bulldozing is effective on sparse stands and medium-sized trees, while disking is limited to small plants. Rootplowing or disking is used when there is a sparse stand of desirable plants and revegetation is needed.

2. *Sprouting or nonsprouting shrubs.* Cabling, chaining, and disking do not give a high degree of kill on shrubs that sprout below the surface of the ground.

3. *Soil conditions.* Cabling or chaining is most effective in areas with lighter-textured sandy or loamy soils. Bulldozing, rootplowing, and disking excessively disturb soil, destroy desirable plants, and may result in soil erosion. Most mechanical methods cannot be used when the soil is excessively wet.

4. *Topography.* Some mechanical methods leave the soil bare, unprotected, and subject to erosion. There should be a minimum of rocks and gullies so that the equipment can operate at relatively high speed. Therefore, most mechanical equipment should be used on relatively level terrain.

Because of the widespread variations in climate, topography, and woody species, there have been many approaches to mechanical plant control. There has been a continuous effort to develop more efficient and effective machinery for control of all types of woody vegetation. These improvements have included both the rudimentary spike drag-type implements used in big sagebrush control and the more advanced tree crushing equipment used in juniper control. Implements that chop or crush woody vegetation are primarily for release of existing desirable vegetation. Sprouting species are not killed by these types of machines. Nonsprouting species are reduced from 50% to 90% depending on the machine used and species treated.

"Brush cutters" are commonly called by their commercial name, Marden or Fleco. These implements resemble the old-fashioned stalk cutter (Figure 15.3). The use of two sections pulled in tandem with angular alignment between the two units allows for greater crushing and cutting effect. Brush cutters of this type have limitations for range work. Best kills of brush, such as sagebrush, are obtained on even-aged mature stands. Kills of uneven-aged brush or small brush seldom reach 50%.

The tree crusher, which weighs 72 tons, uproots, crushes, and splinters juniper trees in one operation. Because most trees are pushed out of the ground before being crushed, the percentage kill is high (about 80%). On fairly level terrain, this machine can crush about 4 ha per hour. The advantage of this machine is its high production. The debris left as mulch

***Figure* 15.3** A brush cutter operated in tandem to reduce palmetto (*Sabal palmetto*) in Florida.

Figure **15.4** A version of the tree crusher that has rotary mower blades mounted in the rear and is shown mowing a stand of juniper-mesquite in northcentral Texas.

on the ground creates a microclimate which is very beneficial in establishing vegetation. The crusher cannot be used on rocky areas or on slopes over 25% (see Figure 15.4).

The business end of the machine known as the "tree eater" is a revolving drum mounted to the rear of the tractor. The tree eater cuts a swath 2.4 m wide. Disadvantages of this machine are its high operating and maintenance costs, plus the noise and dust created by this machine are extremely tough on the operator. The tree eater cannot be used on slopes over 15% because it will not tolerate rock.

The brush shredder and sprayer combination "chem-cut" unit combines the shredding capabilities of a rotary mower with a herbicide application to the stem surfaces as they are cut. The machine contains a specially adapted hollow shaft and cutter bars to allow the herbicide to be applied directly to the freshly cut stump. This combination is effective on sprouting species.

Chaining and cabling are used in control of unwanted vegetation and have been used most successfully in sagebrush and piñon-juniper control projects (Stevens and Walker 1996). The method is also useful in knocking down trees previously killed by aerial spraying, thereby reducing the cost of working livestock. Chains with links weighing over 30 kg are recommended. Wiedemann and Cross (1990) have added a disk to chain links to improve preparation of land for subsequent revegetation with desirable plants. The advantages of this method are

1. Large areas can be treated at low cost.

2. Desirable perennial grasses and forbs are not seriously damaged.

3. Debris and trash protect the soil from erosion.

Limitations are

1. Many of the small or very pliable shrubs and trees are not killed.
2. Undesirable herbaceous plants and sprouting shrubs are damaged only slightly.

Pushing with a dozer is another method of mechanical control. This is suited to light stands of trees (less than 500 stems per hectare). A hinged pusher bar has been developed to aid in uprooting the trees. Bulldozer blades or front-end loaders may be fitted with a "stinger" blade that is pushed under the crown of the plant to ensure uprooting of the bud zone. Experienced operators can lift and push over a shrub or small tree in one operation. Grubbing with a stinger blade is used to control mesquite growing on medium-textured soils.

A rootplow is a horizontal blade attached to a track-type tractor. Rootplowing cuts off the shrub or small tree at depths of approximately 40 cm for resprouting species and 20 cm for nonsprouting species. Rootplowing kills 90% or more of all the vegetation growing on the treated area. The method is best adapted to dense brush areas having little or no residual grass and where seeding of desirable species is possible. It is used to control mesquite, creosotebush, and chaparral.

In disking, the plants are uprooted with a large disk plow or tandem disk. It is limited to small shallow-rooted plants such as tarbush and creosotebush. It also destroys herbaceous plants growing on the treated area, so, like rootplowing, it should be used only in areas where desirable plants can be established.

Chemical Control

Satisfactory control on unwanted plants and considerable improvement in the grazing capacity of rangelands may often be obtained by applications of herbicides (Gibbens et al. 1986). Specific approaches to this problem have been developed for numerous plant species, but information is still needed on others.

Herbicides may be classified as contact, translocated, selective, nonselective, and soil sterilant (Figure 15.5). A contact herbicide kills only those plant parts that are directly exposed to the chemical, for example, diquat (6,7-dihydro-dipyrido[1,2-a:2;1-c]pyrazinediium ion) and paraquat (1,1'-dimethyl-4,4'-bipyridinium ion). A translocated herbicide is applied to one part of a plant but is carried to other parts of the plant by plant tissues (Figure 15.5), for example, 2,4-D [(2,4-dichlorophenoxy)acetic acid], 2,4,5-T [(2,4,5-trichlorophenoxy)acetic acid], silvex [2-(2,4,5-trichlorophenoxy)propionic acid], picloram (4-amino-3,5,6-trichloropicolinic acid), and dicamba (3,6-dichloro-o-anisic acid). A selective herbicide (e.g., the herbicides listed as translocated herbicides) kills or damages a particular species or groups of species with little or no injury to other plants. A nonselective herbicide kills or damages all plant species [e.g., amitrole (3-amino-s-triazole) and paraquat]. A soil sterilant is a herbicide that kills or damages plants when it is present in the soil, for example, bromacil (5-bromo-3-sec-butyl-b-methyluracil), dicamba, monuron [3-(p-chlorophenyl)-1, 1-dimethylurea], picloram, or tebuthiuron (N-[5-(1,1-dimethylethyl)-1,3,4-thiadizol-2-yl]-N,N'-dimethylurea). Most of the latter herbicides are selective at low rates and nonselective at high rates.

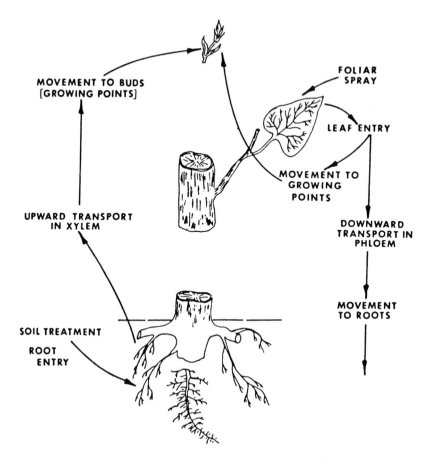

***Figure* 15.5** Translocation of a herbicide applied to the soil and a herbicidal spray applied to the foliage. (From Charles J. Scifres, 1980.)

Broadcast spraying is the method of herbicide application most commonly used on rangelands. Since the herbicide is applied to all plants, desirable as well as undesirable, selective herbicides are generally required. Broadcast sprays can be applied either by ground rigs or by aircraft. Applying granulated or pelleted herbicide is also used to control unwanted plants. The latter method is less dependent than foliar sprays on stage of growth but does require precipitation to dissolve the granules or pellets so that the herbicide may penetrate into the soil. In some areas, excessive herbicide losses may result from leaching beyond the root zone of the target plants, absorption on soil colloids, or decomposition by light or high temperatures (Bovey 1987). Fundamentals to consider are discussed next.

Proper Kind of Herbicide. Herbicides such as 2,4-D, 2,4,5-T, silvex, picloram, and dicamba control a wide variety of plants. Spraying with ground or aerial equipment may be used to control most plants.

Proper Rate of Herbicide. The amounts of herbicide required to provide adequate control vary among plant species. Higher rates than those needed for adequate plant kill cause damage or death to leaves and branches, so that herbicides are not translocated to the proper site and death of the plant does not result. As an example, effective rates are 0.3 to 0.6 kg/ha acid equivalent (a.e.) of 2,4,5-T for mesquite, and 2.2 to 3.3 kg/ha (a.e.) of 2,4-D for big sagebrush. Higher rates than those required for adequate plant control are rarely more effective.

Proper Volume. The volume is dictated by the target species. It is important to obtain adequate coverage but not excessive amounts that will seriously contaminate the adjacent environment. On mesquite, aerial applications of a total volume of 9.3 liters/ha gave as much or slightly better plant kill as total volume of 46.5 liters/ha. This total volume is composed of 1/8 herbicide, 1/8 diesel oil, and 6/8 water.

Proper Time. The phenologic development of the target species, or associated plants, is a reliable index to seasonal susceptibility. Plants are most sensitive to foliar sprays when they are growing vigorously and the leaves are fully expanded. In New Mexico, there was an increased control of mesquite when precipitation for the November to May period before treatment was average or above average. In addition to adequate precipitation, mesquite control in Texas was enhanced when the spring soil temperature at 30 cm depth was at least 24°C. Dry herbicides, applied to individual plants or broadcast, should be applied just before or in the early part of a period of expected precipitation.

Proper Method. Fixed-wing or helicopter aircraft are commonly used to apply herbicidal sprays to large areas. Foliar sprays may also be applied with ground equipment, but the size of the job, the terrain, or the size and density of plants often prevent such operations. Aerial spraying is a specialized job (Figure 15.6a,b). Herbel (1983) listed factors that must be considered:

1. *Application equipment.* The application equipment on the aircraft must be in good condition and the nozzles must be properly placed.

2. *Weather conditions.* Best coverage will result if spraying is done under calm, cool conditions. Spraying should be discontinued when the average wind velocity exceeds 10 km/hr and the temperature exceeds 30°C.

3. *Swath width.* The pesticide dispersal mechanisms must be calibrated and the swath width determined for the proper amount of spray material for a unit area. For fixed-wing aircraft, the swath width is often about 10% greater than the wingspan. Proper marking of the target is necessary to obtain uniform coverage of the spray area.

4. *Flight height.* The aircraft should fly as low as safety will permit, but not higher than the top of the brush.

5. *Mixing-loading equipment.* The equipment must provide adequate agitation to mix emulsions and suspensions properly and rapidly. The equipment must also be large enough and have adequate plumbing to enable quick loading on an aircraft.

(a)

(b)

Figure **15.6** (a) Aerial spraying of an area infested with honey mesquite. The airplane is flying as low as possible so that the herbicidal spray is delivered near the target. Note the mounding of soil around the mesquites and the absence of herbaceous plants. (b) The same vicinity as showing in (a), three years later. There has been a leveling of the soil so that much of the A horizon is spread over the entire treated area, wind erosion is greatly reduced, and there has been a natural revegetation of herbaceous plants.

6. *Spray material.* Recommended mixing instructions must be followed because the herbicide must be mixed with the carrier materials in the proper order to obtain a suitable spray material. In an oil-water emulsion, the oil phase is mixed first and then the water phase is added.

7. *Proximity to nontarget plants.* Some herbicides are toxic to a broad range of species. Drift during application, volatilization from the soil or target plants and subsequent drift of the fumes, or dust blown from treated areas have caused damage to nontarget plants. A nonvolatile herbicide (e.g., a dry material) should be used near sensitive plants, or if the prevailing wind direction poses a problem.

8. *Proximity to livestock.* Most herbicides have a low toxicity to livestock. To assure that livestock are not injured by the herbicide or by grazed plants that develop an increase in toxic properties after spraying, it is desirable to defer grazing of livestock from just before treatment to a period after spraying. In most situations the deferment to avoid poisoning of livestock need not exceed 30 days, but up to 6 months may be required when specific toxins are present.

9. *Directions on the herbicide container.* Herbicides have been developed for certain uses, and specific recommendations are indicated on the label. Some treatments commonly used are

Downy brome (*Bromus tectorum*), also known as cheatgrass and downy chess, is an annual weedy grass that is widely distributed on rangelands of the western United States. Paraquat at 0.6 to 1 kg/ha aerially applied in the spring has controlled downy brome. Applying atrazine (2-chloro-4-ethylamino-6-isopropylamine-*s*-triazine) at 1.1 kg/ha during one fall, followed by seeding to perennial grasses the next fall (chemical fallow), is another approach to improvement of areas infested with downy brome.

Tebuthiuron pellets controlled one-seed juniper (*Juniperus monosperma*), piñon (*Pinus edulis*), waveyleaf oak (*Quercus undulata*), sand sagebrush (*Artemisia filifolia*), skunkbush (*Rhus trilobata*), and algerita (*Berberis* spp.) in New Mexico trails (McDaniel and Duncan 1995).

Locoweeds (*Astragalus* and *Oxytropis* spp.) can be controlled with picloram, dicamba, clopyralid (3,6-dichloropicoliric acid), and triclopyr([(3,5,6-trichloro-2-pyridinyl)oxy] acetic acid) applied aerially (Ralphs and Ueckert 1988).

Aerial spraying of honey mesquite has resulted in plant kills of 8% to 57% in New Mexico. Control is best in years with available soil water before and at the time of spraying, and when the plants were fully leafed and growing vigorously. Control is poor in years with little or no available soil water during the winter-spring prior to spraying. The most effective treatment to control mesquite, considering the price of herbicide, has been 0.6 kg/ha 2,4,5-T in a 1:7 diesel oil water emulsion at a total volume of 9 L/ha. An area aerially sprayed twice for mesquite control during 1958–1961 had an annual average yield of 204 kg/ha of air-dry perennial grass herbage for 1963–1976 compared to 33 kg/ha on an adjacent unsprayed area (Figure 15.7). Since 2,4,5-T is no longer available in the United States, the most effective spray material on honey mesquite is clopyralid or a mixture of clopyralid and picloram (Jacoby et al. 1991; Bovey and Whisemant 1992; Herbel and Gould 1995).

Yucca (*Yucca glauca*) can be controlled with an aerial application of 0.8 to 2.2 kg/ha silvex during the prebloom stage. Effects of this treatment may not become apparent for several months and some resprouting may occur one or more seasons after treatment. The degree of plant kill on shrubby plants generally cannot be ascertained for 2 to 3 years after herbicidal treatment.

The environmental impact of herbicides is discussed in CAST (1987), Deer (1990), Ames and Gold (1990), Seiber (1990), and Bovey (1993). Federally approved herbicides are generally environmentally benign when used at recommended levels and with appropriate application techniques. Weed management in riparian areas is discussed by Sheley et al. (1995).

Biological Control

Biological control has been effective on some weeds and brush in rangelands. It is best used where the low economic return per hectare makes other control methods too expensive. However, biological control alone may be inappropriate against some weeds and may give less control than desired against others. The limitations of its use include situations where complete eradication of the weed is necessary, where important conflicts of interests exist in the use of the weed, or where the target weed is widely scattered within a vegetation complex. In some situations, a combination of control methods are useful. Biological control is generally less expensive than other methods because (1) biological suppressants spread naturally from a few release sites, (2) the biological suppressants become permanently established, and (3) biological suppressants exert a constant pressure on the noxious plant. Often, chemical or mechanical controls must be repeated to part or all of the treated area (De Loach 1978). The reader is referred to Wisdom et al. (1989), Wapshere et al. (1989), Evans (1991), Moffat (1991), and Story (1992) for more recent developments in the use of biological control for unwanted plants.

General Approaches. Two basic methods of biological control are available: (1) introduction of foreign organisms, and (2) augmentation of the effectiveness of organisms already present. The first method is more suitable for use on rangeland because it is usually less expensive. The controlling organisms are released on the range at a few sites; they then multiply and disperse over the infested area, providing permanent control. A disadvantage is that the organisms will probably attack the target weed wherever it grows, including situations where it may be beneficial.

Augmentation is usually too expensive for use on rangelands. Mass rearing and release of insects or pathogens must be applied to all parts of the infested area, often yearly or more frequently. This method has the advantage that it can be applied to specific areas and that it can be terminated at any time.

For a successful biological control program of either native or introduced plants by the introduction of foreign control agents, the following conditions are necessary:

1. Natural enemies must exist somewhere in the world that do not occur where control is desired and that could be introduced. This means that the same species to be controlled or at least other species in the same genus as the unwanted plant must occur in other ar-

eas, and preferably, they should be native to that area so that host-specific natural enemies have evolved.

2. The target plant or species closely related to the target plant should have no substantial
beneficial value and the gains expected from control should be much greater than the expected losses. Highly specific organisms sometimes can be found that do not harm even
closely related plant species, but the chances of finding effective control agents are
greater if rigid host specificity is not required.

In general, biological control of native plants is less likely to succeed than is control of
introduced plants, because of (1) reduced chances of finding safe, yet effective control
agents; (2) a greater possibility of a destructive impact on the ecosystem by the introduced
organism; and (3) a greater chance that the introduced organism would be attacked by native parasites, predators, or pathogens, since there is a possibility that the introduced organism has a close relative already present in the ecosystem.

Examples of Biological Control. There are over 200 insect species that feed on mesquites
in South America. Sixteen of these species seem to be particularly promising for introduction to North America, including foliage feeders, limb and trunk borers, and fruit and seed
feeders. Several insects attack broomweed (*Gutierrezia* spp.) in Argentina. Broomweed is
rarely abundant in Argentina, apparently because of the insects (Cordo and DeLoach 1992).
Several insects in Argentina bore in the stems and leaves of whitebrush (*Aloysia* spp.). A
pathogenic rust of the stem and leaves of whitebrush is also present in Argentina.

In Argentina there are beetles, scale insects, and a grasshopper that inflict considerable
damage on the creosotebush species growing in that country. Numerous larvae bore in the
crowns and stems of tarbush species growing in Argentina. A cerambycid larvae in Brazil kills
baccharis (*Baccharis* spp.) plants and has been introduced into Australia. A stem-boring moth
and a leaf-feeding moth show promise of controlling Russian thistle (*Salsola iberica*).

One of the most famous cases of biological control of rangeland pests has been the use
of the insect *Cactoblastis* to control prickly pear cactus (*Opuntina* spp.) in Australia.
Cactoblastis has not been widely used in the United States because cacti have considerable
beneficial value: (1) the fruits are used as food, (2) the plants are widely used as ornamentals, (3) ranchers burn off the spines and use the pads for supplemental feed during
droughts, and (4) some wildlife species use cacti in their diets. Some biological control of
cacti on Santa Cruz Island was obtained with a cochineal insect. This insect is native to the
mainlands of southern California but is free of its natural enemies on the island.

Use of browsing animals, particularly goats, may control some shrubs. Use of goats to
control brush regrowth following other brush treatments has shown good results in reducing Gambel oak sprouts in Colorado (Davis et al. 1975), in maintaining fuel breaks in southern California chaparral (Green et al. 1978), suppressing wood species after brush control
in Texas (Ueckert 1980; Lopes and Stuth 1984), and in controlling brush species in
Tanzania (Martin and Huss 1981). Spanish goats are more effective in controlling shrubs
than angora goats (Merrill and Taylor 1976). Goats will graze out desirable forage if not
carefully controlled. Grazing by cattle reduced yellow starthistle (*Centaurea solstitialis*) on
California rangeland (Thomsen et al. 1989). Cattle and deer have reduced the rate of brush

invasion on cleared lands (Johnson and Fitzhugh 1990) and forested rangelands in California (Kie and Boroski 1996).

In recent years, domestic sheep have shown potential to reduce infestations of certain noxious range plants. In Montana, a study determined that densities of leafy spurge (*Euphorbia esula*) were effectively reduced by either sheep grazing or the use of picloram (Lacey and Sheley 1995). Mosley (1996) reviewed considerable information indicating sheep grazing, if timing and intensity are carefully controlled, can be effective in cheatgrass (*Bromus tectorum*) suppression in the northwestern United States.

Control of the screw-worm fly in the southwestern United States and northern Mexico has greatly benefited the ranching industry. To eliminate the pest, large numbers of sterile male flies are released where there is a screw-worm infestation. The eradication of the screw-worm has facilitated some livestock practices.

A viral disease, myxomatosis, exhibited some control on the wild rabbit in Australia. However, the production of less viral strains of myxomatosis and the increased resistance of rabbits to myxomatosis has reduced effectiveness. This viral disease has not been introduced into the United States because some rabbits are used for beneficial purposes.

ECONOMIC CONSIDERATIONS

Although brush control has long been one of the quickest ways to increase forage for livestock, ranchers are showing increased concern over its financial effectiveness. The cost of additional land versus the cost and potential increase in forage from range improvement on existing land is an important concern of ranchers. The question of whether brush control is financially sound or unsound depends on a number of criteria. These include potential for natural recovery, rate of return from different brush control options, government subsidies, rate of return from alternative investments, risk of failure, short- and long-term vegetation effects, present livestock prices and ranching costs, and projected livestock prices and ranching costs. We will examine these considerations and develop guidelines that should help range managers in brush control decisions following the discussion in Holechek and Hess (1994). We refer the reader to Workman and Tanaka (1991), and Workman (1995) for more detailed discussions on this subject.

Criteria for Financial Success

Often when brush control projects have been evaluated financially, criteria for success have been vague or undefined with no adjustment for biological risk. We consider a minimum of a 13% return on investment necessary to justify brush control financially. This gives recovery of investment in 10 years plus a 3% premium for illiquidity. These criteria are based on the expected return from the stock market over the past 45 years which has averaged 10%. Stocks in publicly owned companies are liquid investments that can be converted into cash by a phone call. In contrast brush control benefits can only be recovered indirectly through sale of livestock or wildlife and require a higher return to compensate for illiquidity. We have added a 3% illiquid premium although some ranchers might want an even higher premium if there is a good chance they might need the cash for some other purpose within 10 years.

Next is an adjustment to the minimum required rate of return (13%) for the biological risk associated with the particular practice. The literature indicates that generally herbicides involve more risk than burning or mechanical control. However, arid ranges often lack sufficient understory for burning and mechanical control is quite costly. Presently burning costs will be about $1 to $3/acre, herbicide costs $12 to $20/acre, and mechanical control costs $25 to $50/acre.

On southwestern ranges the chances for success in controlling mesquite are no more than 65% with present herbicides. Success might be defined as killing half the mesquite and increasing annual average forage production by 300 lb/acre for 10 years. To adjust for risk, the required rate of return (13%) is divided by 0.65. This means the required liquidity/risk adjusted rate of return necessary to justify the investment is 20%.

Generally big sagebrush is easier to kill than mesquite. Success probabilities of 90% with herbicidal control would be reasonable based on the literature. Therefore, a required risk/liquidity adjusted rate of return of 15% might be reasonable.

We have made no attempt to adjust for inflation because no one is certain what the future might bring. Based on history since the 1870s, inflationary conditions have enhanced returns from range improvements while returns have been diminished under deflationary conditions. Livestock prices have generally been elevated relative to production costs when inflationary spirals occurred. The reverse has been the case under deflationary conditions.

In Table 15.1 we have provided some guidelines on annual per acre returns from different range types when land condition and management are good. As a rule it is financially unsound for a rancher to spend more than 10 times these returns on any range improvement practice.

How Much Increase in Forage Production?

Generally ranchers are most interested in how much they can increase grazing capacity if they control brush. Related to this issue are how long will the forage increases last and how dependable is the increase among years. Increased forage in drought years is more beneficial than increased forage in average or above average years. Under the financial criteria we have developed, most brush control projects require a longevity of 15 to 20 years to be justifiable. The first 10 years are required to recover the investment.

Usually the consistency of forage increase among years is as important as the total increase in forage through time. If the increase is erratic among years, most ranchers will have a hard time stocking their range to use the extra forage when it does occur.

During the first 5 to 6 years big sagebrush control has consistently doubled or tripled forage yields compared to areas without control (Bartolome and Heady 1988; McDaniel et al. 1992). Forage increases of around 300–600 lb/acre per year can be expected for the first 5 years (Table 15.2). After 10 years the effect of big sagebrush control on forage yields diminished on most southeastern Oregon sites (Bartolome and Heady 1988).

In contrast, mesquite control has resulted in more erratic initial increases in forage yields occurring primarily in wet years (Martin and Cable 1974; Dahl et al. 1977)

TABLE 15.1 Forage Production and Financial Returns ($/acre) from Different Range Types in the United States under Good Range Condition and Good Management

RANGE TYPE	TYPE OF OPERATION	STATE	FORAGE PRODUCTION (LB/ACRE)	FINANCIAL RETURNS ($/ACRE)
Southern pine forest	Cattle-cow	Louisiana	2,500–4,000	8–14
Tallgrass prairie	Cattle-cow	Kansas	2,500–3,500	9–12
Coastal prairie	Cattle-cow	Texas	2,500–3,500	9–12
Coastal prairie	Wildlife-cattle	Texas	2,500–3,500	25 (15 wildlife + 10 cattle)
Southern mixed prairie	Cattle-cow	Texas	2,000–3,000	6–8
Southern mixed prairie	Cattle-wildlife	Texas	2,000–3,000	17 (10 wildlife + 7 cattle)
High plains-shinnery	Cattle-cow	New Mexico	800–1,700	3.00–4.00
Oak-savannah	Sheep-goats	Texas	2,000–3,000	8–14
Oak-savannah	Wildlife-cattle	Texas	2,000–3,000	28 (20 wildlife + 8 cattle)
Shortgrass prairie	Cattle-cow	New Mexico	800–1,400	4.50–5.50
Shortgrass prairie	Cattle-yearling	New Mexico	800–1,400	4.00–10.00
Shortgrass prairie	Sheep	Wyoming	600–1,000	3.80–4.50
Desert prairie	Cattle-sheep	New Mexico	500–900	2.50–3.50
Northern mixed prairie	Cattle-cow	Montana	900–1,600	2.50–3.00
California annual grassland	Cattle-cow	California	300–1,500	1.00–3.00
Palouse prairie	Cattle-cow	Oregon	500–800	1.25–2.50
Palouse prairie	Wildlife-cattle	Oregon	500–800	4 (2.50 wildlife + 1.50 cattle)
Chihuahuan desert	Cattle-cow	New Mexico	300–700	0.60–1.00

Sonoran desert	Cattle-cow	Arizona	100–400	0.30–0.60
Salt desert	Sheep	Utah	150–350	0.30–0.70
Salt desert	Cattle-cow	Nevada	150–350	0.15–0.40
Mojave desert	Cattle-cow	California	50–200	0.10–0.30
Big sagebrush	Cattle-cow	New Mexico	250–500	0.50–0.80
Big sagebrush	Cattle-cow	Wyoming	300–800	1.00–2.00
Big sagebrush	Cattle-cow	Nevada	150–400	0.50–1.50
Piñon-juniper	Cattle-cow	New Mexico	100–500	0.25–1.00
Coniferous forest	Cattle-cow	Eastern Oregon	400–800	2.00–3.00
Coniferous forest	Cattle-cow	New Mexico	400–1,000	2.40–3.00

Source: From Holechek and Hess 1994.

TABLE 15.2   Perennial Grass Standing Crop (kg/ha) on Untreated Rangeland and after Big Sagebrush was Controlled by Tebuthiuron on Sites in Northwestern New Mexico, 1985–1990

SITE NUMBER/ LOCATION	TREATMENT	YEAR					
		1985	1986	1987	1988	1989	1990
1. Aztec	T	743	656	839	815	893	485
	C	96	106	173	151	106	102
2. Bloomfield	T	1,508	929	605	461	982	494
	C	327	265	219	203	152	224
3. Gobernator	T	471	377	333	830	858	316
	C	85	97	134	199	175	173
4. Taos	T	46	221	113	264	220	157
	C	19	32	6	0	13	2
5. Navajo Lake	T	945	648	497	NA	1,002	1,181
	C	124	109	157	NA	156	328
6. Tres Piedras-N	T	783	725	519	515	563	NA
	C	110	139	60	61	87	NA
7. Tres Piedras-SA	T	785	1,147	324	475	684	606
	C	368	466	140	215	151	248
8. Tres Piedras-SG	T	749	765	350	358	541	268
	C	259	440	111	74	91	71
9. Lindrith	T	545	570	386	411	152	677
	C	160	245	95	156	72	350

Source: From McDaniel et al. 1992.

NA = Not available.

T = Tebuthiuron at 0.55 kg/ha.

(Table 15.3). A southern New Mexico study on long-term impacts of mesquite control on forage yield indicated areas with 65% kill of mesquite yielded no more than noncontrol areas 20 to 30 years after treatment (Warren et al. 1996). On an experimental area in Arizona, a site with 100% kill of mesquite yielded about 40% more forage than an equivalent site without control 10 years after treatment (Galt et al. 1982). It is important to recognize that for ranchers 100% control of mesquite would be impractical because repeated treatments are necessary, and make the cost excessive.

TABLE 15.3 Perennial Grass Yields in Oven-Dry (lb/acre) Following Aerial 2,4,5-T Application to Honey Mesquite on Texas Rolling Plains Rangeland, 1970–1975

SITE	INITIAL CANOPY COVER (%)	YEAR AFTER SPRAYING							
		1ST		2ND		3RD		4TH	
		UNSPRAYED	2,4,5-T SPRAYED	UNSPRAYED	2,4,5-T SPRAYED	UNSPRAYED	2,4,5-T SPRAYED	UNSPRAYED	2,4,5-T SPRAYED
Shallow redland	5	120	330	820	1,000	1,960	2,080	1,590	1,210
Deep hardland	12	1,070	680	1,060	1,140	1,760	2,040	1,060	1,600
Deep hardland	21	2,040	2,530	1,370	1,640	560	840	—[a]	—[a]
Valley	28	330	1,290	2,320	1,990	2,340	2,670	1,670	1,710
Deep hardland	34	390	1,050	820	990	940	1,420	—[a]	—[a]
Deep hardland	36	480	2,480	1,490	2,030	810	1,220	1,190	1,580
Deep hardland	54	540	770	1,880	1,270	580	990	—[a]	—[a]

Source: From Dahl et al. 1978.

[a]Plots lost due to mechanical disturbance.

In the shinnery oak ranges of Texas and New Mexico, a 600–800 lb/acre/year increase in perennial grass yield occurred the first few years after herbicidal control on the more mesic/sandier sites (Pettit 1979). However, on some of the drier sites there has been little to no increase in perennial grasses. Here brush control can be a disadvantage, since shinnery oak helps to stabilize the site and receives some browsing use by livestock.

CONSIDERATIONS IN SEEDING

Direct planting is an excellent tool for speeding range rehabilitation, but it does have limitations. It is expensive, it is not universally applicable, and there is a calculated risk that success can be achieved. On the other hand, if the following general criteria are carefully observed, chances of success can be high (Herbel 1983; Frasier and Evans 1987; Call and Roundy 1991; Archer and Pyke 1991).

A variety of native and introduced species have been used in reclamation of depleted rangelands, former croplands, and mined lands. Species selection for seeding programs depends on goals, cost, availability, and adaptability. In the past, introduced species have often been used when the goal was provision of more forage for livestock because of their low cost and broad adaptability. However, in recent years there has been great interest in native plants for use on wildlife areas, roadways, ski areas, subdivisions, and former dumping grounds. Many ranchers are shifting to native plants for rangeland revegetation. We believe as rangeland resources shrink in the United States there will be much more interest in featuring native plants in range revegetation programs on public lands. At the same time native range plants are receiving greater emphasis in urban and ranchette landscaping around the new western communities because of their low water and maintenance requirements.

Basic Questions

In deciding whether an area should be seeded, the range manager should ask the following four questions:

1. *Is seeding absolutely needed?* Range can be rehabilitated more positively and at lower cost by better livestock distribution, better systems, or reduced stocking. Only where the desirable native perennial forage plants are almost completely killed out is seeding essential. Such areas will have a forage condition rating of poor or very poor. Where the forage condition rating of a range is fair or better and acceptable forage species are present, a range will generally improve under good grazing management.

2. *Are proven methods available for the site?* Where not available, projects should not be undertaken until satisfactory procedures have been developed.

3. *Can proven methods be used?* On many sites the procedures are known for the general type but cannot be applied because excessive rocks, steep slopes, or other factors prevent use of the types of equipment or procedures needed.

4. *Can the area be given proper grazing management after seeding?* Seeding should not be started until proper grazing management can be assured.

Basic Criteria for Successful Revegetation

1. *Change in plant cover must be necessary and desirable.* The usual goal of developing a useful stand of desirable plants may be achieved by selective plant control or a change in grazing management. Seeding is an expensive and risky undertaking and should be avoided if possible. However, at least one shrub and one desirable herbaceous plant per square meter should be present after revegetation in arid areas. Watershed considerations and soil conditions are important.

2. *Terrain and soil must be suitable for seeding.* Deep fertile soils on level-to-gently sloping land are preferred sites for seeding. Shallow or rocky soils seldom have the potential to justify expensive reclamation measures. Excessive amounts of soluble salts in the soil often require additional attention to ensure adequate plant establishment.

3. *Precipitation and water concentration must be adequate to assure establishment and survival of seeded species.* Average annual precipitation, or equivalent from water concentration, must be adequate for germination and seedling growth. This is dependent on temperatures, but in temperature climates a minimum of 250 mm of precipitation may be needed. Where precipitation is near this limit, only the more drought-resistant species should be used. Existing vegetation is a good indicator of the moisture situation.

4. *Remove or reduce competition from unwanted plants.* Most plants used for revegetation are perennials. Seedlings of these species are often slow growing and cannot compete with existing, unwanted plants. A good seedbed will provide the best possible moisture conditions for germination and plant growth. This requires the control of most existing plants before seeding (Figure 15.7). In addition, it is sometimes necessary to control unwanted plants that are competing with the seedlings of the desirable plants. It has been speculated that livestock grazing may reduce competing vegetation in seedings (and serve as an alternative to herbicides). In Nebraska a variety of grazing treatments were found to be ineffective in controlling weeds on big bluestem (*Andropogon gerardii*) seedings (Lawrence et al. 1995). However, acceptable stands of big bluestem did develop on areas treated with atrazine for weed control.

5. *Use adapted plant materials.* The plant species selected for seeding must be compatible with management objectives (e.g., palatability and growth period). It is important to use only those species and varieties well adapted to the soil, climate, and topography of the specific site being revegetated. If native plants are revegetated, species from local origin are used. Local origin would include species from about the same elevation, and within 320 km north, east, or west, and 480 km south of the area to be seeded. Improved ecotypes, varieties, and introduced species may be available for revegetation and should be used.

***Figure* 15.7** The unwanted shrubs similar to those in the background were killed in the foreground with a rootplow prior to revegetation.

6. *Mixtures of plant types rather than single species should be seeded.* The danger in seeding a monoculture is that a disease or insect infestation can eliminate one species, whereas a mixture would have some survival. Because we often encounter a variable terrain, mixtures will have some survivors on most sites. A variable ground cover will generally result in superior control of soil erosion. Also, mixtures of grasses, forbs, and shrubs will better meet the multiple needs of the land user.

7. *Use seed treatments.* Various microbial treatments (e.g., nitrogen-fixing bacteria or mycorrhizal fungi) may enhance seedling survival. Dormancy of most seeds can be reduced by stratification—subjecting them to temperatures 0° to 4°C for 6 to 20 weeks in moist sand, peat moss, or newspaper. For some shrubs, treatment with thiourea or scarification with sulfuric acid or mechanical abrasion helps overcome dormancy (Hardegree and Emmerich 1992).

8. *Use proper seeding rates.* It is important to use enough seed to get a good stand, but not more than necessary. Too much seed can produce a stand of seedlings so thick that individual plants compete with each other. Species of plants, number of pure live seeds (PLS) per kilogram, and potential productivity of the site are the major factors determining the seeding rate. PLS is determined by multiplying the germination of a lot of seed by its purity. Seeding rates providing 125 to 250 PLS/m^2 should be used when the seed is placed in the soil with a drill. Broadcast seeding is inefficient and not an effective method of revegetation, and should be avoided. Many broadcast seeds are left on the soil surface, where germination and seedling establishment are tenuous. Where broadcast seeding must be used, a rate of 500 PLS/m^2 is recommended.

9. *Use the proper depth.* Proper depth of seeding is determined by the plant species. Optimum depth of seeding is roughly four to seven times the diameter of the seed. Seeding equipment should be used that provides positive seed placement at the desired depth. More stands are lost because seeds are planted too deep than too shallow (Newman and Moser 1988).

10. *Correct seeding dates are important.* The most desirable time to seed nonirrigated areas is immediately before the season of the most reliable rainfall and when temperature is favorable for plant establishment. In some situations, seedings are made prior to snowfall, the seed germinates while the ground is covered with snow, and the seedlings develop after snow melts.

11. *Distribute the seed.* Uniform distribution of seed is essential. Seeding equipment must be checked frequently to assure that it is working properly.

12. *Alter the microenvironment.* Many areas are deficient in soil water for germination and seedling establishment of the desirable plants. In some areas, associated treatment is needed to reduce high soil temperature and provide more soil water (e.g., mulching) or just to provide more soil water (e.g., summer fallow or establishing basins or pits) (Hauser 1989; Roundy et al. 1992).

13. *Seedbed preparation is essential.* The major objectives of preparing seedbeds for seeding are (a) to remove or substantially reduce competing vegetation, (b) prepare a favorable microenvironment for seedling establishment, (c) firm the soil below seed placement and cover the seed with loose soil, and (d) if possible, leave mulch on the soil surface to reduce erosion and to improve the microenvironment (Winkel and Roundy 1991).

14. *Consider fertilization.* Where water is not limiting, supplementation of plant nutrients in bands near the seed zone may be helpful in plant establishment. In harsh areas such as mined lands where soil fertility is low, light rates of nitrogen and phosphorus fertilizer application after seedling emergence can improve short- and long-term plant establishment (Holechek et al. 1981; Holechek 1982).

15. *Revegetated areas must be properly managed.* All seedings must be protected from grazing by animals through the second growing season, or until the seeded species are well established. Spraying to control weeds competing with the new seedlings can prevent the loss of seeding. Rodents, rabbits, insects, and other pests should also be controlled where they are a menace to new seedlings. In the United States, fencing is the most practical method of controlling movement of livestock on rangelands. It is important that fences be in place in advance of revegetation work. For many Third World countries, herding by people on foot or on horseback can be used as a way of protecting new seedings from grazing damage.

Methods of Direct Planting

Drill Seeding. Drilling is by far the superior method of planting seed where site conditions permit. The seed is covered to the proper depth by the drill control, distribution is uniform, the rate of seeding is positively controlled, and compaction can be utilized if needed. There

Figure **15.8** The rugged rangeland drill was designed for seeding rangelands. (Picture is courtesy of U.S. Department of Agriculture–Forest Service.)

are several types of drills available. The rangeland drill was specially adapted for seeding rangelands. This drill is a rugged seeder with high clearance, designed to work on rough sites and it has performed well on rough seedbeds. It can be converted to a deep furrow implement by removing the depth bands. The disks are cupped enough to make good furrows. The depth of the furrow is controlled by adding or taking off disk arm weights. Weights up to 30 kg have been used under some conditions. The feed on this drill will not handle trashy seed unless it is especially designed for that purpose (Figure 15.8). The trashy seed drill was specially designed for this type of seed (Dyck et al. 1994).

Broadcasting. Broadcasting is any method that scatters the seed directly on the soil without soil coverage. The seed, however spread, must be covered in some way if it is to germinate and become established. Size of seed and condition of the seedbed are important factors if the seed is to become covered with soil. A seedbed that has 5 cm to 8 cm of loose soil generally sluffs sufficiently to cover the seed. Covering the seed with a mulch is better than no coverage at all, but mulch coverage is inferior to soil coverage. If mulches are used in conjunction with seeding, best results are obtained by broadcasting the seed, covering the soil, and applying the mulch. Limitations to broadcast seeding are (1) requires a heavier seeding rate; (2) covering of seed is poor compared to drilling; (3) distribution of seed is often poor; (4) loss of seed to rodents and birds can be great; and (5) establishment is generally slower. This method should be avoided if possible.

Transplanting. Transplanting involves the removal of plant material from nursery beds and planting such material in field situations as wind breaks, tree plantations, and so on. Various plant parts can be transplanted from a nursery to a field situation. Planting mater-

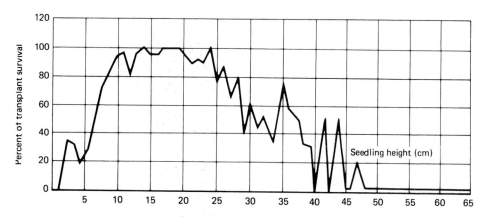

***Figure* 15.9** Survival of transplants of fourwing saltbush (*Atriplex canescens*) in relation to height at time of transplanting. (From Moghaddam and McKell 1975.)

ial includes plants propagated from seed, layering, root suckers, root cuttings, and shoot cuttings. The young plants should be about 75 to 100 days old when they are transplanted; they should not exceed 20 cm to 25 cm in height nor have a diameter of over 5 mm at the neck (Figure 15.9). In the arid zone it is recommended that the resistance of nursery plants be strengthened by low and infrequent irrigation at least a month before planting out. It is also recommended that the aboveground part of the plant be cut back to 20 cm (hardwoods and grasses but not conifers), and where appropriate, the roots be trimmed. A potentially low-cost method of seeding rangelands is to use floodwaters in dry stream channels to spread propagules of desirable plants. The other method is to feed encapsulated seed to live-stock and let them spread the seed around the range unit through dung (Barrow 1992; Barrow and Havstad 1992).

FERTILIZATION

The variety encountered in the world's rangelands, with the diversity of climate, topography, soil types, and vegetation, complicates any attempt to generalize on a range management practice such as fertilization. Seasonal variations in local weather conditions add further complications. In addition, the complex mixture of plants found on rangelands requires more diligent management than does a seeded pasture with one or two species. Each species will respond differently to fertilization.

In some areas, low amounts of available nitrogen, phosphorus, and other soil nutrients limit plant growth. Water is generally the most important factor limiting plant growth, but when the need is satisfied, additional plant nutrients may be useful. Nitrogen was the major growth-limiting plant nutrient on the rangelands of the northern Great Plains, with measurable responses to phosphorus occurring as nitrogen became nonlimiting (Wight and

Black 1979). Fertilizing with the deficient nutrients is economical only where there is adequate moisture and plants respond to the added nutrients. The root systems of range plants often act as nutrient-deficient sinks that have a high potential to immobilize relatively large quantities of applied nitrogen and phosphorus. Infestation of legume roots and nodules with a plant toxin increased plant growth and nitrogen content (Knight and Langston-Unkefer 1988). Some points on range fertilization to consider are as follows:

1. *Soil water.* Response to fertilization is directly related to the availability of soil water. Range fertilization should not be used in areas with low average precipitation, but seasonal distribution of precipitation and evaporative demand may be confounding factors. Range fertilization has been effective in the northern Great Plains in areas where annual precipitation is as low as 280 mm. Annual precipitation of 380 mm or more may be necessary before range fertilization is feasible in warmer regions and where precipitation is more evenly distributed during the year. Areas that have overland flow or are subirrigated may have less precipitation, but they have sufficient water so that the plants can use the added nutrients.

2. *Economics.* Applications of 30 to 50 kg/ha of nitrogen annually or in annual rate equivalents are most efficient in the northern Great Plains. This rate will produce up to 20 kg of additional forage per kilogram of nitrogen applied, or under a grazing situation, about 1 kg of beef per kilogram of nitrogen. Thus, when the price of beef exceeds the cost of applying nitrogen, fertilization become an economical management practice. The total cost of fertilization must be weighed against the benefits.

3. *Ecology and fertilizer timing.* Usually, cool-season species respond most to nitrogen fertilization. However, the effect of fertilization on species composition can be somewhat offset by timing fertilizer applications. Late spring or summer applications tend to benefit warm-season species, whereas late fall or early spring applications tend to benefit cool-season plants. If application rates are high enough to cause a significant carryover of fertilizer nitrogen from one year to the next, cool-season species may use the fertilizer to the detriment of warm-season plants. Nitrifying bacteria, occurring in the soil, are less active in cool weather than in warm weather. Thus plants growing earliest in the season will use the residual nitrogen.

4. *Toxicity.* At nitrogen rates above 200 kg/ha, nitrates accumulate in some plants, especially annual forbs. Caution is required if applying high nitrogen rates on rangelands with nitrate-accumulating plants. Groundwater contamination with nitrates may also result where high nitrogen levels are used and where the groundwater is close to the surface.

5. *Water use efficiency.* Range fertilization increases the efficiency of the limited water supply in plant growth processes. Fertilized range plants extract more water from the soil profile than do unfertilized plants. Thus, if precipitation is adequate to fully recharge the soil profile, fertilized range will use the precipitation more effectively than will unfertilized range.

6. *Drought.* There has been concern that fertilization will compound the effects of droughts, resulting in additional damage to the range vegetation. In some situations, there is a greater loss of desirable plants during drought. However, fertilizer not used during drought years is available for plant use following the drought.

7. *Fertilizer materials.* There have been very little response differences to the inorganic forms of nitrogen and phosphorus. Under some conditions, urea, an organic formulation, will undergo high volatilization losses when broadcast on the soil surface.

8. *Management.* It is generally necessary to fertilize the entire range unit or the animals will concentrate on the fertilized portion and neglect the unfertilized area. Plants that have been fertilized generally are green earlier in the spring and later in the fall if soil water is available. Increasing palatability of fertilized plants may be useful as a management tool to improve animal distribution and forage utilization. However, plants toxic to animals on fertilized areas may also become more palatable and create toxicity problems among the animals using rangeland. Plants growing on fertilized range generally have a higher nutrient content, and this will also affect management decisions.

In an extensive review of research in the Great Plains of North America, Rogler and Lorenz (1974) found that high-yielding, cool-season grasses were most responsive to fertilization with nitrogen. Cool-season species showed a marked early spring response to nitrogen fertilizer, even on soils high in total nitrogen, because low soil temperatures reduced the nitrification rate at the time of the year these species are beginning growth. Soil water often limits plant growth during the summer in the northern Great Plains, but early in the spring, soil water is usually adequate to allow efficient plant use of the additional nitrogen applied by fertilization. In the central and southern parts of the Great Plains, fertilization of warm-season species increased forage production, but weedy species were often favored. The weedy species often slow growth by the warm-season grasses. Rogler and Lorenz (1974) concluded that benefits from fertilization generally outweigh disadvantages in most areas of the Great Plains (semiarid to subhumid climate). Benefits reported in their review included increased forage and livestock production, increased palatability, better livestock distribution, a longer green-feed period, higher forage quality, increased root growth, greater water-use efficiency, greater use of solar energy, and improvement in range condition. The major disadvantages included problems related to increased weed growth or other undesirable changes in species composition, possibility of groundwater pollution, and a remote possibility of metabolic disorders in livestock.

Two years of fertilization with 101 kg/ha nitrogen each year did more to improve deteriorated mixed prairie rangeland near Mandan, North Dakota, than did 6 years of deferment from grazing (Rogler and Lorenz 1974). A deteriorated rangeland is one in which the more productive species have been reduced in vigor or eliminated and have been replaced by less desirable plants. Deferment from grazing, sometimes combined with one or more appropriate manipulative treatments, has been the common means of attempting to correct the situation. However, where applicable, fertilization will hasten the return to a productive condition by stimulating a rapid change in species composition, accompanied by an increase in plant vigor.

The relative effectiveness of deferment from grazing versus use of nitrogen to restore productivity of deteriorated mixed prairie in North Dakota is shown in Table 15.4. The major undesirable species was fringed sage (*Artemisia frigida*). One application of 2,4-D and annual application of 45 kg/ha nitrogen, with grazing continued, did more to increase production of usable forage than did deferment for up to 55 years (Rogler and Lorenz 1974).

TABLE 15.4   Dry-matter Yield of Mixed Prairie near Mandan, North Dakota, Comparing Various Periods of Isolation from Grazing with Fertilization and Weed Control for Improving Deteriorated Range

YEARS OF COMPLETE REST	DRY MATTER (KG/HA)		
	GRASS	FORBS	TOTAL
55	1,353	1,425	2,778
30	1,708	421	2,129
26	1,793	391	2,184
20	2,185	221	2,406
5	2,010	246	2,256
0 (fertilized)[a]	4,926	0	4,926

Source: Adapted from Rogler and Lorenz 1974.

[a]Grazed and fertilized annually with 45 kg/ha nitrogen for 5 years; broadleaf forbs, mostly fringed sage, controlled with one application of 2,4-D.

Residual nitrogen increased forage production for 3 years after the single application of 22, 45, or 90 kg/ha nitrogen on heavily and lightly grazed pastures in South Dakota (Westin et al. 1955). Also, the application of 90 kg/ha nitrogen once in 3 years resulted in more herbage per unit of nitrogen than did 90 kg nitrogen applied once each year for 3 years.

Graves and McMurphy (1969) included burning with fertilization in an attempt to improve poor-condition range in central Oklahoma. Burning and fertilizer increased the desirable grass species, but an increase in undesirable forbs was a major problem. Graves and McMurphy (1969) concluded that rangeland infested with low-quality vegetation should not be fertilized.

Herbel (1963) conducted a 5-year fertilizer study on floodplains in southern New Mexico dominated by tobosa (*Hilaria mutica*). In only 2 years a significant increase in production due to fertilization with nitrogen and/or phosphorus occurred. During one year with available soil moisture for a continuous 60-day period, fertilization with 101 kg/ha nitrogen increased herbage production by 4,664 kg/ha, but in the other 4 years herbage increases were small. Protein content of the herbage at the close of the growing season was generally 20% to 35% higher with 67 or 101 kg/ha nitrogen. One application of 80 kg/ha nitrogen plus 28 kg/ha phosphorus increased forage and beef production in Chihuahua, Mexico.

Rangeland in New Mexico dominated by blue grama (*Boutelous gracilis*) was fertilized with 45 kg/ha nitrogen annually and was grazed by yearling heifers (Donart et al. 1978). Average summer gains were 19 and 45 kg/ha for the unfertilized and fertilized pastures, respectively. The 8-year (1968–1976) average gain per head was 98 kg on the fertilized pasture and 89 kg on the unfertilized pasture.

Average steer gains per grazing season were 41 kg and 104 kg, respectively, on caucasian bluestem (*Bothriochloa caucasica*) unfertilized and fertilized with 84 kg/ha nitrogen in western Kansas (Launchbaugh 1971). Applications of 38 kg/ha nitrogen on weeping lovegrass in the southern Great Plains in Oklahoma increased forage production about 40% and beef production about 31% over the control in a 4-year grazing trail (McIlvain and Shoop 1970). On California annual rangeland, fertilization has increased cattle gains and forage production (Raguse et al. 1988; van Riet and Bailey 1991). Fertilization in this type can be particularly advantageous in drought years.

In Oklahoma, nitrogen fertilization of mixed native warm-season grass stands resulted in substantial increases in herbage yields (Berg 1995). Grass stands in this study were typical of those used in Conservation Reserve plantings on sandy soils.

FORAGE CONSERVATION

The major source of roughage for the livestock industry during the dormant season may be naturally cured forage, particularly when cold temperatures and snow prevent yearlong grazing. This may be in the form of hay chemically cured during a nutritious stage of range forage.

Hay

Hay is one of the major sources of roughages in the dormant season in areas with wet meadows. Runoff from mountains often collects on adjacent meadows that are not grazed during the summer. Fertilizers are sometimes applied to these areas early in the growing season. At the proper growth stage, the plants are harvested, cured and baled. The baled hay is often left on the area and used by livestock during the dormant season. The reader is referred to Heath et al. (1985) for a detailed discussion of haying operations in various regions of the United States.

Chemical Curing

Low concentrations of paraquat can be used to arrest growth of range grasses. These plants then cure in place, and the quality of this forage is preserved for later use by livestock (Sneva 1967). In the northwestern United States, where livestock are wintered on low elevation meadows (hayfields), this strategy provides an alternative to haying. Chemically cured meadow forage is a low-cost alternative to expensive haying operations in this area.

RANGE MANAGEMENT PRINCIPLES

■ Intensive range management practices such as brush control, seeding, and fertilization will often quickly change rangeland botanical composition and greatly increase forage production. However, they involve high cost and high risk and their benefits often diminish within 10 to 20 years after application.

- Burning is one of the cheapest and most natural vegetation manipulation tools. However, herbicides may be the only practical means of manipulating vegetation in degraded desert areas that lack sufficient understory to carry fire.

- Range seeding is quite risky and costly and generally is only practical on productive sites where desirable perennial forage species are no longer present.

Literature Cited

Ames, B. N., and L. S. Gold. 1990. Too many rodent carcinogens: Mitogenesis increases mutagenesis. *Sci.* 249:970–971.

Antognini, J., P. C. Quimby, Jr., C. E. Turner, and J. A. Young. 1995. Implementing effective noxious range weed control on rangelands. *Rangelands* 17:158–163.

Archer, S., and D. A. Pyke. 1991. Plant-animal interactions affecting plant establishment and persistence on revegetated rangeland. *J. Range Manage.* 44:558–565.

Barnitz, J. A., Jr., V. W. Howard, Jr., and G. M. Southward. 1990. Mule deer and rabbit use on areas of piñon-juniper woodland treated by two-way cabling. *N. Mex. Agric. Exp. Sta. Bull.* 752.

Barrow, J. R. 1992. Use of floodwater to dispense grass and shrub seeds on native arid lands. In W. P. Clary, E. D. McArthur, D. Bedunah, and C. L. Wambolt (Compilers). *Proc. Symp., Ecology and Management of Riparian Shrub Communities. U. S. Dep. Agric., Intermountain Res. Sta. Gen. Tech. Rep.* INT-289, pp. 167–169.

Barrow, J. R., and K. M. Havstad. 1992. Recovery and germination of gelatin-incapsulated seed fed to cattle. *J. Arid Environ.* 22:395–399.

Bartolome, J. W., and H. F. Heady. 1988. Grazing Management 1962 to 1986. In The Vale rangeland rehabilitation program: An evaluation. *U.S. Dept. Agr.-Forest Service Resource Bulletin* PNW-R13-157.

Baumgartner, D. M., D. W. Breuer, B. A. Zamora, L. F. Neuenschwander, and R. H. Wakimoto (Eds.). 1989. *Prescribed fire in the intermountain region: Forest site preparation and range improvement.* Wash. State Ext. Serv., Pullman, WA.

Berg, W. A. 1995. Response of a mixed native warm-season grass planting to nitrogen fertilization. *J. Range Manage.* 48:64–67.

Billings, W. D. 1994. Ecological impacts of cheatgrass and resultant fire on ecosystems in the western Great Basin. In S. B. Monsen and S. G. Kitchen (Compilers). Proc. Symp., Ecology and Management of Annual Rangelands. *U.S. Dep. Agric. Intermountain Res. Sta. Gen. Tech. Rep.* INT-313, pp. 22–30.

Bovey, R. W. 1987. Weed control problems approaches and opportunities in rangeland. *Rev., Weed Sci.* 3:57–91.

Bovey, R. W. 1993. Dissipation, movement, and environmental impact of herbicides on Texas rangelands—A 25 year summary. *Tex. Agric. Exp. Stab.* B-1713.

Bovey, R. W. and S. G. Whisemant. 1992. Honey mesquite (*Prosopis glandulosa*) control by synergistic action of clopyralid: Triclopyr mixtures. *Weed Sci.* 40:563–567.

Buffington, L. C., and C. H. Herbel. 1965. Vegetational changes on a semidesert grassland range from 1858 to 1963. *Ecol. Monogr.* 35:139–164.

Call, C. A. and B. A. Roundy. 1991. Perspectives and processes in revegetation of arid and semiarid rangelands. *J. Range Manage.* 44:543–549.

Cordo, H. A., and C. J. DeLoach. 1992. Occurrence of snakeweeds (*Gutierrezia compositae*) and their natural enemies in Argentina: Implications for biological control in the United States. *Biol. Control* 2:143–158.

Council for Agricultural Science and Technology (CAST). 1987. Perspectives on the safety of 2,4-D. *Comments, CAST* 1987-3.

Dahl, B., R. E. Sosebee, J. P. Goen, and C. S. Brumley. 1978. Will mesquite control with 2,4,5-T enhance grass production? *J. Range Manage.* 31:129–131.

Davis, E. G., L. E. Bartel, and C. W. Cook. 1975. Control of Gambel oak sprouts by goats. *J. Range Manage.* 28:216–218.

Deer, H. 1990. Rangeland herbicide doesn't appear to move far. *Utah Sci.* 51(2):60.

De Loach, C. J. 1978. Considerations in introducing foreign biotic agents to control native weeds of rangelands, pp. 39–50. In T. E. Freeman (Ed.). *Proc. 4th International Symposium on the Biological Control of Weeds,* University of Florida, Gainesville, FL.

Donart, G. B., E. E. Parker, R. D. Pieper, and J. D. Wallace. 1978. Nitrogen fertilization and livestock grazing on blue grama rangeland. In D. N. Hyder (Ed.). *Proc. First International Rangeland Congress,* Denver, CO., pp. 614–615.

Dyck, F. B., G. G. Bowes, and J. Waddington. 1994. Drills for rangeland seeding. In S. B. Morsen and S. G. Kitchen (Compilers). *Proc. Symp. Ecology and Management of Annual Rangelands. U.S. Dep. Agric. Intermountain Res. Sta. Gen. Tech. Rep.* INF-313, pp. 323–327.

Engle, D. M., J. F. Stritzke, T. G. Bidwell, and P. L. Claypool. 1993. Late-summer fire and follow-up herbicide treatments in tallgrass prairie. *J. Range Manage.* 46:542–547.

Evans, T. 1991. Several insects promise to control weeds. *Utah Sci.* 52(3):99–101.

Everitt, J. H., D. E. Escobar, and M. R. Davis. 1995. Using remote sensing for detecting and mapping noxious plants. *Weed Abst.* 44:639–649.

Florence, M. 1987. Plant succession on prescribed burn sites in chamise chaparral. *Rangelands* 9:119–122.

Frasier, G. W. and R. A. Evans. 1987. Seed and seedbed ecology of rangeland plants. *Proc. Symp. U. S. Dep. Agric.,* Agric. Res. Serv.

Galt, A. D., B. Theurer, and S. C. Martin. 1982. Botanical composition of steer diets on mesquite and mesquite full desert grassland. *J. Range Manage.* 35:320–325.

Gibbens, R. P., C. H. Herbel, H. L. Morton, W. C. Lindemann, J. A. Ryder-White, D. B. Richman, E. W. Huddleston, W. H. Conley, C. A. Davis, J. A. Reitzel, D. M. Anderson, and A. Guiao. 1986. Some impacts of 2,4,5-T on a mesquite duneland ecosystem in southern New Mexico: A synthesis. *J. Range Manage.* 39:320–326.

Gibbens, R. P., R. A. Hicks, and W. A. Dugas. 1996. Structure and function of C_3 and C_4 Chihuahuan Desert plant communities. Standing crop and leaf area index. *J. Arid Environ.* 34:47–62.

Graves, J. E. and W. E. McMurphy. 1969. Burning and fertilization for range improvement in central Oklahoma. *J. Range Manage.* 22:165–168.

Green, L. R., C. L. Hughes, and W. L. Graves. 1978. Goat control of brush regrowth in southern California chaparral fuel breaks. *Proc. Intel. Rangel. Cong.* 1:451–455.

Hardegree, G. P. and W. E. Emmerich. 1992. Seed germination response of four southwestern range grasses to equilibration at subgermination matric-potentials. *Agron. J.* 84:994–998.

Hauser, V. L. 1989. Improving grass seedling establishment. *J. Soil and Water Conserv.* 44:153–156.

Heady, H. F., R. F. Barnes, and D. S. Metcalfe, 1985. *Forages: The science of grassland agriculture.* 4th ed. Iowa State University Press, Ames, IA.

Heady, H. F., and R. D. Child. 1994. *Rangeland Ecology and Management.* Westview Press, San Francisco.

Heath, M. E., R. F. Barnes, and D. S. Metcalfe. 1985. *Forages: The science of grassland agriculture.* 4th ed. Iowa State University Press, Ames, IA.

Herbel, C. H. 1963. Fertilizing tobosa on flood plains in the semi-desert grassland. *J. Range Manage.* 16:133–138.

Herbel, C. H. 1983. Principles of intensive range improvements. *J. Range Manage.* 36:140–144.

Herbel, C. H., W. L. Gould, W. F. Liefeste, and R. P. Gibbens. 1983. Herbicide treatment and vegetation response to treatment of mesquite in southern New Mexico. *J. Range Manage.* 36:149–151.

Herbel, C. H., and W. L. Gould. 1995. Management of mesquite, creosotebush, and tarbush with herbicides in the northern Chihuahan Desert. *N. Mex. Agric. Exp. Sta. Bull.* 775.

Hobbs, N. T., D. S. Schimel, C. E. Owensby, and D. S. Ojima. 1991. Fire and grazing in the tallgrass prairie: Contingent effects on nitrogen budgets. *Ecology* 72:1374–1382.

Holechek, J. L. 1982. Fertilizer effects on above- and below-ground biomass of four species. *J. Range Manage.* 35:39–42.

Holechek, J. L., E. J. Depuit, J. G. Coenenberg, and R. Valdez. 1981. Fertilizer effects on establishment of two seed mixtures on mined land in southeastern Montana. *J. Soil and Water Cons.* 36:39–42.

Holechek, J. L., K. Hess. 1994. Brush control considerations: A financial perspective. *Rangelands.* 16:193–196.

Jacoby, P. W., R. J. Ansky, and C. H. Meadors. 1991. Late season control of honey mesquite with clopyralid. *J. Range Manage.* 44:56–58.

Johnson, W. H., and E. L. Fitzhugh. 1990. Grazing helps maintain brush growth on cleared land. *Calif. Agric.* 44:31–32.

Jordan, G. L. 1981. Range seeding and brush management for Arizona rangelands. *Univ. Ariz. Bull.* T81121.

Kie, J. G., and B. B. Boroski. 1996. Cattle distribution, habitats, and diets in the Sierra Nevada in California. *J. Range Manage.* 49:482–488.

Klett, W. E., D. Hollingsworth, and J. L. Schuster. 1971. Increasing infiltration by burning. *J. Range Manage.* 24:22–24.

Klinger, R. C., M. J. Kutilek, and H. S. Shellhammer. 1989. Population responses of black-tailed deer to prescribed burning. *J. Wildl. Manage.* 53:863–871.

Knight, T. J., and P. J. Langston-Unkefer. 1988. Enhancement of symbiotic dinitrogen fixation by a toxin-releasing plant pathogen. *Sci.* 241:951–954.

Krammes, J. S. (Coord.). 1990. Effects of fire management of southwestern natural resources. *Proc. Symp. U.S. Dep. Agric., Rocky Mtn. For. & Range Exp. Sta. Gen. Tech. Rep.* RM-191.

Krausman, P. R. (Ed.). 1996. *Rangeland wildlife.* The Society for Range Management, Denver, CO. 440 pp.

Lacey, J. R., and R. L. Sheley. 1995. Leafy spurge and grass response to picloram and intensive grazing. *J. Range Manage.* 49:311–314.

Launchbaugh, J. L. 1971. Upland seeded pastures compared for grazing steers at Hays, Kansas. *Kans. Agric. Exp. Stn. Bull.* 548.

Lawrence, B. K., S. S. Waller, L. E. Moser, B. E. Anderson, and L. L. Larson. 1995. Weed suppression with grazing or atrazine during big bluestem establishment. *J. Range Manage.* 48:376–379.

Lopes, E. L., and J. W. Stuth. 1984. Dietary selection and nutrition of Spanish goats as influenced by brush management. *J. Range Manage.* 37:554–560.

Martin, J. A., and D. L. Huss. 1981. Goats much maligned but necessary. *Rangelands* 3:199–201.

Martin, S. C., and D. R. Cable. 1974. Managing semi-desert grass-shrub ranges: Vegetation responses to precipitation, grazing, soil texture, and mesquite control. *U.S. Dept. Agr. Tech. Bull.* 1480.

Masters, R. A., G. A. Rassmussen, and G. R. McPherson. 1986. Prescribed burning with a helitorch on the Texas Rolling Plains. *Rangelands* 8:173–176.

McDaniel, K. C., and K. W. Duncan. 1995. Juniper control with soil-applied herbicides. *N. Mex. Agric. Exp. Sta. Bull.* 772.

McDaniel, K., D. L. Anderson, and L. A. Torell. 1992. Vegetation change following big sagebrush control with Tebuthiuron. *New Mexico Agr. Exp. Sta. Bull.* 764.

McIlvain, E. H., and M. C. Shoop. 1970. Fertilizing weeping lovegrass in western Oklahoma. *In Proc. First Weeping Lovegrass Symposium, Agriculture Division.* The Samuel Roberts Noble Foundation Inc., Ardmore, OK, pp. 61–70.

Merrill, L. B., and C. A. Taylor. 1976. Take note of the versatile goat. *Rangeman's J.* 3:74–76.

Moffat, A. S. 1991. Research on biological pest control moves ahead. *Sci.* 252:211–212.

Moghaddam, M. R. and C. M. McKell. 1975. Fourwing saltbush for land rehabilitation in Iran and Utah. *Utah Sci.* 36:114–116.

Mosley, J. C. 1996. Prescribed sheep grazing to suppress cheatgrass: A review. *Sheep and Goat Res. J.* 12:74–81.

Newman, P. R., and L. E. Moser. 1988. Grass seedling emergence, morphology, and establishment as affected by planting depth. *Agron. J.* 80:383–387.

Novak, J. L., R. Trimble, P. A. Duffy, and W. Hanselk. 1987. Analyzing the economics of brush control in South Texas using a present value of costs approach. *J. Amer. Soc. Farm Managers & Rural Appraisers* 51:64–67.

Pase, C. R., and A. W. Lindenmuth, Jr. 1971. Effects of prescribed fire on vegetation and sediment in oak-mountain mahogany chaparral. *J. For.* 69:800–805.

Pettit, R. 1979. Effects of picloram and tebuthiuron pellets on sand shinnery oak communities. *J. Range Manage.* 32:196–201.

Raguse, C. A., J. L. Hull, M. R. George, J. A. Morris, and K. L. Taggard. 1988. Foothill range management and fertilization improve beef cattle gains. *Calif. Agric.* 42(3):4–8.

Ralphs, M. H., and D. N. Ueckert. 1988. Herbicide control of locoweeds: A review. *Weed Tech.* 2:460–465.

Riggs, R. A., S. C. Bunting, and S. E. Daniels. 1996. Prescribed fire. In P. R. Krausman (Ed.), *Rangeland wildlife,* pp. 295–321. The Society for Range Management, Denver, CO.

Rogler, G. A., and R. J. Lorenz. 1974. Fertilization of mid-continent range plants. In *Proc. Symposium on Forage Fertilization,* Muscle Shoals, AL, pp. 231–254.

Rollins, D., F. C. Bryant, D. D. Waid, and L. C. Bradley. 1988. Deer response to brush management in central Texas. *Wildl. Soc. Bull.* 16:277–284.

Roundy, B. A., V. K. Winkel, H. Khalifa, and A. D. Matthias. 1992. Soil water availability and temperature dynamics after one-time heavy cattle trampling and land imprinting. *Arid Soil Res. & Rehab.* 6:53–69.

Scifres, C. J. 1980. *Brush management.* Texas A&M University Press, College Station, TX.

Scifres, C. J. 1987. Economic assessment of tebuthiuron-fire systems for brush management. *Weed Tech.* 1:22–28.

Seiber, J. 1990. Food safety and chemical contamination: Fact vs. fantasy. *Utah Sci.* 51(4):156–161.

Sheley, R. L., B. H. Mullin, and P. K. Fay. 1995. Managing riparian weeds. *Rangelands* 17:154–157.

Sneva, F. A. 1967. Chemical curing of range forage. *J. Range Manage.* 20:389–394.

Stevens, R., and S. C. Walker. 1996. Juniper-piñon population dynamics over 30 years following anchor chaining. In J. R. Barrow, E. D. McArthur, R. E. Sosebee, and R. J. Tausch (Compilers). Proc. Symp. Shrublands Ecosystem Dynamics in a Changing Environment. *U.S. Dep. Agric. Intermountain Res. Sta. Gen. Tech. Rep.* INT-338, 125–128.

Story, J. M. 1992. Biological control of weeds: Selective, economical and safe. *West. Wildlands* (summer): 18–23.

Thomsen, C. D., W. A. Williams, M. R. George, W. B. McHenry, F. L. Bell, and R. S. Knight. 1989. Managing yellow starthistle on rangeland. *Calif. Agric.* 43(5):4–7.

Thompson, M. W., M. A. Shaw, R. W. Umber, J. E. Skeen, and R. E. Thackston. 1991. Effects of herbicides and burning on overstory defoliation and deer forage production. *Wildl. Soc. Bull.* 19:163–170.

Ueckert, D. N. 1980. Manipulating range vegetation with prescribed burning. *Symposium on Prescribed Range Burning in the Edwards Plateau of Texas Proc.,* pp. 27–44.

Ueckert, D. N., J. L. Petersen, R. L. Potter, J. D. Whipple, and M. W. Wagner. 1988. Managing prickly pear with herbicides and fire. *Tex. Agric. Exp. Sta.* PR-4570.

Vallentine, J. F. 1989. *Range development and improvements.* 3d ed. Brigham Young University Press, Provo, UT.

Van Riet, W. J., and R. Bailey. 1991. Fertilizers produce more range forage in drought than normal years. *Calif. Agric.* 45(3):28–30.

Wade, D. D., and C. E. Lewis. 1987. Managing southern grazing ecosystems with fire. *Rangelands* 9:115–119.

Wapshere, A. J., E. S. Delfosse, and J. M. Cullen. 1989. Recent developments in biological control of weeds. *Crop Protection* 8:227–250.

Warren, A., J. Holechek, and M. Cardenas. 1996. Honey mesquite influences on Chihuahuan Desert vegetation. *J. Range Manage.* 49:46–52.

Watts, M. J., and C. J. Wambolt. 1989. Economic evaluation of Wyoming big sagebrush (*Artemisia tridentata*) control methods. *Weed Tech.* 3:640–645.

Westin, F. C., G. J. Buntley, and B. C. Brage. 1955. Soil and weather. Agricultural research. *S. Dak. Agric. Exp. Stn. Circ.* 116, pp. 6–18.

White, L. D., and C. W. Hanselka. 1989. *Prescribed range burning in Texas.* Tex. Agric. Ext. Serv. B-1310.

Whitson, T. D., M. A. Ferrell, and H. P. Alley. 1988. Changes in rangeland canopy cover seven years after tebuthiuron application. *Weed Tech.* 2:486–489.

Wiedemann, H. T., and B. T. Cross. 1990. Disk-chain-diker implement selection and construction. *Tex. Agric. Exp. Sta. Center Tech. Rep.* 90-1.

Wiedemann, H. T., B. T. Cross, and C. E. Fisher. 1977. Low-energy grubber for controlling brush. *Trans. Am. Soc. Agric. Eng.* 20:210–213.

Wight, J. R., and A. L. Black. 1979. Range fertilization: Plant response and water use. *J. Range Manage.* 32:345–348.

Winkel, V. K., and B. A. Roundy. 1991. Effects of cattle trampling and mechanical seedbed preparation on grass seedling emergence. *J. Range Manage.* 44:176–180.

Wisdom, C. S., C. S. Crawford, and E. F. Aldon. 1989. Influence of insect herbivory on photosynthetic area and reproduction in *Gutierrezia* species. *J. Ecol.* 77:685–692.

Wood, G. W. 1988. Effects of prescribed fire on deer forage and nutrients. *Wildl. Soc. Bull.* 16:180–186.

Workman, J. P. 1995. The value of increased forage from improved rangeland condition. *Rangelands* 17:46–48.

Workman, J. P., and J. A. Tanaka. 1991. Economic feasibility and management considerations in range revegetation. *J. Range Manage.* 44:566–573.

Wright, H. A., and A. W. Bailey. 1982. *Fire ecology.* John Wiley & Sons, Inc., New York.

CHAPTER 16

RANGE MANAGEMENT IN DEVELOPING COUNTRIES

R angelands represent an important resource in many countries around the world. About 30 to 40 million people in arid and semiarid regions have "animal-based" economies (Sandford 1983). Over 50% of these people live on the continent of Africa, and they are commonly referred to as "pastoralists" (Sandford 1983). They derive most of their income and sustenance from livestock grazing in arid and semiarid areas. In developing countries, pastoralists are more dependent on rangelands than in other countries because there are seldom other employment opportunities, such as in industry. Rangelands in many developing countries are being stressed as animal numbers expand to meet a growing human population dependent on a shrinking resource base.

Developing countries face multifaceted problems in range resource management. Some of these problems are somewhat unique to developing countries, whereas others are more general and apply to rangelands everywhere. The following discussion will be an analysis of these problems.

PROBLEMS RELATING TO LIVESTOCK NUMBERS

Problems with livestock and adequate rangeland for grazing date to Biblical times, when Abraham and Lot were forced to separate their herds to provide adequate forage. Despite the long-standing nature of problems with livestock numbers, solutions to these problems are as elusive as ever. In some cases problems relate to livestock distribution, while in others they are more pervasive.

Determining optimum stocking rate is complicated by extreme temporal and spatial variability in herbage and browse production on rangelands. Coping with these variations is a challenge to the most progressive operator as well as the nomadic pastoralist in developing countries.

Data in Table 16.1, taken from exclosures in arid rangeland in Niger, exemplify the type of variation found in herbage production on rangelands the world over. At Gadebeji herbage production declined from nearly 1,200 kg/ha in 1980 to less than 200 kg/ha the following year. In many developing countries supplemental feed, complementary pastures, and alternative feed sources are limited. Consequently, herders in this vicinity were faced with a feed resource in 1981 representing only 16% of that present the previous year. These fluctuations represent "normal" conditions, although a single year may be considered a drought (which is covered in more detail later in the chapter).

Spatial variation in herbage production is also pronounced (Table 16.1). Average production was lowest at Aderbissinat, but in 1982 it was lowest at Ibecetene. Figure 16.1 shows differences between two areas of the Sahel region of northern Africa south of the Sahara desert.

Providing flexibility in livestock operations to meet these variations is difficult under the best conditions. There may be incentives to maximize herd size in terms of prestige, wealth, and higher family security. Herders or livestock operators deal with these yearly fluctuations in herbage production by adopting one of several stocking plans (Pieper 1981). Variations in herbage production from year to year are shown in Table 16.1 while stocking plans to meet these fluctuations are discussed in Chapter 8.

Most developing countries have experienced destructive grazing to varying degrees (Strange 1980; Thomas 1980). In many cases, range deterioration is most pronounced around watering points and other areas of livestock concentration. Mixed herds of camels, cattle, sheep, and goats are very efficient in exploiting range resources (Figure 16.2).

TABLE 16.1   Annual Herbage Standing Crop (kg/ha) at Four Locations Within the Pastoral Zone of Niger, 1980–1982

LOCATION	1980	1981	1982	AVERAGE
North Dakorro	1,500	750	1,200	1,150
Ibecetene	968	637	586	730
Gadabeji	1,173	188	1,132	831
Aderbissinat	657	499	728	628
Average	1,075	519	911	835

Source: From Wylie et al. 1983.

***Figure* 16.1** Differences in herbage production from annual grasses in the Sahel. Top photo represents high production in deciduous woodland in eastern Senegal, while bottom photo represents virtually no herbaceous vegetation on a range in Niger.

The immediate prospects of reducing destructive grazing in developing countries is not bright. The reasons for heavy stocking are many and complicated. However, heavy stocking and deterioration of basic resources remain to be the most serious problems facing the range livestock sector worldwide. In developing countries, the destructive cycle (Figure 16.3) was initiated some time ago (Strange 1980). However, in recent years human population increase has accelerated the spiral in which the people are involved. Population increases are among the highest in developing countries. Many of these countries have only

Figure **16.2** Mixed groups of livestock are often able to exploit range vegetation to a greater extent than one species. Photo taken in Mali by Rex Pieper.

limited industries or other avenues of employment for people outside of agriculture. In the Sahel, for example, large portions of the population are directly dependent on their live-stock for food and cash to purchase alternative food and other necessities (Simpson and Evangelou 1984). Two examples will be used to show the extent of the problem. In 1970 there were about 261 million people in sub-Saharan Africa. This number had expanded to 347 million by 1980 and is expected to nearly double to 639 million by 2000 (Simpson 1984). Cattle numbers would have to increase from 147 million head in 1980 to 280 million head by 2000 and sheep and goats from 230 million head to 410 million to support this human population. If, as many ecologists believe, the Sahel is already fully stocked or over-stocked, such livestock increases could result in disaster. In Niger, the number of tropical livestock units (TLU = 250 kg live weight) for each family member was 4 to 5 prior to the drought of the late 1960s. This number was sufficient to meet minimal human requirements. During the drought nearly 50% of the national herd died, but by 1983 herds were reconsti-tuted. However, because of the annual human population growth rate (2.5%) each family member had only 3.5 TLU, a subsistence level (Pase et al. 1985). In contrast, the 1984 drought was much more severe than the previous one.

Vegetation of the Sahel is now mainly savannas with annual grass understory or annual grassland with scattered shrubs. Concepts of proper use based on physiological require-ments of perennial plants may have little relevance for these annual grasslands since there seems to be little opportunity for restoring perennial grasses under present livestock num-bers. Indeed, heavy stocking during certain periods may increase primary production (Wylie et al. 1983). However, when stocking is heavy enough to restrict livestock intake and to allow accelerated erosion, it is considered excessive.

Heavy grazing may also take a toll on other basic resources in many rangelands. Shrubby vegetation often serves a critical role in dry-season grazing. In the Sahel, grass vegetation is often depleted after 5 to 6 months of dormancy and shrubs provide an impor-

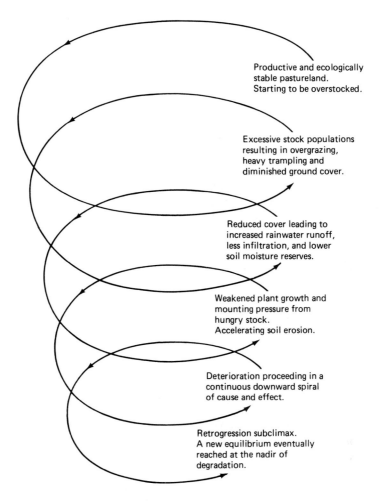

Productive and ecologically
stable pastureland.
Starting to be overstocked.

Excessive stock populations
resulting in overgrazing,
heavy trampling and
diminished ground cover.

Reduced cover leading to
increased rainwater runoff,
less infiltration, and lower
soil moisture reserves.

Weakened plant growth and
mounting pressure from
hungry stock.
Accelerating soil erosion.

Deterioration proceeding in a
continuous downward spiral
of cause and effect.

Retrogression subclimax.
A new equilibrium eventually
reached at the nadir of
degradation.

***Figure* 16.3** Retrogressive spiral on rangeland under heavy grazing. (From
Strange 1980.)

tant source of protein, vitamin A, certain minerals, and so forth. Heavy utilization can be
very detrimental to these shrubs (Bille 1978). Bille's (1978) data indicate that it may require
30 years of complete protection for these shrubs to regenerate their productivity. The range
livestock industry would be severely limited in many range areas if the woody resources
were lost (Le Houerou 1980).

Remedies for problems relating to excessive livestock numbers on rangelands in developing countries are complex and not easily resolved. Often, they are compounded by lack of
land tenure regulations and expanding human populations. Herders have few alternatives
in terms of opportunities in other agricultural segments or industry. However, continued
exploitation of the basic range resources may lead eventually to elimination of livestock

production for thousands of these nomadic and seminomadic people. The reader is referred to Foster (1992) for a more complete discussion on food problems in developing countries.

LAND TENURE AND COMMUNAL GRAZING

The primary type of land tenure for extensive range areas around the world is open grazing. Herders who graze on these unrestricted ranges are sometimes divided into three classes:

1. *Nomadic.* Herders have no permanent base; they take all their provisions with them as they move with their livestock.
2. *Transhumance.* Herders have a permanent base to which they return each year. They move with their livestock during certain portions of the year.
3. *Sedentary.* These people are often farmers who also raise stock on the side. They have a permanent home and graze livestock in the vicinity of their permanent base.

Examples of nomadic and transhumance patterns abound in developing countries. Most tribes in Africa are nomadic to some degree (Abercrombie 1974; Breman et al. 1978; El Hassan 1978; Hickey 1978; Thomas 1980). Bedouin herders are prominent in the Middle East (Noy-Meir and Seligman 1979), while Mongol herders raise cattle, sheep, horses, and camels in certain areas in Asia.

It is easy to visualize nomadic grazing systems in arid and semiarid areas as a mechanism with maximum flexibility for herders to provide feed and water for livestock. Often, these patterns were seasonal, with movements away from permanent water during the rainy season and with the reverse in the dry season. The impact of local droughts can be minimized by altering some of these movements to those areas less influenced by the drought. If the dry season extends longer than normal, herders can move to other areas in search of forage.

Breman et al. (1978) provide an example of movement patterns for pastoralists who live mostly in Mali. Zebu herds feed mainly on perennial grass species on the floodplain of the Niger River during the long dry season (8 to 9 months). In the early dry season the cattle graze on pastures owned by the village of Diafarabe. Later they graze on communal pastures in the middle of the central delta. During the rainy season, the delta is flooded and livestock make the long trek to native rangeland, composed mainly of annual grasses and forbs in southeast Mauritania. At the end of the rainy season, they return to the area of the Niger near Diafarabe.

These nomadic and seminomadic systems were well adapted for many rangelands around the world when human populations were relatively low and forage and water plentiful for the livestock supporting the human population. However, as the human population increased, the number of livestock needed to support the human population also increased. These increases in livestock numbers put additional stress on fragile vegetation and resulted in considerable range deterioration.

Many western range people consider communal grazing as one of the prime factors responsible for destructive grazing with excessive livestock. Hardin's (1968) "Tragedy of the Commons" argument added fuel to the argument. According to Hardin's analysis of the case, it is in the interest of the herder to increase his herd size since he will gain all the benefit of having a larger herd while sharing any adversity with all others grazing the common rangeland. The example given by Hardin is as follows: An individual herder is considering adding one more animal to his herd. The action has two components, one positive and one negative. The positive component is represented by the value of one more animal. It might be increased sale value of the animal, increased milk production for his family, increased prestige in the view of other herders, and so on. The negative value of adding one more animal to the herd comes from the added grazing pressure on the resource base. However, this degree of exploitation, -1, must be shared by all herders using the range and is only a small fraction for the herder who added one animal to his herd. Consequently, it is to the advantage of all herders to increase their herds, with resultant range deterioration. These and similar arguments have often led U.S. range advisors to recommend against communal grazing and to advocate greater control of livestock movements and numbers.

However, not all authors agree completely with Hardin's argument. Artz (1985), for example, argues that in most communal systems there are various regulations within the group to control abuse. Not all herders make decisions in a vacuum; they do consider others who also use the commons. Sandford (1983) also believes that the argument is too strict and not based on sound assumptions in all cases. Shifts from communal grazing to private ownership have not alleviated all problems of destructive grazing (Runge 1981; Sandford 1983). On the other hand, Gilles and Jamtgaard (1982) have listed examples in Peru and Africa where communal rangelands have been grazed for years without serious range deterioration. Other critical resources, under the control of a family group, grazing association, and so on, may indirectly control grazing pressure (Gilles and Jamtgaard 1982). In Kenya, Ellis and Swift (1988) argue that development policies should facilitate and build on traditional pastoral strategies rather than constrain them.

PROBLEMS RELATING TO DRY-SEASON GRAZING

In most range areas precipitation patterns are unimodal, with a fairly short rainy period followed by an extended dry period with limited opportunities for plant growth. This alternate pattern of plant growth and dormancy presents two problems with respect to grazing animals:

1. Declining forage quantity
2. Declining forage quality

At the start of the rainy season, forage for livestock on rangelands is at a minimum. As the plants begin new growth, herbage weight increases and becomes plentiful, until it reaches a peak. Even without grazing by large herbivores (livestock and large game animals), there is a sharp decline in available herbage (Pieper et al. 1974). With livestock grazing the decline is even sharper. In a tropical area of India, the decline during the dormant season was nearly 64% without large grazing animals (Singh and Yadava 1974). On many ranges, declines of 25% to 35% are not uncommon (Brockington 1960; Ratliff and Heady 1962; Pieper et al. 1974).

The decline in nutritive quality of range forage throughout the dormant season has been well documented (Chapter 11). For example, annual grasses in Niger lose about 70% of their crude protein content during the dry period (Table 16.2). Digestible protein and vitamin A may be difficult to detect in standing dead plants during this dormant period (Louis et al. 1983).

Consequently, at the end of the dry season and early in the rainy season, livestock are under the maximum stress from the standpoint of intake and nutritive quality of their diet. In many cases, pastoralists have little flexibility to adjust to these stress conditions and must suffer outright mortality of their animals and reduced production of their surviving animals. In most cases, individual pastoralists may not be able to alter their operations to adjust to these problems. Some type of herder organizations or governmental interventions may be necessary to develop programs to alleviate problems associated with dry-season grazing.

The suggestions listed below represent only a few possibilities for reducing stress on both livestock and human populations during dry seasons in developing countries. Some of these may be impractical under most situations, but a few may have relevance in specific cases.

1. *Seasonal grazing.* Traditional nomadic or transhumance grazing was largely a response to seasonal variation in available forage supplies and enables herders the flexibility necessary to maintain their herds. Increasing human and livestock populations have reduced the opportunities for these movements in many cases. However, these patterns have evolved over years and appear to be the most feasible response by the herders.

2. *Complementary forage.* In many cases, both in developing and developed countries, crop residue is utilized in a grazing program to supplement native range. Crop residue

TABLE 16.2 Content of Certain Nutrients in Three Abundant Grass Species in the Pastoral Zone of Central Niger

SPECIES	SEASON	CRUDE PROTEIN	DIGESTIBLE PROTEIN	PERCENTAGE PHOSPHORUS	VITAMIN A (IU/KG)[a]
Aristida mutabilis	Rainy	8.9	4.7	0.22	7,600
	Dry	2.3	0	0.10	ND
Cenchrus biflorus	Rainy	7.3	3.3	0.24	5,400
	Dry	2.6	0	0.09	ND
Schoenfeldia gracilis	Rainy	8.6	4.5	0.19	5,400
	Dry	2.4	0	0.10	ND

Source: From Louis et al. 1983.

[a]ND = Not detectable.

may add energy to the diet and in some cases protein. Improved pastures may also be used in conjunction with native range to provide a yearlong balanced program. Nomads in the Niger River delta in Mali use improved pasture as part of their grazing program, as discussed earlier (Breman et al. 1978). The use of seasonal suitability grazing is discussed further in Chapter 9.

3. *Supplemental feed.* In the United States, supplementing livestock on native range during periods of nutritional stress is a common practice (Chapter 10). However, supplements are not so readily available in most developing countries. As supplements are more readily considered, their availability in developing countries may increase. Protein supplements such as peanut meal and cottonseed meal are now available in some African countries. Supplements are often supplied to meet livestock needs for protein, energy, minerals (e.g., phosphorus, trace elements), and vitamin A. Vitamin A can also be supplied by injections.

4. *Stored feed.* The type of stored feed depends, again, upon availability. Examples include peanut straw, silage, and hay (alfalfa or grass hay). Native grass hay was used to supplement native range in tropical areas of Tanzania (Sullivan et al. 1980) (Table 16.3). This system proved beneficial to both the livestock and human population. There may be many opportunities for herders to harvest grass hay by hand, but provisions for storage and subsequent feeding may be a problem.

5. *Forage reserves.* Introduced or native plants can be protected from grazing until late in the dry season. Ideally, they should be plants that retain their nutritive value well into the dry season. The International Livestock Center for Africa has been conducting such trials in Nigeria. The forage reserves might be used for livestock at greatest risk. Palatable and nutritious shrubs have potential for this type of operation. Direct seeding or transplanting of shrubs in small exclosures near water may provide forage reserves as well as augmenting the valuable but often depleted browse supplies. Bille (1978) has outlined the importance and use of shrubs in the Sahel region of Africa.

6. *Centripetal grazing.* The normal pattern of grazing around permanent water is to graze close to water first and then to work out progressively. This pattern results in repeated grazing of herbage close to water and a "sacrifice" area around the water point with scattered plant cover and considerable bare ground. During extended dry-season grazing, animals must spend considerable periods trailing to and from water when their physiological condition is at its lowest. As the forage is depleted at greater and greater distances from water, weakened animals must walk greater and greater distances. With centripetal grazing, the process is reversed. Animals are trailed to areas at greatest distances early in the dry season and graze progressively closer to water late in the dry season. Thus when the animals are in the poorest condition, they have the least distance to walk between water and forage. For such a system to work, control of animals would be mandatory, however, livestock herders do have such control in many countries. Procedures would have to be developed to prevent trampling of herbage close to water when livestock were moved away from water. Livestock numbers would need to be in balance with available herbage to prevent destructive grazing patterns.

TABLE 16.3   Forage Intake and Milk Production (1,000 kg) by Cattle for a Traditional Village and a Supplement System in Tanzania

| MONTH | FORAGE INTAKE | | | MILK YIELD | | | |
| | TRADITIONAL | HAY SUPPLEMENT | | TRADITIONAL | | HAY SUPPLEMENT | |
		FORAGE	HAY	TOTAL	TO HUMANS	TOTAL	TO HUMANS
January	844	873	0	86	19	88	21
February	838	863	0	88	17	90	20
March	851	877	0	85	14	87	16
April	828	858	0	74	9	78	13
May	634	622	0	60	6	63	8
June	592	626	0	39	3	43	7
July	580	602	18	18	2	21	6
August	515	540	15	10	2	15	7
September	445	460	16	10	2	25	12
October	525	531	22	18	4	26	13
November	688	677	37	36	8	43	16
December	856	888	0	76	14	80	17
				600	102	648	150

Source: From Sullivan et al. 1980.

DROUGHT

The serious drought in the Sahel in Africa in the late 1960s and early 1970s and the recent situation in many countries in Africa have brought worldwide attention to this recurring problem. Much has been written about the famine and mass starvation of human populations. Direct food aid has been contributed in massive proportions. Yet little has been done to prepare or plan for long-term solutions to handle drought problems.

Some authors have decried the lack of planning for drought (Wallen and Gwynne 1978). Each drought tends to be treated as if it were an isolated event. Once the drought is broken, everything is okay and little is done to prepare for the next one. Perhaps one of the problems is that droughts are unpredictable, whereas dry-season grazing occurs every year.

Drought should be distinguished from aridity since the two are confused at times. Aridity refers to the general lack of rainfall and is a chronic or continuous condition. Drought, on the other hand, represents a departure from average or normal conditions. Drought must be evaluated in relation to average conditions that reduce vegetational growth. Droughts may be short-term (Pieper and Donart 1973) or long-term, such as the drought of the early 1970s in the Sahel (Wallen and Gwynne 1978) and of the 1950s in the southwestern United States (Herbel et al. 1972). The initial impact of drought is a reduction of available forage for livestock (Pieper and Donart 1975; Swanson and Sellars 1978). Livestock performance may be reflected in weight gains, milk production, and reproduction (Wallace and Foster 1975). Eventually, mortality of perennial grass and shrubs will occur, depending on the time and severity of the drought (Chamrad and Box 1965; Pieper and Donart 1973). Livestock mortality is often a result of drought when all other options have been exhausted. Livestock mortality during the Sahelian drought was substantial for all countries. Many herders believe that maintaining a large herd is protection against drought, however, the limited data in Table 16.4 do not support this idea completely.

Development of early-warning systems over large areas to locate forage-deficient areas and possible areas of relatively high forage will aid governments and planners to prepare for shortages late in the dry season. Satellite imagery is one technique that has shown promise in this area (Tucker et al. 1985). A network of rain gauges could also be used in this conjunction, provided that they can be read at appropriate times. Such information does not provide relief from the drought, but does provide time for planning emergency action.

TABLE 16.4   Comparison of Livestock Numbers in Both Pre- and Post-drought Years in Mali

VILLAGE	NAME OF PASTORALIST	HEAD OF CATTLE PRIOR TO DROUGHT	LOST	REMAINING IN 1974
Koinse	Souaibo dit Kouira	40	40	0
	Guibrila Gouranke	40	40	0
	Ousmane Kaissara	13	13	0
Ziguiberi	Azida ag Gouhoun	65	60	5
	Abdoulaye ag Takamades	70	63	7
	Manguiel	81	70	11
	Rali ag Higoum	50	48	2
	El Moustapha	13	5	8
	Hohammed	40	40	0
Goumgam	Sita Baye	80	40	40
	Tazoudi	160	130	30
Tangmole	Moussa Hamidou	65	36	29
	Sedou Dibi	10	4	6

Source: From Franke and Chasin 1980.

Basically, there are two approaches to drought:

1. Reduce livestock numbers.
2. Provide supplementary feed.

Reducing livestock numbers is not popular or economically feasible in most cases. Herders often depend directly on their livestock for a livelihood and cannot reduce numbers without having immediate impact on their families.

Wallen and Gwynne (1978) recommend utilization of the boundary zone between rainfed agriculture and rangeland as an area for production of livestock feed. This zone can be one alternative between grazing and crop agriculture, depending on climatic patterns.

Several practices discussed under dry-season grazing apply to drought conditions. Management of any supplemental feed or forage is more complex for drought conditions because these feeds are perishable and cannot be stored until a drought occurs. Drought planning is difficult because of unknown length and severity.

The following quotation from Glantz and Katz (1977) summarizes many of the problems of drought rangelands:

> The mistaken idea that drought in the Sahel is an unexpected event has often been used to excuse the fact that long-range planning has failed to take rainfall variability into account. People blame the climate for agricultural failures in semi-arid regions making it a scapegoat for faulty population and agricultural policies. It will therefore be crucial that decision makers know what statistical and quasi-statistical measures actually mean: no single number can adequately describe the climate regime of an arid or semi-arid region. Decision makers must supplement such terms as the "means" with more informative statistical measures to characterize adequately the variability of the climate. An understanding of this high degree of variability will serve to remove one of the major obstacles to resolving the perennial problems of the Sahel and of other arid or semi-arid regions.

The reader is referred to Rietkerk (1996) for a recent discussion on the dynamics of Sahelian rangeland vegetation with management implications.

DEVELOPMENT OF LIVESTOCK WATER

Arid and semiarid rangelands are often lacking in water for livestock and game, resulting in some areas being underutilized. Early range development projects often included water development to provide additional land available for grazing (Sandford 1983; Child et al. 1984). These developments often were dirt tanks for the temporary storage of water following rains or boreholes which tapped groundwater sources. Such projects were highly visible and often favored by herders and local governments. However, often these water development projects were initiated without any plan for livestock grazing in the new area made available by water development.

Areas around water points serve as areas of concentration for livestock and game. Destructive grazing is common in these "sacrifice zones." The size of the sacrifice area depends on stocking rate and grazing period. Water point development without grazing management often resulted in exploitation of new areas and further range deterioration. Heavy grazing results in a depletion of the area around the well. Consequently, livestock water development has not been a qualified success, even though many areas are now well watered.

Another problem with water developments in developing countries is maintenance. Often there is no provision for maintenance at the end of the project or in training of local people for maintenance. Table 16.5 shows that many wells developed in Africa were never used, and many became inoperative after several years.

Thus is appears that water point development in developing countries should be evaluated carefully. Adequate grazing management and provision for maintenance should be requirements for water point development.

RANGE BURNING

Prescribed or controlled burning can be used as a tool for various land management objectives. However, in developing countries, indiscriminant burning may destroy valuable forage and leave bare areas subject to severe erosion. Burning is done for a variety of reasons, but in many cases there is no plan to keep the fire confined to a specific location.

The magnitude of the burning problem in many countries can be illustrated by Sudan. Obeid (1978) estimated that 34% of the annual forage crop in Sudan is lost through indiscriminate burning. In some provinces, as much as 45% of the herbage is burned each year. Considerable livestock could graze this forage to help support the human population.

TABLE 16.5   Fate of Water Developments in Several African Countries

COUNTRY	TYPE OF DEVELOPMENT	OUTCOME
Northeastern Kenya	54 boreholes in 1969	25% operating by 1976
Kenya	New water points	62% functioning after 10 years
Botswana	Boreholes	40% never functioned
	Water points	65% functioning
		19% abandoned
		16% temporarily nonfunctioning
Niger	Boreholes	15 of 23 functioning after 11 years
Sudan	145 boreholes	44% never operated
		28% broken pumps

Source: From Sandford 1983.

AGRICULTURAL VERSUS INDUSTRIAL DEVELOPMENT

Presently there is much debate over whether meaningful improvement can occur in living conditions in developing countries without seven basic sociopolitical factors being in place. These factors, discussed by Miller (1990), Foster (1992), and Schiller (1994), include:

1. National unity

2. Market oriented economy

3. Democratic form of government

4. Sound education system

5. Protection of private property rights

6. Opportunity for social and economic mobility

7. Level of economic growth exceeds level of population growth

Organizations that administer assistance to developing countries increasingly have chosen to focus on those countries where major improvements are being made in the above factors. Once the institutional foundation is put in place for economic development various tradeoffs must be made regarding whether to emphasize agricultural or industrial production. Narjisse (1995) provided an assessment of the role of rangelands in the development of African countries. Both problems and prospects were discussed in considerable detail. He emphasized that rangeland development should not occur in isolation from the rest of agriculture and the national economy. He considered fair trade policies on the part of industrialized countries to be indirectly important to African rangeland sustainability. New development strategies based on a decentralized, multidisciplinary approach were advocated. Under his suggested strategy the role of the central government would be limited to setting goals for the country's long term food security, assessing land suitability in different ecozones of the country, and identifying mechanisms to promote appropriate land uses. These mechanisms would include credit and other incentives. For other detailed discussions on economic and agricultural development approaches in Latin American, Asian, and African countries we refer readers to Brown and Wolf (1985), Miller (1990), Foster (1992), and Schiller (1994).

RANGE MANAGEMENT PRINCIPLES

- Careful study of successes and failures of range management policies and programs in developed and developing countries will permit managers to better identify approaches and practices likely to be successful in developing countries.

- Programs that do not take into consideration human customs and cultures and do not educate tribal herdsmen on program objectives will probably be failures.

- In areas where large increases are occurring in human populations, improvement in rangeland and human conditions will in most cases be difficult unless there is rapid economic development.

- Planning and preparation for drought is a critical aspect of range management in both developing and developed countries.

Literature Cited

Abercrombie, F. D. 1974. *Range development and management in Africa.* Agency for International Development, Office of Developmental Services, Washington, DC.

Artz, N. E. 1985. Must communal grazing lead to tragedy? In L. D. White and J. A. Tiedeman (Eds.). *Proc. 1985 International Rangeland Resources Development Symposium.* Cooperative Extension, Department of Forestry and Range Management, Washington State University, Pullman, WA.

Bille, J. C. 1978. Woody forage species in the Sahel: Their biology and use. *Proc. Int. Rangel. Congr.* 1:392–395.

Breman, H., A. Diallo, G. Traore, and M. M. Djiteye. 1978. The ecology of the annual migrations of cattle in the Sahel. *Proc. Int. Rangel. Congr.* 1:592–595.

Brockington, N. R. 1960. Studies of the growth of *Hyparrhenia*—Dominant grassland in northern Rhodesia. I. Growth and reaction to cutting. *J. Br. Grassl. Soc.* 15:323–338.

Brown, L. R., and E. C. Wolf. 1985. Reversing Africa's decline. *Worldwatch Paper 65.* Worldwatch Institute, Washington DC.

Chamrad, A. D., and T. W. Box. 1965. Drought-associated mortality of range grasses in south Texas. *Ecology* 46:780–785.

Child, R. D., H. F. Heady, W. C. Hickey, R. A. Peterson, and R. D. Pieper. 1984. *Arid and semiarid lands—Sustainable use and management in developing countries.* Winrock International Institute, Morrilton, AR.

El Hassan, B. A. 1978. Nomadism and range management in the Sudan. *Proc. Int. Rangel. Congr.* 1:127–129.

Ellis, J. E., and D. Swift. 1988. Stability of African pastoral ecosystems: Alternative paradigms and implications for development. *J. Range Manage.* 41:450–459.

Foster, P. 1992. *The world food problem.* Lynne Rienner Publishing, Inc., Boulder, CO.

Franke, R. W., and B. H. Chasin. 1980. *Seeds of famine.* Allanheld, Osmun & Co., Publishers, Inc., Montclair, NJ.

Gilles, J. L., and K. Jamtgaard. 1982. The commons revisited. *Rangelands* 4:51–54.

Glantz, M., and R. W. Katz. 1977. When is a drought a drought? *Nature* 267:192–193.

Hardin, G. 1968. The tragedy of the commons. *Sciences* 162:1243–1248.

Herbel, C. H., F. N. Ares, and R. A. Wright. 1972. Drought effects on a semidesert grassland range. *Ecology* 53:1084–1093.

Hickey, J. V. 1978. Fulani nomadism and herd maximization: A model for government mixed farming and ranching schemes. *Proc. Int. Rangel. Congr.* 1:95–99.

Le Houerou, H. N. (Ed.). 1980. *Browse in Africa—The current state of knowledge.* International Livestock Centre for Africa, Addis Ababa, Ethiopia.

Louis, S. L., A. Dankintafo, B. Bookary, and N. Goumeye. 1983. Seasonal influence on the nutritive value of the rangeland in Niger pastoral zone. Ministry of Rural Development. *US AID Tech. Bull.* 2. Tahoua, Niger.

Miller, G. T. 1990. *Resource conservation and management.* Wadsworth Publ., Co., Belmont, CA.

Narjisse, H. 1995. The range livestock industry in developing countries: Current assessment and prospects. *Intn'l Rangel. Cong.* 5:14–21.

Noy-Meir, I., and N. G. Seligman. 1979. Management of semi-arid ecosystems in Israel. In B. H. Walker (Ed.), *Management of semi-arid ecosystems.* Elsevier Science Publishing Co., Inc., New York.

Obeid, M. M. 1978. The impact of human activities and land use practices on the grazing lands in the Sudan. *Proc. Int. Rangel. Congr.* 1:48–51.

Pase, C. P., R. D. Pieper, and M. Bagoudou. 1985. Range management problems and opportunities in a transhumance pastoral region of Niger. In L. D. White and J. A. Tiedeman (Eds.). *Proc. 1985 International Rangeland Resources Development Symposium.* International Affairs Committee, Society for Range Management, Cooperative Extension Department of Forestry and Range Management, Washington State University, Pullman, WA.

Pieper, R. D. 1981. The stocking rate decision. *Proc. First International Rancher's Round-Up.* Texas Agriculture Extension Service, College Station, TX, pp. 199–204.

Pieper, R. D., and G. B. Donart. 1973. Drought effects on blue grama rangeland. *N. Mex. State Univ. Livestock Feeders Rep.*

Pieper, R. D., C. H. Herbel, D. D. Dwyer, and R. E. Banner. 1974. Management implications of herbage weight changes on native rangeland. *J. Soil Water Conserv.* 29:227–229.

Pieper, R. D., and G. B. Donart. 1975. Drought effects on southwestern vegetation. *Rangeman's J.* 2:176–178.

Ratliff, R. D., and H. F. Heady. 1962. Seasonal changes in herbage weight in an annual grass community. *J. Range Manage.* 15:146–149.

Rietkerk, M., P. Ketner, L. Strooshijder, and H. H. T. Prins. 1996. Sahelian rangeland development: A catastrophe. *J. Range Manage.* 49:512–519.

Runge, C. F. 1981. Common property externalities: Isolation, assurance, and resource depletion in a traditional grazing context. *Am. J. Agric. Econ.* 63:595–606.

Sandford, S. 1983. *Management of pastoral development in the Third World.* John Wiley & Sons, Inc., New York.

Schiller, B. 1994. *The economy today.* 6th ed. McGraw-Hill, Inc., New York.

Simpson, J. R. 1984. Problems and constraints, goals and policy: Conflict resolution in development of subsaharan Africa's livestock industry. In J. R. Simpson and P. Evangelou (Eds.), *Livestock Development in Subsaharan Africa. Constraints, prospects, policy.* Westview Press, Inc., Boulder, CO.

Simpson, J. R., and P. Evangelou (Eds.). 1984. *Livestock development in Subsaharan Africa. Constraints, prospects, policy.* Westview Press, Inc., Boulder, CO.

Singh, J. S., and P. S. Yadava. 1974. Seasonal variation in composition, plant biomass, and net primary productivity of a tropical grassland at Kurukshetra, India. *Ecol. Monogr.* 44:351–376.

Strange, L. R. N. 1980. *Africa pastureland ecology.* Food and Agriculture Organization of the U.N. Rome, Italy.

Sullivan, G. M., K. W. Stokes, D. E. Farris, T. C. Nelsen, and T. C. Cartwright. 1980. Transforming a traditional forage/livestock system to improve human nutrition in tropical Africa. *J. Range Manage.* 33:174–178.

Swanson, J. D., and D. V. Sellars. 1978. The record drought on California's annual grass rangeland. *Proc. Int. Rangel. Congr.* 1:212–215.

Thomas, G. W. 1980. *The Sahelian/Sudanian zones of Africa. Profile of a fragile environment.* Report to the Rockefeller Foundation. New Mexico State Univ., Las Cruces, NM.

Tucker, C. J., J. R. G. Townhead, and T. E. Goff. 1985. African ground cover classification using satellite data. *Science* 225:1047–1051.

Wallace, J. D., and L. Foster. 1975. Drought and range cattle performance. *Rangeman's J.* 2:178–180.

Wallen, C. C., and M. D. Gwynne. 1978. Drought—A challenge to rangeland management. *Proc. Int. Rangel. Congr.* 1:21–32.

Wylie, B., R. Senock, L. Snyder, L. Roettgen, and S. Porter. 1983. *Niger range and livestock project. Range research and results.* Report of Nigerian Ministry of Rural Development, Niamey, Niger.

CHAPTER 17

COMPUTER APPLICATIONS AND THE FUTURE

Successful ranching increasingly depends on current knowledge of biological and financial outcomes of various management practices. In tomorrow's world any successful ranch management program will increasingly depend on highly skilled personnel who can use computers to assess the risk/reward ratios of various management options. Each ranch is a complete system that must be managed as a business. Modern ranchers must have understanding of biology, weather, economics, politics, and the interactions among these factors to successfully manage their operations. Considerable emphasis has been placed on the biological factors in the past but increasingly it is being recognized that other factors cannot be ignored. Recent developments such as a 30% drop in cattle prices (1993–1996), drought in the southwestern United States, discontinuation of USDA emergency feed and wool/mohair subsidies, and increased land use regulations have greatly impacted the western range livestock industry and are important considerations in selection of range management practices. The rapid increase in technology and information has made it imperative that today's ranchers be skilled in the use of computers. We consider a strong background in business management and economics as important to today's rancher as knowledge in range management, animal husbandry, ecology, agronomy and wildlife management. Some sources we have found quite invaluable for the rancher interested in improving his or her business capabilities include Gitman and Joehnk (1984), Workman (1986), Lynch (1989), Pring (1992), and Schiller (1994).

Rangeland managers usually must consider large amounts of information to predict the outcome of different range management practices. These managers can and do benefit from computer data compilation and synthesis systems geared toward practical problem solving. Since the 1970s, computers have been used increasingly by the range profession in both data storage and data analysis. George Van Dyne is considered the pioneer in development of computer applications to range management problems. During the 1960s and 1970s, he developed computerized models that integrate knowledge on various aspects of rangeland ecosystems into a framework for management decisions (Van Dyne 1966a). The approach developed by Van Dyne and others is referred to as *systems analysis* and is discussed in detail by Van Dyne (1966b) and Shugart and O'Neill (1979).

SYSTEM ANALYSIS

Systems analysis is the process of defining goals for a range unit and discovering procedures for accomplishing those goals most efficiently. The most important attributes of systems analysis are that it provides a unifying structure for interrelating facts and observations, and that it serves as a dynamic force as new facts and observations become available.

Model

A systems model is a dynamic hypothesis describing the function of the range system (Figure 17.1). A model is a plan or method of organizing information about a range system. It is an abstraction of a real system. It does not represent the system in its entirety, but it does simulate the range system in aspects regarded as essential by the modeler. For example, no conceivable model could possibly represent all the species present in a range system, but it must truly represent the behavior of that portion of the system being simulated. The medium of the abstraction is the various symbols of mathematics and the real system variables have analogous mathematical variable counterparts. Thus a model is a mathematical representation of the portion of the range system under study.

Models can be classified as predictive or theoretical, based on their intended use. Predictive models predict the future behavior of variables, while theoretical models provide insight into how the system functions. Predictive models are validated by establishing, with a reasonable degree of accuracy, the degree of accuracy to which the model simulates the system's behavior and the conditions over which the model can be used. The more complex the range system, the more difficult it is to predict the results of manipulative treatments: grazing treatments, brush management, revegetation, and so on. The model provides us with a tool for examining the effects of many manipulations of the ecosystem and determining the results without doing field trials. However, enough field checking must be done to ensure that the model represents the behavior of the system.

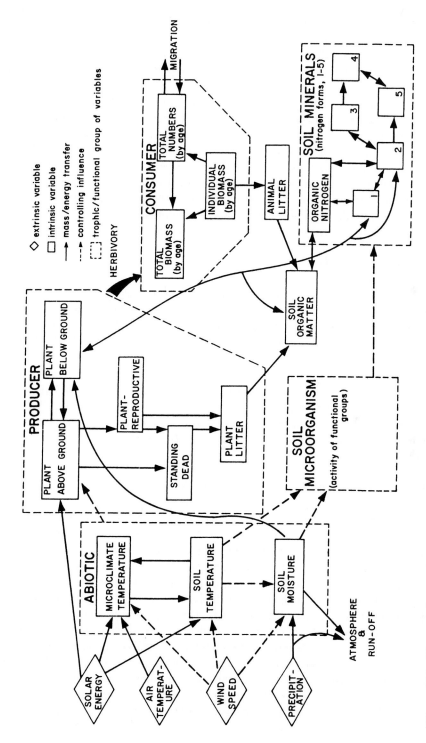

Figure 17.1 Diagram of a grassland ecosystem. Each arrow corresponds to one or more equations. (From Bledsoe and Jameson 1969.)

Validation of theoretical models consists of attempting to disprove the model (i.e., to find a single instance in which the model is not similar to data from the real world) so that confidence in theory can be established. Most ecosystem models combine predictive and theoretical functions. A computer is needed to handle the volume of data acquired for each range unit.

One of the models developed for use on rangelands is the SPUR model (Figure 17.2). SPUR is an acronym for Simulation of Production and Utilization of Rangelands (Wight and Skiles 1987). SPUR is a physically based rangeland simulation model developed to assist resource managers and scientists. SPUR is composed of five basic components: (1) climate, (2) hydrology, (3) plant, (4) animal (domestic and wildlife), and (5) economic. SPUR is driven by daily inputs of precipitation, minimum and maximum temperatures, solar radiation, and wind run. The hydrology component calculates upland surface runoff volumes, peak flow, snowmelt, upland sediment yield, channel streamflow, and sediment. It also calculates a daily soil water balance that is used to generate soil water regimes that control plant growth. Net photosynthesis is the basis for predicting herbage production. Photosynthesis is controlled by temperature and soil water availability. The animal component considers both domestic livestock and wildlife as consumers of plants. Animal production is used by the economic component to estimate benefits and costs of alternative grazing practices, range improvements, and animal management options.

Some of the other useful software includes a forage dynamics model (Blackburn and Kothmann 1989), Simple Model to Assess Range Technology (SMART), Rangeland Plant Profiles (RAPPS), Water Erosion Prediction Project (WEPP) (Foster 1987; Savabi et al. 1995), Generation of Weather Elements for Multiple Applications (GEM), Annual Planning Stock Adjustments Templates (Kothmann and Hinnant 1992), IMAGES (Hacker et al. 1991), KINEROS (Osborn and Simanton 1990), and prescribed burning (Wright et al. 1992). A combination of imagery and models was suggested to determine livestock distribution (Pickup and Chewings 1988; Smith 1988; Coughenour 1991).

Artificial Intelligence

The use of artificial intelligence (AI) in management of natural resources began with the development of expert systems for problem solving and decision making. An *expert system* is a type of artificial intelligence program that follows a few general procedures for solving problems. It uses facts (often obtained from written reports), experience, and models stored in the memory of a computer by human experts. There are two ways in which the computer can arrive at conclusions. It can reason forward, going from facts to a solution, or it can work from a hypothetical solution to seek supporting evidence. Using the two reasoning approaches to solve a particular problem, the computer suggests a set of hypotheses based on data input by the user. The system then considers each hypothesis in turn, attempting to find a specific solution. The artificial intelligence programs analyze decisions, can interpret the meanings, or can ask appropriate questions. Thus the computer aids managers who often reach conclusions from partial or uncertain evidence by following possible lines of reasoning. Also, the system can add to its data base as it gains experience. AI

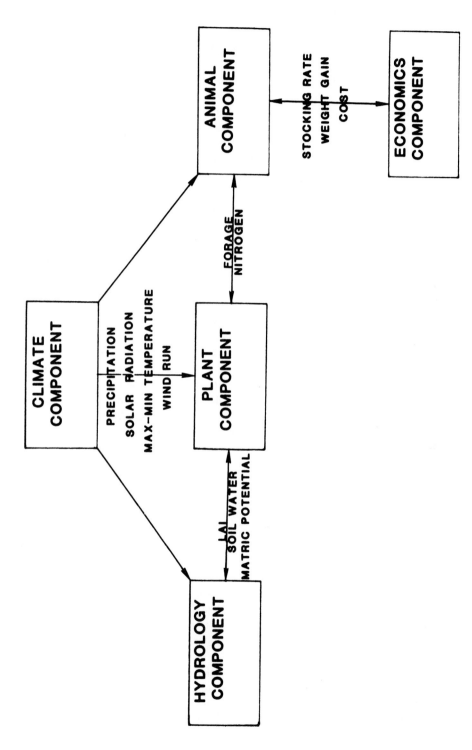

Figure 17.2 Simple flow diagram of SPUR model. (From Wight 1983.)

applications offer flawless memory, a huge information-processing potential, and are indefatigable. They will determine how to perform a requested task, perform it, and avoid performing the tasks with inadequate information. An expert system was used to develop a rangeland simulation model (Ritchie 1989). Simulation allows the manager to examine the entire ranch, and to alter various parameters and determine effects on the ranch. Major management adjustments may be recommended as a result of even minor adjustments to variables. Maintaining a high degree of accuracy in decision-support functions requires a commitment to collect, develop, and verify the experts' knowledge.

The use of expert systems led to the development of other artificial intelligence procedures pertinent to management of natural resources. Some of these are (1) integrated expert systems, which link management models with natural resource models; (2) intelligent Geographic Information Systems (GIS) files; and (3) artificial intelligence modeling of animal behavior and interaction with the environment (Coulson et al. 1987). Expert systems have led to knowledge systems, intelligent decision systems, creativity systems, and literature research systems (Truett 1989).

VIRTUAL REALITY

Virtual reality (VR) is a computer simulation usually experienced through headgear, goggles, and sensory gloves that lets the user feel they are in another place. VR devices let you see, hear, feel, and interact with real or abstract data from computer-generated models. Perspective views may be generated by combining digital terrain data with satellite imagery, scanned aerial photographs or maps, airborne video, ground-penetrating radar, and GIS files (Truman et al. 1988; Everitt et al. 1990; Smith et al. 1991; Hart and Laycock 1996). The three-dimensional views are used to visualize and verify the results of the rangeland models. Database visualization in 3-D enables the decision-makers and the general public to more readily understand complex data. With GIS/GPS (Global Positioning System) rainfall, soil condition, temperatures, and plant conditions can be gathered throughout a season and integrated into data layers useful in managing rangelands. GIS is used to analyze factors in the ecosystem such as quality and quantity of ground and surface water distributions and types of human activities and landscape patterns that affect animal and plant habitats (Whitford et al. 1996). These systems quickly and easily mix text, graphics, sound, and moving objects so that decisions can be made on a timely basis.

Historically, a single soil sampling frequency has been used throughout a region, without regard for varying local conditions or needs. Within GIS/GPS technology, variable sample frequency is a modern alternative (Berry 1996). Time-series satellite sensor data may be used to monitor vegetation change due to various practices (Eve and Peters 1996). In the Great Basin, wildfire scars were identifiable using aerial photography and satellite imagery (Tueller 1994).

High-resolution satellite imagery of the near future will offer error-free topographic data, continuous coverage, frequent update, and digital data instead of cumbersome photography. The map of the future will be 3-D (three dimensional), multispectral with automatically derived overlays of land features. Why stop at 3-D? Animations can be used to add a time dimension to multivariate data. Spatial analysis will be used to create maps representing abstract surfaces, such as cost and risk. Such maps will be dynamic, flexible entities and be user-friendly. Visualization often makes data errors more obvious (Hargrove 1996).

ECONOMIC ANALYSIS

Range economics is concerned with improving efficiency, increasing equity, and managing risk. Efficiency involves allocation of scarce resources among competing uses for defined periods to produce the greatest quantity of net output or products from a given amount of rangeland. Often, maximum profits occur at some productivity level below maximum. The manager must determine the level of production that yields the greatest net returns. Equity concerns the distribution of products among competing consumers and owners of rangelands (Workman et al. 1984). Resource allocation affects both efficiency and equity (Gardner 1984). In some situations, efficient production is a more desirable goal than maximizing gross production. Optimum resource allocation is necessary to achieve this goal. Ranching involves biological, financial, climatic, and political risks. It is one of the riskiest of all businesses. Biological risks can vary from uncertainty regarding outcome of a grazing or range improvement practice to the possibility of disease infecting a rancher's livestock. Climatic risks center around drought and severity of winter. Although there is some repeatability to climatic patterns for given areas (Roberts and Lansford 1979; Holechek 1996), great uncertainty exists in rainfall timing and amount for any ranch in any particular year. Although definite cycles occur for livestock prices, the length and magnitude of the cycles can vary substantially (Holechek et al. 1994). Many outside forces regarding the economy at large such as actions by the Federal Reserve Banking System can influence the financial outcome of range management and ranching decisions (Holechek et al. 1994; Holechek and Hess 1994). Finally, the political policies regarding subsidies, taxes, and environmental regulations are never constant. In recent years ranchers have had to contend with much greater political uncertainty at local, state, and federal levels than ever before. It is probable this uncertainty will increase in the twenty-first century. Range economics attempts to account for all these risks when various management options are evaluated and selected.

Few, if any, range improvement practices yield acceptable monetary returns under standard economic evaluations used by lending institutions. However, a range manager must consider other factors. For example, what is the future value of a stand of forage plants (a renewable resource)? If the stand is lost and renovation is possible, renovation would often cost more than the value of the land. The encroachment of undesirable species adversely affects future generations through increased soil erosion, air pollution, and a depletion of the resource. Soil on rangelands may be considered a nonrenewable resource because it takes hundreds of years to make soil through geologic processes. However, it can be lost within a few years on deteriorated sites. Therefore, a depleted stand of forage plants represents a cost to future generations as well as the present landowners.

Effective range management depends on selecting practices that will keep rangelands in good condition and have favorable cost/return ratios. Computer models have been developed that fulfill this requirement. Figures 17.2 and 17.3 shows the SPUR model integrating range ecosystem components with economics. Grazing Land Alternative Analysis Tool (GAAT) is a decision support system that empowers managers to rapidly analyze economic implications of alternative management strategies, for example, prescribed burning (Kreuter et al. 1996).

COMPUTERS AND THE FUTURE

In this section we will attempt to look into the future, considering implications of how new technologies and other changes could impact rangeland and range management. However, we acknowledge it is quite possible we will overlook some important event or development that would change our scenario.

To develop our look into the future, we will use a typical ranch located in the middle of the western United States. During the night, the ranch computer automatically assembles several databases to obtain current information on availability and prices of fuel, fencing, pump leathers, and vaccines; weather forecasts; and livestock markets. At the manager's command, a digest of this information appears on a monitor (Holt 1985). Sensors in ear tags and implanted devices in the livestock are scanned for health, nutritional status, and estrus.

The range unit will be scanned frequently by remote-sensing devices. Range condition and trend will be monitored. Runoff and soil water will be predicted for each site (Shik et al. 1991). Plant condition will be determined by monitoring the stand and environment. Strategically placed sensors in combination with near infrared spectroscopy (NIRS) will analyze the quality of forages. This information is used to make decisions on supplemental feed for range animals (Holechek et al. 1982; Lyons and Stuth 1992). Distribution of livestock will be controlled by implantable electric stimulators linked to geographic positioning systems (GIS) (Walker 1995).

The computer has also been scanning several weather stations and sensors in key plants scattered in strategic locations on the ranch. It has weather and soil water data, short- and long-range weather forecast information from the network, and plant condition to predict the status of range plant communities. Remote sensing techniques (aerial photography, airborne videography, and satellite sensor imagery) is used to distinguish and measure infestations. Reflectance measurements are used to determine spectral characteristics of the target plants and their phenological stages (Everitt et al. 1995). Portions of a certain unit are scheduled for herbicide treatment to control unwanted shrubs. Accurate prediction of El Niño, warm water temperatures in the tropical Pacific affecting global weather, will assist in strategic planning. A week before treatment, a gate to an ungrazed unit 2 adjacent to unit 1 is opened by computer command at ranch headquarters to the hydraulic unit on the gate. Most livestock move themselves to the fresh forage in unit 2, but two livestock helpers on motorbikes move the remaining livestock the next day. Each animal has an electronic eartag to facilitate movement and record keeping. During the short breeding and parturition seasons, animals are confined to relatively small portions of a range unit by laser beams. Automated feeding systems are used as required by individual animals to supplement the quality of range forage. All the data on an individual animal are

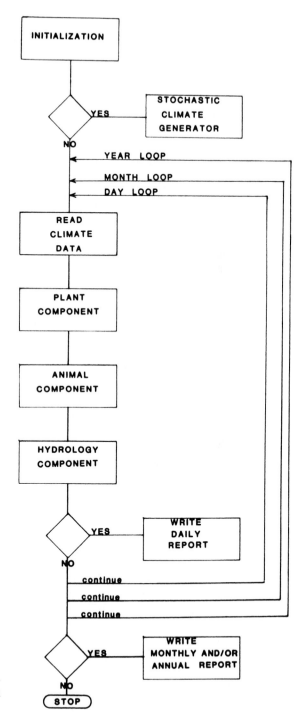

***Figure* 17.3** Interaction between various components of SPUR. (From Wight 1983.)

updated daily so that an animal may be marketed as soon as it is not profitable to keep that animal on the ranch. Life-cycle production will be evaluated to match genetic, nutrition, behavior, and management resources (Koch and Algeo 1983). Reproductive efficiency will be enhanced by hormonal control and multiple births. Breeds will be matched with a certain environment. The proper proportion of sheep, goats, cattle, llamas, and wildlife for a profitable enterprise will be attained for each ranch unit. Rumen-regulating drugs and genetic engineering to produce more useful rumen microflora will provide increased potential for metabolizing range forage into useful nutrients. Instead of chemicals for noxious plant control, new herbicides will be developed that are isolated from plant allelopathic toxins that do not contaminate the soil or water, and do not leave residues harmful to humans, livestock, or wildlife. Range management systems will be designed to match the needs of animals and thereby increase efficiency of production (Stricklin and Kautz-Scanavy 1984).

Supplies and parts are ordered by computer linkage to stores. The manager employs his or her custom-designed master ranch management program, an enormously complex combination of data bases and artificial intelligence, to make decisions on livestock marketing schedule and compliance with regulations. Included in the master ranch program are a complete description of the ranch, including unit size and shape; location of watering points, fences, and roads; soil types; vegetation types; and stocking schedule. Up-to-the-minute information on manipulative treatments is available from national experts. Costs and benefits are calculated for each situation and the results included in the master ranch program. Also included in the master ranch program is a highly automated accounting system providing for complete depreciation schedules, labor records and payroll management, and automatic generation of all necessary tax records and forms for this ranch operation. The ranch is a business and must be handled in a businesslike manner.

Those who desire may register for courses at the university that are taught at home through the computer. The instructor is a national leader in the field as well as an excellent teacher. The knowledge, appearance, and personality of the teacher as well as the training aids are captured on the optical disk by a unique programming approach that makes all interaction as realistic as if the teacher were present. As the student progresses, he or she will spend a few days on the campus for examinations, laboratory experience, counseling, seminars, and simulator training. In general, range education programs will incorporate instruction in practical law, economics, government relations, communication skills, and development of environmental impact statements. There will be an increasing demand for adult-education programs.

Information is becoming increasingly important. Technological and political developments throughout the world are dramatically increasing the number of management options. Multiple media-text, graphics, and sound have been integrated through digital technology to deliver a communications experience that dramatically alters the way people give and receive information. In business, government, education, and social realms, tremendous breakthroughs in communications and interactive information retrieval are being achieved.

In the highly competitive environment of agriculture in the future, those who survive will be those who are able to acquire the most accurate and best organized information and use it most effectively (Holt 1985). Several agricultural databases are presently accessible to ranchers and others with proper equipment. Past productivity improvements resulted from relatively low-priced land, water, fuel, chemicals, and capital inputs. Since costs of these inputs have risen, and will continue to rise, alternative production technologies will be developed to keep food and fiber prices comparable to present levels.

The alteration of living cells plus advances in microculture, cell fusion, regeneration of plants from single cells, and embryo recovery and transfer will create new range plants and animals. Spraying critical range areas with ice-nucleation-negative bacteria will prevent frost damage to plants and thus maintain forage quality during winter months. Many range plants will be inoculated to fix nitrogen so that this important element will not limit production. Other bacteria will be used to render other elements, such as phosphorus and iron, more readily available to range plants. Through genetic engineering, improved range plants will be resistant to environmental stresses. Improved soil- and water-resource technologies will be used to save our valuable resources. Different plants and animals will be developed for the various conditions prevalent on rangelands.

The technology exists for many of the practices described in this scenario. Some of the computer applications are commercially available, including most of the automation equipment, the equipment-monitoring features, and the accounting and inventory management software. Robotics and integrated ranch management and control packages, including expert systems, are in the process of development.

Formerly, the university and similar institutions were producers, collectors, and dispensers of information. However, information is now widely available on the Internet. Most branches of science show an exponential growth, doubling about every 10 years. There will be a migration away from classic campus-based higher education. The alternatives will be video servers with lectures by experts; electronic access to interactive reading materials and study exercises; interactivity with faculty and electronic textbooks; and video-conferencing. These techniques will soon be available around the world (Noam 1995).

A futuristic look at rangelands is provided by Walker (1995). Trends he identifies include (1) increased abundance of resources and food, (2) increased emphasis on environmental quality, (3) more use of open space for recreational endeavors, and (4) a large increase in the use of genetic engineering. Environmental concerns include weather change, loss of biodiveristy, and the sustainability of commodity production systems.

There is a trend toward using fewer and cheaper raw materials in building construction and consumer goods. Modern telecommunication equipment uses fiber optics made of silica instead of expensive copper (Naisbitt and Aburdene 1990). New super plastics are being developed that can be used in automobile engines, bridges, building, etc.

In his vision of the near future, Drexler et al. (1991) stated that nanotechnology (extreme miniaturization) will permit the human race to feed itself with ordinary, naturally grown, pesticide-free foods while returning most of today's croplands back to nature. They cautioned that people working with natural resources could lose their credibility if they continue to use the gloom and doom predictions of the last 200 years to justify future research and management. Molecular biology is causing a revolution in agriculture. The new industry of genetic engineering is expected to drastically increase production from crops and livestock.

The interaction of technological development, information transfer, and the application of improved systems for production of food and material goods should cause a tremendous improvement in our quality of life. While the future is uncertain, we look forward to it with great optimism. For the twenty-first century we foresee greatly increased use of:

1. Robotics to decrease or eliminate manual labor.
2. Artificial intelligence to communicate with ranch animals.
3. Super conductivity to improve transportation and computer use.
4. Genetic engineering to improve range plants and animals.
5. Energy produced on the ranch by solar and wind power with the sale of the surplus to urban areas.

We refer the reader to Walker (1995) for additional discussion on new technologies and range management in the twenty-first century.

RANGE MANAGEMENT PRINCIPLES

- The ability to find and acquire knowledge may be more important to range managers in the twenty-first century than their depth of knowledge. This is because of the rapid increase in information and new technologies that is now occurring.

- Substitution, miniaturization, genetic engineering, and nanotechnology have the potential to largely overcome problems of food scarcity and pollution in the twenty-first century. Preservation of open space and environmental diversity could become the biggest rangeland problems of the twenty-first century.

Literature Cited

Berry, J. K. 1996. Beyond mapping, what's the point? *GIS World* 9(12):30.

Blackburn, H. D., and M. M. Kothmann. 1989. A forage dynamics model for use in range or pasture environments. *Grassl. For. Sci.* 44:283–294.

Bledsoe, L. J., and D. A. Jameson. 1969. Model structure for a grassland ecosystem. In R. L. Dix and R. G. Beidleman (Eds.). The grassland ecosystem: A preliminary synthesis. *Range Science Series No. 2*, p. 417. Colorado State University, Fort Collins, CO.

Coughenour, M. B. 1991. Spatial components of plant-herbivore interactions in pastoral, ranching, and native ungulate ecosystems. *J. Range Manage.* 44:530–542.

Coulson, R. N., L. J. Folse, and D. K. Loh. 1987. Artificial intelligence and natural resource management. *Science* 237:262–267.

Drexler, K. E., C. Petersen, and G. Pergait. 1991. *Unbounding the future: The Nanotechnology revolution.* William Morrow and Co., Inc, New York.

Eve, M. D., and A. J. Peters. 1996. Using high temporal resolution satellite data to assess shrub control effectiveness. Proc. Symp. Shrubland Ecosystem Dynamics in a Changing Environment. J. R. Barrow, E. D. McArthur, R. W. Sosebee, and R. J. Tausch (Compilers). *U.S. Dep. Agric. Intermountain Res. Sta. Gen. Tech. Rep.* INT-338, pp. 88–94.

Everitt, J. H., K. Lulla, D. E. Escobar, and A. J. Richardson. 1990. Aerospace video-imaging systems for rangeland management. *Photo. Engi. & Rem. Gens.* 56:343–349.

Everitt, J. D., D. E. Escobar, and M. R. Davis. 1995. Using remote sensing for detecting and mapping noxious plants. *Weed Abstr.* 44:639–649.

Foster, G. R. (Compiler). 1987. User requirements: USDA—Water erosion prediction project (WEPP). *U.S. Dep. Agric., Nat. Soil Erosion Res. Lab. Rep.* 1. West Lafayette, IN.

Gardner, B. D. 1984. The role of economic analysis in public range management. *Developing strategies for rangeland management.* National Research Council/National Academy of Sciences (Eds.). Westview Press, Inc., Boulder, CO, pp. 1441–1466.

Gitman, L. J., and M. D. Joehnk. 1984. *Fundamentals of investing.* 2d ed. Harper & Row New York.

Hacker, R. B., K. M. Wang. G. S. Richmond, and R. K. Lindner. 1991. IMAGES: An integrated model of an arid grazing ecological system. *Agric. Systems* 37:119–163.

Hargrove, W. W. 1996. Perspectives on future directions in GIS. *GIS World* 9(3):28.

Hart, R. H., and W. A. Laycock. 1996. Repeat photography on range and forest lands in the western United States. *J. Range Manage.* 49:60–67.

Holechek, J. L. 1996. Drought in New Mexico: Prospects and management. *Rangelands* 18:225–227.

Holechek, J. L., J. S. Shenk, M. Vavra, and D. Arthun. 1982. Prediction of forage quality using near infrared reflective spectroscopy on esophageal fistula samples from cattle on mountain range. *J. Anim. Sci.* 55:971–975.

Holechek, J. L., and K. Hess. 1994. Brush control considerations: A financial perspective. *Rangelands* 16:193–196.

Holechek, J. L., J. Hawkes, and T. Darden. 1994. Macroeconomics and cattle ranching. *Rangelands* 16:118–123.

Holt, D. A. 1985. Computers in production agriculture. *Science* 228:422–427.

Koch, R. M., and J. W. Algeo. 1983. The beef cattle industry: Changes and challenges. *J. Anim. Sci.* 57(Suppl. 2):29–43.

Kothmann, M. M., and R. T. Hinnant. 1992. APSAT: An operational level grazing management model. *Proc. Wes. Sec. Amer. Soc. Anim. Sci.* 43:354–357.

Krueter, U. P., R. C. Rowan, J. R. Conner, J. W. Stuth, and W. T. Hamilton. 1996. Decision support software for estimating the economic efficiency of grazingland production. *J. Range Manage.* 49:464–469.

Lynch, P. 1989. *One up on Wall Street.* Penguin Books, New York.

Lyons, R. K., and J. W. Stuth. 1992. Fecal NIRS equation's for predicting diet quality of free-ranging cattle. *J. Range Manage.* 45:238–244.

Naisbitt, J., and P. Aburdene. 1990. *Megatrends 2000, ten new directions for the 1990s.* William Morrow and Co., Inc., New York.

Noam, E. M. 1995. Electronics and the dim future of the university. *Science* 270:247–249.

Osborn, H. B., and J. R. Simanton. 1990. Hydrologic modeling of a treated rangeland watershed. *J. Range Manage.* 43:474–481.

Pickup, G., and V. H. Chewings. 1988. Estimating the distribution of grazing and patterns of cattle movement in a large arid zone paddock. *Int. J. Remote Sensing* 9:1469–1490.

Pring, M. J. 1992. *The all-season investor.* John Wiley & Sons, Inc., New York.

Ritchie, J. R. 1989. An expert system for a rangeland simulation model. *Ecol. Mod.* 46:91–105.

Roberts, W. O., and H. Lansford. 1979. *The climate mandate.* W. H. Freeman and Company, San Francisco, p. 197.

Savabi, M. R., W. J. Rawls, and R. W. Knight. 1995. Water erosion prediction project (WEPP) rangeland hydrology component on a Texas range site. *J. Range Manage.* 48:535–541.

Schiller, B. S. 1994. *The economy today.* 6th ed. Random House, New York.

Shik, S. F., D. S. Harrison, A. G. Smajstrla, and F. S. Zazueta. 1991. Infrared thermometry to estimate soil water content in pasture areas. *Soil & Crop Sci. Soc. Fla., Proc.* 50:158–162.

Shugart, H. H., and R. V. O'Neill (Eds.). 1979. *Systems ecology.* Dowden, Hutchinson & Ross, Inc., Stroudsburg, PA.

Smith, M. S. 1988. Modeling: Three approaches to predicting how herbivore impact is distributed in rangelands. *N. Mex. Agric. Exp. Sta. Res. Rep.* 628.

Smith, S. M., H. E. Schreier, and S. Brown. 1991. Spatial analysis of forage parameters use geographic information system and image-analysis techniques. *Grass & For. Sci.* 46:183–189.

Stricklin, W. R., and C. C. Kautz-Scanavy. 1984. The role of behavior in cattle production: A review of research. *Appl. Anim. Ethol.* 11:359–390.

Truett, W. L. 1989. Artificial intelligence and applications across the 1990s: An applied approach. *Amer. Lab.* 21(4):40–47.

Truman, C. C., H. F. Perkins, L. F. Asmeissen, and H. D. Allison. 1988. Using ground-penetrating radar to investigate variability in soil properties. *J. Soil & Water Conserv.* 43:341–345.

Tueller, P. T. 1994. Great Basin annual vegetation patterns assessed by remote sensing. Proc. Ecology and Management of Annual Rangelands. S. B. Morsen and S. G. Ktichen (Eds.). *U.S. Dept. Agric. Intermountain Res. Sta. Gen. Tech. Rep.* INT-313, pp. 126–131.

Van Dyne, G. M. 1966a. Application and integration of multiple linear regression and linear programming in renewable resource analyses. *J. Range Manage.* 19:356–62.

Van Dyne, G. M. 1966b. *Ecosystems, systems ecology and systems ecologists.* ORNL-3957. Oak Ridge National Laboratory, Oak Ridge, TN.

Walker, J. W. 1995. Viewpoint: Grazing management and research now and in the next millennium. *J. Range Manage.* 48:250–357.

Whitford, W. G., D. J. Rapport, and R. M. Groothouser. 1996. The central Rio Grande Valley. *GIS World* 9(12):60–62.

Wight, J. R. (Ed.). 1983. *SPUR*—Simulation of production and utilization of rangelands: A rangeland model for management and research. *U.S. Dep. Agric. Mis. Publ.* 1431.

Wight, J. R., and J. W. Skiles (Eds.). 1987. SPUR: Simulation of production and utilization of rangelands—Documentation and user guide. *U.S. Dep. Agric., Agric. Res. Serv.* ARS-63.

Workman, J. P., S. K. Fairfax, and W. Burch. 1984. Applying socio-economic techniques to range management decision making: Summary and recommendations. In National Research Council/National Academy of Sciences (Eds.). *Developing strategies for rangeland management*, pp. 1427–1440. Westview Press, Inc., Boulder, CO.

Workman, J. P. 1986. *Ranch economics.* MacMillan Publishing Co., New York.

Wright, H. A., J. R. Burns, H. Chang, and K. Blair. 1992. An expert system for prescribed burning of rangelands. *Rangelands.* 14:286–292.

INDEX